WAR AGAINST THE WEAK

WAR AGAINST THE WEAK

Eugenics and America's Campaign to Create a Master Race

Edwin Black

FOUR WALLS EIGHT WINDOWS
New York / London

PUBLISHED IN THE UNITED STATES BY
Four Walls Eight Windows
39 West 14th Street
New York, NY 10011

U.K. OFFICES:
Four Walls Eight Windows/Turnaround
Unit 3, Olympia Trading Estate
Coburg Road, Wood Green
London N22 6TZ, England

http://www.4w8w.com

First printing September 2003.

LIBRARY OF CONGRESS CATALOGING-IN-PUBLICATION DATA
Black, Edwin.
    War against the weak : eugenics and America's campaign to create a master race / by Edwin
Black.
    p. cm.
    Includes bibliographical references and index.
    ISBN: 1-56858-258-7 (hardcover)
1. Eugenics—United States—History. 2. Sterilization (Birth Control)—United States.
3. Human reproduction—Government policy—United States. 4. United States—Social
policy. 5. United States—Moral conditions. I. Title.

HQ755.5.U5B53 2003
363.9'7—dc21   2003048857

Book design by Sara E. Stemen

10 9 8 7 6 5 4 3 2 1

PRINTED IN THE UNITED STATES

To my mother
who at present is unable to read this book,
but who still remembers when
American *principles of eugenics*
came to Nazi-occupied Poland.

Contents

Acknowledgments   ix

Introduction   xv

PART ONE: FROM PEAPOD TO PERSECUTION

CHAPTER 1: Mountain Sweeps   3

CHAPTER 2: Evolutions   9

CHAPTER 3: America's National Biology   21

CHAPTER 4: Hunting the Unfit   43

CHAPTER 5: Legitimizing Raceology   63

CHAPTER 6: The United States of Sterilization   87

CHAPTER 7: Birth Control   125

CHAPTER 8: Blinded   145

CHAPTER 9: Mongrelization   159

PART TWO: EUGENICIDE

CHAPTER 10: Origins   185

CHAPTER 11: Britain's Crusade   207

CHAPTER 12: Eugenic Imperialism   235

CHAPTER 13: Eugenicide   247

CHAPTER 14: *Rasse und Blut*   261

CHAPTER 15: Hitler's Eugenic Reich   279

CHAPTER 16: Buchenwald   319

CHAPTER 17: Auschwitz   337

PART THREE: NEWGENICS

CHAPTER 18: From Ashes to Aftermath   375

CHAPTER 19: American Legacy   385

CHAPTER 20: Eugenics Becomes Genetics   411

CHAPTER 21: Newgenics   427

Notes   445

Major Sources   505

Index   519

Acknowledgments

Where do I begin to express gratitude, when so many people in so many places have lent so many hands to advance the cause of this years-long project? More than fifty researchers in fifteen cities in four countries, assisted by scores of archivists and librarians at more than one hundred institutions, combined to ingather and organize some 50,000 documents, together with hundreds of pages of translation, as well as to review hundreds of books and journals, all to collectively tear away the thickets of mystery surrounding the eugenics movement around the world. I cannot name all who need naming because of space limitations. In many cases, I do not even know them all. Many helped behind the scenes. But if great projects depend upon great efforts by a vast network, then *War Against the Weak* is greatly indebted indeed.

I must begin by thanking my corps of skilled researchers, mostly volunteers. Because the information needed for *War Against the Weak* resided in many out-of-the-way archives as well as major repositories, the challenge was to locate the right person in the right place at the right time, from the hilly back country of southern Virginia to Berlin. Recruits came from the Internet, organizational bulletin boards, word of mouth and my personal website, as well as the devoted research team involved with my previous books, *IBM and the Holocaust* and *The Transfer Agreement*. Some worked for a few days in a strategic location to extract vital information; others worked for months at a time in archives or my office.

Thanks are due to at least eight people in Germany, including Dennis Riffel, Christina Herkommer and Jakob Kort, who worked tirelessly in Berlin, Munich, Heidelberg, Koblenz and Münster at the archives and libraries of the Max Planck Institutes (successor to the Kaiser Wilhelm Institutes), Bundesarchiv, Heidelberg University, Münster University, the Frei University and many other locations reviewing and summarizing thousands of pages. The laser-like ability of Riffel, Herkommer and their colleagues to identify connections spanning decades between Germans and Americans was indispensable.

In London, Jane Booth, Julie Utley, Diane Utley and several others spent months checking numberless documents, reviewing pamphlets and squinting at microfiche at the Public Record Office, Wellcome Library, University College of London Archives, British Library, Cambridge and other repositories to uncover links across the Atlantic.

In New York, more than a dozen researchers including Max Gross assisted me at the New York Public Library, the archives of New York University, Columbia University, and the Planned Parenthood Foundation. In Virginia, Susan Fleming Cook, Bobby Holt and Aaron Crawford dug through special and restricted library collections, archives, little-known museums, courthouse and institutional records, as well as the files of the ACLU. In California I was assisted by Lorraine Ramsey who worked in Chico, Sacramento and the University of California at Berkeley; Joanne Goldberg at the archives of the Hoover Institution and Stanford University; and others.

No fewer than eight researchers, including Christopher Reynolds and David Keleti, spent long hours at the American Philosophical Society archives in Philadelphia, the country's most precious eugenic resource. I owe a debt to Ashley and Jodie Hardesty who, among a team of four, scoured the valuable files of Vermont eugenicists, which in many cases were still waiting to be processed. At Truman State University in Kirksville, Missouri, I recruited a cadre of students to scrutinize thousands of pages of documents from the files of Harry Laughlin in the Pickler Memorial Library and its archive, and two of the most helpful were Benjamin Garrett and Courtney Carter. The project was also aided when attorney Charles Bradley volunteered to provide follow-up at the Rockefeller Archives.

Of special importance was Phyllis Bailey of Montreal, who labored at university libraries in Montreal, the Public Records of Vermont, the American Philosophical Society in Philadelphia, and the Rockefeller Archives in Sleepy Hollow, New York. Bailey drove from archive to archive displaying extraordinary research skill and keen intellectual understanding of the injustices she was investigating.

My Washington, D.C., research staff—about a dozen individuals—displayed unflagging tenacity in researching at numerous archives, analyzing and organizing thousands of documents, as well as delivering incomparable research and manuscript detail work. No research project could ask for more. Here I include Kate Hanna, who worked at the National Library of

Medicine and the Library of Congress, and, wielding her uncanny memory could recall almost every line of thousands of pages of eugenics journals she reviewed. Once Kate even corrected the date on an archival photograph.

Paul Dwyer displayed a special acumen for locating obscure volumes at numerous libraries, including American University, Catholic University, George Washington University, George Mason University, the University of the District of Columbia and others; he was also among a team of a dozen that pored through record groups at the National Archives. Eve Jones searched files at the National Archives and the Carnegie Institution, and my own considerable archival holdings.

John Corrado, assisted by Eve Jones, led the four-person fact and footnote verification team whose chore it was to cross-examine every fact and bit of fact context and then create the documentation trail, footnote by footnote, folder by folder. Corrado is also an exceptional researcher. Often, as I pounded my keyboard, I would call out an obscure name from decades past; within moments, Corrado was able to report the details. He is a researcher's researcher.

Corrado, Jones, Hanna and Dwyer were augmented and assisted by Patricia Montesinos, Alexandra Carderelli, Greg Greer, Eric Smith, Erica Ashton and several others. Numerous translators worked arduously and often with little notice; chief among them was Susan Steiner, and Karl Lampl also helped.

War Against the Weak could never have been completed without the exceptional cooperation of literally scores of archivists and librarians. Some archivists helped by producing as many as five thousand photocopies from a single institution, often making an exception to their copying regulations, and with special file and fact searches, as well as fellowship.

In England, those who deserve thanks include Anne Lindsay, Helen Wakely, Tracy Tillotson, Chris Hilton and many others at the Wellcome Library; Stephen Wright and Julie Archer at University College of London; and numerous staffers at the Public Record Office and the British Library.

In Germany, the list is long and represents the best of Germany's unparalleled archival services, as well as its dedication to understanding its own history. At the top of the list is Matthias M. Weber, archivist at the Max Planck Institute for Psychiatry and an expert on German eugenics, who spent many hours assisting my project. Wilhelm Lenz and Annegret Neupert at Bundesarchiv both in Berlin and Koblenz greatly expedited our work. Hans Ewald Kessler gave good advice and facilitated our access at

Heidelberg University archives and Robert Giesler did the same at the university archives at Münster. Harry Stein at Buchenwald Archive was indispensable in locating and providing copies of Katzen-Ellenbogen trial materials lost at the National Archives in Washington. Helmut Freiherr von Verschuer granted permission to freely examine his father's records. Many more German librarians and archivists are not named for lack of space and I apologize.

In the United States, I worked with dozens of repositories, many holding local and seemingly innocuous materials and unaware of their international value. The list stretched from community historical societies and corporate libraries to the major eugenic archives. Four institutions rendered profound assistance and their archivists reside at the apex of archival personalities preserving the history of eugenics. Judith Sapko, archivist extraordinaire at Pickler Memorial Library, labored more than I am permitted to say; Sapko was in constant contact with me during months of research. James Byrnes and Jennifer Johnsen at Planned Parenthood's McCormack Library displayed unrivalled and unflinching cooperation by continuously faxing materials—often within minutes of my request—to verify or disprove information about Margaret Sanger. At Cold Spring Harbor, Clare Bunce was a champion of research assistance, helping even as her own archives were in flux; Mila Pollock was also an important help. Valerie Lutz and Rob Cox, undermanned and greatly taxed, did their utmost to respond to pressing needs at the American Philosophical Society for more than a year.

There were many more in America. Marie Carpenti at the National Archives, Amy Fitch and Tom Nussbaum at the Rockefeller Archives and John Strom at the Carnegie Institution Archives all helped continuously.

Several people at the United States Holocaust Memorial Museum assisted greatly, including archivist Henry Mayer, several librarians, Tom Cooney and Andy Hollinger. Unfortunately, the executive staff at USHMM refused to open its records regarding IBM and certain other corporations, as well as the topic of American corporations and eugenics, and even rebuffed a Freedom of Information Act request, claiming the museum was immune to FOIA requests. But this did not stop others at the museum from doing their best to provide other traditional historical materials, and I thank them.

In addition, dozens of librarians helped, finding and copying rare newspapers, journals and other special materials in their collections. At the top of the list is Janice Kaplan at the New York Academy of Medicine

Library, and David Smith, a reference librarian of the New York Public Library; both worked with me for months. Anne Houston at Tulane University and the staff of the Rockville Public Library also deserve special mention, as does Charles Saunders and the staff at the *Richmond Times-Dispatch* newspaper morgue. I apologize to many more who cannot be listed for lack of space.

Numerous state officials went above and beyond. These include Margaret Walsh, Judith Dudley and James S. Reinhard at Virginia's Department of Hygiene, for allowing me to be the first to receive documents from the files of the Central Virginia Training School regarding Carrie Buck. I also thank state of Vermont officials for helping with important archival documents relating to the Hitler regime. Many more state officials worked with me on a confidential basis to reveal closed records. Their names cannot be revealed, but they know who they are.

Literally dozens of experts, eyewitnesses and other sources gave of their time to provide documentation in their possession, help trace facts or exchange ideas. In some instances the exchanges were brief, and in some cases the consultations were extensive and spanned weeks of effort. Among them were Sam Edelman, Nancy Gallagher, Daniel Kevles, Paul Lombardo, Barry Mehler, K. Ray Nelson, Diane Paul, Steve Selden, and Stephen Trombley.

Great guidance, page by page, stretching over many weeks, was rendered by Max Planck archivist Matthias Weber and geneticist Benno Müller-Hill in Germany; health policy historian Paul Weindling in England; eugenics author J. David Smith at the University of Virginia; and National Archives Nazi historian and archivist Robert Wolfe in the U.S. I am also grateful to the many other draft readers whose comments were so essential, including S. Jay Olshansky, a health issues expert at the University of Illinois; William Seltzer, a demographic and census expert at Fordham University; archivist Piotr Setkiewicz at Auschwitz Museum; William Spriggs of the National Urban League; Ariel Szczupak in Jerusalem; Abraham H. Foxman of the Anti-Defamation League; Malcolm Hoenlein at the Conference of Presidents of Major American Jewish Organizations; and more than a dozen others.

In each of my books, I have paid tribute to the musical talents of those who have inspired and energized me. Crowning the playlist are Danny Elfman, Jerry Goldsmith and Hans Zimmer. To this I add John Barry, BT, Moby, Afro Celt Sound System, John Williams and, of course, Dmitri Shostakovich.

Polishing a manuscript is a never-ending process, and here I extend special recognition to Elizabeth Black, Eve Jones, Phyllis Bailey and many others who devoted endless hours to the numerous revisions, tweaks and updates. In particular, Jones's deft understanding of both the historical facts and editorial fine-tuning will be felt on every page.

No author could have asked for a better publishing team. I have been blessed with great editors during the past three decades and among the very finest was Jofie Ferrari-Adler, who melded with every sentence and brought great skill and polish to the finished product. Four Walls Eight Windows publisher John Oakes believed in the book from the beginning and mobilized the entire company behind it; it has been an honor to travel with him. Publicist Penny Simon, who worked with me on *IBM and the Holocaust*, aided this project with her ceaseless energies. Most of all, this project would not have been possible without the steadfast support of my agent and manager, Lynne Rabinoff, who embodies the best of literary representation here and abroad; few authors are fortunate enough to have an agent so dedicated and energized, and Lynne's imprint will be felt throughout this book.

A special word must be written for my family, robbed of my presence for two years while I was holed up amidst stacks of documents. Their indulgence was indispensable.

EDWIN BLACK
Washington, D.C.
June 1, 2003

Introduction

Voices haunt the pages of every book. This particular book, however, speaks for the never-born, for those whose questions have never been heard—for those who never existed.

Throughout the first six decades of the twentieth century, hundreds of thousands of Americans and untold numbers of others were not permitted to continue their families by reproducing. Selected because of their ancestry, national origin, race or religion, they were forcibly sterilized, wrongly committed to mental institutions where they died in great numbers, prohibited from marrying, and sometimes even unmarried by state bureaucrats. In America, this battle to wipe out whole ethnic groups was fought not by armies with guns nor by hate sects at the margins. Rather, this pernicious white-gloved war was prosecuted by esteemed professors, elite universities, wealthy industrialists and government officials colluding in a racist, pseudoscientific movement called eugenics. The purpose: create a superior Nordic race.

To perpetuate the campaign, widespread academic fraud combined with almost unlimited corporate philanthropy to establish the biological rationales for persecution. Employing a hazy amalgam of guesswork, gossip, falsified information and polysyllabic academic arrogance, the eugenics movement slowly constructed a national bureaucratic and juridical infrastructure to cleanse America of its "unfit." Specious intelligence tests, colloquially known as IQ tests, were invented to justify incarceration of a group labeled "feebleminded." Often the so-called feebleminded were just shy, too good-natured to be taken seriously, or simply spoke the wrong language or were the wrong color. Mandatory sterilization laws were enacted in some twenty-seven states to prevent targeted individuals from reproducing more of their kind. Marriage prohibition laws proliferated throughout the country to stop race mixing. Collusive litigation was taken to the U.S. Supreme Court, which sanctified eugenics and its tactics.

The goal was to immediately sterilize fourteen million people in the United States and millions more worldwide—the "lower tenth"—and then continuously eradicate the remaining lowest tenth until only a pure Nordic super race remained. Ultimately, some 60,000 Americans were coercively sterilized and the total is probably much higher. No one knows how many marriages were thwarted by state felony statutes. Although much of the persecution was simply racism, ethnic hatred and academic elitism, eugenics wore the mantle of respectable science to mask its true character.

The victims of eugenics were poor urban dwellers and rural "white trash" from New England to California, immigrants from across Europe, Blacks, Jews, Mexicans, Native Americans, epileptics, alcoholics, petty criminals, the mentally ill and anyone else who did not resemble the blond and blue-eyed Nordic ideal the eugenics movement glorified. Eugenics contaminated many otherwise worthy social, medical and educational causes from the birth control movement to the development of psychology to urban sanitation. Psychologists persecuted their patients. Teachers stigmatized their students. Charitable associations clamored to send those in need of help to lethal chambers they hoped would be constructed. Immigration assistance bureaus connived to send the most needy to sterilization mills. Leaders of the ophthalmology profession conducted a long and chilling political campaign to round up and coercively sterilize every relative of every American with a vision problem. All of this churned throughout America years before the Third Reich rose in Germany.

Eugenics targeted all mankind, so of course its scope was global. American eugenic evangelists spawned similar movements and practices throughout Europe, Latin America and Asia. Forced sterilization laws and regimens took root on every continent. Each local American eugenic ordinance or statute—from Virginia to Oregon—was promoted internationally as yet another precedent to be emulated by the international movement. A tightly-knit network of mainstream medical and eugenical journals, international meetings and conferences kept the generals and soldiers of eugenics up to date and armed for their nation's next legislative opportunity.

Eventually, America's eugenic movement spread to Germany as well, where it caught the fascination of Adolf Hitler and the Nazi movement. Under Hitler, eugenics careened beyond any American eugenicist's dream. National Socialism transduced America's quest for a "superior Nordic race" into Hitler's drive for an "Aryan master race." The Nazis were fond of saying "National Socialism is nothing but applied biology," and in 1934 the *Richmond Times-Dispatch* quoted a prominent American eugenicist as saying, "The Germans are beating us at our own game."

Nazi eugenics quickly outpaced American eugenics in both velocity and ferocity. In the 1930s, Germany assumed the lead in the international movement. Hitler's eugenics was backed by brutal decrees, custom-designed IBM data processing machines, eugenical courts, mass steriliza-tion mills, concentration camps, and virulent biological anti-Semitism—all of which enjoyed the open approval of leading American eugenicists and their institutions. The cheering quieted, but only reluctantly, when the United States entered the war in December of 1941. Then, out of sight of the world, Germany's eugenic warriors operated extermination centers. Eventually, Germany's eugenic madness led to the Holocaust, the destruc-tion of the Gypsies, the rape of Poland and the decimation of all Europe.

But none of America's far-reaching scientific racism would have risen above ignorant rants without the backing of corporate philanthropic largess.

Within these pages you will discover the sad truth of how the scientific rationales that drove killer doctors at Auschwitz were first concocted on Long Island at the Carnegie Institution's eugenic enterprise at Cold Spring Harbor. You will see that during the prewar Hitler regime, the Carnegie Institution, through its Cold Spring Harbor complex, enthusiastically propagandized for the Nazi regime and even distributed anti-Semitic Nazi Party films to American high schools. And you will see the links between the Rockefeller Foundation's massive financial grants and the German sci-entific establishment that began the eugenic programs that were finished by Mengele at Auschwitz.

Only after the truth about Nazi extermination became known did the American eugenics movement fade. American eugenic institutions rushed to change their names from *eugenics* to *genetics*. With its new identity, the remnant eugenics movement reinvented itself and helped establish the modern, enlightened human genetic revolution. Although the rhetoric and the organizational names had changed, the laws and mindsets were left in place. So for decades after Nuremberg labeled eugenic methods genocide and crimes against humanity, America continued to forcibly sterilize and prohibit eugenically undesirable marriages.

I began by saying this book speaks for the never-born. It also speaks for the hundreds of thousands of Jewish refugees who attempted to flee the Hitler regime only to be denied visas to enter the United States because of the Carnegie Institution's openly racist anti-immigrant activism. Moreover, these pages demonstrate how millions were murdered in Europe precisely because they found themselves labeled lesser forms of life, unworthy of existence—a classification created in the publications and academic research rooms of the Carnegie Institution, verified by the research grants of the Rockefeller Foundation, validated by leading scholars from the best Ivy League universities, and financed by the special efforts of the Harriman railroad fortune. Eugenics was nothing less than corporate philanthropy gone wild.

Today we are faced with a potential return to eugenic discrimination, not under national flags or political credos, but as a function of human genomic science and corporate globalization. Shrill declarations of racial dominance are being replaced by polished PR campaigns and patent protections. What eugenics was unable to accomplish in a century, newgenics may engineer in a generation. The almighty dollar may soon decide who stands on which side of a new genetic divide already being demarcated by the wealthy and powerful. As we speed toward a new biological horizon, confronting our eugenic past will help us confront the bewildering newgenic future that awaits.

I first became interested in eugenics while researching my previous books, *The Transfer Agreement* and *IBM and the Holocaust*. *The Transfer Agreement*, published in 1984, documented the tempestuous worldwide anti-Nazi boycott, which included vigorous efforts to stop American organizations from funding medical research. At the time I could not understand why Nazi medical research was so important to American corporate philanthropists. The scope of eugenics escaped me. Then in 2000, while researching *IBM and the Holocaust*—which revealed IBM's role in automating Germany's eugenic institutions—I finally came to see that eugenics was a life and death proposition for Europe's Jews. Yet I still didn't realize that this bizarre cult of Nazi race science was organically linked to America.

As I explored the history of eugenics, however, I soon discovered that the Nazi principle of Nordic superiority was not hatched in the Third Reich but on Long Island decades earlier—and then actively transplanted to Germany. How did it happen? Who was involved? To uncover the story I did as I have done before and launched an international investigation. This time, a network of dozens of researchers, mostly volunteers, working in the United States, England, Germany and Canada unearthed some 50,000 documents and period publications from more than forty archives, dozens of library special collections and other repositories (see Major Sources). But unlike the Holocaust field, in which the documentation is centralized in a number of key archives, the information on eugenics is exceedingly decentralized and buried deep within numerous local and niche repositories.

In the United States alone, the investigation brought my team to the archival holdings of the American Philosophical Society in Philadelphia, to the Cold Spring Harbor Laboratory on Long Island, to Truman State University in northeastern Missouri, to numerous obscure community colleges in the Appalachian states, and a long list of state archives, county historical files and institutional archives where personal papers and period materials are stored. I also spent much time in many small, private libraries and archives, such as the one maintained by Planned Parenthood. We examined records at the Rockefeller Foundation and the Carnegie Institution. There are probably two hundred important repositories in America, many of them special collections and manuscript departments of local libraries or universities. Because eugenics was administered on the local level, every state probably possesses three to five sites hosting important eugenic documentation. I only accessed a few dozen of these across America. Much more needs to be done and American researchers will surely be kept busy for a decade mining the information.

In England I visited the British Library, the Wellcome Library, the University College of London, the Public Record Office and other key archives. These not only provided the information on Britain's eugenic campaigns, but also yielded copies of correspondence with American eugenic organizations that are simply not available in the American holdings. For example, strident propaganda pamphlets long cleansed from American files are still stored in the British records.

Because the German and American wings collaborated so closely, the German archives clearly traced the development of German race hygiene as it emulated the American program. More importantly, because the American and German movements functioned as a binary, their leaders bragged to one another and exchanged information constantly. Therefore I learned much about America's record by examining Reich-era files. For instance, although the number of individuals sterilized in Vermont has eluded researchers in that state, the information is readily available in the files of Nazi organizations. Moreover, obscure Nazi medical literature reveals the Nazis' understanding of their American partners. Probing the prodigious files of Nazi eugenics took my project to the Bundesarchiv in Berlin and Koblenz, the Max Planck Institute in Berlin, Heidelberg University and many other repositories in Germany.

When it was finished, the journey to discover America's eugenic history had taken me from an austere highway warehouse in Vermont, where the state's official files are stacked right next to automotive supplies and retrieved by forklift, to the architectonic British Library, to the massive Bundesarchiv in Berlin—and every type of research environment in between. Sometimes I sat on a chair in a reading room. Sometimes I poked through boxes in a basement.

Even still, I was not prepared for the many profound built-in challenges to eugenic research. My experiences are rooted in Holocaust investigation, where a well-developed infrastructure is in place. Not so with eugenics. In Holocaust research, archives facilitate unlimited speedy photocopying of documents. The Public Record Office in London produces copies within hours. The National Archives in Washington, D.C., allows self-service photocopying. But the most important eugenic archive in Britain, storing thousands of important documents, limits users to just one hundred copies per year. America's largest eugenic archive, housing vast numbers of papers in numerous collections, limits researchers to just four hundred copies per year. Often the beleaguered and understaffed copy departments in these archives needed between three and four weeks to produce the copies. One archive asked for three months to copy a ten-page document. Fortunately, I was able to circumvent these restrictions by deploying teams of five and ten researchers at these archives, and by virtue of the gracious and indispensable flexibility of archivists who continuously assisted me in this massive project (see Acknowledgments). Only by their special efforts and indulgence was I able to secure as many as five thousand copies from a single archive, and reasonably quickly—thus allowing me to gain a comprehensive view of the topic and shorten my work by years.

Another profound obstacle has been the fallacious claim by many document custodians, in both state and private archives, that the records of those sterilized, incarcerated and otherwise manipulated by the eugenics movement are somehow protected under doctor-patient confidentiality stretching back fifty to one hundred years. This notion is a sham that only dignifies the crime. Legislation is needed to dismantle such restrictions. No researcher should ever accept assertions by any document custodian that such records are covered by confidentiality protections accorded to medical procedures—whether in Nazi Germany or the United States. The people persecuted by eugenics were not patients, they were victims. No doctor-patient relationship was established. Most of the unfortunate souls snared by eugenics were deceived and seized upon by animal breeders, biologists, anthropologists, raceologists and bureaucrats masquerading as medical men. Mengele's victims were not patients. Nor were those in America who were caught up in the fraudulent science of eugenics.

In some instances, records were initially denied to me on this basis. Fortunately, the investigative reporter only gets started when he hears the word *no*. I demanded full access and was grateful when I received it. I applaud the State of Virginia for allowing me to be the first to receive files on the infamous sterilization of Carrie Buck; copies of those files are now in my office.

The international scope of the endeavor created a logistical nightmare that depended on devoted researchers scouring files in many cities. For months, I functioned as a traffic cop, managing editor and travel coordinator while simultaneously dispatching researchers to follow leads on both sides of the Atlantic. On the same day that one group might be interviewing mountain people in the hills of Virginia, another might be examining the personal papers of a police chief in California, while another in Berlin scanned the financial records of the Kaiser Wilhelm Institute to identify American financial assistance, while still others reviewed the pamphlets of the Eugenics Society in London.

We were as likely to scrutinize the visitor registers at the Kaiser Wilhelm Institute's guest facility, Harnack House, to see which Americans visited Berlin, as we were to review the mailing lists of Carnegie scientists to see who in Germany was receiving their reports. Progress among my researchers was exchanged by continuous use of the Internet and by the extensive use of faxed and scanned documents. Eventually all of the documents came together in my office in Washington. They were then copied and arranged in chronological folders—one folder for every month of the twentieth century. The materials were then cross-filed to trace certain trends, and then juxtaposed against articles published month-by-month in journals such as *Eugenical News*, *Journal of Heredity* and *Eugenics Review*, as well as numerous race science publications in Nazi Germany. By pulling any one monthly folder I could assemble a snapshot of what was occurring worldwide during that month.

When we were done, we had assembled a mountain of documentation that clearly chronicled a century of eugenic crusading by America's finest universities, most reputable scientists, most trusted professional and charitable organizations, and most revered corporate foundations. They had collaborated with the Department of Agriculture and numerous state agencies in an attempt to breed a new race of Nordic humans, applying the same principles used to breed cattle and corn. The names define power and prestige in America: the Carnegie Institution, the Rockefeller Foundation, the Harriman railroad fortune, Harvard University, Princeton University, Yale

University, Stanford University, the American Medical Association, Margaret Sanger, Oliver Wendell Holmes, Robert Yerkes, Woodrow Wilson, the American Museum of Natural History, the American Genetic Association and a sweeping array of government agencies from the obscure Virginia Bureau of Vital Statistics to the U.S. State Department.

Next came an obsessive documentation process. Every fact and fragment and its context was supported with black and white documents, then double-checked and separately triple-checked in a rigorous multistage verification regimen by a team of argumentative, hairsplitting fact-checkers. Only then was the manuscript draft submitted to a panel of known experts in the field from the United States, Germany, England and Poland, for a line-by-line review. The result: behind each of the hundreds of footnotes, there is a folder that contains the supporting documentation.

To ensure that all of our information was accurate, we also set about verifying the work of numerous other scholars by checking their documentation. We often asked them to provide documents from their files. In other words, we not only documented my book, we verified other works as well. Most of the authors graciously complied, readily faxing copies of their documents or explaining precisely where the information could be found. During this process, however, we discovered numerous errors in many prior works.

For example, in one book an important speech on the value of heredity is attributed to Woodrow Wilson, president of the United States—the speech was actually given by Jim Wilson, president of the American Breeders Association. I can understand how errors like this occur. Many scholars rely on other scholars' works. Summaries of summaries of summaries yield a lesser truth with every iteration. Except for the work of a few brilliant world-class documenters, such as Daniel J. Kevles, Benno Müller-Hill, Paul Weindling and Martin Pernick, I largely considered published works little more than leads. What's more, there is boundless information on eugenics accumulating on the Internet, some of it very prettily presented, much of it hysterical, and unfortunately, most of it filled with profound errors. Hence whenever possible, I acquired primary source material so I could determine the provable facts for myself.

When the research phase was over, I realized that less than half the information I had assembled would even make it into the book. Frankly, I had amassed enough information to write a freestanding book for each of the twenty-one chapters in this volume. It was painful to pick and choose which information would be included, but I am confident that with so many journalists throughout America now aggressively delving into eugenics, the field will soon be as broad and diversified as the investigations of the Holocaust and American slavery. At least one book could be written for each state, starting with California, which was America's most energetic eugenic state. Critical biographies are needed for the key players. In-depth examinations of the links between Germany and the Pioneer Fund, the Rockefeller Foundation, the Carnegie Institution as well as numerous state officials would be welcome. The role of the Chicago Municipal Court must be further explored.

When I began this project in 2001, many in the public were not even aware of eugenics. Indeed, for a while my publisher did not even want me to include the word *eugenics* in the title of this book. In reality, however, the topic has been continuously explored over the past decades by several extremely talented academics and students hailing from a range of disciplines from biology to education. Although most were gracious and supportive, I was surprised to find that many tended to guard their information closely. One such author told me she didn't believe another book on eugenics was necessary. ("It depends on how nuanced," she said with some discomfort.) Another professor astonished me by asking for money to answer some questions within his expertise—the first time I had encountered such a request in thirty-five years of historical research. When I contacted a Virginia professor who had written a dissertation decades earlier, she actually told me she didn't think a member of the media was "qualified" to read her dissertation. One collaborative scholarly eugenic website, ironically funded by a federal grant, restricts media usage while permitting unrestricted scholarly usage.

As I was completing my work, the public was beginning to discover the outlines of eugenics. The *Richmond Times-Dispatch*, *Winston-Salem Journal*, and several other publications and radio stations, as well as the *Los Angeles Times*, *New York Times* and *American Heritage* magazine, all produced exemplary articles on various aspects of eugenics. The *Winston-Salem Journal* series was a feat of investigative journalism. As the manuscript was being typed, the governors of Virginia, Oregon, California, North Carolina and South Carolina all publicly apologized to the victims of their states' official persecution. Others will follow. The topic is now where it belongs, in the hands of hard-driving journalists and historians who will not stop until they have uncovered all the facts.

Now that newspaper and magazine articles have placed the crime of eugenics on the front burner, my book explains in depth exactly how this fraudulent science infected our society and then reached across the world and right into Nazi Germany. I want the full story to be understood in context. Skipping around in the book will only lead to flawed and erroneous conclusions. So if you intend to skim, or to rely on selected sections, please do not read the book at all. This is the saga of a century and can easily be misunderstood. The realities of the twenties, thirties and forties were very different from each other. I have made this request of my readers on prior books and I repeat it for this volume as well.

Although this book contains many explosive revelations and embarrassing episodes about some of our society's most honored individuals and institutions, I hope its contents will not be misused or quoted out of context by special interests. Opponents of a woman's right to choose could easily seize upon Margaret Sanger's eugenic rhetoric to discredit the admirable work of Planned Parenthood today; I oppose such misuse. Detractors of today's Rockefeller Foundation could easily apply the facts of their Nazi connections to their current programs; I reject the linkage. Those frightened by the prospect of human engineering could invoke the science's eugenic foundations to condemn all genomic research; that would be a mistake. While I am as anxious as the next person about the prospect of out-of-control genomics under the thumb of big business, I hope every genetic advance that helps humanity fight disease will continue as fast and as furiously as possible.

This is the right place to note that virtually all the organizations I investigated cooperated with unprecedented rigor, because they want the history illuminated as much as anyone. This includes the Rockefeller Foundation, the Carnegie Institution, Cold Spring Harbor Laboratory, and the Max Planck Institute, successor to the Kaiser Wilhelm Institute. All gave me unlimited access and unstinting assistance. These organizations have all worked hard to help the world discover their pasts and must be commended. Planned Parenthood worked with me closely day after day, searching for and faxing documents, continually demonstrating their interest in the unvarnished truth. The same can be said for numerous other corporations and organizations. This is a book of history, and corporate and philanthropic America must be commended when they cooperate in an investigation as aggressive and demanding as mine.

Indeed, of the scores of societies, corporations, organizations and governmental agencies I contacted around the world, only one obstructed my work. IBM refused me access to its files. Despite this obstruction, I was able to demonstrate that the race-defining punch card used by the SS in

Nazi Germany was actually derived from one developed for the Carnegie Institution years before Hitler came to power.

This project has been a long, exhausting, exhilarating odyssey for me, one that has taken me to the darkest side of the brightest minds and revealed to me one reason why America has been struggling so long to become the country it still wants to be. We have a distance to go. Again, I ask how did this happen in a progressive society? After reviewing thousands upon thousands of pages of documentation, and pondering the question day and night for nearly two years, I realize it comes down to just one word. It was more than the self-validation and self-certification of the elite, more than just power and influence joining forces with prejudice. It was the corrupter of us all: it was *arrogance*.

EDWIN BLACK
Washington, D.C.
March 15, 2003

A Note on the Text

War *Against the Weak* utilized published and private sources spanning a century, and in several languages, and as such presented numerous textual challenges. We relied upon established style conventions as often as possible, and, when required, adapted and innovated styles. Readers may notice certain inconsistencies. Some explanation follows.

Every phrase of quoted material has remained as true as possible to the original terminology, punctuation and capitalization, even to the point of preserving archaic and sometimes offensive terms when used by the original source. No attempt was made to filter out ethnic denigrations when they appeared in period materials. Eugenicists in America called themselves *eugenicists*, but in Britain referred to themselves as *eugenists*, and sometimes the usage crossed; we used *eugenicists* in narrative but *eugenists* whenever it appeared in a specific quotation. In several instances we quoted from profoundly misspelled handwritten letters, and it was our decision to transcribe these as authentically as possible.

When referring to materials originally published in German, journals and magazines are cited by their legal name in German, such as *Archiv für Rassen- und Gesellschaftsbiologie*, with the first usage including a translation in parentheses. Titles of books are referred to by their English translations; the first usage includes the original German title in parentheses. When multiple translations of a book title or organization name exist, we selected the most appropriate. We made an exception when a book's title rose to the public awareness of a *Mein Kampf*. We used the German *für* whenever possible but were compelled to use the variant *fuer* when it was used in American headlines.

For most points of style, this book has followed *The Chicago Manual of Style*. Unfortunately, not even the near-thousand pages of standards set forth in *Chicago* could cover all the varied forms in which primary information was received. This is especially true when dealing with electronic sources such as Internet web pages, and actual documents—new and old—reproduced in PDF formats, electronic books and other Internet sources. This is one of the first history books to incorporate widespread use of legitimate materials on the Internet. For example, we obtained copies of Papal encyclicals from the Vatican's website, PDFs of original historical programs, and electronic books—all on the Internet. These are legitimate materials when used with extreme caution.

Citing the Internet is a profound challenge. Given the lack of style consensus, and the fact that websites are continuously updated and rearranged, it was necessary to create a new style for Internet citations. We decided to include just two key elements: the website's home page address and the title of the document. General search engines such as Google and site-specific search engines will be the best means of locating the content of these cited pages. Naturally we retained printouts of all cited web materials.

PART ONE

From Peapod
to Persecution

CHAPTER 1

Mountain Sweeps

When the sun breaks over Brush Mountain and its neighboring slopes in southwestern Virginia, it paints a magical, almost iconic image of America's pastoral splendor. Yet there are many painful stories, long unspoken, lurking in these gentle hills, especially along the hiking paths and dirt roads that lead to shanties, cabins and other rustic encampments. Decades later, some of the victims have been compelled to speak.

In the 1930s, the Brush Mountain hill folk, like many of the clans scattered throughout the isolated Appalachian slopes, lived in abject poverty. With little education, often without running water or indoor plumbing, and possessing few amenities, they seemed beyond the reach of social progress. Speaking with the indistinct drawls and slurred vestigial accents that marked them as hillbillies, dressed in rough-hewn clothing or hand-me-downs, and sometimes diseased or poorly developed due to the long-term effects of squalor and malnutrition, they were easy to despise. They were easily considered alien. Quite simply, polite Virginia society considered them white trash.

Yet Brush Mountain people lived their own vibrant rural highlands culture. They sang, played mountain instruments with fiery virtuosity to toe-tapping rhythms, told and retold engaging stories, danced jigs, sewed beautiful quilts and sturdy clothing, hunted fox and deer, fished a pan full and fried it up.[1] Most of all, they hoped for better—better health, better jobs, better schooling, a better life for their children. Hill people did produce great men and women who would increasingly take their places in modern society. But hopes for betterment often became irrelevant because these people inhabited a realm outside the margins of America's dream. As such, their lives became a stopping place for America's long biological nightmare.

A single day in the 1930s was typical. The Montgomery County sheriff drove up unannounced onto Brush Mountain and began one of his many

raids against the hill families considered socially inadequate. More precisely, these hill families were deemed "unfit," that is, unfit to exist in
nature. On this day the Montgomery County sheriff grabbed six brothers
from one family, bundled them into several vehicles and then disappeared
down the road. Earlier, the sheriff had come for the boys' sister. Another
time, deputies snared two cousins.[2]

"I don't know how many others they took, but they were after a lot of
them," recalled Howard Hale, a former Montgomery County supervisor,
as he relived the period for a local Virginia newspaper reporter a half century later. From Brush Mountain, the sheriff's human catch was trucked to
a variety of special destinations, such as Western State Hospital in
Staunton, Virginia. Western State Hospital, formerly known as the Western Lunatic Asylum, loomed as a tall-columned colonial edifice near a hill
at the edge of town. The asylum was once known for its so-called "moral
therapy," devised by Director Dr. Francis T. Stribling, who later became
one of the thirteen founding members of the American Psychiatric
Association. By the time Brush Mountain hillbillies were transported
there, Western housed not only those deemed insane, but also the so-
called "feebleminded."[3]

No one was quite sure how "feebleminded" was defined.[4] No matter.
The county authorities were certain that the hill folk swept up in their raids
were indeed mentally—and genetically—defective. As such, they would
not be permitted to breed more of their kind.

How? These simple mountain people were systematically sterilized
under a Virginia law compelling such operations for those ruled unfit.
Often, the teenage boys and girls placed under the surgeon's knife did not
really comprehend the ramifications. Sometimes they were told they were
undergoing an appendectomy or some other unspecified procedure.
Generally, they were released after the operation. Many of the victims did
not discover why they could not bear children until decades later when the
truth was finally revealed to them by local Virginia investigative reporters
and government reformers.[5]

Western State Hospital in Staunton was not Virginia's only sterilization mill. Others dotted the state's map, including the Colony for
Epileptics and the Feebleminded near Lynchburg, the nation's largest
facility of its kind and the state's greatest center of sterilization. Lynchburg
and Western were augmented by hospitals at Petersburg, Williamsburg
and Marion. Lower-class white boys and girls from the mountains, from
the outskirts of small towns and big city slums were sterilized in assembly

line fashion. So were American Indians, Blacks, epileptics and those suffering from certain maladies—day after day, thousands of them as though orchestrated by some giant machine.[6]

Retired Montgomery County Welfare Director Kate Bolton recalled with pride, "The children were legally committed by the court for being feebleminded, and there was a waiting list from here to Lynchburg." She added, "If you've seen as much suffering and depravity as I have, you can only hope and pray no one else goes through something like that. We had to stop it at the root."[7]

"Eventually, you knew your time would come," recalled Buck Smith about his Lynchburg experience. His name is not really Buck Smith. But he was too ashamed, nearly a half century later, to allow his real name to be used during an interview with a local Virginia reporter. "Everybody knew it. A lot of us just joked about it....We weren't growed up enough to think about it. We didn't know what it meant. To me it was just that 'my time had come.'"[8]

Buck vividly recounted the day he was sterilized at Lynchburg. He was fifteen years old. "The call came over the dormitory just like always, and I knew they were ready for me," he remembered. "There was no use fighting it. They gave me some pills that made me drowsy and then they wheeled me up to the operating room." The doctor wielding the scalpel was Lynchburg Superintendent Dr. D. L. Harrell Jr., "who was like a father to me," continued Buck. Dr. Harrell muttered, "Buck, I'm going to have to tie your tubes and then maybe you'll be able to go home." Drowsy, but awake, Buck witnessed the entire procedure. Dr. Harrell pinched Buck's scrotum, made a small incision and then deftly sliced the sperm ducts, rendering Buck sterile. "I watched the whole thing. I was awake the whole time," Buck recalled.[9]

Buck Smith was sterilized because the state declared that as a feebleminded individual, he was fundamentally incapable of caring for himself. Virginia authorities feared that if Buck were permitted to reproduce, his offspring would inherit immutable genetic traits for poverty and low intelligence. Poverty, or "pauperism," as it was called at the time, was scientifically held by many esteemed doctors and universities to be a genetic defect, transmitted from generation to generation. Buck Smith was hardly feebleminded, and he spoke with simple eloquence about his mentality. "I've worked eleven years at the same job," he said, "and haven't missed more than three days of work. There's nothing wrong with me except my lack of education."[10]

"I'll never understand why they sterilized me," Buck Smith disconsolately told the local reporter. "I'll never understand that. They [Lynchburg] gave me what life I have and they took a lot of my life away from me. Having children is supposed to be part of the human race."[11]

The reporter noticed a small greeting card behind Buck Smith. The sterilized man had eventually married and formed a lasting bond with his stepchildren. The card was from those stepchildren and read: "Thinking of you, Daddy." Through tears, Buck Smith acknowledged the card, "They call me *Daddy*."[12]

Mary Donald was equally pained when she recalled her years of anguish following her sterilization at Lynchburg when she was only eleven. Several years later, she was "released" to her husband-to-be, and then enjoyed a good marriage for eighteen years. But "he loved kids," she remembered. "I lay in bed and cried because I couldn't give him a son," she recounted in her heavily accented but articulate mountain drawl. "You know, men want a son to carry on their name. He said it didn't matter. But as years went by, he changed. We got divorced and he married someone else." With these words, Mary broke down and wept.[13]

Like so many, Mary never understood what was happening. She recalled the day doctors told her. "They ask me, 'Do you know what this meeting is for?' I said, 'No, sir, I don't.' 'Well this is a meeting you go through when you have to have a serious operation, and it's for your health.' That's the way they expressed it. 'Well,' I said, 'if it's for my health, then I guess I'll go through with it.' See, I didn't know any difference." Mary didn't learn she had been sterilized until five years after her operation.[14]

The surgeon's blade cut widely. Sometimes the victims were simply truants, petty thieves or just unattended boys captured by the sheriffs before they could escape. Marauding county welfare officials, backed by deputies, would take the youngsters into custody, and before long the boys would be shipped to a home for the feebleminded. Many were forced into virtual slave labor, sometimes being paid as little as a quarter for a full week of contract labor. Runaways and the recalcitrant were subject to beatings and torturous ninety-day stints in a darkened "blind room." Their release was generally conditional on family acquiescence to their sterilization.[15]

Mary Donald, "Buck Smith," the brothers from Brush Mountain and many more whose names have long been forgotten are among the more than eight thousand Virginians sterilized as a result of coercion, stealth and deception in a wide-ranging program to prevent unwanted social, racial and ethnic groups from propagating. But the agony perpetrated against these people was hardly a local story of medical abuse. It did not end at the Virginia state line. Virginia's victims were among some sixty thousand who were forcibly sterilized all across the United States, almost half of them in California.[16]

Moreover, the story of America's reproductive persecution constitutes far more than just a protracted medical travesty. These simple Virginia people, who thought they were isolated victims, plucked from their remote mountain homes and urban slums, were actually part of a grandiose, decades-long American movement of social and biological cleansing determined to obliterate individuals and families deemed inferior. The intent was to create a new and superior mankind.

The movement was called eugenics. It was conceived at the onset of the twentieth century and implemented by America's wealthiest, most powerful and most learned men against the nation's most vulnerable and helpless. Eugenicists sought to methodically terminate all the racial and ethnic groups, and social classes, they disliked or feared. It was nothing less than America's legalized campaign to breed a super race—and not just any super race. Eugenicists wanted a purely Germanic and Nordic super race, enjoying biological dominion over all others.[17]

Nor was America's crusade a mere domestic crime. Using the power of money, prestige and international academic exchanges, American eugenicists exported their philosophy to nations throughout the world, including Germany. Decades after a eugenics campaign of mass sterilization and involuntary incarceration of "defectives" was institutionalized in the United States, the American effort to create a super Nordic race came to the attention of Adolf Hitler.

Those declared unfit by Virginia did not know it, but they were connected to a global effort of money, manipulation and pseudoscience that stretched from rural America right into the sterilization wards, euthanasia vans and concentration camps of the Third Reich. Prior to World War II, the Nazis practiced eugenics with the open approval of America's eugenic crusaders. As Joseph DeJarnette, superintendent of Virginia's Western State Hospital, complained in 1934, "Hitler is beating us at our own game."[18]

Eventually, out of sight of the world, in Buchenwald and Auschwitz, eugenic doctors like Josef Mengele would carry on the research begun just years earlier with American financial support, including grants from the Rockefeller Foundation and the Carnegie Institution. Only after the secrets of Nazi eugenics horrified the world, only after Nuremberg declared compulsory sterilization a crime against humanity, did American eugenics recede, adopt an enlightened view and then resurface as "genetics" and "human engineering."[19] Even still, involuntary sterilization continued for decades as policy and practice in America.

True, the victims of Virginia and hundreds of thousands more like them in countries across the world were denied children. But they did give birth to a burning desire to understand how the most powerful, intelligent, scholarly and respectable individuals and organizations in America came to mount a war against the weakest Americans to create a super race. Just as pressing is this question: Will the twenty-first-century successor to the eugenics movement, now known as "human engineering," employ enough safeguards to ensure that the biological crimes of the twentieth century will never happen again?

CHAPTER 2

Evolutions

Mankind's quest for perfection has always turned dark. Man has always existed in perpetual chaos. Continuously catapulted from misery to exhilaration and back, humanity has repeatedly struggled to overcome vulnerability and improve upon its sense of strength. The instinct is to "play God" or at least mediate His providence. Too often, this impulse is not just to improve, but to repress, and even destroy those deemed inferior.

Eventually, the Judeo-Christian world codified the principle that all human life should be valued. A measure of our turbulent civilization and even of our humanity has always been how well people have adhered to that precept. Indeed, as societies became more enlightened, they extended respect for life to an ever-widening circle of people, including the less fortunate and the less strong.

Racism, group hatred, xenophobia and enmity toward one's neighbors have existed in almost every culture throughout history. But it took millennia for these deeply personal, almost tribal hostilities to migrate into the safe harbor of scientific thought, thus rationalizing destructive actions against the despised or unwanted.

Science offers the most potent weapons in man's determination to resist the call of moral restraint. To forge the new science of human oppression—a race science—several completely disconnected threads of history twined. Indeed, it took centuries of development for three disciplines—socioeconomics, philosophy and biology—to come together into a resilient and fast-moving pseudoscience that would change the world forever.

Perhaps the story truly begins with the simple concept of charity. Charity is older than the Bible.[1] Organized refuges for the poor and helpless date to the Roman era and earlier.[2] The concept of extending a helping hand was established in the earliest Judeo-Christian doctrine. "There will always be poor people in the land, therefore, I command you to be open-handed toward your brothers and toward the poor and needy in your land," declared Deuteronomy.[3] Jesus Christ based his ministry on helping the helpless—the lame, the blind, lepers, the mentally deranged, and social outcasts such as thieves and prostitutes. He proclaimed, "The meek... shall inherit the earth."[4]

After the Roman Empire adopted Christianity, the Canones Arabici Nicaeni of 325 A.D. mandated the expansion of hospitals and other monastic institutions for the sick and needy.[5] During medieval times, the church was chiefly responsible for "houses of pity."[6] In England, such charitable institutions for the poor were abundantly required.

The Black Death killed millions across Europe between 1348 and 1350. Labor shortages motivated bands of itinerant workers and beggars to wander from town to town in search of the highest paying pittance. As they wandered, many resorted to petty thievery, highway robbery, and worse. With their impoverished existence came the associated afflictions of illiteracy, poor health, rampant disease and physical disability.[7]

During the early and mid-1500s, economic upheavals took their toll on all but the richest of the nobility. Silver from the New World and official coinage debasements caused prices to rise, increasing the suffering of the poor. Tribes of vagrants migrated from the countryside to villages. Later, in response to the booming wool market, England's landowners switched from estate farming to vast sheep breeding enterprises. Consequently, great numbers of farm workers were evicted from their peasant domiciles, bloating the hordes of the unemployed and destitute. This teeming hardship only increased the church's role in tending to a multitude of the wretched and poor.[8]

Everything changed in the 1530s when Pope Clement VII refused to annul Henry VIII's marriage to Catherine of Aragon. Furious, King Henry seized church property and monasteries in England, and charitable institutions slowly became a governmental responsibility.[9] Tending to the poor was expensive but the alternative was food riots.[10]

By the early sixteenth century, the first poor laws were enacted in England. Such measures categorized the poor into two groups. The *deserving poor* were the very young and the very old, the infirm and families who fell on financial difficulties due to a change in circumstances. The *undeserving poor* were those who had turned to crime—such as highwaymen, pickpockets, and professional beggars—and also included paupers who roamed the country looking for a day's work. The undeserving poor were considered an affliction upon society, and the law laid out harsh punishment. Poverty, or more precisely, *vagrancy*, was criminalized. Indeed, the concept of criminal vagrancy for those with "no visible means of support" has persisted ever since.[11]

Despite all attempts to contain welfare spending, England's enormous expenditures only escalated. In 1572, compulsory poor law taxes were assessed to each community to pay for poor houses and other institutions that cared for the deranged, diseased and decrepit among them. These taxes created a burden that many resented.[12] Now it was the poor and helpless against the rest of society.

Indeed, a distinct pauper class had emerged. These people were perceived by the establishment as both an arrogant lot who assumed an inherited "right to relief," and as seething candidates for riot and revolution. Overcrowded slums and dismal poorhouses caused England to reform its poor laws and poverty policies several times during the subsequent three hundred years. The urbanization of poverty was massively accelerated by the Industrial Revolution, which established grim, sunless sweatshops and factories that in turn demanded—and exploited—cheap labor. Appalling conditions became the norm, inspiring Charles Dickens to rouse the public in novels such as *Oliver Twist*. Despite progress, by the mid-1800s the state was still spending £1,400 a year (equivalent to about $125,000 in modern money) per 10,000 paupers. The ruling classes increasingly rebelled against "taxing the industrious to support the indolent."[13]

Soot-smeared and highly reproductive, England's paupers were looked down upon as a human scourge. The establishment's derogatory language began to define these subclasses as subhumans. For example, a popular 1869 book, *The Seven Curses of London*, deprecated "those male and female pests of every civilized community whose natural complexion is dirt, whose brow would sweat at the bare idea of earning their bread, and whose stock-in-trade is rags and impudence."[14]

England's complex of state-sponsored custodial institutions stretched across a distant horizon. Over time, the proliferation of poor houses, lunacy asylums, orphanages, health clinics, epilepsy colonies, rescue shelters, homes for the feebleminded and prisons inevitably turned basic Christian charity into what began to be viewed as a social plague.

While Britain's perceived social plague intensified, a new social philosophy began evolving in Europe. In 1798, English economist Thomas Malthus published a watershed theory on the nature of poverty and the controlling socioeconomic systems at play. Malthus reasoned that a finite food supply would naturally inhibit a geometrically expanding human race. He called for population control by moral restraint. He even argued that in many instances charitable assistance promoted generation-to-generation poverty and simply made no sense in the natural scheme of human progress. Many who rallied behind Malthus's ideas ignored his complaints about an unjust social and economic structure, and instead focused on his rejection of the value of helping the poor.[15]

In the 1850s, agnostic English philosopher Herbert Spencer published *Social Statics*, asserting that man and society, in truth, followed the laws of cold science, not the will of a caring, almighty God. Spencer popularized a powerful new term: "survival of the fittest." He declared that man and society were evolving according to their inherited nature. Through evolution, the "fittest" would naturally continue to perfect society. And the "unfit" would naturally become more impoverished, less educated and ultimately die off, as well they should. Indeed, Spencer saw the misery and starvation of the pauper classes as an inevitable decree of a "far-seeing benevolence," that is, the laws of nature. He unambiguously insisted, "The whole effort of nature is to get rid of such, and to make room for better.... If they are not sufficiently complete to live, they die, and it is best they should die." Spencer left no room for doubt, declaring, "all imperfection must disappear." As such, he completely denounced charity and instead extolled the purifying elimination of the "unfit." The unfit, he argued, were predestined by their nature to an existence of downwardly spiraling degradation.[16]

As social and economic gulfs created greater generation-to-generation disease and dreariness among the increasing poor, and as new philosophies suggested society would only improve when the unwashed classes faded away, a third voice entered the debate. That new voice was the voice of hereditary science.

In 1859, some years after Spencer began to use the term "survival of the fittest," the naturalist Charles Darwin summed up years of observation in a lengthy abstract entitled *The Origin of Species*. Darwin espoused "natural selection" as the survival process governing most living things in a world of limited resources and changing environments. He confirmed that his theory "is the doctrine of Malthus applied with manifold force to the whole animal and vegetable kingdoms; for in this case, there can be no artificial increase of food, and no prudential restraint from marriage."[17]

Darwin was writing about a "natural world" distinct from man. But it wasn't long before leading thinkers were distilling the ideas of Malthus, Spencer and Darwin into a new concept, bearing a name never used by Darwin himself: *social Darwinism*.[18] Now social planners were rallying around the notion that in the struggle to survive in a harsh world, many humans were not only less worthy, many were actually destined to wither away as a rite of progress. To preserve the weak and the needy was, in essence, an unnatural act.

Since ancient times, man has understood the principles of breeding and the lasting quality of inherited traits. The Old Testament describes Jacob's clever breeding of his and Laban's flocks, as spotted and streaked goats were mated to create spotted and streaked offspring. Centuries later, Jesus sermonized, "A good tree cannot bear bad fruit, and a bad tree cannot bear good fruit. Every tree that does not bear good fruit is cut down and thrown into the fire."[19]

Good stock and preferred traits were routinely propagated in the fields and the flocks. Bad stock and unwanted traits were culled. Breeding, whether in grapes or sheep, was considered a skill subject to luck and God's grace.

But during the five years between 1863 and 1868, three great men of biology would all promulgate a theory of evolution dependent upon identifiable hereditary "units" within the cells. These units could actually be seen under a microscope. Biology entered a new age when its visionaries proclaimed that good and bad traits were not bestowed by God as an inscrutable divinity, but transmitted from generation to generation according to the laws of science.

Spencer, in 1863, published *Principles of Biology*, which suggested that heredity was under the control of "physiological units."[20]

Three years later, the obscure Czech monk Gregor Mendel published his experiments with smooth-skinned and wrinkled peas; he constructed a predictable hereditary system dependent on inherited cellular "elements."[21]

Finally, in 1868, Darwin postulated the notion that "the units throw off minute granules which are dispersed throughout the entire system.... They are collected from all parts of the system to constitute the sexual elements, and their development in the next generation forms a new being; but they are likewise capable of transmission in a dormant state to future generations." Darwin named these minute granules *gemmules*.[22]

By any name, science had now pulled away the shroud covering the genetic realities of mankind.

Far-flung notions of social planning, philosophy and biology—centuries in the making—now gravitated toward each other, culminating in a fascinating new ideology that sought to improve the human race—not by war or charity, but by the progressive logic of science and mathematics. The driving force behind this revelation was not really a scientist, although his scientific methodology influenced many scientists. He was not really a philosopher, although his ability to weave scientific principles into social philosophy spawned fiery movements of dogma. He was not really a physician, although his analyses of human physiology ultimately governed much of the surgical and medical profession. The man was Francis J. Galton. He was above all a clever and compulsive counter—a counter of things, of phenomena, of traits, of all manner of occurrences, obvious and obscure, real and conjured. If any pattern could be discerned in the cacophony of life, Galton's piercing ratiocination could detect it and just maybe systemize it to the level of predictability.

Galton never finished his studies at London's King College Medical School and instead studied math at Cambridge, where he quickly became an aficionado of the emerging field of statistics.[23] He joyously applied his arithmetic prowess and razor-like powers of observation to everyday life, seeking correlation. Galton distinguished himself by his ability to recognize patterns, making him an almost unique connoisseur of nature—sampling, tasting and discerning new character in seemingly random flavors of chaos.

More than correlation, Galton's greatest quest was prediction. To his mind, what he could predict, he could outwit—even conquer. And so Galton's never-ending impulse was to stand before life and defy its mysteries, one by one, with his indomitable powers of comprehension.

Perhaps counting relieved the throbbing of his constant headaches or was an intellectual consequence of his insatiable desire to excel. More than once, he succumbed to palpitations and even a nervous breakdown amidst the fury of his cogitations. Even his visage seemed sculpted to seek and measure. A pair of bushy eyebrows jutted out above his orbits almost like two hands cupped over the brow of a man peering into an unfathomable distance. At the same time, his dense windswept sideburns swerved back dramatically just behind his earlobes, as though his mind was speeding faster than the rest of his head.[24]

Galton counted the people fidgeting in an audience and tried to relate it to levels of interest. He tried to make sense of waves in his bathtub. He gazed from afar at well-endowed women, using a sextant to record their measurements. "As the ladies turned themselves...to be admired," wrote Galton, "I surveyed them in every way and subsequently measured the distance of the spot where they stood...and tabulated the results at my leisure." He even tried to map the concentration of beauty in Britain by noting how many lovely women were located in different regions of the country.[25]

Galton's favorite adage was, "Whenever you can, count."[26]

Much of Galton's quantitative musings amounted to little more than distraction. But some of it became solid science. In 1861, he distributed a questionnaire to the weather stations of Europe, asking the superintendents to record all weather details for the month of December. He found a pattern. Analyzing the data, Galton drew up the world's first weather maps, peppering them with his own idiosyncratic symbols for wind direction, temperature and barometric pressure. His maps, revealing that counterclockwise wind currents marked sudden changes in pressure, eventually made isobaric charts possible. Galton's 1863 publication, *Meteorographica: or Methods of Mapping the Weather*, greatly advanced the science of meteorology.[27]

Later, he discovered that the raised ridges on human fingertips were each unique. No two were alike. He devised a system for analyzing and categorizing the distinctive sworls, and inking them into a permanent record. Galton simply called these fingerprints. The new discipline permitted the identification of criminals—this at a time when a wave of crime by unidentifiable felons gripped London and Jack the Ripper prowled the East End. Galton's book, *Finger Prints*, featured the author's own ten arched across the page as a personal logotype.[28]

About the time Darwin, Spencer and Mendel began explaining the heredity of lower species, Galton was already looking beyond those theories. He began to discern the patterns of various qualities in human beings. In 1865, Galton authored a two-part series for *Macmillan Magazine* that he expanded four years later into a book entitled *Hereditary Genius*. Galton studied the biographical dictionaries and encyclopedias, as well as the genealogies of eminent scholars, poets, artists and military men. Many of them were descendants of the same families. The frequency was too impressive to ignore. Galton postulated that heredity not only transmitted physical features, such as hair color and height, but mental, emotional and creative qualities as well. Galton counted himself among the eminent, since he was Darwin's cousin, and both descended from a common grandfather.[29]

Galton reasoned that talent and quality were more than an accident. They could be calculated, managed and sharpened into a "highly gifted race of men by judicious marriages during several consecutive generations." Far from accepting any of Malthus's notions of inhibited procreation, Galton suggested that bountiful breeding of the best people would evolve mankind into a superlative species of grace and quality. He actually hoped to create a regulated marriage process where members of the finest families were only wed to carefully selected spouses.[30]

Galton did not worry that inbred negative qualities would multiply. He said there was "no reason to suppose that, in breeding for the higher order of intellect, we should produce . . . a feeble race." He explained his own incapacitating physical frailties away as a manifestation of hereditary distinction. "Men who leave their mark on the world," wrote Galton, "are very often those who, being gifted and full of nervous power, are at the same time haunted and driven by a dominant idea, and are therefore within a measurable distance of insanity."[31]

Galton struggled to find the pattern, the predictability, the numerical formula that governed the character of progeny. Mathematics would be the key to elevating his beliefs from an observation to a science. He didn't have the answer yet, but Galton was certain that the secret of scientific breeding could be revealed—and that it would forever change humankind. "Could not the undesirables be got rid of and the desirables multiplied?" he asked.[32]

In 1883, Galton published *Inquiries into Human Faculty and Development* and created a new term for his discipline. He played with many names for his new science. Finally, he scrawled Greek letters on a hand-sized scrap of paper, and next to them the two English fragments he would join into one. The Greek word for *well* was abutted to the Greek word for *born*.[33]

In a flourish, Galton invented a term that would tantalize his contemporaries, inspire his disciples, obsess his later followers and eventually slash through the twentieth century like a sword. The finest and the fiendish would adopt the new term as their driving mantra. Families would be shattered, generations would be wiped away, whole peoples would be nearly erased—all in the name of Galton's word. The word he wrote on that small piece of paper was *eugenics*.[34]

* * *

Eugenics was a protoscience in search of vindicating data. Galton had described the eugenically well-born man as a trend in science, but he desperately sought to quantify the biological process. After all, if Galton could advance from merely discovering the scientific mechanism controlling human character to actually predicting the quality of the unborn, his knowledge would become almost divine. In theory, the master of any enforced eugenics program could play God—deciding who would be born and who would not. Indeed, the notion of constructing a brave new world by regimented reproduction has never receded.

Numbers were needed. In 1884, Galton opened his Anthropometric Laboratory at London's International Health Exhibition. Using questionnaires—just as he had in quantifying weather—Galton asked families to record their physical characteristics, such as height, weight and even lung power. Later Galton even offered cash rewards for the most comprehensive family history. The data began to accrue. It wasn't long before nine thousand people, including many complete families, offered their physical details for Galton's calculations.[35] He began pasting numbers together, sculpting formulas, and was finally able to patch together enough margins of error and coefficients of correlation into a collection of statistical eugenic probabilities.

At the same time, German cellular biologist August Weismann, using more powerful microscopes, announced that something called "germ plasm" was the true vehicle of heredity. Weismann observed what he termed a "nucleus." He theorized, "The physical causes of all apparently unimportant hereditary habits... of hereditary talents, and other mental peculiarities must all be contained in the minute quantity of germ-plasm which is possessed by the nucleus of a germ cell."[36] Others would later identify character-conveying threads termed "chromatic loops" or "chromosomes."

Superseding Darwinian precepts of descent and Weismann's germ plasm, Galton, in his essays and an 1889 book entitled *Natural Inheritance*, tried to predict the precise formulaic relationship between ancestors and their descendants. He concluded, "The influence, pure and simple, of the mid-parent may be taken as $\frac{1}{2}$, of the mid-grandparent $\frac{1}{4}$, of the mid-great-grandparent $\frac{1}{8}$, and so on. That of the individual parent would therefore be $\frac{1}{4}$, of the individual grandparent $\frac{1}{16}$, of an individual in the next generation $\frac{1}{64}$, and so on." In other words, every person was the measurable and predictable sum of his ancestors' immortal germ plasm. Inheritable traits included not only physical characteristics, such as eye color and height, but subtle qualities, such as intellect, talent and personality. Galton ultimately reduced all notions of heritage, talent and character to a series of complex, albeit fatally flawed, eugenic equations.[37]

Above all, Galton concluded that the caliber of progeny always reflected its distant ancestry. Good lineage did not improve bad blood. On the contrary, in any match, undesirable traits would eventually outweigh desirable qualities.[38] Hence, when eugenically preferred persons mated with one another, their offspring were even more valuable. But mixing eugenically well-endowed humans with inferior mates would not strengthen succeeding generations. Rather, it would promote a downward biological spiral. What was worse, two people of bad blood would only create progressively more defective offspring.

It was all guesswork, ancestral solipsism and mathematical acrobatics—some of it well-founded and some of it preposterous—forged into a self-congratulatory biology and social science. Scholarly kudos and celebration abounded. Yet Galton himself was forced to admit in 1892, in the preface to the second edition of *Hereditary Genius*, that his theories and formulae were still completely unprovable. "The great problem of the future betterment of the human race is confessedly, at the present time, hardly advanced beyond the state of academic interest."[39]

Years later, in a preface to a eugenic tract about gifted families, Galton again warned that musing about "improved breeds" of the human race were still nothing more than "speculations on the theoretical possibility."[40]

Nonetheless, Galton remained convinced that germ-plasm was the ultimate, elusive governing factor. As such, environment and the quality of existence were by and large irrelevant and actually an impediment to racial improvement. No amount of social progress or intervention could help the unfit, he insisted. Qualifying his sense of charity with a biological imperative, Galton asserted, "I do not, of course, propose to neglect the sick, the feeble or the unfortunate. I would do all...for their comfort and happiness, but I would exact an equivalent for the charitable assistance they receive, namely, that by means of isolation, or some other drastic yet adequate measure, a stop should be put to the production of families of children likely to include degenerates."[41]

Galton called for a highly regulated marriage licensing process that society at large would endorse. By prohibiting eugenically flawed unions and promoting well-born partners, Galton believed "what Nature does blindly, slowly and ruthlessly, man may do providently, quickly and kindly."[42]

Galton believed that eugenics was too broad a societal quest to be left to individual whim. He espoused a new definition of eugenics that wed the biology to governmental action. "Eugenics," asserted Galton, "is the study of all agencies under social control which can improve or impair the racial quality of future generations."[43]

Galton's ideas ultimately became known as "positive eugenics," that is, suggesting, facilitating, predicting and even legally mandating biologically conducive marriages. Every family hopes its offspring will choose wisely, and Galton hoped his scientific, equation-filled epistles would encourage families and government bureaus to require as much. His convictions, even those involving legislation and marriage regimentation, were, within his own utopian context, deemed noninvasive and nondestructive.

But a few years later, by the dawn of the twentieth century, Galton's notions of voluntary family planning and positive governmental structures would be transmogrified into an entirely different constellation of negative and coercive thought. The new faithful called it "negative eugenics." Galton died in 1911. With his passing, his positive eugenic principles of marriage regimentation also disappeared from the eugenics main stage. Certainly his name lived on as a rallying call, stamped on the plaques of societies and academic departments. But before long others would come along to chew up his ideas and spit them out as something new and macabre, barely resembling the original.

What Galton hoped to inspire in society, others were determined to force upon their fellow man. If Galton was correct—and these new followers were certain he was—why wait for personal choice or flimsy statutory power? In their minds, future generations of the genetically unfit—from the medically infirm to the racially unwanted to the economically impoverished—would have to be wiped away. Only then could genetic destiny be achieved for the human race—or rather, the white race, and more specifically, the Nordic race. The new tactics would include segregation, deportation, castration, marriage prohibition, compulsory sterilization, passive euthanasia—and ultimately extermination.

As the twentieth century opened for business, the eugenic spotlight would now swing across the ocean from England to the United States. In America, eugenics would become more than an abstract philosophy; it would become an obsession for policymakers. Galton could not have envisioned that his social idealism would degenerate into a ruthless campaign to destroy all those deemed inadequate. But it would become nothing less than a worldwide eugenic crusade to abolish all human inferiority.

CHAPTER 3

America's National Biology

America was ready for eugenics before eugenics was ready for America. What in England was the biology of class, in America became the biology of racial and ethnic groups. In America, class was, in large measure, racial and ethnic.

Everything Galtonian eugenics hoped to accomplish with good matrimonial choices, American eugenicists preferred to achieve with draconian preventive measures designed to delete millions of potential citizens deemed unfit. American eugenicists were convinced they could forcibly reshape humanity in their own image. Their outlook was only possible because American eugenicists believed the unfit were essentially subhuman, not worthy of developing as members of society. The unfit were diseased, something akin to a genetic infection. This infection was to be quarantined and then eliminated. Their method of choice was selective breeding—spaying and cutting away the undesirable, while carefully mating and grooming the prized stock.

Breeding was in America's blood. America had been breeding humans even before the nation's inception. Slavery thrived on human breeding. Only the heartiest Africans could endure the cruel middle passage to North America. Once offloaded, the surviving Africans were paraded atop auction stages for inspection of their physical traits.[1]

Notions of breeding society into betterment were never far from post–Civil War American thought. In 1865, two decades before Galton penned the word *eugenics*, the utopian Oneida Community in upstate New York declared in its newspaper that, "Human breeding should be one of the foremost questions of the age...." A few years later, with freshly expounded Galtonian notions crossing the Atlantic, the Oneida commune began its first selective human breeding experiment with fifty-three female and thirty-eight male volunteers.[2]

21

Feminist author Victoria Woodhull expressed the growing belief that both positive and negative breeding were indispensable for social improvement. In her 1891 pamphlet, *The Rapid Multiplication of the Unfit*, Woodhull insisted, "The best minds of today have accepted the fact that if superior people are desired, they must be bred; and if imbeciles, criminals, paupers and [the] otherwise unfit are undesirable citizens they must not be bred."[3]

America was ready for eugenic breeding precisely because the most established echelons of American society were frightened by the demographic chaos sweeping the nation. England had certainly witnessed a mass influx of foreigners during the years leading up to Galton's eugenic doctrine. But the scale in Britain was dwarfed by America's experience. So were the emotions.

America's romantic "melting pot" notion was a myth. It did not exist when turn-of-the-century British playwright Israel Zangwill optimistically coined the term.[4] In Zangwill's day, America's shores, as well as the three thousand miles in between, were actually a cauldron of undissolvable minorities, ethnicities, indigenous peoples and other tightly-knit groups— all constantly boiling over.

Eighteen million refugees and opportunity-seeking immigrants arrived between 1890 and 1920. German Lutherans, Irish Catholics, Russian Jews, Slavic Orthodox—one huddled mass surged in after another.[5] But they did not mix or melt; for the most part they remained insoluble.

But ethnic volatility during the late 1800s arose from more than the European influx. Race and group hatred crisscrossed the continent. Millions of Native Americans were being forced onto reservations. Mexican multitudes absorbed after the Mexican-American War, in which Mexico lost fully half its land to United States expansion, became a clash point in the enlarged American West and Southwest. Emancipated African slaves struggled to emerge across the country. But freed slaves and their next generation were not absorbed into greater society. Instead, a network of state and local Jim Crow laws enforced apartheid between African Americans and whites in much of the nation, especially in the South. The Chinese Exclusion Act of 1882 temporarily halted the immigration through California of any further Chinese laborers, and blocked the naturalization of those already in the country; the measure was made permanent in 1902.[6]

"Race suicide" was an alarum commonly invoked to restrict European immigration, as 1880 Census Bureau Director Francis Walker did in his 1896 *Atlantic Monthly* article, "Restriction of Immigration." Walker lamented the statistical imbalance between America's traditional Anglo-

Saxon settlers and the new waves flowing in from southern Europe. Eminent sociologist E. A. Ross elevated the avoidance of "race suicide" to a patriotic admonishment, decrying "the beaten members of beaten breeds" from Croatia, Sicily and Armenia flooding in through Ellis Island. Ross warned that such groups "lack the ancestral foundations of American character, and even if they catch step with us they and their children will nevertheless impede our progress."[7]

As the nineteenth century closed, women still could not vote, Native Americans who had survived governmental genocide programs were locked onto often-barren reservations, and Blacks, as well as despised "white trash," were still commonly lynched from the nearest tree—from Minnesota to Mississippi. In fact, 3,224 Americans were lynched in the thirty-year period between 1889 and 1918—702 white and 2,522 black. Their crimes were as trivial as uttering offensive language, disobeying ferry regulations, "paying attention to [a] white girl," and distilling illicit alcohol.[8]

The century ahead was advertised as an epoch for social progress. But the ushers of that progress would be men and women forged from the racial and cultural fires of prior decades. Many twentieth-century activists were repelled by the inequities and lasting scars of racial and social injustice; they were determined to transform America into an egalitarian republic. But others, especially American eugenicists, switched on the lights of the new century, looked around at the teeming, dissimilar masses and collectively declared they had unfinished business.

Crime analysis moved race and ethnic hatred into the realm of heredity. Throughout the latter 1800s, crime was increasingly viewed as a group phenomenon, and indeed an inherited family trait. Criminologists and social scientists widely believed in the recently identified "criminal type," typified by "beady eyes" and certain phrenological shapes. The notion of a "born criminal" became popularized.[9] Ironically, when robber barons stole and cheated their way into great wealth, they were lionized as noble leaders of the day, celebrated with namesake foundations, and honored by leather-bound genealogies often adorned with coats of arms. It was the petty criminals, not the gilded ones, whom polite society perceived as the great genetic menace.

Petty criminals and social outcasts were abundant in Ulster County, New York. Little did these seemingly inconsequential people know they were making history. In the first decades of the nineteenth century, this rustic Catskill Mountain region became a popular refuge for urban dropouts who preferred to live off the land in pastoral isolation. Fish and game were abundant. The lifestyle was lazy. Civilization was yonder. But as wealthy New Yorkers followed the Hudson River traffic north, planting opulent Victorian mansions and weekend pleasure centers along its banks, the very urbanization that Ulster's upland recluses spurned caught up to them. Pushed from their traditional fishing shores and hillside hunting grounds, where they lived in shanties, the isolated, unkempt rural folk of Ulster now became "misfits." Not a few of them ran afoul of property and behavior laws, which became increasingly important as the county's population grew.[10] Many found themselves jailed for the very lifestyle that had become a local tradition.

In 1874, Richard Dugdale, an executive of the New York Prison Association, conducted interviews with a number of Ulster County's prisoners and discovered that many were blood relatives. Consulting genealogies, courthouse and poorhouse records, Dugdale documented the lineages of no fewer than forty-two families heavily comprised of criminals, beggars, vagrants and paupers. He claimed that one group of 709 individuals were all descendants of a single pauper woman, known as Margaret and crowned "mother of criminals." Dugdale collectively dubbed these forty-two troubled families "the Jukes." His 1877 book, *The Jukes, a Study in Crime, Pauperism, Disease and Heredity*, calculated the escalating annual cost to society for welfare, imprisonment and other social services for each family. The text immediately exerted a vast influence on social scientists across America and around the world.[11]

While Dugdale's book spared no opportunity to disparage the human qualities of both the simple paupers and the accomplished criminals among the Jukes family, he blamed not their biology, but their circumstances. Rejecting notions of heredity, Dugdale instead zeroed in on the adverse conditions that created generation-to-generation pauperism and criminality. "The tendency of heredity is to produce an environment which perpetuates that heredity," he wrote. He called for a change in social environment to correct the problem, and predicted that serious reform could effect a "great decrease in the number of commitments" within fifteen years. Dugdale cautioned against statistics that inspired false conclusions. He even reminded readers that not a few wealthy clans made their fortunes by cheating the masses—yet these scandalous people were considered among the nation's finest families.[12]

But Dugdale's cautions were ignored. His book was quickly hailed as proof of a hereditary defect that spawned excessive criminality and poverty—even though this was the opposite of what he wrote. For example, Robert Fletcher, president of the Anthropological Society of Washington, insisted in a major 1891 speech that germ plasm ruled, that one criminal bred another. "The taint is in the blood," Fletcher staunchly told his audience, "and there is no royal touch which can expel it.... Quarantine the evil classes as you would the plague."[13]

The Jukes was the first such book, but not the last. Tribes of paupers, criminals and misfits were tracked and traced in similar books. The Smokey Pilgrims of Kansas, the Jackson Whites of New Jersey, the Hill Folk of Massachusetts and the Nam family of upstate New York were all portrayed as clans of defective, worthless people, a burden to society and a hereditary scourge blocking American progress. Most convincing was a presentation made in 1888 to the Fifteenth National Conference of Charities and Correction by the social reformer Reverend Oscar McCulloch. McCulloch, a Congregationalist minister from Indianapolis, presented a paper entitled *Tribe of Ishmael: A Study of Social Degeneration.* The widely-reported speech described whole nomadic pauper families dwelling in Indianapolis, all related to a distant forefather from the 1790s.[14]

Ishmael's descendants were in fact bands of roving petty thieves and con artists who had victimized town and countryside, giving McCulloch plenty of grist for his attack on their heredity. He compared the Ishmael people to the Sacculina parasites that feed off crustaceans. Paupers were inherently of no value to the world, he argued, and would only beget succeeding generations of paupers—and all "because some remote ancestor left its independent, self-helpful life, and began a parasitic, or pauper life." His research, McCullouch assured, "resembles the study of Dr. Dugdale into the Jukes and was suggested by that."[15]

Many leading social progressives devoted to charity and reform now saw crime and poverty as inherited defects that needed to be halted for society's sake. When this idea was combined with the widespread racism, class prejudice and ethnic hatred that already existed among the turn-of-the-century intelligentsia—and was then juxtaposed with the economic costs to society—it created a fertile reception for the infant field of eugenics. Reformers possessed an ingrained sense that "good Americans" could be bred like good racehorses.

Galton had first pronounced his theory of the well-born in 1883. For the next twenty years, eugenics bounced around America's intellectual circles as a perfectly logical hereditary conclusion consistent with everyday observations. But it lacked specifics. Then, as one of the first sparks of the twentieth century, Gregor Mendel's theory of heredity was rediscovered.

True, between 1863 and 1868, various theories of heredity had been published by three men: Spencer, Darwin and Mendel. But while Darwin and Spencer presided with great fanfare in London's epicenter of knowledge, Mendel was alone and overlooked by the world of science he aspired to.

The son of simple mountain peasants, Mendel was not socially adept. Combative exchanges with those in authority made him prefer solitude. "He who does not know how to be alone is not at peace with himself," he wrote. Originally, he had hoped to devote himself to the natural sciences. But he failed at the university and retreated to an Augustinian monastery in Brno, Moravia. There, while tending the gardens, he continued the work of a long line of students of plant hybridization.[16]

Mendel preferred peas. Peering through flimsy wire-rim glasses into short tubular microscopes and scribbling copious notes, Mendel studied over ten thousand cross-fertilized pea plants. Key differences in their traits could be predicted, depending upon whether he bred tall plants with short plants, or plants yielding smooth pods with plants yielding wrinkled pods. Eventually, he identified certain governing inheritable traits, which he called "dominating" and "recessive." These could be expressed in mathematical equations, or traced in a simple genealogical chart filled with line-linked A's and B's. Among his many conclusions: when pea plants with wrinkled skins were crossed with plants yielding smooth skins, the trait for wrinkled skin dominated.[17] In other words, the smooth peapod skin was corrupted by wrinkled stock. Wrinkled peapods ultimately became a powerful image to those who found the human simile compelling.

Mendel's scientific paper, describing ten years of tedious work, was presented to a local scientific society in Brno and mailed to several prominent biologists in Europe, but it was ignored by the scientific world. Mendel grew more unhappy with the rejection. His combative exchanges with local officials on unrelated issues were so embarrassing to the order that when Mendel died in 1884, the monastery burned all his notes.[18]

In May of 1900, however, the esteemed British naturalist and Darwin disciple William Bateson unexpectedly discovered references to Mendel's laws of heredity in three separate papers. The three papers were independently researched and simultaneously submitted by three different students. Amazed at Mendel's findings, an excited Bateson announced to the world through the Royal Horticultural Society that he had "rediscovered" Mendel's crucial studies in plant heredity. The science that Bateson called *genetics* was born. Mendel's laws became widely discussed throughout the horticultural world.[19]

But Galton's eugenic followers understood that the biological arithmetic of peapods, cattle and other lower species did not ordain the futures of the most complex organism on earth: *Homo sapiens*. Height, hair color, eye color and other physical attributes could be partially explained in Mendelian terms. But intelligent, thought-driven humans beings were too subtle, too impressionable, too variable and too unpredictable to be reduced to a horticultural equation. Man's environment and living conditions were inherent to his development. Nutrition, prenatal and childhood circumstances, disease, injury, and upbringing itself were all decisive, albeit not completely understood, factors that intervened in the development of any individual. Some of the best people came from the worst homes, and some of the worst people came from the best homes.

Hence, during the first decade of the twentieth century, as Mendel was being debated, most Galtonian eugenicists admitted that their ideas were still too scantily clad to be called science, too steeped in simple statistics rather than astute medical knowledge, too preliminary to even venture into the far-reaching enterprise of organized human breeding. Eugenics was all just theory and guesswork anyway. For example, in 1904 Galton wrote to his colleague Bateson seeking any initial evidence of "Mendelianism in Man," suggesting that any data could contribute to what he still called a "theoretical point of view." In another 1904 letter, Galton reminded Bateson, "I do indeed fervently hope that exact knowledge may be gradually attained and established beyond question, and I wish you and your collaborators all success in your attempts to obtain it."[20]

As late as 1910, Galton's most important disciple, mathematician Karl Pearson, head of the Eugenics Laboratory, admitted just how thin their knowledge was. In a scientific paper treating eugenics and alcoholism, Pearson confessed, "The writers of this paper are fully conscious of the slenderness of their data; they have themselves stated that many of their conclusions are probabilities ... rather than demonstrations. They will no doubt be upbraided with publishing anything at all, either on the ground that what they are dealing with is 'crude and worthless material' or that as 'mathematical outsiders,' they are incapable of dealing with a medico-social problem." Pearson added in a footnote that he also understood why some would find the linkage of eugenics and alcoholism an act of inebriation in itself. He went on to quote a critic: "The educated man and the scientist is as prone as any other to become the victim ... of his prejudices. ... He will in defense thereof make shipwreck of both the facts of science and the methods of science ... by perpetrating every form of fallacy, inaccuracy and distortion."[21]

Galton himself dismissed the whole notion of human breeding as socially impossible—with or without the elusive data he craved. "We can't mate men and women as we please, like cocks and hens," Galton quipped to Bateson in 1904. At the time, Galton was defending his recently published *Index to Achievements of Near Kinfolk*, which detailed how talent and skill run in the same celebrated families. Wary of being viewed as an advocate of human breeding, Galton's preface cautioned Mendelian devotees with strong conditionals, ifs and buts. "The experience gained in establishing improved breeds of domestic animals and plants," he wrote, "is a safe guide to speculations on the theoretical possibility of establishing improved breeds of the human race. It is not intended to enter here into such speculations but to emphasize the undoubted fact that members of gifted families are...more likely...to produce gifted offspring."[22]

Nor did Galton believe regulated marriages were a realistic proposition in any democratic society. He knew that "human nature would never brook interference with the freedom of marriage," and admitted as much publicly. In his published memoir, he recounted his original error in suggesting such utopian marriages. "I was too much disposed to think of marriage under some regulation," he conceded, "and not enough of the effects of self-interest and of social and religious sentiment."[23]

Unable to achieve a level of scientific certainty needed to create a legal eugenic framework in Britain, Galton hoped to recast eugenics as a religious doctrine governing marriages, a creed to be taken on faith without proof. Indeed, faith without proof constitutes the essence of much religious dogma. Eugenical marriage should be "strictly enforced as a religious duty, as the Levirate law ever was," wrote Galton in a long essay, which listed such precedents in the Jewish, Christian and even primitive traditions. He greeted the idea of a religion enthusiastically, suggesting, "It is easy to let the imagination run wild on the supposition of a whole-hearted acceptance of eugenics as a national religion."[24]

Many of Galton's followers agreed that founding a national religion was the only way eugenics could thrive. Even the playwright George Bernard Shaw, a eugenic extremist, agreed in a 1905 essay that "nothing but a eugenic religion can save our civilization." Late in his life, in 1909, Galton declared that eugenics in a civilized nation would succeed only as "one of its religious tenets."[25]

But in America, it did not matter that Galton and his followers found themselves fighting for intellectual acceptance with little evidence on their side. Nor did it matter that British eugenic leaders themselves admitted that eugenics did not rise to a level of scientific certainty sufficient to formulate public policy. Nor did it matter that Mendel's newly celebrated laws of heredity might make good sense for peapods, but not for thinking, feeling men, women and children.

In America, racial activists had already convinced themselves that those of different races and ethnic backgrounds considered inferior were no more than a hereditary blight in need of eugenic cleansing. Many noted reformers even joined the choir. For example, in a 1909 article called "Practical Eugenics," the early twentieth-century education pioneer John Franklin Bobbitt insisted, "In primal days was the blood of the race kept high and pure, like mountain streams." He now cautioned that the "highest, purest tributaries to the stream of heredity" were being supplanted by "a rising flood in the muddy, undesirable streams."[26]

Bobbitt held out little value for the offspring of "worm-eaten stock." Although considered a social progressive, he argued that the laws of nature mandating "survival of the fittest" were constantly being countermanded by charitable endeavors. "Schools and charities," he harangued, "supply crutches to the weak in mind and morals... [and] corrupt the streams of heredity." Society, he pleaded, must prevent "the weaklings at the bottom from mingling their weakness in human currents."[27]

Defective humans were not just those carrying obvious diseases or handicaps, but those whose lineages strayed from the Germanic, Nordic and/or white Anglo-Saxon Protestant ideal. Bobbitt made clear that only those descended from Teutonic forefathers were of pure blood. In one such remonstration, he reminded, "One must admit the high purity of their blood, their high average sanity, soundness and strength. They were a well-born, well-weeded race." Eugenic spokesman Madison Grant, trustee of the American Museum of Natural History, stated the belief simply in his popular book, *The Passing of the Great Race*, writing that Nordics "were the white man par excellence."[28]

Indeed, the racism of America's first eugenic intellectuals was more than just a movement of whites against nonwhites. They believed that Germans and Nordics comprised the supreme race, and a typical lament among eugenic leaders such as Lothrop Stoddard was that Nordic populations were decreasing. In *The Rising Tide of Color Against White World Supremacy*, Stoddard wrote that the Industrial Revolution had attracted squalid Mediterranean peoples who quickly outnumbered the more desirable Nordics. "In the United States, it has been much the same story. Our country, originally settled almost exclusively by Nordics, was toward the close of the nineteenth century invaded by hordes of immigrant Alpines and Mediterraneans, not to mention Asiatic elements like Levantines and Jews. As a result, the Nordic native American has been crowded out with amazing rapidity by these swarming, prolific aliens, and after two short generations, he has in many of our urban areas become almost extinct." Madison Grant agreed: "The term 'Caucasian race' has ceased to have any meaning."[29]

By no means did the eugenics movement limit its animus to non-English speaking immigrants. It was a movement against non-Nordics regardless of their skin color, language or national origin. For example, Stoddard denigrated the "swart cockney" in Britain "as a resurgence of the primitive Mediterranean stock, and probably a faithful replica of his ancestors of Neolithic times." All mixed breeds were vile. "Where the parent stocks are very diverse," wrote Stoddard, "as [in] matings between whites, Negroes and Amerindians, the offspring is a mongrel—a walking chaos, so consumed by his jarring heredities that he is quite worthless."[30]

Grant's tome lionized the long-headed skulls, blue eyes and blond hair of true Nordic stock, and outlined the complex history of Nordic migrations and invasions across Eurasia and into Great Britain. Eventually, these Nordic settlements were supplanted by lesser breeds, who adopted the Nordic and Anglo-Saxon languages but were in fact the carriers of corrupt human strains.[31]

Indeed, those Americans descended from lower-class Scottish and Irish families were also viewed as a biological menace, being of Mediterranean descent. Brunette hair constituted an ancestral stigma that proved a non-Nordic bloodline. Any claims by such people to Anglo-Saxon descent because of language or nationality were considered fraudulent. Grant railed, "No one can question . . . on the streets of London, the contrast between the Piccadilly gentleman of Nordic race and the cockney coster-monger [street vendor] of the Neolithic type."[32] Hence, from Ulster County to the Irish slums of Manhattan, to the Kentucky and Virginia hills, poor whites were reviled by eugenicists not for their ramshackle and destitute lifestyles, but for a heredity that supposedly made pauperism and criminality an inevitable genetic trait.

Even when an individual of the wrong derivation was healthy, intelligent and successful, his existence was considered dangerous. "There are many parents who, in many cases, may themselves be normal, but who produce defective offspring. This great mass of humanity is not only a social menace to the present generation, but it harbors the potential parenthood of the social misfits of our future generations."[33]

Race mixing was considered race suicide. Grant warned: "The cross between a white man and an Indian is an Indian; the cross between a white man and a Negro is a Negro; the cross between a white man and a Hindu is a Hindu; and the cross between any of the three European races and a Jew is a Jew."[34]

The racial purity and supremacy doctrines embraced by America's pioneer eugenicists were not the ramblings of ignorant, unsophisticated men. They were the carefully considered ideals of some of the nation's most respected and educated figures, each an expert in his scientific or cultural field, each revered for his erudition.

So when the facts about Mendel's peapods appeared in America in 1900, these influential and eloquent thinkers were able to slap numbers and a few primitive formulas on their class and race hatred, and in so doing create a passion that transcended simple bigotry. Now their bigotry became science—race science. Now Galtonian eugenics was reborn, recast and redirected in the United States as a purely and uniquely American quest.

To succeed, all American eugenics needed was money and organization. Enter Andrew Carnegie.

Steel made Andrew Carnegie one of America's wealthiest men. In 1901, the steel magnate sold out to J. P. Morgan for $400 million and retreated from the world of industry. The aging Scotsman would henceforth devote his fortune to philanthropy. The next year, on January 28, 1902, the millionaire endowed his newly created Carnegie Institution with $10 million in bonds, followed by other endowments totaling $12 million. The entity was so wealthy that in 1904, Washington agreed to reincorporate the charity by special act of Congress, chartering the new name "Carnegie Institution of Washington." This made the Carnegie Institution a joint incarnation of the steel man's money and the United States government's cachet.[35]

The Carnegie Institution was established to be one of the premier scientific organizations of the world, dedicated by charter "to encourage, in the broadest and most liberal manner, investigation, research, and discovery, and the application of knowledge to the improvement of mankind." Twenty-four of America's most respected names in science, government and finance were installed as trustees. The celebrated names included National Library of Medicine cofounder John Billings, Secretary of War Elihu Root and philanthropist Cleveland Dodge. Renowned paleontologist John C. Merriam became president. Merriam and his staff were required under the bylaws to closely scrutinize and preapprove all activities, audit all expenditures and regularly publish research results.[36]

Several principal areas of scholarly investigation were identified from the worthy realms of geophysics, astronomy and plant biology. Now another scientific endeavor would be added: negative eugenics. The program would quickly become known as "the practical means for cutting off defective germ-plasm" and would embrace a gamut of remedies from segregation to sterilization to euthanasia.[37] This radical human engineering program would spring not from the medical schools and health clinics of America, but from the pastures, barns and chicken coops—because the advocates of eugenics were primarily plant and animal breeders. Essentially, they believed humans could be spawned and spayed like trout and horses.

America's formless eugenics movement found its leader in zoologist Charles Davenport, a man who would dominate America's human breeding program for decades. Davenport, esteemed for his Harvard degrees and his distinguished background, led the wandering faithful out of the wilderness of pure prejudice and into the stately corridors of respectability. More than anyone else, it was Davenport who propelled baseless American eugenics into settled science—wielding a powerful sociopolitical imperative.

Who was Charles Benedict Davenport?

He was a sad man. No matter how celebrated Davenport became within his cherished circles, throughout his career he remained a bitter and disconsolate person boxing shadows for personal recognition. Even as he judged the worthiness of his fellow humans, Davenport struggled to prove his own worthiness to his father and to God. Ironically, it was his mother who inspired the conflict between devotion to science and subservience to God that Davenport would never bridge.[38]

Davenport grew up in Brooklyn Heights as the proud descendent of a long line of English and Colonial New England Congregationalist ministers. His authoritarian father, Amzi Benedict Davenport, did not join the clergy, but nonetheless cloaked his family's world in the heavy mantle of puritanical religion. The elder Davenport's business was real estate. But as a cofounder of two Brooklyn churches—ruling elder of one and a longtime deacon of the other—Amzi Davenport infused his household with pure fire and brimstone, along with the principles of commerce and market value. He demanded from his family impossible levels of Bible-thumbing rectitude and imposed an unyielding disdain for joy.[39]

A close friend described the father's face as one of "bitter unhappiness," and characterized his parental manner as "harsh masterfulness." Charles Davenport was the last of eleven children. Siblings were born like clockwork in the Davenport home, every two years. Rigorous and often punishing Gospel studies intruded into every aspect of young Davenport's upbringing, morning and night. The boy's diary records one typical entry about grueling Sunday school lessons. Using personal shorthand and misspelling as a boy would, young Davenport scribbled, "stuiding S.S. lesson from 8:30 A.M. to 9:30 P.M. All day!" Once, it was the day after Christmas, he jotted, "Woke at 6:30 A.M. and was late for prayers. After breakfast father sent me to bed for that reason for two hours."[40]

Ancestry was a regular theme in the Davenport household. The elder Davenport organized two extensive volumes of family genealogy, tracing his Anglo-Saxon tree back to 1086. That was the year William the Conqueror compiled his massive Domesday census book.[41] Shades of Davenport's glorified forebearers must have pursued the boy at every moment.

Yet in the midst of young Davenport's dour, patriarchal domination, his mother Jane was somehow permitted to live a life of irrepressible brightness. A Dutch woman, Jane offered unconditional affection to her children, a wonderful flower garden to delight in, and a fascination with natural history. Young Davenport's refuge from the severe and unapproachable man he trepidatiously called "Pa" was the world of beauty his mother represented.[42]

When Davenport as a young man escaped from theology into academia, it was to the world of measurable mysteries: science, math and engineering. In doing so, he declared that God's work was not infinite—it could indeed be quantified. That surely spurned the absolutist precepts of his father's sermonizing. Later, Davenport dedicated his first scientific book, *Experimental Morphology*, "to the memory of the first and most important of my teachers of Natural History—my mother." Such inscriptions were not a sign of intellectual liberation. Davenport was never quite comfortable with his defection to the world of nature. At one point, he formally requested his father's written permission to study the sciences; seven weeks later he finally received an answer permitting it. His father's written acquiescence hinged on "the question of prime importance, [that] is how much money can you make for yourself and for me."[43]

After his graduation from Brooklyn Polytechnic, Davenport became a civil engineer. His love of animals and natural history led Davenport to Harvard, where he enrolled in nearly every natural science course offered and quickly secured his doctorate in biology. In the 1890s, he became a zoology instructor at Harvard. Later, he held a similar position at the University of Chicago.[44]

Long-headed and mustachioed, Davenport always looked squeezed. His goatee created a slender but dense column from chin to lower lip; as he aged, it would fade from black to white. With a deeply parted haircut hanging high above his ears, Davenport's face tapered from round at the top to a distinct point at the inverted apex of his beard.[45]

Davenport married Gertrude Crotty in 1894. A fellow biologist, Gertrude would continually encourage him to advance in personal finance and career. However, Davenport never escaped his upbringing. Puritanical in his sexual mores, domineering in his own family relationships, inward and awkward in most other ways, Davenport was described by a close lifelong colleague as "a lone man, living a life of his own in the midst of others, feeling out of place in almost any crowd." Worse, while Davenport's thirst for scholarly validation never quenched, he could not tolerate criticism. Hearing adverse comments, reading them, just sensing that rejection might dwell between the lines of a simple correspondence caused Davenport so much distress, he could blurt out the wrong words, sometimes the exact opposite of his intent. Criticism paralyzed him.[46] Yet this was the scientist who would discover and deliver the evidence that would decide the biological fate of so many.

Davenport's pivotal role as eugenic crusader-in-chief began taking shape at the very end of the nineteenth century. He found a modicum of professional and personal success directing the Brooklyn Institute of Arts and Sciences's biological laboratory on Long Island. There, he could apply his precious Harvard training. The quiet, coveside facility at Cold Spring Harbor was located about an hour's train and carriage ride from Manhattan. Situated down the road from the state fish hatchery, and ensconced in a verdant, marshy inlet ideal for marine and mammal life, the biological station allowed Davenport to concentrate on the lowest species. He investigated such organisms as the Australian marine pill bug, which clings to the underside of submerged rocks and feeds on rotted algae. He employed drop nets to dredge for oysters and other mollusks. Flatfish and winter flounder were purchased for spawning studies.[47]

To supplement his income during school breaks, Davenport, aided by botany instructors from other institutions, offered well-regarded summer courses at Cold Spring Harbor. Students in bacteriology, botany and animal biology from across the nation were attracted to these courses.[48] Davenport also corresponded with other academic institutions, which pleased him greatly.

While at the Brooklyn Institute's biological station, Davenport became fascinated with Galton. In a series of fawning missives to Galton during the spring of 1897, Davenport praised the British scientist's work, requested his photograph, and ultimately tried to schedule a meeting in London that summer. Galton hardly knew what to make of the unsolicited admiration. "I am much touched," Galton replied to Davenport's earliest praise, "by the extremely kind expression in your letter, though curious that you ascribe to me more than I deserve."[49] The two exchanged brief notes thereafter. Davenport's were formal and typed. Galton's were scrawled on monarch stationery.

Davenport incorporated the statistical theories of Galton and Galton's disciple, Pearson, into an 1899 book, *Statistical Methods with Special Reference to Biological Variation*. He wanted the volume to be a serious scientific publication of international merit, and he proudly mailed a copy to Galton for his inspection. Galton penned back a short word of thanks for "your beautiful little book with its kindly and charming lines." Later, Galton sent Davenport some sample fingerprints to examine.[50] But meteorology, statistics and fingerprints were only the threshold to the real body of Galtonian knowledge that riveted Davenport. The precious revelation for the American biologist was the study of superiority and ancestry, the principle Galton called eugenics.

Eugenics appealed to Davenport not just because his scientific mind was shaped by a moralized world choked with genealogies and ancestral comparisons, but because of his racial views and his obsession with race mixture.[51] Davenport saw ethnic groups as biologically different beings—not just physically, but in terms of their character, nature and quality. Most of the non-Nordic types, in Davenport's view, swam at the bottom of the hereditary pool, each featuring its own distinct and indelible adverse genetic features. Italians were predisposed to personal violence. The Irish had "considerable mental defectiveness," while Germans were "thrifty, intelligent, and honest."[52]

Social reformers may have held out hope that America's melting pot might one day become a reality, but eugenicists such as Davenport's outspoken ally Lothrop Stoddard spoke for the whole movement when he declared, "Above all, there is no more absurd fallacy than the shibboleth of 'the melting pot.' As a matter of fact, the melting pot may mix but does not melt. Each race-type, formed ages ago, and 'set' by millenniums of isolation and inbreeding, is a stubbornly persistent entity. Each type possesses a special set of characters: not merely the physical characters visible to the naked eye, but moral, intellectual and spiritual characters as well. All these characters are transmitted substantially unchanged from generation to generation."[53]

When Mendel's laws reappeared in 1900, Davenport believed he had finally been touched by the elusive but simple biological truth governing the flocks, fields and the family of man. He once preached abrasively, "I may say that the principles of heredity are the same in man and hogs and sun-flowers."[54] Enforcing Mendelian laws along racial lines, allowing the superior to thrive and the unfit to disappear, would create a new superior race. A colleague of Davenport's remembered him passionately shaking as he chanted a mantra in favor of better genetic material: "Protoplasm. We want more protoplasm!"[55]

Shortly after the Carnegie Institution appeared in 1902, in its pre-Congressional form, Davenport acted to harness the institution's vast financial power and prestige to launch his eugenic crusade. The Carnegie Institution was just months old, when on April 21, 1902, Davenport outlined a plan for the institution to establish a Biological Experiment Station at Cold Spring Harbor "to investigate...the method of Evolution." Total initial cost was estimated to be $32,000.[56]

By the time Davenport penned his formal proposal to Carnegie trustees two weeks later on May 5, 1902, his intent was unmistakably racial: "The aims of this establishment would be the analytic and experimental study of...race change." He explained how: "The methods of attacking the problem must be developed as a result of experience. At present, the following seems the most important: Cross-breeding of animals and plants to find the laws of commingling of qualities. The study of the laws and limits of inheritance." Davenport tantalized the trustees with the prospect: "The Carnegie fund offers the opportunity for which the world has so long been waiting."[57]

Hence from the very start, the trustees of the Carnegie Institution understood that Davenport's plan was a turning-point plan for racial breeding.

Redirecting human evolution had been a personal mission of Davenport's for years, long before he heard of Mendel's laws. He first advocated a human heredity project in 1897 when he addressed a group of naturalists, proposing a large farm for preliminary animal breeding experiments. Davenport called such a project "immensely important." With the Carnegie Institution now receptive to his more grandiose idea, Davenport knew it was important to continue rallying support from the scientific establishment. He convinced the Brooklyn Institute of Arts and Science, which controlled the lab site at Cold Spring Harbor, to form a prestigious scientific committee to press the "plan for a permanent research laboratory...in connection with the Carnegie Institution at Washington."[58]

Knowing Carnegie officials would refer the question to the institution's Zoological Committee, Davenport elicited support from prominent zoologists.[59] In May of 1902, he sent a letter of tempting intrigue to his friend Professor Henry Fairfield Osborn, director of the New York Zoological Society and the American Museum of Natural History. "I do not think this is the place to tell in detail what I should expect to do," wrote Davenport, adding only, "The station should undertake to do what is impracticable elsewhere."[60]

Osborn, a like-minded eugenicist, wrote back with encouragement, reporting that Carnegie's committee had considered the general topic before. British eugenicists had already approached Andrew Carnegie directly. But Osborn assured, "I know of no one better qualified to do this work than you."[61]

Shoring up his knowledge and enlisting wider consensus, Davenport traveled to Europe for four months, where he briefly visited with Galton. The founding eugenicist warned Davenport that any such effort must be a serious scientific enterprise, not just "any attempt at showy work, for the sake of mere show." Untroubled, Davenport traveled to several European marine life research centers gathering academic accord for his project.[62]

Fresh from his European travels, and fortified with the latest international views on eugenics, Davenport dispatched to the Carnegie Institution a more detailed letter plus a lengthy report on the state of human evolution studies to date. The documents made clear that far-reaching American race policy could not be directed without supportive scientific data based on breeding experiments with lower species. The results of those experiments would be applied in broad strokes to humans. "Improvement of the human race can probably be effected only by understanding and applying these methods," he argued. "How appalling is our ignorance, for example, concerning the effect of a mixture of races as contrasted with pure breeding; a matter of infinite importance in a country like ours containing numerous races and subspecies of men."[63]

Davenport hoped to craft a super race of Nordics. "Can we build a wall high enough around this country," he asked his colleagues, "so as to keep out these cheaper races, or will it be a feeble dam . . . leaving it to our descendants to abandon the country to the blacks, browns and yellows and seek and an asylum in New Zealand."[64]

Man was still evolving, he reasoned, and that evolution could and should be to a higher plane. Carnegie funds could accelerate and direct that process. "But what are these processes by which man has evolved," posited

Davenport, "and which we should know...in hastening his further evolution." He disputed the value of improved conditions for those considered genetically inferior. He readily admitted that with schooling, training and social benefits, "a person born in the slums can be made a useful man." But that usefulness was limited in the evolutionary scheme of things. No amount of book learning, "finer mental stuff" or "intellectual accumulation" would transfer to the next generation, he insisted, adding that "permanent improvement of the race can only be brought about by breeding the best."[65]

Drawing on his belief in raceology, Davenport offered the Carnegie trustees an example he knew would resonate: "We have in this country the grave problem of the negro," he wrote, "a race whose mental development is, on the average, far below the average of the Caucasian. Is there a prospect that we may through the education of the individual produce an improved race so that we may hope at last that the negro mind shall be as teachable, as elastic, as original, and as fruitful as the Caucasian's? Or must future generations, indefinitely, start from the same low plane and yield the same meager results? We do not know; we have no data. Prevailing 'opinion' says we must face the latter alternative. If this were so, it would be best to export the black race at once."[66]

Proof was needed to fuel the social plans the eugenicists and their allies championed. Davenport was sure he could deliver the proof. "As to a person to carry out the proposed work," he wrote Carnegie, "I am ready at the present moment to abandon all other plans for this." To dispel any doubt of his devotion, Davenport told the institution, "I propose to give the rest of my life unreservedly to this work."[67]

The men of Carnegie were impressed. They said *yes*.

*　*　*

During 1903, while the esteemed men of the Carnegie Institution were readying their adventure into eugenics, Davenport worked to broaden support for the perception of American eugenics as a genuine science. Since the great men of medicine were, for the most part, devoted to improving individual health, not stunting it, few of them wanted to be affiliated with the nascent movement. So Davenport instead turned to the great men of the stable, the field and the barnyard.

He found a willing ear at the newly established American Breeders Association. The ABA was created in 1903 by the Association of Agricultural Colleges and Experimental Stations, after four years of preparatory effort spurred by a request from the U.S. Secretary of Agriculture. The

American government urged animal breeders and seed experts to "join hands." The idea of bringing the two groups together was first suggested to Washington in 1899 by the Hybridizer's Conference in London meeting under the auspices of the Royal Horticultural Society. In light of Mendel's discoveries about peapods, the American government pushed the plan.[68]

Many breeders were convinced that their emerging Mendelian knowledge about corn and cattle was equally applicable to the inner quality of human beings. A typical declaration came from one New York State breeder: "Every race-horse, every straight-backed bull, every premium pig tells us what we can do and what we must do for man.... The results of suppressing the poorest and breeding from the best would be the same for them as for cattle and sheep."[69]

At the ABA's first annual meeting in St. Louis during the chilly final days of December 1903, Davenport was well received and elected to the permanent five-man oversight committee. Two organizational sections were established: Plants and Animals. But Davenport prevailed upon the ABA to add a third group, a so-called Eugenics Committee. The establishing resolution declared the committee should "devise methods of recording the values of the blood of individuals, families, people and races." The resolution specified that the goal was to "emphasize the value of superior blood and the menace to society of inferior blood."[70]

Eventually, Davenport bluntly confessed to an ABA audience: "Society must protect itself; as it claims the right to deprive the murderer of his life, so also it may annihilate the hideous serpent of hopelessly vicious protoplasm." A report to the committee called for broad public awareness through "popular magazine articles, in public lectures ... in circular letters to physicians, teachers, the clergy and legislators." The report decried "such mongrelization as is proceeding on a vast scale in this country.... Shall we not rather take the steps ... to dry up the springs that feed the torrent of defective and degenerate protoplasm?" In the process, the report claimed, the United States would curtail the $100 million in annual expenditures for the destitute, insane, feebleminded, defective and criminal elements—a group comprised of at least two million people. How? The report, circulated to the entire ABA membership and the federal government, was explicit: "By segregation during the reproductive period or even by sterilization."[71]

Once defectives were eliminated in America, the same methods could be employed worldwide. ABA president Willet Hays, who also served as assistant secretary of agriculture, authored an article entitled "Constructive Eugenics" for *American Breeders Magazine*, in which he proposed a global solution to all unwanted races. "Eugenic problems are much the same throughout as the problems of plant breeding and animal improvement," wrote Hays, adding, "May we not hope to...lop off the defective classes below, and also increase the number of the efficient at the top?" His suggestion? A massive international numbering convention, assigning descriptive eleven-digit "number names" to every man, woman and child on earth using census bureaus. By creating a series of nearly 100 billion numbers, for an estimated world population of only 1.5 billion, Hays hoped to enroll "every person now living, any person of whom there is any history, and any person who might be born in the next thousand years.... No two persons would have the same number." These eleven-digit "number names" would not only identify the individual, they would trace his lineage and assign a genetic rating, expressed as a percentage. Methodically, one nation after another would identify its population and eliminate the unwanted strains. "Who, except the prudish, would object if public agencies gave to every person a lineage number and genetic percentage ratings, that the eugenic value of every family and of every person might be available to all who have need of the truth as to the probable efficiency of the offspring."[72]

On January 19, 1904, the Carnegie Institution formally inaugurated what it called the Station for Experimental Evolution of the Carnegie Institution at bucolic Cold Spring Harbor. Davenport's annual salary was fixed at $3,500 plus travel expenses. It was a significant compensation package for its day. For example, in 1906, the president of the University of Florida received only $2,500 per year, and Northwestern University's librarian earned only $1,200.[73]

A new building for the experimental station costing $20,000 was approved. Everything would be first class, as it should be, endowed by Andrew Carnegie's fortune. The undertaking was not merely funded by Carnegie, it was an integral part of the Carnegie Institution itself. Letterhead prominently made it clear at the top that the station was wholly part of the Carnegie Institution. Moreover, the purse strings would be tightly held with the smallest activity being considered in advance and authorized after approval. "The sum of $300 [shall] be paid to Prof. Davenport to enable him to procure certain animals for the proposed laboratory," instructed Carnegie's chairman, John Billings, "... provided that he shall furnish properly acceptable vouchers for the expenditure of this money."[74]

Billings was fastidious about record keeping and supervision. He was one of America's most distinguished citizens. Some would eventually call him "the father of medical and vital statistics" in the United States. He ensured that medical statistics were included in the United States Census of 1880, and he took a leadership role in drawing up the nation's vital statistics for the censuses of 1890 and 1900. During Billings's tenure in the Surgeon General's Office, he was considered America's foremost expert on hygiene.[75]

Billings and the Carnegie Institution would now mobilize their prestige and the fortune they controlled to help Davenport usher America into an age of a new form of hygiene: racial hygiene. The goal was clear: to eliminate the inadequate and unfit. Now it was time to search the nation, from its busiest metropolises to its most remote regions, methodically identifying exactly which families were qualified to continue and which were not.

CHAPTER 4

Hunting the Unfit

The Carnegie Institution's Station for Experimental Evolution at Cold Spring Harbor opened for business in 1904. But in the beginning, little happened. The experimental station's first years were devoted to preparatory work, mostly because Davenport was fundamentally unsure of just how he would go about reshaping mankind in his image. "I have little notion of just what we shall do," Davenport confided in a note. "We shall reconnoiter the first year." [1]

So Davenport focused on the basics. Lab animals were purchased: a tailless Manx cat, long-tailed fowl, canaries and finches for breeding experiments. Hundreds of seeds were acquired for Mendelian exercises. A staff was hired, including an animal keeper from Chicago, several research associates, an expert in botany and entomology, plus a gardener and a librarian. The librarian assembled shelf after shelf of the leading English, German and French biology publications: 2,000 books, 1,500 pamphlets, and complete sets of twenty-three leading journals, including *American Journal of Physiology*, *Canadian Entomologist*, *Der Zoologische Garten* and *L'Annee Biologique*. Associates were recruited from the scholarly ranks of Harvard, the University of Chicago, Columbia University and other respected institutions to actively research and consult. Corresponding scientists were attracted from Cambridge, Zurich, Vienna, Leipzig and Washington, D.C. to share their latest discoveries from the fields of entomology, zoology and biology.[2]

Davenport was so busy getting organized that the Carnegie Institution did not issue its official announcement about the experimental station until more than a year later, in March of 1905.[3]

Indeed, only after Davenport had recruited enough scholars and amassed enough academic resources to create an aura of eugenic preeminence, did he dispatch a letter to Galton, in late October of 1905, inviting him to become a so-called "correspondent." Clearly, Davenport wanted

Galton's name for its marquee value. "Acceptance of this invitation," Davenport wrote, "[is] implying only [a] mutual intention to exchange publications and occasionally ideas by letter." But Galton was reluctant. "You do me honor in asking," Galton scribbled back, ". . . but I could only accept in the understanding that it is an wholly honorary office, involving no duties whatever, for I have already more on my head than I can properly manage." That said, Galton asked Davenport to "exercise your own judgment" before using his name "under such bald restriction."[4]

During the next two years, Davenport's new experimental station confined its breeding data to the lower life forms, such as mice, canaries and chickens, and he contributed occasional journal articles, such as one on hereditary factors in human eye color.[5]

But how could Davenport translate his eugenic beliefs into social action?

Talk and theories gave way to social intervention at the December 1909 American Breeders Association meeting in Omaha, Nebraska. Subcommittees had already been formed for different human defects, such as insanity, feeblemindedness, criminality, hereditary pauperism and race mongrelization. Davenport encouraged the ABA to escalate decisively from pure hereditary research into specific ethnic and racial investigation, propaganda and lobbying for legislation. He convinced his fellow breeders to expand the small Eugenics Committee to a full-fledged organizational section. ABA members voted yes to Davenport's ideas by a resounding 499 to 5. Among his leading supporters was Alexander Graham Bell, famous for inventing the telephone and researching deafness, but also a dedicated sheep breeder and ardent eugenicist.[6]

Now the real work began. Davenport and Bell had already devised a so-called "Family Record" questionnaire. Bell agreed to use his influence and circulate the forms to high schools and colleges. The ABA also agreed to distribute five thousand copies. Davenport's eugenic form asked pointed questions about eye defects, deafness and feeblemindedness in any of a suspect family's ancestry. Bell wondered why Davenport would not also trace the excellence in a suspect family, as well as its defects.[7]

But Davenport was only interested in documenting human defects in other races and ethnic groups, not their achievements. He believed that inferiority was an inescapable dominant Mendelian trait. Even if a favorable environment produced a superior individual, if that individual derived from inferior ethnic or racial stock, his progeny would still constitute a biological "menace."[8]

Davenport's scientific conclusion was already set in his mind; now he craved the justifying data. Even with the data, making eugenics a practical and governing doctrine would not be easy. American demographics were rapidly transforming. Political realities were shifting. Davenport well understood that as more immigrants filed into America's overcrowded political arena, they would vote and wield power. Race politics would grow harder and harder to legislate. It mattered not. Davenport was determined to prevail against the majority—a majority he neither trusted nor respected.

The inspiration to persevere against a changing world of ethnic diversity would come weeks later, during a visit to Kent, England. Davenport called the experience "one of the most memorable days of my life." That morning, the weather was beautiful and Davenport could not help but walk several miles through the bracing English countryside. He found himself at Downe House, Darwin's longtime residence. For an hour, the American eugenicist pondered Darwin's secluded walking paths and gardens. "It is a wonderful place," Davenport wrote, "and seems to me to give the clue to Darwin's strength—solitary thinking out of doors in the midst of nature. I would give a good deal for such a walk.... Then I would build a brick wall around it.... I know you will laugh at this," he continued, "but it means success in my work as opposed to failure. I must have a convenient, isolated place for continuous reflection."[9]

Davenport returned to America and began constructing his scientific bastion, impervious to outside interference. The first step would be to establish the so-called Eugenics Record Office to quietly register the genetic backgrounds of all Americans, separating the defective strains from the desired lineages. Borrowing nomenclature and charting procedures from the world of animal breeding, these family trees would be called pedigrees. Where would the ERO obtain the family details? "They lie hidden," Davenport told his ABA colleagues, "in records of our numerous charity organizations, our 42 institutions for the feebleminded, our 115 schools and homes for the deaf and blind, our 350 hospitals for the insane, our 1,200 refuge homes, our 1,300 prisons, our 1,500 hospitals and our 2,500 almshouses. Our great insurance companies and our college gymnasiums have tens of thousands of records of the characters of human bloodlines. These records should be studied, their hereditary data sifted out and properly recorded on cards, and [then] the cards sent to a central bureau for study... [of] the great strains of human protoplasm that are coursing through the country."[10]

At the same time, Davenport wanted to collect pedigrees on eminent, racially acceptable families, that is, the ones worth preserving.[11]

The planned ERO would also agitate among public officials to accept eugenic principles even in the absence of scientific support. Legislation was to be pressed to enable the forced prevention of unwanted progeny, as well as the proliferation by financial incentives of acceptable families. Whereas the experimental station would concentrate on quotable genetic research, the ERO would transduce that research into governing policy in American society.

In early 1910, just after the impetus for the new eugenics section of the American Breeders Association, Davenport swiftly began making his Eugenics Record Office a reality. Once more, the undertaking would require a large infusion of money. So once again he turned to great wealth. Reviewing the names in Long Island's *Who's Who*, Davenport searched for likely local millionaires. Going down the list, he stopped at one name: "Harriman."[12]

E. H. Harriman was legendary. America's almost mythic railroad magnate controlled the Union Pacific, Wells Fargo, numerous financial institutions and one of the nation's greatest personal fortunes. Davenport knew that Harriman craved more than just power and wealth; he fancied himself a scientist and a naturalist. The railroad man had financed a famous Darwin-style expedition to explore Alaskan glaciers. The so-called "Harriman Expedition" was organized by famous botanist and ornithologist C. Hart Merriam, a strong friend of eugenics. In 1907, Merriam had single-handedly arranged a private meeting between Davenport's circle of eugenicists and President Theodore Roosevelt at the president's Long Island retreat.[13]

Harriman died in 1909, leaving a fabulous estate to his wife, Mary.[14]

Everything connected in Davenport's mind. He remembered that three years earlier, Harriman's daughter, also named Mary, had enrolled in one of Cold Spring Harbor's summer biology courses. She was so enthusiastic about eugenics, her classmates at Barnard College had nicknamed her "Eugenia." Mrs. Harriman was the perfect candidate to endow the Eugenics Record Office to carry on her husband's sense of biological exploration, and cleanse the nation of racial and ethnic impurity.[15]

Quickly, Davenport began cultivating a relationship with the newly widowed Mrs. E. H. Harriman. Her very name invoked the image of wealth and power wielded by her late husband, but identified her as now possessing the power over that purse. Even though the railroad giant's wife was now being plagued by philanthropic overtures at every turn, Davenport knew just how to tug the strings. Skilled in the process, it only took about a month.[16]

In early 1910, just days after the ABA elected to launch the Eugenics Record Office, Davenport reconnected with his former student about saving the social and biologic fabric of the United States. Days later, on January 13, Davenport visited Mary to advance the cause. On February 1, Davenport logged an entry in his diary: "Spent the evening on a scheme for Miss Harriman. Probably time lost." Two days later, the diary read: "Sent off letter to Miss Harriman." By February 12, Davenport had received an encouraging letter from the daughter regarding a luncheon to discuss eugenics. On February 16, Davenport's diary entry recorded: "To Mrs. Harriman's to lunch" and then several hours later, the final celebratory notation: "All agreed on the desirability of a larger scheme. A Red Letter Day for humanity!"[17]

Mrs. E. H. Harriman had joined the eugenic crusade. She agreed to create the Eugenics Record Office, purchasing eighty acres of land for its use about a half mile from the Carnegie Institution's experimental station at Cold Spring Harbor. She also donated $15,000 per year for operations and would eventually provide more than a half million dollars in cash and securities.[18]

Clearly, the ERO seemed like an adjunct to the Carnegie Institution's existing facility. But in fact it would function independently, as a joint project of Mrs. Harriman and the American Breeders Association's eugenic section. "As the aims of the [ABA's] Committee are strongly involved," Davenport wrote Mrs. Harriman on May 23, 1910, "it is but natural that, on behalf of the Committee, I should express its gratitude at the confidence you repose in it."[19]

Indeed, all of Davenport's numerous and highly detailed reports to Mrs. Harriman were written on American Breeders Association eugenic section letterhead. Moreover, the ABA's eugenics committee letterhead itself conveyed the impression of a semiofficial U.S. government agency. Prominently featured at the top of the stationery were the names of ABA president James Wilson, who was also secretary of the Department of Agriculture, and ABA secretary W. M. Hays, assistant secretary of the Department of Agriculture. In fact, the words "U.S. Department of Agriculture, Washington D.C." appeared next to Hays's name, as a credential.[20] The project must have seemed like a virtual partnership between Mrs. Harriman and the federal government itself.[21]

Although the establishment of the Eugenics Record Office created a second eugenics agency independent of the Carnegie Institution, the two facilities together with the American Breeders Association's eugenic section in essence formed an interlocking eugenic directorate headquartered at Cold Spring Harbor. Davenport ruled all three entities. Just as he scrupulously reported to Carnegie trustees in Washington about the experimental station, and ABA executives about its eugenic section, Davenport continuously deferred to Mrs. Harriman as the money behind his new ERO. Endless operational details, in-depth explanations regarding the use of cows to generate milk for sale at five cents per quart to defray the cost of a caretaker, plans to plant small plots of hay and corn, and requests to spend $10 on hardware and $50 on painting—they were all faithfully reported to Mrs. Harriman for her approval.[22] It gave her the sense that she was not only funding a eugenic institution, but micromanaging the control center for the future of humanity.

While the trivialities of hay and hardware consumed report after report to Mrs. Harriman, the real purpose of the facility was never out of anyone's mind. For example, in his May 23, 1910 report to Mrs. Harriman, Davenport again recited the ERO's mission: "The furtherance of your and its [the ABA's] ideal to develop to the utmost the work of the physical and social regeneration of our beloved country [through] the application ... of ascertained biological principles." Among the first objectives, Davenport added, was "the segregation of imbeciles during the reproductive period." No definition of "imbeciles" was offered. In addition, he informed Mrs. Harriman, "This office has addressed to the Secretary of State of each State a request for a list of officials charged with the care of imbeciles, insane, criminals, and paupers, so as to be in a position to move at once ... as soon as funds for a campaign are available. I feel sure that many states can be induced to contribute funds for the study of the blood lines that furnish their defective and delinquent classes if only the matter can be properly brought to their attention."[23]

Referring to the increase in "defective and delinquent classes" that worried so many of America's wealthy, Davenport ended his May 23 report by declaring, "The tide is rising rapidly; I only regret that I can do so little."[24]

Davenport could not do it alone. Fundamentally, he was a scientist who preferred to remain in the rarefied background, not a ground-level activist who could systemize the continuous, around-the-clock, county-by-county and state-by-state excavation of human data desired. He could not prod the legislatures and regulatory agencies into proliferating the eugenic laws envisioned. The eugenics movement needed a lieutenant to work the trenches—someone with ceaseless energy, a driven man who would never be satisfied. Davenport had the perfect candidate in mind.

"I am quite convinced," Davenport wrote Mrs. Harriman, "that Mr. Laughlin is our man."[25]

* * *

Fifty-five miles west of where northeast Missouri meets the Mississippi River, rolling foothills and hickory woodlands veined with lush streams finally yield to the undulating prairie that seats the town of Kirksville. In colonial times, mound-building Indians and French trappers prowled this region's vast forests hunting beaver, bear and muskrat pelts. After the Louisiana Purchase in 1803, only the sturdiest pioneers settled what became known as the state of Missouri. Kirksville was a small rural town in its northeast quadrant, serving as the intellectual and medical center of its surrounding agricultural community.[26]

In 1891, the Laughlin clan was among the tough middle-class pioneer families that settled in Kirksville, hoping to make a life. George Laughlin, a deeply religious college professor, migrated from Kansas to become pastor at Kirksville's Christian Church. The next year, the classically trained Laughlin was hired as chairman of the English Department of the Normal School, the area's main college.[27] Quickly, the Laughlins became a leading family of Kirksville.

In a modest home on East Harrison Street, the elder Laughlin raised ten children including five sons, one of whom was Harry Hamilton Laughlin. Young Harry was expected to behave like a "preacher's kid," even though his father was a college professor and no longer a clergyman. Preacher's kid or not, Harry was prone to youthful pranks and was endearingly nicknamed "Hi Yi" by his siblings. Once, on a sibling dare, Harry swung an axe at his younger brother Earl's hand, which was poised atop a chopping block. One of Earl's fingers was nearly severed, but was later reattached.[28]

Ancestry and social progress were both important in the Laughlin household. Reverend Laughlin could trace his lineage back to England and Germany, and it included U.S. President James Madison. His mother, Deborah, a Temperance League activist, acknowledged that her great-grandfather was a soldier in the English Light Dragoons during colonial times.[29]

When a well-educated Harry Laughlin graduated from college, he saw himself destined for greater things. Unfortunately, opportunity did not approach. So Laughlin became a teacher at a desolate one-room school-house in nearby Livonia, Missouri. Life in Livonia was an unhappy one for Laughlin. He had to walk through a small stream just to reach the front door of the schoolhouse. Laughlin referred to his ramshackle school as being "20 miles from any civilized animal." Sneering at the locals, he wrote, "People here are 75 years behind the times." Laughlin denigrated his students as "very dull" and admitted to "a forced smile" when he wasn't grumbling.[30]

Laughlin returned to Kirksville at his first chance. Initially, he hired on as principal of the local high school in 1900. However, he soon advanced to the Department of Agriculture, Botany and Nature at his college alma mater, the Normal School. His wife Pansy had also graduated from there. Hence, it was where Laughlin felt most comfortable. Indeed, despite the wide travels and illustrious circles he ultimately attained, Laughlin always considered simple Kirksville his true home and refuge.[31]

Still, Laughlin was convinced his days at Normal were temporary. A political dreamer, Laughlin had already drafted the first of numerous outlines for a one-world government comprised of six continental jurisdictions, complete with an international parliament apportioning seats in favor of the hereditarily superior nations. In Laughlin's world scheme, the best stocks would rule. Laughlin submitted his detailed plans to heads of state and opinion makers, but to no avail. No one paid attention.[32]

Highfalutin proposals for a personally crafted world order were only the outward manifestations of a man who desperately sought to make a mark, and not just any mark, but an incandescent mark visible to all. In pursuit of this, Laughlin spent a lifetime submitting his writings on everything from politics to thoroughbred horseracing to world leaders and influential personalities, seeking favorable comments, approval and recognition. And if none of that was possible, just a simple "thank you" would do.

It was not unusual for Laughlin to mail an obscure journal article or scientific paper to dozens of perfect strangers in high places, soliciting any measure of written approbation. These reply letters typed on important letterheads were then filed and cherished. Many were little more than polite but depthless two-sentence acknowledgments written by well-placed people who scarcely understood why they had been contacted. For example, Laughlin sent one immigration study to dozens of embassies, newspaper editors, business tycoons and private foundation leaders seeking comment. The Columbian Ambassador to Washington formally wrote back: "I take pleasure in acknowledging receipt of... the books... which I will be glad to look over." The editor of *Foreign Affairs* magazine issued a curt two-sentence thank you, indicating, "It will be useful in our reference files." An assistant in Henry Ford's office dashed off a two-sentence pro forma note, "We...wish to take this opportunity of thanking you on behalf of Mr. Ford for the copy of your work...."[33]

Self-promotion was a way of life for Laughlin.[34] But no matter how high his station, it was never high enough. "If I can't be great," Laughlin once confessed to his mother, at least "I can certainly do much good."[35]

Laughlin's desperate quest for greatness turned a historic corner on May 17, 1907, when he wrote to Davenport asking to attend one of Cold Spring Harbor's continuing summer biology courses. His application was immediately approved.[36] The relationship between Davenport and Laughlin finally ignited in January of 1909 when both men attended the American Breeders Association meeting in Columbia, Missouri.[37] The next year, after Mrs. Harriman approved the ERO, Laughlin was Davenport's number one choice.

Within Davenport's grandiose ideas about reshaping mankind, Laughlin could both find a niche and secure personal gratification. Working in the eugenics movement, with his notions of a one-world government, Laughlin might achieve a destiny he could barely imagine in any other endeavor.

Davenport understood Laughlin's deeply personal needs. As such, he structured Laughlin's employment to be more than just a career. The Eugenics Record Office would become Laughlin's life—from morning to night and into the next morning. Laughlin found such rigor comforting; it represented a personal acceptance he'd never known. Davenport had certainly chosen the right man.

Stressing to Mrs. Harriman that the ERO's task was a long-term project, Davenport proposed that Laughlin be hired for at least ten years. Laughlin's residence would actually be on the grounds of the Eugenics Record Office, and his title would be "superintendent." Davenport understood human nature. The very title "superintendent" was reminiscent of railroad station managers, the kind who had catered to Mrs. Harriman's late husband's steel-tracked empire. "Do you wish first to see Mr. Laughlin," Davenport asked Mrs. Harriman with apparent deference, but quickly added, "or do you authorize me to offer Mr. Laughlin $2,400 for the first year?"[38]

Mrs. Harriman approved. Davenport notified Laughlin. The campaign to create a superior race would soon be launched.

* * *

By late 1910 the Laughlins had arrived at Cold Spring Harbor to open the facility. They lived on the second floor of the ERO's main building, where they enjoyed four large rooms and a fifth smaller one. Laughlin would have continuous access to the library, dining room and kitchen adjacent to the main business area on the first floor. He would eat and sleep eugenics. Working fastidiously on the smallest details of the ERO's establishment, it was not uncommon to find him in the office seven days a week including most holidays.[39]

The Eugenics Record Office went into high gear even before the doors opened in October of 1910. Its first mission was to identify the most defective and undesirable Americans, estimated to be at least 10 percent of the population. This 10 percent was sometimes nicknamed the "submerged tenth" or the lower tenth. At the time, this amounted to millions of Americans. When found, they would be subjected to appropriate eugenic remedies to terminate their bloodlines. Various remedies were debated, but the leading solutions were compulsory segregation and forced sterilization.[40]

No time was wasted. During the ERO's preparatory summer months, a dozen field workers, mainly women, were recruited to canvas prisons and mental institutions, establishing good working relationships with their directors. The first junket on July 15, 1910, proved to be typical. First, field workers visited the notorious prison at Ossining, New York, known as Sing Sing, where they were granted a complete tour of the "hereditary criminals" they would be studying. After Sing Sing, the group traveled to the State Asylum at Matteawan, New York, where Superintendent Lamb promised to open all patient records to help "demonstrate at once the hereditary basis of criminal insanity." An albino family was then examined in nearby Millerton, New York. The eugenic investigators ended their outing at a school for the feebleminded in Lakeville, Connecticut. In Lakeville, once again, "the records were turned over to us," Davenport reported to Mrs. Harriman, enabling the "plotting on a map of Connecticut the distribution of birth-places of inmates." None of the institutions hesitated to turn over their confidential records to the private ERO—even before the agency opened its doors.[41]

After a few weeks of training in eugenic characteristics and principles, Laughlin's enthusiastic ERO field investigators swept across the eastern seaboard. Their mission was to identify those perceived as genetically inferior, as well as their extended families and their geographic concentrations. By pegging hotspot origins of defectives, eugenic cleansing priorities could be established. By no means was this a campaign directed solely against racial groups, but rather against any individual or group—white or black—considered physically, medically, morally, culturally or socially inadequate in the eyes of Davenport and Laughlin. Often there was no racial or cultural consistency to the list of those targeted. The genuinely lame, insane and deformed were lumped in with the troubled, the unfortunate, the disadvantaged and those who were simply "different," thus creating a giant eugenic underclass simply labeled "the unfit."

The hunt began.

ERO researcher A. H. Estabrook traveled to western Massachusetts and Connecticut to collect family trees on albino families. He was then "attached" to the State Asylum at Matteawan to research criminal insanity. Thereafter, Laughlin assigned him to search for "degenerates in the isolated valleys around the upper Hudson [River]." Estabrook developed 35 pages of pedigrees and 168 pages of personal descriptions in his first forays, but Laughlin became most interested in one "large family with much intermarriage that promises to be as interesting as the Juke or Zero family."[42]

Mary Drange-Graebe was assigned to Chicago where she worked with the Juvenile Psychopathic Institute under Dr. William Healy. After four months in Chicago, she was reassigned to track down the so-called Ishmael clan of nomadic criminals and vagabonds in and around Indianapolis. The tribe of racially mixed white gypsies, Islamic blacks and American Indians had been described years earlier in the study *The Tribe of Ishmael: a Study in Social Degeneration*, as a prime example of genetic criminality. This book had become a fundamental text for all eugenics. Now the ERO considered the book, written a generation earlier, as "too advanced for the times." So Drange-Graebe would resume tracing the family lineages of the infamous Ishmaelites. Within months, she had assembled 77 pages of family pedigrees and 873 pages of individual descriptions.[43]

Criminal behavior was hardly a prerequisite for the ERO's scrutiny. Field worker Amey Eaton was assigned to Lancaster County, Pennsylvania, to report on the Amish. Buggy-riding Amish folk, the most conservative wing of Mennonite Christians, were among the most law-abiding, courteous and God-fearing people in America. But they were also known for their unshakable pacifism, their peculiar refusal to adopt industrial technology and their immutable clannishness. This made them different. "In this small sect," Laughlin reported, "considerable intermarriage has occurred. These people kindly cooperated in our efforts to learn whether...these consanguineous [family-linked] marriages had resulted in defective offspring."[44]

The ERO's sights were broad, so their workers continued fanning out. Helen Reeves sought records of so-called feebleminded patients in various New Jersey institutions. Another researcher was sent to trawl the files of the special genealogy collection of the New York Public Library, looking for family ties to unfit individuals. Various hospitals around the country were scoured, yielding records on eighty immigrant families with Huntington's chorea, a devastating disease of the central nervous system. Even when Davenport vacationed in Maine, he used the occasion to visit the area's islands and peninsulas to record the deleterious effects of intermarriage in groups considered unfit. Idyllic Washington and Hancock counties in Maine were of particular interest.[45]

Epileptics were a high-priority target for Laughlin and the ERO. Field worker Florence Danielson was dispatched to collect the family trees of epileptics at Monson State Hospital for Epileptics in Massachusetts. Monson had previously been an almshouse or poorhouse. In line with eugenic thought, Monson's administrators believed that epilepsy and poverty were genetically linked.[46]

Laughlin dispatched a second ERO investigator, Sadie Deavitt, to the New Jersey State Village for Epileptics at Skillman to chart individual pedigrees. At Skillman, Deavitt deftly interviewed patients and their families about the supposed traits of their relatives and ancestors. The ERO's scientific regimen involved ascribing various qualities and characteristics to epileptic patient family members, living or dead. These qualities included medical characteristics such as "deaf" or "blind," as well as strictly social factors such as "wanderer, tramp, confirmed runaway" and "criminal."[47] The definition of "criminal" was never delineated; it included a range of infractions from vagrancy to serious felony.

Miss Deavitt employed warmth and congeniality to extract family and acquaintance descriptions from unsuspecting patients, family members and friends. A New Jersey State instructive report explained, "The investigator visits the patients in their cottages. She does this in the way of a friendly visit and leads the patient on to tell all he can about his friends and relatives, especially as to addresses. Often they bring her their letters to read and from these she gleans considerable information. Then comes the visit to the [family's] home. It is the visitor's recent and personal knowledge of the patient that often assures her of a cordial welcome." By deftly gaining the confidence of one family member and friend after another, Miss Deavitt was able to map family trees with various social and medical qualities penned in with special codes. "Sx" meant "sexual pervert"; "im" stood for "immoral."[48] None of the hundreds of people interviewed knew they were being added to a list of candidates for sterilization or segregation in special camps or farms.

Laughlin and the ERO focused heavily on the epileptic menace because they believed epilepsy and "feeblemindedness" were inextricably linked in human nature. Indeed, they often merged statistics on epileptic patients with those of the feebleminded to create larger combined numbers. The term "feeblemindedness" was never quite defined; its meaning varied from place to place, and even situation to situation. The eugenically damning classification certainly included genuine cases of severely retarded individuals who could not care for themselves, but it also swept up those who were simply shy, stuttering, poor at English, or otherwise generally nonverbal, regardless of their true intellect or talent.[49] Feeblemindedness was truly in the eye of the beholder and frequently depended upon the dimness or brightness of a particular moment.

But there was little room for gray in Laughlin's world. To accelerate the campaign against epileptics, Laughlin distributed to hospital and institutional directors a special thirty-page bulletin, filled with dense scientific documentation, number-filled columns, family charts and impressive Mendelian principles warning about the true nature of epilepsy. The bulletin, entitled "A First Study of Inheritance of Epilepsy," and first published in the *Journal of Nervous and Mental Diseases*, was authored by Davenport and a doctor employed by New Jersey's epileptic village. The treatise asserted conclusively that epilepsy and feeblemindedness were manifestations of a common defect, due to "the absence of a protoplasmic factor that determines complete nervous development." The bulletin emphasized that the genetic menace extended far beyond the family into the so-called genetic "fraternity," or the lineages of everyone related to every person who was considered epileptic. The more such "tainted" defectives were allowed to reproduce, the more numerous their epileptic and feebleminded descendants would become. In one example, the research declared that "in 28 families of normal parents of epileptic children every one shows evidence of mental weakness."[50]

The ERO dismissed the well-known traumatic causes of epilepsy or insanity, such as a fall or severe blow to the head, in favor of hereditary factors. In one typical insanity case originally blamed on a fall, the bulletin explained, "This defect may be purely traumatic but, on the other hand, he has an epileptic brother and a feeble-minded niece so there was probably an innate weakness and the fall is invoked as a convenient 'cause.'"[51]

Strikingly, the ERO's definition of epilepsy itself was so sweeping that it covered not only people plagued by seizures, but also those suffering from migraine headaches and even brief fainting spells possibly due to exhaustion, heat stroke or other causes. "Epilepsy is employed in this paper," Davenport wrote, "in a wide sense to include not only cases of well-marked convulsions, but also cases in which there has been only momentary loss of consciousness."[52]

The prospect of epileptics in the population would haunt Laughlin for decades as he feverishly launched every effort to identify them. Once he identified them, Laughlin wanted to neutralize their ability to reproduce. The ERO's epilepsy bulletin concluded: "The most effective mode of preventing the increase of epileptics that society would probably countenance is the segregation during the reproductive period of all epileptics."[53]

America's geography was diverse. Since the western regions of the United States were still being settled, the ERO understood that many family trees in those regions would be incomplete. Indeed, many people moved out West precisely because they wanted to begin a new life detached from their former existence. Public records in western locales often lacked information about extended family and ancestry. Overcoming the challenge of documenting the population of a vast continent with only broken bits of family data, the ERO promised that "the office is now prepared to index any material, no matter how fragmentary or how extensive, concerning the transmission of biological traits in man; and it seeks to become the depository of such material." To that end, the ERO contacted "the heads of all institutions in the United States concerned with abnormal individuals."[54]

Extending beyond the reach of his field workers, Laughlin promised the eugenics movement that the ERO would register information on all Americans no matter where they lived to "[prevent] the production of defective persons." While defectives were to be eliminated, the superior families were to be increased. The eugenics movement would seek out and list "men of genius" and "special talents," and then advocate that those families receive special entitlements, such as financial rewards and other benefits for increased reproduction.[55] Eventually, the superior race would be more numerous and would control American society. At some point, they alone would comprise American society.

The eugenic visions offered by Davenport and Laughlin pleased the movement's wealthy sponsors. On January 19, 1911, Andrew Carnegie doubled the Carnegie Institution's endowment with an additional ten million dollars for all its diverse programs, including eugenics. Mrs. Harriman increased her enthusiastic grants. John D. Rockefeller's fortune also contributed to the funding. A Rockefeller philanthropic official became "much interested in eugenics and seems willing to help Dr. Davenport's work," reported one eugenic leader to Mrs. Harriman in a handwritten letter. "His preference is to give a small sum at first... raising the amount as the work advances." Initial Rockefeller contributions amounted to just $21,650 in cash and were earmarked to defray field worker expenses. But the highly structured Rockefeller philanthropic entities donated more than just cash; they provided personnel and organizational support, as well as the visible name of Rockefeller.[56]

Clearly, eugenics and its goal of purifying America's population was already more than just a complex of unsupported racist theorems and pronouncements. Eugenics was nothing less than an alliance between biological racism and mighty American power, position and wealth against the most vulnerable, the most marginal and the least empowered in the nation. The eugenic crusaders had successfully mobilized America's strong against America's weak. More eugenic solutions were in store.

* * *

On May 2 and May 3, 1911, in Palmer, Massachusetts, the research committees of the ABA's eugenic section adopted a resolution creating a special new committee. "Resolved: that the chair appoint a committee commissioned to study and report on the best practical means for cutting off the defective germ-plasm of the American population." Laughlin was the special committee's secretary. He and his colleagues would recruit an advisory panel from among the country's most esteemed authorities in the social and political sciences, medicine and jurisprudence. The advisory panel eventually included surgeon Alexis Carrel, M.D., of the Rockefeller Institute for Medical Research, who would months later win the Nobel Prize for Medicine; O. P. Austin, chief of the Bureau of Statistics in Washington, D.C.; physiologist W. B. Cannon and immigration expert Robert DeCourcy Ward, both from Harvard; psychiatrist Stewart Paton from Princeton; public affairs professor Irving Fisher from Yale; political economist James Field from the University of Chicago; renowned attorney Louis Marshall; and numerous other eminent men of learning.[57]

Commencing July 15, 1911, Laughlin and the main ABA committee members met at Manhattan's prestigious City Club on West Forty-fourth Street. During a number of subsequent conferences, they carefully debated the "problem of cutting off the supply of defectives," and systematically plotted a bold campaign of "purging the blood of the American people of the handicapping and deteriorating influences of these anti-social classes." Ten groups were eventually identified as "socially unfit" and targeted for "elimination." First, the feebleminded; second, the pauper class; third, the inebriate class or alcoholics; fourth, criminals of all descriptions including petty criminals and those jailed for nonpayment of fines; fifth, epileptics; sixth, the insane; seventh, the constitutionally weak class; eighth, those predisposed to specific diseases; ninth, the deformed; tenth, those with defective sense organs, that is, the deaf, blind and mute. In this last category, there was no indication of how severe the defect need be to qualify; no distinction was made between blurry vision or bad hearing and outright blindness or deafness.[58]

Not content to eliminate those deemed unfit by virtue of some malady, transgression, disadvantage or adverse circumstance, the ABA committee targeted their extended families as well. Even if those relatives seemed perfectly normal and were not institutionalized, the breeders considered them equally unfit because they supposedly carried the defective germ-plasm that might crop up in a future generation. The committee carefully weighed the relative value of "sterilizing all persons with defective germ-plasm," or just "sterilizing only degenerates." The group agreed that "defective and potential parents of defectives not in institutions" were also unacceptable.[59]

Normal persons of the wrong ancestry were particularly unwanted. "There are many others of equally unworthy personality and hereditary qualities," wrote Laughlin, "who have . . . never been committed to institutions." He added, "There are many parents who, in many cases, may themselves be normal, but who produce defective offspring. This great mass of humanity is not only a social menace to the present generation, but it harbors the potential parenthood of the social misfits of our future generations." Davenport had consistently emphasized that "a person who by all physical and mental examinations is normal may lack in half his germ cells the determiner for complete development. In some respects, such a person is more undesirable in the community than the idiot, (who will probably not reproduce), or the low-grade imbecile who will be recognized as such."[60]

How many people did the eugenics movement target for countermeasures? Prioritizing those in custodial care—from poor houses to hospitals to prisons—the unfit totaled close to a million. An additional three million people were "equally defective, but not under the state's care." Finally, the group focused on the so-called "borderline," some seven million people, who "are of such inferior blood, and are so interwoven in kinship with those still more defective, that they are totally unfitted to become parents of useful citizens." Laughlin insisted, "If they mate with a higher level, they contaminate it; if they mate with the still lower levels, they bolster them up a little only to aid them to continue their unworthy kind." The estimated first wave alone totaled nearly eleven million Americans, or more than 10 percent of the existing population.[61]

Eleven million would be only the beginning. Laughlin readily admitted that his first aim was at "ten percent of the total population, but even this is arbitrary." Eugenics would then turn its attention to the extended families deemed perfectly normal but still socially unfit.[62] Those numbers would add many million more.

Indeed, the eugenicists would push further, attempting a constantly upward genetic spiral in their insatiable quest for the super race. The movement intended to constantly identify the lowest levels of even the acceptable population and then terminate those families as well. "It will always be desirable," wrote Laughlin on behalf of the committee, "in the interests of still further advancement to cut off the lowest levels, and encourage high fecundity among the more gifted."[63]

The committee was always keenly aware that their efforts could be deemed unconstitutional. Legal fine points were argued to ensure that any eugenical countermeasure not "be considered as a second punishment... or as a cruel or unusual punishment." The eugenic committee hoped to circumvent the courts and due process, arguing that "sterilization of degenerates, or especially of criminals, [could] be legitimately effected through the exercise of police functions." In an ideal world, a eugenics board or commission would unilaterally decide which families would be the targets of eugenic procedures. The police would simply enforce their decisions.[64]

Human rights attorney Louis Marshall, the committee's main legal advisor, opined that eugenic sterilization might be legal if ordered by the original sentencing judge for criminals. But to venture beyond criminals, he wrote, targeting the weak, the diseased and their relatives, would probably be unconstitutional. "I understand that the operation of vasectomy is painless," wrote Marshall, "... other than to render it impossible for him to have progeny.... The danger, however, is that it might be inflicted upon one who is not a habitual criminal, who might have been the victim of circumstances and who could be reformed. To deprive such an individual of all hope of progeny would approach closely to the line of cruel and unusual punishment. There are many cases where juvenile offenders have been rendered habitual criminals who subsequently became exemplary citizens...

the very fact that they exist would require the exercise of extreme caution in determining whether such a punishment is constitutional."[65]

Marshall added with vagueness, "Unless justified by a conviction for crime, it [eugenical sterilization] would be a wanton and unauthorized act and an unwarranted deprivation of the liberty of the citizen. In order to justify it, the person upon whom the operation is to be performed has, therefore, the right to insist upon his right to due process of law. That right is withheld if the vasectomy is directed…by a board or commission, which acts upon its own initiative.…I fear that the public is not as yet prepared to deal with this problem."[66]

But Laughlin and his fellow breeders envisioned eugenical measures beyond mere sterilization. To multiply the genetically desired bloodlines, they suggested polygamy and systematic mating. Additional draconian remedies that were proposed to cut off defective germ-plasm included restrictive marriage laws, compulsory birth control and forced segregation for life—or at least until the reproductive years had passed. Davenport believed mass segregation or incarceration of the feebleminded during their entire reproductive years, if "carried out thoroughly" would wipe out most defectives within fifteen to thirty years. All the extra property acquired to incarcerate the inmates could be sold off for cash. As part of any long-term incarceration program, the patient could be released if he or she willingly submitted to sterilization "just prior to release." This was viewed as a central means of bypassing the need for a court order or even a commission decision. These sterilizations could then be called "voluntary."[67]

One option went further than any other. It was too early to implement. However, point eight of the American Breeders Association plan called for euthanasia.[68]

Despite the diversity of proposals, the group understood that of the various debated remedies, the American public was only ready for one: sterilization. The committee's tactic would be to convince America at large that "eugenics is a long-time investment" appealing to "far-sighted patriots." The agenda to terminate defective bloodlines was advocated and its underlying science was trumpeted as genuine, even as the committee confessed in their own summary report, "our knowledge is, as yet, so limited." Laughlin and his colleagues pursued their mission even as the original Galtonian eugenicists in London publicly declared they were "fully conscious of the slenderness of their data." American eugenicists pressed on even as Pearson of the Eugenics Laboratory openly quoted criticism by a fellow of the Royal Statistical Society, "The educated man and the scientist is as prone as any other to become the victim... of his prejudices.... He will in defence thereof make shipwreck of both the facts of science and the methods of science... by perpetrating every form of fallacy, inaccuracy and distortion."[69] America's eugenicists continued even as their elite leaders acknowledged, "public sentiment demanding action was absent."[70]

Laughlin and the American eugenics movement were undeterred by their own lack of knowledge, lack of scientific evidence, and even the profound lack of public support. The crusade would continue. In their eyes, the future of humanity—or their version of it—was at stake.

Moreover, America's eugenicists were not satisfied with merely cleansing the United States of its defectives. The movement's view was global. The last of eighteen points circulated by Laughlin's committee was entitled "International Cooperation." Its intent was unmistakable. The ERO would undertake studies "looking toward the possible application of the sterilization of defectives in foreign countries, together with records of any such operations." Point eighteen made clear that Laughlin's ERO and the American eugenics movement intended to turn their sights on "the extent and nature of the problem of the socially inadequate in foreign countries."[71]

CHAPTER 5

Legitimizing Raceology

When Galton's eugenic principles migrated across the ocean to America, Kansas physician F. Hoyt Pilcher became the first in modern times to castrate to prevent procreation. In the mid-1890s, Dr. Pilcher, superintendent of the Kansas Home for the Feebleminded, surgically asexualized fifty-eight children. Pilcher's procedure was undertaken without legal sanction. Once discovered, Kansas citizens broadly condemned his actions, demanding he stop. The Kansas Home's embattled board of trustees suspended Pilcher's operations, but staunchly defended his work. The board defiantly proclaimed, "Those who are now criticizing Dr. Pilcher will, in a few years, be talking of erecting a monument to his memory." Later, Pilcher's national association of institution directors praised him as "courageous" and as a "pioneer, strong [enough] to face ignorance and prejudice."[1]

Enter Dr. Harry Clay Sharp, physician at the Indiana Reformatory at Jeffersonville. Sharp earned his medical degree in 1893. Two years later, he was hired by the Indiana Reformatory as its doctor. The Indiana Reformatory, the state's first prison, was proud of its progressive sanitation and medical policies. Sharp was already performing extralegal medical castrations to cure convicts of masturbation. In early 1899, he read an article in the *Journal of the American Medical Association (JAMA)* by distinguished Chicago physician Albert John Ochsner, who later cofounded the American College of Surgeons. Dr. Ochsner advocated compulsory vasectomy of prisoners "to eliminate all habitual criminals from the possibility of having children." In this way, Ochsner hoped to reduce not only the number of "born criminals" but also "chronic inebriates, imbeciles, perverts and paupers."[2]

Sharp combined Ochsner's idea with a second suggestion by another Chicago doctor, Daniel R. Brower. Brower read a paper before the American Medical Society, reprinted in *JAMA*, similarly urging that someone employ vasectomy on convicts to prevent the propagation of a criminal class.[3]

Sharp was willing to be that someone. In October of 1899, he became the first in the world to impose vasectomy on a person in custody. A nineteen-year-old Indiana Reformatory prisoner complained of excessive masturbation, and Sharp used the opportunity. After disinfecting the prisoner's scrotum, the doctor made a one-inch incision, severed the ducts, and then buried a stitch. Sharp was pleased with his work. During the next several years, he performed the same operation on scores of additional inmates, becoming the world expert in human sterilization. Each operation took about three minutes. Anesthetic was not used for subsequent operations.[4]

The Indiana prison doctor proudly lectured his colleagues about the procedure's advantages in a 1902 article in the *New York Medical Journal*. He presented the surgery strictly as a tool for human breeding. Quoting an old essay, Sharp railed: "We make choice of the best rams for our sheep... and keep the best dogs... how careful then should we be in begetting of children!"[5]

Sharp's article described his method in instructive, clinical detail. Yet involuntary sterilization was still not legal, and was thought by many to be unconstitutional. So he urged his fellow institutional doctors to lobby for both restrictive marriage laws and legal authority for every institutional director in every state to "render every male sterile who passes its portals, whether it be an almshouse, insane asylum, institute for the feeble minded, reformatory or prison." Sharp declared that widespread sterilization was the only "rational means of eradicating from our midst a most dangerous and hurtful class.... Radical methods are necessary."[6]

It is no wonder that the world was first prompted to embrace forced sterilization by Indiana. Within the state's mainly rural turn-of-the-century population existed a small but potent epicenter of radical eugenic agitation. For decades, Indiana law provided for the compulsory servitude of its paupers. They could be farmed out to the highest bidder. Unwashed homeless bands wandering through Indiana were reviled by many within charitable circles as genetically defective, and beyond help.[7]

Reverend Oscar McCulloch, pastor of Indianapolis's Plymouth Congregational Church, was known as a leading reformer and advocate of public charity. Ironically, McCulloch actually harbored an intense hatred of paupers and the displaced. He was greatly influenced by the publication of Dugdale's *The Jukes*, which traced a Hudson Valley family of paupers and criminals as a living example of the need to improve social conditions. But McCulloch was foremost among those who twisted Dugdale's work from a cry for social action into a vicious hereditary indictment.[8]

McCulloch went even farther, adding his own genealogical investigation of Indiana's thieving vagabonds, the so-called Tribe of Ishmael. He proffered their stories as further scientific proof of degeneration among the impoverished. McCulloch preached to his fellow reformers at the 1888 National Conference of Charities and Corrections that paupers were nothing more than biologically preordained "parasites" suffering from an irreversible hereditary condition. By 1891, McCulloch had become president of the National Conference of Charities and Corrections, further ingraining his degeneracy theories upon the nation's charity and prison officials, who were only too quick to accept.[9]

Reverend McCulloch's outspoken sermons and investigations of the Ishmael tribe drew the attention of another leading Indianian, biologist David Starr Jordan, president of the University of Indiana. Convinced that paupers were indeed parasites, as McCulloch so fervently claimed, Jordan lectured his students and faculty to accept that some men were "dwarfs in body and mind." Quickly, Jordan became America's first eminent eugenic theorist. His 1902 book, *Blood of a Nation*, first articulated the concept of "blood" as the immutable basis for race. He readily proclaimed, "The pauper is the victim of heredity, but neither Nature nor Society recognizes that as an excuse for his existence." Jordan left Indiana in 1891 to become the first president of the newly created Stanford University, founded by the estate of wealthy railroad entrepreneur Leland Stanford. While at Stanford, Jordan used his position to further champion the eugenic cause, damning paupers in his writings and leading the like-minded elite in national eugenic organizations.[10]

Among the staunchest of Indiana's radical eugenicists was Dr. J. N. Hurty, who quickly rose from his insignificant station as the proprietor of an Indianapolis drug store to become the secretary of Indiana's State Board of Health. A close colleague of Hurty's once recalled for a eugenic audience: "It was not until Hurty had become the State Health Officer and had observed the stupidity of mankind, the worthlessness and the filthiness of certain classes of people, that he became really greatly interested in the subject [eugenics]." Once, when a prominent minister argued that all human beings were God's children, subject not to the laws of Mendel, but to the laws of grace, Hurty retorted, "Bosh and nonsense! Men and woman are what they are largely because of the stock from which they sprang." Hurty was eventually elected president of the American Public Health Association.[11]

By 1904, Sharp had performed 176 vasectomies as a eugenic solution designed to halt bloodlines. But the procedure was still not legal. So for three years, Drs. Sharp and Hurty lobbied the Indiana legislature to pass a bill for mandatory sterilization of all convicts. No distinction was made betwen lesser or graver crimes. There was no groundswell of public support for the measure, just the private efforts of Sharp, aided by Hurty and a few colleagues. The men stressed the social cost to the state of caring for its existing degenerates, and promised the new procedure would save Indiana from caring for future degenerates.[12] Drs. Sharp and Hurty were not immediately successful. But they did not give up.

It was an uphill battle. Indiana was not the first state to consider reproductive intervention, but until now, the idea had been rebuffed. In 1897, in the wake of Dr. Pilcher's first castrations, Michigan's legislature rejected a proposed law to make such actions legal. From 1901 through 1905, a key Pilcher supporter, Dr. Martin Barr, director of the Pennsylvania Training School for the Feebleminded, pushed for compulsory sterilization of mental defectives and other degenerates. Barr was undoubtedly among those responding to Sharp's early call to seek legislation. In 1905, both houses of Pennsylvania's legislature finally passed an "Act for the Prevention of Idiocy." The bill mandated that if the trustees and surgeons of the state's several institutions caring for feebleminded children determined "procreation is inadvisable," then the surgeon could "perform such operation for the prevention of procreation as shall be decided safest and most effective."[13]

Pennsylvania Governor Samuel Pennypacker's veto message denounced the very idea: "It is plain that the safest and most effective method of preventing procreation would be to cut the heads off the inmates," wrote Pennypacker, adding, "and such authority is given by the bill to this staff of scientific experts.... Scientists, like all other men whose experiences have been limited to one pursuit ... sometimes need to be restrained. Men of high scientific attainments are prone ... to lose sight of broad principles outside their domain.... To permit such an operation would be to inflict cruelty upon a helpless class ... which the state has undertaken to protect." Governor Pennypacker ended his incisive veto with five words: "The bill is not approved." No effort was made to override.[14]

What failed in Michigan and Pennsylvania found greater success in Indiana. Throughout 1906, Sharp ramped up his campaign. But the Indiana legislature was still resistant. So Sharp reminded Indiana's governor, J. Frank Hanley, that he was constantly performing vasectomies anyway, and his total had by now surged to 206. "I therefore wish to urge you," Sharp wrote the governor, "to insist upon the General Assembly [that] passing such a law or laws . . . will provide this as a means of preventing pro-creation in the defective and degenerate classes."[15]

On January 29, 1907, Indiana Representative Horace Reed introduced Sharp's bill. The measure's phrasing was an almost verbatim rendering of the previously vetoed Pennsylvania bill. Three weeks later, with little debate, Indiana's House approved the eugenic proposal, 59 in favor and 22 opposed. About two weeks later, again with virtually no debate, Indiana's Senate ratified the bill, 28 voting aye and 16 nay. This time, there was no governor's veto.[16] Indiana thereby made its mark in medical history, and became the first jurisdiction in the world to legislate forced sterilization of its mentally impaired patients, poorhouse residents and prisoners. Sharp's knife would now be one of a multitude, and the practice would crisscross the United States.

*   *   *

In 1907, most Americans were unaware that sterilization had become legal in Indiana. Nor did they comprehend that a group of biological activists were trying to replicate that legislation throughout the country. Frequently, the dogged state lobbying efforts were mounted by just one or two individuals, generally local physicians who carried the eugenic flame.[17]

In February of 1909, Oregon's first woman doctor, Bethenia Owens-Adair, promoted Bill 68, sporting provisions virtually identical to Indiana's law, but vesting the sterilization decision in a committee of two medical experts. Both Oregon houses ratified and Governor George Chamberlain had promised to sign the bill into law. But when Chamberlain finally com-prehended the final text, he vetoed the bill. In a letter to Dr. Owens-Adair, the governor explained, "When I first talked to you about the matter, with-out knowing the terms of the Bill in detail, I was disposed to favor it." But, he added, there were too few safeguards to prevent abuse.[18]

In early 1909, several additional attempts in other states also failed. Illinois's Senate Bill 249 authorized either castration or sterilization of con-firmed criminals and imbeciles when a facility doctor felt procreation was "inadvisable"; it failed to pass. Wisconsin's Bill 744 to sterilize the feeble-minded, criminals, epileptics and the insane on the recommendation of two experts was also rejected despite an amendment.[19]

But three states did ratify eugenic sterilization in 1909. Washington targeted "habitual criminals" and rapists, mandating sterilization as addi-tional punishment for the "prevention of procreation." Connecticut enacted a law permitting the medical staff at two asylums, Middletown and

Norwich, to examine patients and their family trees to determine if feeble-minded and insane patients should be sterilized; the physicians were permitted to perform either vasectomies on males or ovariectomies on women.[20]

California was the third state to adopt forced sterilization in 1909; Chapter 720 of the state's statutory code permitted castration or sterilization of state convicts and the residents of the California Home for the Care and Training of Feebleminded Children in Sonoma County. Two institutional bureaucrats could recommend the procedure if they deemed it beneficial to a subject's "physical, mental or moral condition."[21]

During the next two years, more states attempted to enact eugenic sterilization laws. Efforts in Virginia to pass House Bill 96, calling for the sterilization of all criminals, imbeciles and idiots in custody when approved by a committee of experts, died in the legislature. But efforts in other states were successful. Nevada targeted habitual criminals. Iowa authorized the operation for "criminals, idiots, feebleminded, imbeciles, drunkards, drug fiends, epileptics," plus "moral or sexual perverts" in its custody. The Iowa act was tacked onto a prostitution law.[22]

New Jersey's legislation was passed in 1911. Chapter 190 of its statutory code created a special three-man "Board of Examiners of Feebleminded, Epileptics and Other Defectives." The board would systematically identify when "procreation is inadvisable" for prisoners and children residing in poor houses and other charitable institutions. The law included not only the "feebleminded, epileptic [and] certain criminals" but also a class ambiguously referred to as "other defectives." New Jersey's measure added a veneer of due process by requiring a hearing where evidence could be taken, and a formal notice served upon a so-called "patient attorney." No provision permitted a family-hired or personally selected attorney, but only one appointed by the court. The administrative hearing was held within the institution itself, not in a courtroom under a judge's gavel. Moreover, the court-designated counsel for the patient was given only five days before the sterilization decision was sealed. Thus the process would be swift, and certainly beyond the grasp of the confused children dwelling within state shelters. New Jersey's governor, Woodrow Wilson, signed the bill into law on April 21, 1911. The next year, he was elected president of the United States for his personal rights campaign known as the "New Freedoms." Stressing individual freedoms, Wilson helped create the League of Nations. President Wilson crusaded for human rights for all, including the defenseless, proclaiming to the world the immortal words: "What we seek is the reign of law, based upon the consent of the governed, and sustained by the organized opinion of mankind."[23]

New York was next. In April of 1912, New York amended its Public Health Law with Chapter 445, which virtually duplicated New Jersey's eugenic legislation. New York law created its own "Board of Examiners for feebleminded, epileptics and other defectives," comprised of a neurologist, a surgeon and a general physician. Any two of the three examiners could rule whether family history, feeblemindedness, "inherited tendency" or other factors proved that procreation was inadvisable for the patients or prisoners they reviewed. Once again, a so-called "patient attorney" was to be appointed by the court. Vasectomies, salpingectomies (tubal ligations), and full castrations were authorized, at the discretion of the board.[24]

Despite the spreading patchwork of state eugenic sterilization laws, by late 1911 and early 1912, the Cold Spring Harbor stalwarts of the American Breeders Association, its Eugenic Record Office and the Carnegie Institution's Experimental Station remained frustrated. Their joint Committee to Study and Report the Best Practical Means of Cutting off the Defective Germ-plasm of the American Population knew that few Americans had actually undergone involuntary sterilization. True, in the years since 1907, when Indiana legalized such operations, Sharp had vasectomized scores of additional prisoners and even published open appeals to his professional colleagues to join his eugenic crusade. More than two hundred had been forcibly sterilized in California. Connecticut's Norwich Hospital had performed the operation on fewer than ten, mostly women. But only two eugenic sterilizations had been ordered in Washington state, and both were held in abeyance. An extralegal vasectomy had been performed on one Irish patient in a Boston hospital constituting a juridical test. However, none were authorized in Nevada, Iowa, New Jersey, or New York.[25]

Many state officials were clearly reluctant to enforce the laws precisely because the results were radical and irreversible. The legality of the operations and the question of due process had never been satisfactorily answered. The Eugenics Section of the American Breeders Association admitted in a report that the prior legislation had been pushed by "some very small energetic groups of enthusiasts, who have had influence in the legislatures...[but] it was a new and untried proposition. Public sentiment demanding action was absent. Law officers of the state were not anxious to undertake defense of a law the constitutionality of which was questioned."[26]

Moreover, the whole concept of eugenic solutions, such as marriage restriction, forced segregation and involuntary sterilization was still disdained by most Americans. Catholics by and large considered the termination of reproductive capability to be an act against God. "It is evident," the report continued, "that active hostility and opposition will arise as soon as there is any attempt to carry out the laws in a through-going manner." The report concluded, "So we must frankly confess that... this movement for race betterment is as yet little more than a hobby of a few groups of people."[27]

The Eugenics Section declared, "It is, therefore, easy to see why little has been actually done. The machinery of administration has to be created.... Much more extensive education of the public will be necessary before the practice of sterilization can be carried out to the extent which will make it a factor of importance."[28]

Clearly, the eugenics movement needed scientific validation, standards to identify exactly who was feebleminded and unfit, and most importantly, society's acceptance of the need to cut off defective families. Eugenicists in other countries, who had been corresponding together for some years, also felt the need to broaden acceptance of their beliefs. All of them wanted eugenic solutions to be applied on a global basis. Their mission, after all, was to completely reshape humanity, not just one corner of it. Toward this end, the Americans, working closely with their counterparts in Germany and England, scheduled an international conference in London. July of 1912 was selected because it coincided with a visit to London by Stanford University's Jordan and other eugenic leaders.[29]

Galton had died in January of 1911. By that time, his original theories of positive marriage, as well as his ideas on biometric study, had been circumvented by a more radical London group, the Eugenics Education Society. The Eugenics Education Society had adopted American attitudes on negative eugenics. By now, America's negative eugenics had also been purveyed to like-minded social engineers throughout Europe, especially in Germany and the Scandinavian nations, where theories about Nordic superiority were well received. Hence, this first conference was aptly called the First International Congress on Eugenics, bringing together some several hundred delegates and speakers from across America, Belgium, England, France, Germany, Italy, Japan, Spain and Norway.[30]

Not a few of the conferees would attend simply to investigate the emerging field of eugenics. But many of the Europeans attended because they harbored their own racial or ethnic biases against their nations' indigenous, immigrant or defective populations. For example, Jon Alfred Mjøen of Norway was that country's leading raceologist and eugenicist. He believed that crossing blond-haired Norwegians with native dark-haired Lapps produced a defective mulatto-like breed. Another major delegate was Alfred Ploetz, the spiritual father of Germany's race hygiene and eugenics movement.[31]

Organizers draped the conference with some of the most prestigious names in the world. Major Leonard Darwin, son of Charles Darwin, was appointed president. Britain's First Lord of the Admiralty, Winston Churchill, would represent the king. Churchill was alarmed at Britain's growing population of "persons...of mental defect" and advocated a eugenic solution. The vice presidents would include David Starr Jordan, Davenport, Ploetz and Alexander Graham Bell. To impress American governors and scientific organizations, the Eugenics Congress leadership wanted the U.S. State Department to send an official American delegate. Missouri's representative on the all-powerful House Appropriations Committee proffered the request. However, the State Department could not comply because the meeting was nongovernmental; therefore the U.S. government could not participate.[32]

Instead, Secretary of State P. C. Knox agreed to write the invitations on official letterhead and mail them to distinguished Americans in the realms of science, higher learning and state government all across the country. The U.S. State Department invitations would be officially extended on behalf of Alfred Mitchell Innes, the British Embassy's *chargé d'affaires* in Washington, who in turn was submitting them on behalf of the Eugenics Education Society in London. Hence the invitations bore the clear imprimatur of the U.S. Secretary of State, yet technically Secretary Knox was merely conveying the invitation. The Knox letter also promised "to be the medium of communication to the Embassy" for any reply.[33]

Knox's official-looking invitations were each virtually alike. "At the request of the British Embassy at this capital, I have the honor to send you herewith an invitation extended to you by the Organizing Committee of the First International Eugenics Congress." Kansas Governor Walter Stubbs received one. Kentucky Governor James McCreary received one. Maryland Governor Phillip L. Goldsborough received one. Every governor of every state received one. Invitations were also sent to the presidents of the National Academy of Sciences, the American Academy of Political and Social Sciences, the American Economic Association at Yale University, the American Philosophical Society, and many other esteemed organizations of science and academic study. Knox also mailed an invitation to every president of every leading medical society, including the American Gynecological Society, the American Neurological Association, the American Pediatric Society and, of course, the American Medical Association. Hundreds of such letters were posted on a single day—June 20, 1912.[34]

Because the invitations were distributed just a few weeks before the London congress, few if any of the invitees could actually attend. This fact must have been understood in advance. After all, many received the invitation quite late, often only after their summer travels were complete. Nonetheless, nearly every recipient issued a gracious decline, and a personal note of thanks expressing their regret at missing an important event. All but one, that is. Secretary of War Henry Stimson dashed off a stern rebuff reminding Secretary of State Knox that such official involvement in a private conference was precluded by law. Stimson quoted the law in his reply: "No money...shall be expended...for expenses of attendance of any person at any meeting or convention of members of any society or association" unless authorized by statutory appropriation.[35]

The message was clear. Knox had, for all intents and purposes, turned the State Department into a eugenics post office and invitation bureau. From Knox's point of view, however, he was undoubtedly only too happy to help the eugenics program of the Carnegie Institution. Prior to his service as secretary of state, Knox had been an attorney for the Carnegie Steel Company, and was once called by Carnegie "the best lawyer I have ever had."[36]

Proper or not, eugenics had overnight been packaged into an officially recognized and prestigious science in the eyes of those who counted.

*   *   *

Some four hundred delegates from America and Europe gathered at the University of London in late July of 1912, where for five days a diverse assemblage of research papers were presented exploring the social science and heredity of man. Two French doctors reviewed Parisian insanity records for the previous half-century. Alcoholism as an inheritable trait was debated. But the proceedings were dominated by the U.S. contingent and their theories of racial eugenics. Galton's hope of finding the measurable physical qualities of man, an endeavor named biometrics, had become passé. One leading eugenicist reported, "'Biometry'...might have never existed so far as the congress was concerned." Indeed, Galton's chief disciple, Karl Pearson, declined to even attend the congress.[37]

Instead, the racial biology of America's ERO, and its clarions for sterilization, dominated. The preliminary ABA report from what was dubbed "the American Committee on Sterilization" was heralded as a highlight of the meeting. One prominent British eugenicist, writing in a London newspaper, identified Davenport as an American "to whom all of us in this country are immensely indebted, for the work of his office has far outstripped anything of ours."[38]

One key British eugenicist added that if Galton were still alive and could "read the recent reports of the American Eugenics Record Office, which have added more to our knowledge of human heredity in the last three years than all former work on that subject put together, [he] would quickly seek to set our own work in this country upon the same sure basis."[39]

The medical establishment began to take notice as well, presenting eugenics as a legitimate medical concept. The *Journal of the American Medical Association*'s coverage glowed. *JAMA*'s headline rang out: "The International Eugenics Congress, An Event of Great Importance to the History of Evolution, Has Taken Place." Its correspondent enthusiastically portrayed the eugenicists' theory of social Darwinism, spotlighting the destructive quality of charity and stressing the value of disease to the natural order. "The unfit among men," the *JAMA* correspondent reported from a key congress speech, "were no longer killed by hunger and disease, but were cherished and enabled to reproduce their kind. It was true, they [society] could not but glory in this saving of suffering; but they must not blind themselves to the danger of interfering with Nature's ways. Cattle breeders bred from the best stocks.... Conscious selection must replace the blind forces of natural selection."[40]

Legitimacy, recognition and proliferation were only the beginning. In 1911, Davenport had authored a textbook entitled *Heredity in Relation to Eugenics*. It had been published by the prestigious Henry Holt & Co. The volume blended genuine biological observation with bizarre pseudoscientific postulations on personal habits and even simple preferences commanded by one's heredity. "Each 'family' will be seen to be stamped with a peculiar set of traits depending 'upon the nature of its germ plasm," wrote Davenport. "One family will be characterized by political activity, another by scholarship, another by financial success, another by professional success, another by insanity in some members with or without brilliancy in others, another by imbecility and epilepsy, another by larceny and sexual immorality, another by suicide, another by mechanical ability, or vocal talent, or ability in literary expression."[41]

Davenport's book promulgated a law of heredity that condemned the marriage of cousins as prohibited consanguinity, or marriage of close relatives. "[Should] a person that belongs to a strain in which defect is present...marry a cousin or other near relative...such consanguineous marriages are fraught with grave danger." Nonetheless, Davenport and his colleagues extolled the marriage of cousins among the elite as eugenically desired; for example, they commonly pointed to great men, such as Darwin, who married his first cousin.[42]

In the same textbook, Davenport insisted that if immigration from southeastern Europe continued, America would "rapidly become darker in pigmentation, smaller in stature, more mercurial, more attached to music and art, more given to crimes of larceny, kidnapping, assault, murder, rape and sex-immorality." He added a scholarly note about Jews: "There is no question that, taken as a whole, the horde of Jews that are now coming to us from Russia and the extreme southeast of Europe, with their intense individualism and ideals of gain at the cost of any interest, represent the opposite extreme from the early English and the more recent Scandinavian immigration with their ideals of community life in the open country, advancement by the sweat of the brow, and the uprearing of families in the fear of God and the love of country."[43]

Davenport's textbook concluded, "In other words, immigrants are desirable who are of 'good blood'; undesirable who are of 'bad blood.'"[44]

The volume declared that, without question, Mendel's laws governed all human character: "Man is an organism—an animal; and the laws of improvement of corn and of race horses hold true for him also." In Davenport's mind, this axiom spawned far-reaching social consequences. Applying Mendelian formulas to pauperism, for example, Davenport cited "shiftlessness" as a genuine genetic trait, which could be rated for severity. On page 80 of his textbook, Davenport explained with mathematical authority, "Classifying all persons in these two families as *very shiftless, somewhat shiftless*, and *industrious*, the following conclusions are reached. When both parents are *very shiftless*, practically all children are *very shiftless* or *somewhat shiftless*....When both parents are shiftless in some degree, about 15 percent of the known offspring are recorded as *industrious*." Not even the sudden onset of a prolonged disease incapacitating or killing the family breadwinner, and thereby creating financial woes for widows and orphans, was an excuse for poverty. "The man of strong stock," Davenport's textbook explained, "will not suffer from prolonged disease."[45]

As a solution to society's eugenic problem, Davenport's textbook strongly advocated for mass compulsory sterilization and incarceration of the unfit, a proliferation of marriage restriction laws, and plenty of government money to study whether intelligence testing would justify such measures against a mere 8 percent of America's children or as many as 38 percent.[46]

But could Davenport's eugenic textbook, and two or three others like it, become accepted doctrine at the nation's universities? American eugenicists were firmly entrenched in the biology, zoology, social science, psychology and anthropology departments of the nation's leading institutions of higher learning. Methodically, eugenic texts, especially Davenport's, were integrated into college coursework and, in some cases, actually spurred a stand-alone eugenics curriculum. The roster was long and prestigious, encompassing scores of America's finest schools. Harvard University's two courses were taught by Drs. East and Castle. Princeton University's course was taught by Dr. Schull and Laughlin himself. Yale's by Dr. Painter. Purdue's by Dr. Smith. The University of Chicago's by Dr. Bisch. Northwestern University, a hotbed of radical eugenic thought, offered a course by Dr. Kornhauser, who had interned at Cold Spring Harbor.[47]

Each school wove eugenics into its own academics. At the University of California, Berkeley, Dr. Holmes's semester-long sociology course was simply named "Eugenics." At New York University, Dr. Binder's fifteen-week sociology course was named "Family and Eugenics," and was attended by some twenty-five male and female students. At Stanford University, Dr. V. L. Kellogg taught a course covering zoology and eugenics. Even tiny schools inaugurated eugenics courses. At Alma College in Michigan, the biology department offered Dr. MacCurdy's "Heredity and Eugenics" as an eighteen-week course. At tiny Bates College in Maine, Dr. Pomeroy's eighteen-week biology course was called "Genetics."[48]

Eugenics rocketed through academia, becoming an institution virtually overnight. By 1914, some forty-four major institutions offered eugenic instruction. Within a decade, that number would swell to hundreds, reaching some 20,000 students annually.[49]

High schools quickly adopted eugenic textbooks as well. Typical was George William Hunter's high school biology book, published by the nation's largest secondary school book publisher, the American Book Company. Hunter's 1914 textbook, *A Civic Biology: Presented in Problems*, echoed many of Davenport's principles. For example, in one passage Hunter railed against unfit families "spreading disease, immorality, and crime to all parts of this country." His text added, "Largely for them, the poorhouse and the asylum exist. They take from society but they give nothing in return. They are true parasites." Before long, the overwhelming majority of high schools employed eugenic textbooks that emphasized clear distinctions between "superior families" and "inferior families."[50]

But impeding Davenport and Laughlin's campaign for eugenic programs of sterilization, segregation and social restriction was the lack of easy-to-apply standards to earmark the inferior. Measuring man's intelligence had always been a eugenic pursuit. In 1883, Galton established what amounted to an intelligence test center in London, charging applicants three pence each to be evaluated. He measured physical response time to auditory, tactile and visual cues. In 1890, Galton's idea was refined by his associate, the psychologist James Cattell, who devised a series of fifty tests he called "Mental Tests and Measurements." Like Galton's intelligence examinations, these "mental tests" logged physical reaction time to sounds and pressures.[51]

French psychologist Alfred Binet was not a eugenicist; he believed that one's environment shaped one's mind. In 1905, at the request of the French education ministry, Binet and physician Theodor Simon published the first so-called "intelligence test" to help classify the levels of retarded children, allowing them to be placed in proper classes. The Binet-Simon Test offered students thirty questions of increasing difficulty from which the test grader could calculate a "mental level." But Binet insisted that his test did not yield fixed numbers. With assistance, special educational methods and sheer practice a child could improve his score, "helping him literally to become more intelligent than he was before." To this end, Binet developed mental and physical exercises designed to raise his students' intelligence levels. These exercises actually yielded improved scores.[52] Heredity was in no way a predeterminer of intelligence, he insisted.

But Binet's intent was turned upside down by American eugenicists. The key instrument of that distortion was psychologist Henry Goddard, an ardent eugenic crusader who became the movement's leading warrior against the feebleminded. In 1906, the year after Binet published his intelligence test, Goddard was hired to direct the research laboratory at the Vineland Training School for Feebleminded Girls and Boys in Vineland, New Jersey. When the ERO was created a few years later, Goddard routinely made his patients available for assessment and family tracing.[53]

In 1913, Goddard published an influential book in the eugenics world, *The Kallikak Family: A Study in the Heredity of Feeblemindedness*. In the tradition of *The Jukes* and *The Tribe of Ishmael*, Goddard traced the ancestry, immorality and social menace of a large family he named the Kallikaks. He created the surname by combining the Greek words for "beauty" and "bad." The story of the Kallikaks presented more than just another defective genealogy. The book spun a powerful eugenic lesson and moral warning.[54]

Family patriarch Martin Kallikak, from the Revolutionary War era, was actually a splendid eugenic specimen who fathered an illustrious line of American descendants by his legitimate and eugenically sound Quaker wife. But Goddard claimed that the same Martin Kallikak had also engaged in an illicit affair with a feebleminded girl, which spawned "a race of defective degenerates."[55]

Foreshadowing a philosophy that low intelligence was a hereditary curse, Goddard wrote that the bad Kallikaks were "feebleminded, and no amount of education or good environment can change a feebleminded individual into a normal one, any more than it can change a red-haired stock into a black-haired stock." To drive his point home, Goddard included a series of photographs of nefarious-looking and supposedly defective Kallikak family members. These photos had been doctored, darkening and distorting the eyes, mouths, eyebrows, nose and other facial features to make the adults and children appear stupid. Although retouching published photos was common during this era, the consistent addition of sinister features allowed Goddard to effectively portray the Kallikaks as mental and social defectives.[56]

Added to the ominous photos were highly detailed descriptions of the Kallikak family tree. Goddard had anticipated that some might question how such meticulous biographical information about Kallikak ancestors— often hailing back nearly a century and a half—could be reliably extracted from feebleminded descendants. His answer: "After some experience, the field worker becomes expert in inferring the condition of those persons who are not seen, from the similarity of the language used in describing them to that used in describing persons whom she has seen."[57]

For example, Goddard's assistant asked one farmer, "Do you remember an old man, Martin Kallikak, who lived on the mountain edge yonder?" The book's text quotes the exchange: "'Do I?' he answered. 'Well, I guess! Nobody'd forget him. Simple,' he went on; 'not quite right here,' tapping his head, 'but inoffensive and kind. All the family was that.'" Goddard recited this documentation in a chapter entitled "Further Facts."[58]

Mass sterilization, in Goddard's view, was merely the first step in corralling the feebleminded. Sterilization did not diminish sexual function, just reproductive capability. Therefore, Goddard asked, "What will be the effect upon the community in the spread of debauchery and disease through having within it a group of people who are thus free to gratify their instincts without fear of consequences in the form of children? ... The feebleminded seldom exercise restraint in any case."[59]

His answer: mass incarceration in special colonies. "Segregation through colonization seems in the present state of our knowledge to be the ideal and perfectly satisfactory method."[60]

Davenport and Goddard both craved a more scientific measurement to identify the feebleminded they targeted. To that end, Goddard translated Binet's intelligence test into English to create a new American tool for intelligence testing. Binet had originally labeled the highest class of retarded child *débile*, French for "weak." Goddard changed that, coining a new word: *moron*. It was derived from *moros*, Greek for "stupid and foolish."[61]

Financing would be needed to prove Goddard's new test reliable in the field. "It would be very valuable for the general problem of Eugenics," Goddard outlined to Davenport in a July 25, 1912 letter, "... in connection with the heredity of feeble-mindedness because ... we could judge the probable development of the child from the mental condition of the parents." The problem? "Our finances have failed us," wrote Goddard. "I trust you will be able to provide for some such work as this."[62]

Goddard was provided for. By 1913, he had taken his new intelligence test and a team of testers to Ellis Island to conduct experiments. American eugenicists long believed that the majority of immigrants, especially brown-haired Irish, Eastern European Jews and southeastern Italians, were genetically defective. As such, they could be expected to contribute a disproportionate number of feebleminded to American shores. At Ellis Island's massive intake centers, Goddard's staff initially selected just twenty Italians and nineteen Russians for assessment because they "appeared to be feebleminded." He believed in the "unmistakable look of the feebleminded," bragging that to spot the feebleminded, just "a glance sufficed." Ultimately, 148 Jews, Hungarians, Italians and Russians were chosen for examination.[63]

Predictably, Goddard's version of the Binet test showed that 40 percent of immigrants tested as feebleminded. Moreover, he wrote, "60 percent of the [Jewish immigrants] classify as morons." In reporting his results in the *Journal of Delinquency*, Goddard further argued that an improved test would reveal even greater numbers of feebleminded immigrants. "We cannot escape feeling," wrote Goddard, "that this method is too lenient ... too low for prospective American citizens." He explained, "It should be noted that the immigration of recent years is of a decidedly different character from the earlier immigration. It is no longer representative of the respective races. It is admitted on all sides that we are now getting the poorest of each race."[64]

Goddard's version of Binet's test, and the new term *moron*, began to proliferate throughout eugenic, educational, custodial, psychological and other scientific circles as a valid—if still developing—form of intelligence testing. Mental testing, under different names and on different scales, quickly emerged as a fixture of social science, frequently linked to eugenic investigation and sterilization efforts. Such tests were invariably exploited by the ERO for its eugenic agenda. In 1915, for example, Detroit's superintendent of schools tested 100 teenagers who had attended special classes. The Eugenics Record Office circulated a note in connection with the test: "It would be very interesting to secure the family history of those children who improve and did not markedly improve." Mental examinations as a condition of a marriage licenses were advocated by the president of New York's Association of County Superintendents of Poor and Poor Law Officers; moreover, the association president also urged the sterilization of any children who could be shown as feebleminded or epileptic by age twelve.[65]

Chicago's central jail, the House of Correction, studied the "practicality of the Binet Scale and the question of the border line case." By including the so-called "borderline," who tested near but not within the moron range, more persons could be classed as feebleminded or "nearly feebleminded." Chicago Municipal Chief Judge Harry Olson, responsible for sentencing prisoners to the House of Correction, was a revered leader of the eugenics movement. At the time of the House of Correction study, he reminded colleagues, "We have laid too great importance on the environmental factors and paid too little attention to the problem of heredity."[66]

Mental tests applied to Blacks led to an article in the *Archives of Psychology* reporting that when 486 whites and 907 Blacks were examined, Blacks scored only three-fourths as well as their white counterparts. The article noted that pure Blacks tested the lowest, about 60 percent lower than whites. But as the amount of white blood increased in their ancestry, so did the test scores. The authors concluded, "In view of all the evidence it does not seem possible to raise the scholastic attainments of the negro.... It is probable that no expenditure of time or of money would accomplish this end, since education cannot create mental power."[67]

In 1916, a conference on feeblemindedness and insanity assembled in Indiana to an overflowing attendance, where, as eugenicists reported, "The keynote of the whole conference was *prevention* rather then *cure*." The group heard many papers on "mental tests and their value." Even though many conferees claimed these mental tests were still in their infancy, eugenicists insisted the examinations did not need to be judged because they were merely "short-cuts" to "the final test of the person's mentality."[68]

Nonetheless, many openly disputed the validity of Goddard's intelligence test. In one case, the Magdalen Home for the Feebleminded commenced an involuntary commitment of a slow-learning twenty-one-year-old New York woman, based on her low Binet scores. The woman's fervent protest against incarceration was vindicated by a New York judge, who ruled in her favor, declaring: "All criteria of mental incapacity are artificial and the deductions therefrom must necessarily lack verity and be, to a great extent, founded on conjecture."[69]

More sophisticated tests than Goddard's began to appear. The Yerkes-Bridge Point Scale for Intelligence, for instance, was employed by ERO field workers "measuring the intelligence of members of pedigrees that are being investigated." The ERO printed special rating forms for the test. The test's creator, Harvard psychologist Robert Yerkes, was a leading eugenic theorist and a former student of Davenport's. Yerkes was a member of many elite eugenic committees, including the Committee on the Inheritance of Mental Traits and the Committee on the Genetic Basis of Human Behavior. Two years after helping invent the Point Scale, Yerkes became president of the American Psychological Association.[70]

Europe exploded into war in 1914. America did not join the fray until 1917, but when it did, Washington struggled to classify more than three million drafted and enlisted soldiers. American Psychological Association president Yerkes pleaded for intelligence testing. He gathered Goddard and Stanford University eugenic activist Lewis Terman and others to help develop standardized examinations. Working from May to July of 1917 at Goddard's laboratory at the Vineland Training School for Feebleminded Girls and Boys in New Jersey, these eugenic psychologists and others jointly developed what they portrayed as scientifically designed army intelligence tests. These were submitted to the army, and the surgeon general soon authorized mass testing.[71]

Two main tests were devised: the written Army Alpha test for English-speaking literate men, and the pictoral Army Beta test for those who could not read or speak English. The Alpha test's multiple-choice questions could certainly be answered by sophisticated urbanites familiar with the country's latest consumer products, popular art and entertainment. Yet most of America's draftees hailed from an unsophisticated, rural society. Large numbers of them had "never been off the farm."[72] Many came from insular religious families, which disdained theater, slick magazines and smoking. No matter, the mental capacity of everyone who could read and write was measured by the same pop culture yardstick.

> *Question*: "Five hundred is played with…" *Possible answers*: rackets, pins, cards, dice. *Correct response*: cards.
> *Question*: "Becky Sharp appears in…" *Possible answers*: Vanity Fair, Romola, The Christmas Carol, Henry IV. *Correct response*: Vanity Fair.
> *Question*: "The Pierce Arrow car is made in…" *Possible answers*: Buffalo, Detroit, Toledo, Flint. *Correct response*: Buffalo.
> *Question*: "Marguerite Clark is known as a…" *Possible answers*: suffragist, singer, movie actress, writer. *Correct response*: movie actress.
> *Question*: "Velvet Joe appears in advertisements for…" *Possible answers*: tooth powder, dry goods, tobacco, soap. *Correct response*: tobacco.
> *Question*: "'Hasn't scratched yet' is used in advertising a…" *Possible answers*: drink, revolver, flour, cleanser. *Correct response*: cleanser.[73]

Americans and naturalized immigrants who could neither read nor write English were administered the Beta picture exam. For example, Beta Test 6 offered twenty simple sketches with something missing. "Fix it," the subject was instructed. He was then expected to pencil in the missing element. Bowling balls were missing from a bowling lane. The center net was subtracted from a tennis court. The incandescent filament was erased from a lightbulb. A stamp was missing from a postcard. The upper left diamond was missing from a sketch of the jack of diamonds on a playing card.[74]

A third test was administered to those who could not score appreciably on either the Alpha or Beta tests. Dr. Terman of Stanford had created a so-called Stanford revision of the Binet test, later named the Stanford-Binet Test. This test was only an update of Goddard's work.[75]

Predictably, Yerkes's results from all three tests identified vast numbers of morons among the eugenically inferior groups—so many that Yerkes asserted the army could not afford to reject all of them and still go to war. "It would be totally impossible to exclude all morons," reported Yerkes, because "47 percent of whites and 89 percent of Negroes" were shown to have a mental capacity below that of a thirteen-year-old. By contrast, the tests verified that feeblemindedness among eugenically cherished groups was indeed miniscule: Dutch people, a tenth of a single percent; Germans, just two-tenths of one percent; English, three-tenths; Swedes, less than half of one percent.[76]

In 1912, the German psychologist William Stern had begun referring to Binet's original "intelligence level" as an "intelligence age." Stern went further, dividing the intelligence age by the chronological age to create a ratio. In doing so, he coined the term *intelligence quotient*. Four years later, after Terman created the Stanford version of Goddard's Binet test, Terman and Yerkes wanted a more identifiable number, one that could be popularized. In 1916, using the Stanford-Binet test, Terman divided mental age by chronological age, and then multiplied by 100. This became the American version of the intelligence quotient. Terman nicknamed it IQ. The moniker became an instant icon of intelligence. Scales and rankings were devised. Those classified below a certain level, 70 scale points, were graded as either "morons," "imbeciles," or "idiots."[77]

Feeblemindedness now had a number. Soon everyone would receive one. Terman knew how such a number could be used. While studying California public school children, he argued, "If we would preserve our state for a class of people worthy to possess it, we must prevent, as far as possible, the propagation of mental degenerates."[78]

Yerkes's work was advanced by another eugenic activist, Princeton psychologist Carl Brigham. A radical raceologist, Brigham analyzed Yerkes's findings for the world at large, casting them as eugenic evidence of Nordic supremacy and the racial inferiority of virtually everyone else. Brigham's 1922 book, *A Study of American Intelligence*, published by no less than Princeton University Press, openly conceded that the volume was based on two earlier raceological books, Madison Grant's virulently racist *Passing of the Great Race*, and William Ripley's equally biased *Races of Europe*. Before Brigham's book was published, a team of prestigious colleagues from the surgeon general's office, Harvard, Syracuse University and Princeton pored over his manuscript, verifying his conclusions, as did Yerkes himself, who also wrote the foreword.[79]

"We still find tremendous differences between the non-English speaking Nordic group and the Alpine and Mediterranean groups," wrote Brigham. "The underlying cause of the nativity differences we have shown is race and not language." Moreover, "The decline in intelligence is due to two factors: the change in the races migrating to this country, and to the additional factor of the sending of lower and lower representatives of each race.... The conclusion [is] that our test results indicate a genuine intellectual superiority of the Nordic group over the Alpine and Mediterranean groups."[80]

According to Brigham, Negro intelligence was predestined by racial heredity, but could be improved by "the greater amount of admixture of white blood."[81]

Brigham concluded, "According to all evidence available, then, American intelligence is declining, and will proceed with an accelerating rate as the racial admixture becomes more and more extensive. The decline of American intelligence will be more rapid than the decline of the intelligence of European national groups," he warned, "owing to the presence here of the negro." He added, "The results which we obtain by interpreting the Army data...support Mr. Madison Grant's thesis of the superiority of the Nordic type...."[82]

Quickly, *A Study of American Intelligence* became a scientific standard. Shortly after its publication, Brigham adapted the Army Alpha test for use as a college entrance exam. It was first administered to Princeton freshman and applicants to Cooper Union. Later the College Board asked Brigham to head a committee to create a qualifying test for other private colleges in the Northeast and eventually across the country. Brigham's effort produced the Scholastic Aptitude Test, administered mainly to upper middle-class white students. The test quickly became known as the SAT and was eventually employed at colleges across the country. Over time, more and more colleges required high school students to take the test and score high enough to qualify for application.[83]

The deeply flawed roots of the IQ test, the SAT and most other American intelligence tests were more than apparent to many thinking people of the period. It became glaringly obvious that the tests were vehicles for cultural exclusion. Poor-scoring southern Italian immigrants would not have known who the latest Broadway stars were or which brands of flour were popular. They were, however, steeped in the arias of operatic masters, the arts in general, and had discovered the secrets of fine cooking centuries before. Jews—who overwhelmingly scored as moronic—were often only literate in Yiddish. But they enjoyed a rich tradition of Talmudic scholarship that debated to abstraction the very essence of life and God's will. Farm boys may not have been aware that Velvet Joe was a cigarette advertising character, but they grasped the intricate agrarian tenets of growing and curing tobacco leaves to produce the perfect smoke.

Blacks might not have been able to decipher the reading, writing and arithmetic denied to them by a discriminatory educational system intent on keeping them illiterate. They may not have been able to comprehend the first thing about tennis nets, bowling lanes or incandescent bulbs. But the descendants of men and women ripped from Africa had cultivated a rich oral storytelling tradition, an intense, almost enraptured scripture-quoting religion, and as a group they would originate the revolutionary music that would dominate the twentieth century. Perhaps most remarkably, they were smart enough to stay alive in a world where an uppity black man with too much on the ball, or too much spring in his step, could be lynched for looking in the wrong direction or asking too many questions.[84]

Brigham's book would be circulated to all the state legislatures, congressional committees and throughout the marble halls of Washington as proof positive that the inferior were not just poor or uneducated, but genetically defective. This notion was welcome news to many. Now the pages of polished scholarship could be held up as justification for the draconian measures the movement advocated.

But dissident schools of psychologists and social works emerged. Common sense rejected the numbers. Resistance grew.

The U.S. Army never acted on Yerkes's voluminous findings, declining to classify its inductees according to his data. Indeed, three independent investigations of the project were launched, one by the army's general staff, one by the surgeon general and one by the secretary of war. The general staff's investigation derisively concluded, "No theorist may…ride it [the test scores] as a hobby [horse] for the purpose of obtaining data for research work and the future benefit of the human race." Nor would military planners utilize the information in the next war.[85]

Vituperative attacks upon the objectivity and credibility of the Alpha and Beta tests were widespread and highly publicized. Typical were the public denunciations of syndicated journalist Walter Lippmann in the *New Republic*. "The danger of the intelligence tests," warned Lippmann, "is that in a wholesale system of education, the less sophisticated or the more prejudiced will stop when they have classified and forget that their duty is to educate. They will grade the retarded child instead of fighting the causes of his backwardness. For the whole drift of the propaganda based on intelligence testing is to treat people with low intelligence quotients as congenitally and hopelessly inferior." Terman's answer to Lippmann was simply, "Some members of the species are much stupider than others." But Lippmann summed it up for many when he declared that the Stanford-Binet and other IQ tests were "a new chance for quackery in a field where quacks breed like rabbits, and…doped evidence to the exponents of the New Snobbery."[86]

Eventually, even some of the architects of the IQ, SAT and kindred intelligence tests could no longer defend their creations from the growing rejection in their own professions. In 1928, Goddard grudgingly retreated from his hereditarian stance. "This may surprise you, but frankly when I see what has been made out of the moron by a system of education, which as a rule is only half right, I have no difficulty in concluding that when we get an education that is entirely right there will be no morons who cannot manage themselves and their affairs and compete in the struggle for existence. If we could hope to add to this a social order that would literally give every man a chance, I should be perfectly sure of the result."[87]

As for the compulsion to sterilize, Goddard eventually abandoned the eugenic creed entirely, at least publicly. "It may still be objected that moron parents are likely to have imbecile or idiot children. [But] there is not much evidence that this is the case. The danger is probably negligible." Aware he had recanted his whole life's work, Goddard confessed in exasperation, "As for myself, I think I have gone over to the enemy."[88]

In 1929, Brigham finally rejected those scholarly publications that asserted a racial basis for intelligence—including his own. Whether out of shame or embarrassment, the Princeton scholar submitted, "Comparative studies of various national and racial groups may not be made with existing tests...one the most pretentious of these comparative racial studies—the writer's own—was without foundation."[89]

Meaningful as they were to the history of science, the several quiet recantations were published in obscure medical and scholarly journals. Academia could relish the debate and savor the progress. But the system hewed in stone by the eugenics movement's intelligence warriors has stubbornly remained in place to this day. By the time some scientists saw the folly of their fiction, the politicians, legislators, educators and social workers who had adopted eugenic intelligence notions as firm science had enacted laws, procedures, systems and policies to enforce their tenets. Quiet apologies came too late for thousands of Americans who would be chased down by the quotients, scales and derisive labels eugenics had branded upon them.

No longer constrained by newness or lack of scientific proof, the eugenic crusade blitzed across America. The weak, the socially maligned, the defenseless and the scientifically indefensible of America's lowest biological caste would now be sterilized by the thousands, and in some cases euthanized.

CHAPTER 6

The United States of Sterilization

It didn't matter that the majority of the American people opposed sterilization and the eugenics movement's other draconian solutions. It didn't
matter that the underlying science was a fiction, that the intelligence measurements were fallacious, that the Constitutionality was tenuous, or that
the whole idea was roundly condemned by so many. None of that mattered
because Davenport, Laughlin and their eugenic constellation were not
interested in furthering a democracy—they were creating a supremacy.

Of course, American eugenicists did not seek the approbation of the
masses whose defective germ plasm they sought to wipe away. Instead, they
relied upon the powerful, the wealthy and the influential to make their war
against the weak a conflict fought not in public, but in the administrative
and bureaucratic foxholes of America. A phalanx of shock troops sallied
forth from obscure state agencies and special committees—everyone from
the elite of the academic world to sympathetic legislators who sought to
shroud their racist beliefs under the protective canopy of science. In tandem, they would hunt, identify, label and take control of those deemed
unfit to populate the earth.

During the years bracketing World War I, a potent, if unsound, intelligence classification system was taking root. A patchwork of largely inert
state sterilization laws awaited greater validation. The elite thinkers of
American medicine, science and higher education were busy expanding the
body of eugenic knowledge and evangelizing its tenets. However, the
moment had still not arrived for eugenic rhetoric to massively impact the
country. During these percolating years, Davenport and Laughlin continued to prepare the groundwork. They knew humanity could not be recre-
ated overnight. They were patient men.

During the war years, eugenic organizations proliferated in America.
Like-minded citizens found ethnic solace and even self-vindication in the
idea of biological superiority. The Race Betterment Foundation was

among the leading eugenic organizations that sprouted around the country to augment the work at Cold Spring Harbor. The society was founded by yet another wealthy American, Dr. John Harvey Kellogg of Battle Creek, Michigan. Dr. Kellogg was a member of the state board of health and operated a health sanitarium renowned for its alternative and fanciful food regimens. He had developed for his patients a natural product, a cereal made of wheat flakes. In 1898, Dr. Kellogg's brother, Will, created the corn flake, and in 1906 he began selling it commercially through a company that would ultimately become the cereal giant known as Kellogg Company. In that same year, Dr. Kellogg founded the Race Betterment Foundation to help stop the propagation of defectives.[1]

The Race Betterment Foundation attracted some of the most radical elements of the eugenics community. The organization wanted to compile its own eugenic registry, listing the backgrounds of as many Americans as possible, this to augment the one being developed by the Eugenics Record Office. In 1914, Dr. Kellogg organized the First Race Betterment Conference in Battle Creek, Michigan. The conference's purpose was to lay the foundations for the creation of a super race, amid an atmosphere of lavish banquets, stirring calls to biological action, and scientific grandiloquence. "We have wonderful new races of horses, cows, and pigs," argued Dr. Kellogg. "Why should we not have a new and improved race of men?" He wanted the "white races of Europe ... to establish a Race of Human Thoroughbreds."[2]

Davenport told the Battle Creek conferees that this could be accomplished by working quietly with the heads of state institutions. "The superintendents of state institutions," he explained, "were very desirous of assistance. We were able to give it to them, and they to us." Davenport relied upon institutional figures to authenticate his findings. "We have found that a large proportion of the feeble-minded, the great majority of them, are such because they belong to defective stock."[3]

Whatever restraint Laughlin used in his formal writings was absent from his speeches to the eugenic vanguard. Laughlin boldly put the Battle Creek gathering on notice: "To purify the breeding stock of the race at all costs is the slogan of eugenics." His three-pronged program was based on sterilization, mass incarceration, and sweeping immigration restrictions. "The compulsory sterilization of certain degenerates," affirmed Laughlin, "is therefore designed as a eugenical agency complementary to the segregation of the socially unfit classes, and to the control of the immigration of those who carry defective germ-plasm."[4]

The mothers of unfit children should be relegated to "a place comparable to that of the females of mongrel strains of domestic animals," said
Laughlin. He complained that although twelve states had enacted laws,
only a thousand people had been sterilized. "A halfway measure will never
strike deeply at the roots of evil," he railed.[5]

At the Second Race Betterment Conference held the next year, ERO
Scientific Director Irving Fisher, a Yale University economist, was equally
blunt. "Gentlemen and ladies," Fisher sermonized, "you have not any idea
unless you have studied this subject mathematically, how rapidly we could
exterminate this contamination if we really got at it, or how rapidly the
contamination goes on if we do not get at it."[6]

Eugenic extremism enjoyed layer upon layer of scientific veneer not
only because eminent scholars enunciated its doctrine and advocated its
solutions, but also by virtue of its numerous respected "research bodies."
The Eugenics Record Office had inaugurated a Board of Scientific
Directors in December of 1912. The board was initially comprised of
Davenport, plus eminent Harvard neuropathologist E. E. Southard,
Alexander Graham Bell and renowned Johns Hopkins University patholo-
gist William Welch. Welch enjoyed impeccable qualifications; he had
served as both the first scientific director of the Rockefeller Institute for
Medical Research and as a trustee of the Carnegie Institution. Moreover,
before and during his term on the ERO's scientific board, Welch was also
elected president of the American Association for the Advancement of
Sciences, the American Medical Association and the National Academy of
Science. Understandably, Laughlin and Davenport felt it only fitting that
he should serve as chairman of the ERO's Board of Scientific Directors.[7]

Among the biological issues the board identified as vital were "the consequences of marriages between distinct races—miscegenation," "the study
of America's most effective bloodlines," as well as "restricting the strains
that require state care." The board also sought to examine the ancestral caliber of immigrants being allowed into the country. As usual, feeblemindedness took the spotlight. Several key regions of the East Coast were targeted
for investigation.[8]

Among the directors, only Bell became uncomfortable with the ERO's
direction. He immediately voiced consternation over eugenics' constant
focus on inferior traits. "Why not vary a little from this program and investigate the inheritance of some desirable characteristics," Bell wrote
Davenport on December 27, 1912, just days after the board's first meeting.
For emphasis, Bell reiterated over and over in his letter that the ERO's sub-

stantial funding might be better "devoted to the study of...*desirable* characteristics rather than undesirable. The whole subject of eugenics has been too much associated in the public mind with fantastical and impractical schemes for restricting marriage and preventing the propagation of undesirable characteristics, so that the very name 'Eugenics' suggests, to the average mind...an attempt to interfere with the liberty of the individual in his pursuit of happiness in marriage."[9]

Perhaps the most militant of the eugenic research bodies was the Eugenics Research Association, created in June of 1913 at Cold Spring Harbor. Like many other eugenic groups, this association was also dominated by Davenport and Laughlin. But unlike the other eugenic bodies, the Eugenics Research Association was determined to go far beyond family investigations and position papers. The body was determined to escalate its "research" into legislative and administrative action, and public propaganda for the causes of eugenics, raceology and Nordic race supremacy. As such, the Eugenics Research Association brought together America's most esteemed eugenic medical practitioners, the field's most respected university professors, the movement's most intellectual theorists and the nation's most rabid eugenic racists.[10]

Only fifty-one charter members created the ERA, and its ranks did not exceed five hundred in later years. Those fifty-one charter members included men and women from the senior echelons of psychology, such as Yerkes and Adolf Meyer; later, Goddard, Brigham, Terman and other intelligence measurement authorities would join up. Professors from the medical schools and life science departments of Harvard, Columbia, Yale, Emory, Brown and Johns Hopkins were counted among the ranks.[11]

Two race hatred fanatics, Madison Grant and Lothrop Stoddard, achieved leadership roles within the organization. Grant was internationally known for his bestseller, *The Passing of the Great Race*, which promoted Nordic whites as the superior race. Grant's book, revered by eugenicists, lamented that America had been infested by "a large and increasing number of the weak, the broken and the mentally crippled of all races drawn from the lowest stratum of the Mediterranean basin and the Balkans, together with hordes of the wretched, submerged populations of the Polish Ghetto." Grant called these "human flotsam." Among America's genetic enemies, Grant singled out Irishmen, whom he insisted "were of no social importance." As a eugenic remedy, he preached: "A rigid system of selection through the elimination of those who are weak or unfit—in other words, social failures—would solve the whole question in a century...."

Grant held numerous leadership roles in the Eugenics Research Association, including its presidency, and ultimately sat with Davenport on the three-man executive committee.[12]

Stoddard would write an equally belligerent bestseller, published by Scribner's, entitled *The Rising Tide of Color Against White World Supremacy.* Harvard-educated Stoddard defiantly summarized his science in these words: "You cannot make bad stock into good...any more than you can turn a cart-horse into a hunter by putting it into a fine stable, or make a mongrel into a fine dog by teaching it tricks." He urged widespread segregation and immigration restrictions to combat the unfit races, which Stoddard compared to infectious bacteria. "Just as we isolate bacterial invasions and starve out the bacteria by limiting the area and amount of their food-supply, so we can compel an inferior race to remain in its native habitat...[which will] as with all organisms, eventually limit...its influence." Stoddard was one of the early members of the Eugenics Research Association, joining in response to the association's official invitation.[13]

The ranks of the ERA included eugenic activists of all sorts, but of the fifty-one original members, none was more enigmatic than charter member #14. His name was Dr. Edwin Katzen-Ellenbogen.[14]

Dr. Katzen-Ellenbogen had distinguished himself in the field of psychology, mostly though his work with epileptics. In the years just prior to his charter membership, Katzen-Ellenbogen served as the director of the Psychopathological Laboratory at New Jersey's State Village for Epileptics at Skillman. Before that he had been an assistant physician at Danvers Hospital in Massachusetts, as well as a clinical assistant at a medical school in New York and a lecturer in abnormal psychology at Harvard. Just a year before joining the ERA, he had presented a paper on the mental capacity of epileptics before the National Association for the Study of Epilepsy at Goddard's Vineland Training School for Feebleminded Girls and Boys in New Jersey. He was considered an up-and-coming talent. Although just twenty-seven years of age, Katzen-Ellenbogen was listed as a leading psychologist in the distinguished biographical volume, *American Men of Science.*[15]

Who was Katzen-Ellenbogen, really? He spelled his last name numerous ways, hyphenated and unhyphenated. He was an American citizen, but he was actually born in Stanislawow, in Austrian-occupied Poland; he immigrated to the United States in 1905. He settled in Fitchburg, Massachusetts. Shortly after arriving in Fitchburg, the twenty-four-year-old Katzen-Ellenbogen married Marie A. Pierce, an American woman six years his junior. Two months later, he traveled to Paris for further studies, but returned to the U.S. in 1907 when he was naturalized. He boasted credentials from Harvard and was a member of that university's postgraduate teaching staff, but he had actually received his primary education in Poland and his secondary schooling in Germany. He assumed the middle name "Maria," perhaps after his wife's name, but his real middle name was Wladyslaw. He claimed to be Roman Catholic, but was actually Jewish.[16]

Long-skulled, with bushy eyebrows, a thin mustache and a semicircular receding hairline topped by a very high brow, Katzen-Ellenbogen's head seemed almost too large for his body. As one who had worked with epileptics, disturbed children and the insane, Katzen-Ellenbogen had become accustomed to tinkering with the extremes of human frailty and the limits of will. He was attracted to the mysteries of the mind, but was convinced that the field of psychology was still in its infancy as it probed those mysteries. "Psychology is a discipline of undue hopes and uncritical skepticism," he wrote, adding, "It has been a hard battle, which in forty years time has elevated psychology from a cinderella science domiciled in one room at the Leipzig University to palace-like institutions, such as for instance the Harvard Psychological Institute...."[17]

In 1915, two years after he joined the Eugenics Research Association, Katzen-Ellenbogen sailed again to Europe. He would never return to America. He traveled first to Russia, but ended up in Germany. By then, Europe was embroiled in a bloody World War. But Katzen-Ellenbogen remained an "active member" of the organization even while abroad. Then America entered the war against Germany, and on March 21, 1918, the association's executive committee dropped Katzen-Ellenbogen from its rolls.[18]

Katzen-Ellenbogen studied troubled minds but was also familiar with intense personal pain and the fire of his own considerable mental anguish. In 1920, his only son, still in America, fell from a roof garden and was killed. The boy's death destroyed Katzen-Ellenbogen's sense of personal existence. There would be no male heir to carry on his bloodline, which contradicted the central aspiration of eugenics. But beyond any tenet of science, the untimely death would haunt Katzen-Ellenbogen for the rest of his life. He was in Europe when it occurred, yet he did not return for the funeral. The doctor's wife slid into profound depression. Katzen-Ellenbogen never forgave himself for staying away. Suicidal impulses would grip him for years.[19]

Bitter but also philosophical, purely scientific yet overwhelmingly ambitious, Katzen-Ellenbogen wandered from mental place to mental place. He emerged with the disconnected sense of a man with nothing to lose. Abortionist, drug peddler, informer, medical theorist, murderer—Katzen-Ellenbogen eventually drifted into all of these realms.[20] This American eugenicist would disappear from America, but his biological vision of humanity would eventually shock the world. Nor would he be alone in his crimes.

*   *   *

Eugenics found allies not just among the nation's learned men, but also among the affluent and influential. In 1912, shortly before the Eugenics Record Office installed its board of scientific directors, the New York State legislature had created the Rockefeller Foundation, which boasted fabulous assets. John D. Rockefeller donated $35 million the first year, and $65 million more the next year.[21] Davenport was keen to funnel Rockefeller's money into eugenics. As he had done with Mrs. Harriman, Davenport cultivated a personal connection with Rockefeller's son, John D. Rockefeller Jr. The younger Rockefeller controlled the foundation's millions.[22]

Shy and intensely private, the oil heir seemed to enjoy corresponding with Davenport about sundry eugenic topics. On January 27, 1912, using his personal 26 Broadway stationery, the young Rockefeller wrote Davenport a letter about a plan to incarcerate feebleminded criminal women for an extra length of time, so they "would...be kept from perpetuating [their] kind...until after the period of child bearing had been passed." Two months later, Rockefeller Jr. sent Davenport a copy of a *Good Housekeeping* article referencing Pearson and British eugenicists. Rockefeller asked, "Will you be good enough to return the article with your reply, which I shall greatly appreciate." On April 2, Rockefeller sent Davenport a formal thank you for answering a letter just received. About a month later, Rockefeller sent another note of personal thanks, this time for answering questions about the *Good Housekeeping* article.[23]

At its first meeting, the ERO's board of scientific directors "voted to recommend to Mr. John D. Rockefeller the support of the following investigations." The ERO's board, chaired by William Welch (who doubled as Rockefeller's own scientific director), compiled a short list: first, "an analysis of feeblemindedness"; second, "a study of a center of heavy incidence of insanity in Worcester County, Massachusetts"; third, a well-financed "preliminary study of the sources of the better and the poorer strains of immigrants" to be conducted overseas. They also petitioned Rockefeller to fund a statistician who would compile the data.[24]

Welch found his work with the ERO satisfying, and did not mind becoming vice-chairman when Alexander Graham Bell was appointed to the top post. Two years after Welch joined the board of scientific directors, Davenport used the connection to secure additional Rockefeller financial support. On March 1, 1915, Davenport told Welch, "It seems to me a favorable time to approach the Rockefeller Foundation on the subject of giving a fund for investment to the Eugenics Record Office." Davenport skillfully played Mrs. Harriman's wealth against Rockefeller's vastly superior fortune. To date, Rockefeller's foundation had "given us $6,000 a year, whereas Mrs. Harriman has given us $25,000" as well as funds for construction and other general expenses. Davenport's new plan called for an annual investment fund, as well as money to establish a better indexing operation to link surnames, traits and geographic locales. After adding up the columns, itemizing the projects and totaling the results, Davenport wrote Welch, "I would suggest that we should ask for $600,000 [$10.1 million in modern money] from the Rockefeller Foundation."[25]

If Rockefeller agreed to the $600,000 subvention, Davenport planned to go back to Mrs. Harriman and ask her to go one better. "We should then ask Mrs. Harriman to consider an endowment of $800,000 to $1 million." That would almost double her annual tithe.[26]

As expected, Davenport lunched with Mrs. Harriman just days later. Their discussion was fruitful. "She is, I understand, ready to turn over some property to [the Eugenics Record Office]," Davenport happily reported to Bell. Mrs. Harriman's financial support would ultimately grow to hundreds of thousands of dollars.[27]

Big money made all the difference for eugenics. Indeed, biological supremacy, raceology and coercive eugenic battle plans were all just talk until those ideas married into American affluence. With that affluence came the means and the connections to make eugenic theory an administrative reality.

Providing her opulent 1 East Sixty-ninth Street home as a meeting place, Mrs. Harriman bestowed her prestige as well as her wealth on the eugenic crusade. At one meeting in her home on April 8, 1914, more than a dozen experts gathered to plan action against those considered feebleminded. Most offered short presentations. Goddard, fresh from his intelligence-testing accomplishments, began the meeting with a proposed definition of "feebleminded." Another outlined ideas on "segregation of the feebleminded." A third offered "new and needed legislation in re: the feebleminded." Laughlin presented a fifteen-minute talk on "sterilization of the feebleminded." Davenport spoke on county surveys of the feebleminded.[28]

Mrs. Harriman wielded great power. When she made a request of New York State officials, it was difficult for them to say no. Davenport's proposed county surveys in search of the unfit, for example, were implemented by state officials. Eugenic agencies were established, often bearing innocuous names. Robert Hebberd, secretary of the New York State Board of Charities, reported to Mrs. Harriman that "our Eugenics Bureau is known officially as the Bureau of Analysis and Investigation." In describing the agency's work, Hebberd's letter reflected the usual eugenic parlance, "The study of groups of defective individuals is so closely related to the welfare of future generations that the lessons drawn from the histories of abnormal families... [can] prevent the continuance of conditions which foster social evils." He added that to this end, the records of some 300,000 people had already been tabulated in twenty-four of New York State's counties. Hebberd promised to coordinate his agency's work with privately financed eugenic field surveys "in Rockland County, under your direction." He deferentially added, "Permit me to say that it is gratifying to know of your deep interest in this branch of the work of the State Board of Charities."[29]

Rockefeller also financed private county surveys. His foundation would cover the $10,000 cost of a hunt for the unfit in New York's Nassau County. Davenport and several Nassau County appointees formed an impromptu "Committee on the Enumeration of Mental Defectives," which worked closely with local school authorities in search of inferior students. Eight field workers would assist the search.[30]

Some ordinary New York State agencies changed their focuses from benign to eugenic. One such agency operated under the innocuous-sounding name of the Bureau of Industries and Immigration. Originally established to protect disadvantaged immigrants, the bureau began employing investigators to identify "defectives," the feebleminded and the insane. One typical report on fifteen feebleminded newcomers began with Case #258, which focused on Teresa Owen, a forty-year-old woman from Ireland who was classified as insane. The case note on Owen read, "Has been released to her husband and is cohabiting with him, with what disastrous results to posterity... no one can foretell. She is a menace... [and] should be removed and segregated pending removal." Case #430 treated Eva Stypanovitz, an eighteen-year-old Russian Jew who was classified as feebleminded. The file on Stypanovitz noted, "Case diagnosed by relatives. Is of marriageable age, and a menace to the community." Case #918 dealt with Vittorio Castellino, a thirty-five-year-old from Italy, and recorded, "Such a case cannot be too extravagantly condemned from a eugenic and economic point of view."[31]

Another such agency was the organization that became known as the National Committee on Prison and Prison Labor, first organized in 1910 by the New York State Department of Labor to investigate the exploitation of convict-manufactured goods. Four years later, the body changed its name amid a "widening of its activities." Judge Olson, the stalwart eugenic activist who also directed the Municipal Court of Chicago Psychopathic Laboratory, steered his colleagues on the prison committee to create similar municipal psychopathic labs to document hereditary criminality in their cities. The New York City Police Department did indeed establish a psychopathic laboratory for eugenic investigations, utilizing Eugenics Record Office field workers supplied by Mrs. Harriman. Davenport himself headed up the prison group's special committee on eugenics, which was established "to get at the...heredity factors in anti-social behavior...with the aid of a careful family history." Prisoners at Sing Sing were the first to be examined by Davenport's researchers under a year-long joint project with the Eugenics Record Office.[32]

In 1916, New York's Senate Commission to Investigate Provision for the Mentally Deficient held hearings and published a 628-page special report, including a 109-page bibliography of eugenic books and articles. The commission's purview included imposed sterilization. Among its cited resources were eugenic county surveys in Westchester County supervised by Dr. Gertrude Hall, one of the eugenic experts in Mrs. Harriman's circle and the director of the Bureau of Analysis and Investigation.[33]

Many officials were easily swayed by the stacks of scientific documentation eugenicists could amass. New York's State Hospital Commission—comprised of a coterie of leading physicians—emerged from meetings with Davenport at the Eugenics Record Office in July of 1917 expressing a new determination to concentrate on the feebleminded—even though there was not yet a definition for feeblemindedness. After the meeting, the commission announced it would recommend that the state legislature allocate $10 to $20 million during the next decade to eugenically address the insane and feebleminded. The ERO pledged its assistance in the effort.[34]

New York State was hardly alone. Indiana's legislature appropriated $10,000 for a Committee on Mental Defectives in 1917. Initial research was completed by ERO field workers Clara Pond (in Jasper, Wabash and Elkhart counties) and Edith Atwood (in Shelby, Vanderburgh and Warrick

THE UNITED STATES OF STERILIZATION || 97

counties). A commission to investigate the feebleminded was empanelled in Utah. Arkansas did the same. One ERO field worker, Ethel Thayer, traveled some 10,000 miles during six months in 1917, interviewing 472 individuals to produce what the ERO termed "more or less complete histories of 84 [families]."[35]

There was no way for the public to know if a seemingly unrelated government agency was actively pursuing a eugenic agenda. The United States Department of Agriculture maintained an active role in America's eugenics movement by virtue of its quasi-official domination of the American Breeders Association. Various Department of Agriculture officials either sponsored or officially encouraged eugenic research. Agricultural department meetings went beyond the bounds of simple agronomy; they often encompassed human breeding as well. On November 14, 1912, Professor C. L. Goodrich, at the Washington office of the Department of Agriculture, was asked by a colleague in the USDA's Columbia, South Carolina, office whether two Negro siblings, both with six fingers on each hand, should be brought to an ABA meeting at the National Corn Exposition for eugenic evaluation. Professor Goodrich, who controlled the presentations of the ABA's Eugenic Section, replied a few days later, "Have the children brought. . . . I will put you on the program for a paper before the Eugenics section. . . ."[36]

On November 26, 1912, the USDA's Office of Farm Management wrote to Davenport on official government letterhead suggesting that the ERO assign "a eugenic worker on the case and develop the facts in relation to the negro's family by the time of the meeting of the Breeder's Association in Columbia [South Carolina] in February." Receptive to the idea, Davenport replied three days later, "Perhaps he can present one or more of the polydactyls to the eugenics section."[37]

On January 3, 1913, Davenport wrote to George W. Knorr at the USDA in Washington asking, "If not too late, please add two titles to the eugenics program." One of these would be Davenport's own last-minute entry, "A Biologist's View of the Southern Negro Problem." Knorr wrote back asking for a lecturer on eugenic immigration issues. On January 8, Davenport referred Knorr to a Harvard eugenicist specializing in immigration, and reminded the department to make sure "the meeting of the eugenics section [was all arranged] at the Insane Asylum." That same day, Davenport wrote his colleague at Harvard, asking him to contact the USDA to get on the program. On January 10, Davenport asked Knorr to approve yet another eugenics paper entitled "Heredity of Left-handedness."[38]

Secretary of Agriculture James Wilson doubled as president of the ABA. At the group's 1913 convention, he rallied the forces. In his presidential address, Wilson declared, "You have developed in your eugenics section a great experiment station and institution of research, with a splendid building called the Eugenics Record Office.... Your laboratory material is the heredity that runs through the veins of the good, bad, and indifferent families of our great country... assembling the genetic data of thousands of families... making records of the very souls of our people, of the very life essence of our racial blood.... Those families which have in them degenerate blood will have new reason for more slowly increasing their kind. Those families in whose veins runs the blood of royal efficiency, will have added reason for that pride which will induce them to multiply their kind." Wilson also encouraged the ERO to seek even greater funding. "I observe that you are publicly asking for a foundation of half a million dollars," he said. "Twenty times that sum, or ten millions, would come nearer the mark."[39]

The speeches presented at obscure agricultural meetings in South Carolina, the eugenic surveys in small Indiana counties or by major New York State agencies, the eugenics courses taught in small colleges or in prestigious universities—none of this eugenic activity remained a local phenomenon. It quickly accumulated and became national news for a movement hungry for the smallest advance in its crusade. Therefore in January of 1916, the ERO launched a new publication, *Eugenical News*, which was edited by Laughlin and reported endless details of the movement's vicissitudes. Approximately 1,000 copies of each issue were distributed to activists. From the most important research to the most obscure minutia, an eager audience of committed eugenic devotees would read about it in *Eugenical News*. Almost every administrative proposal, every legislative measure, every academic course, every speech and organizational development was reported in this publication.[40]

When field worker Clara Pond began her eugenic duties at the New York Police Department on January 15, 1917, it was reported in the February issue. When the ERO received records of 128 family charts from Morgan County, Indiana, it was reported. When the Village for Epileptics at Skillman, New Jersey, contributed 798 pages of data on its patients, it was reported. When Laughlin spoke before the Illinois Corn Growers Convention at the University of Illinois, it was reported. When Dr. Walter Swift of the Speech Disorder Clinic wrote on inherited speech problems in the *Review of Neurology and Psychiatry*, his article was reviewed in depth.

When Yerkes paid a courtesy visit to the Eugenics Record Office in Cold Spring Harbor, it was reported. When Congress overrode President Wilson's veto of an immigration bill, the vote tallies were reported. When the state of Delaware appropriated $10,000 for an institution for the feeble-minded, it was reported. When eugenic field worker Elizabeth Moore took up gardening at her home in North Anson, Maine, this too was reported.[41]

No legislative development was too small, nor was any locale too obscure for coverage. Indeed, the more obscure the eugenic development, the more enthusiastic the reportage seemed. The more significant the research or legislative effort, the more readers looked to *Eugenical News* for information and guidance. In effect, *Eugenical News* offered the movement organizational, scientific, legislative and theoretical cohesion.

Eventually, the eugenics movement and its supporters began to speak a common language that crept into the general mindset of many of America's most influential thinkers. On January 3, 1913, former President Theodore Roosevelt wrote Davenport, "I agree with you ... that society has no busi-ness to permit degenerates to reproduce their kind. ... Some day, we will realize that the prime duty, the inescapable duty, of the good citizen of the right type, is to leave his or her blood behind him in the world; and that we have no business to permit the perpetuation of citizens of the wrong type." Episcopalian Bishop John T. Dallas of Concord, New Hampshire, issued a public statement: "Eugenics is one of the very most important subjects that the present generation has to consider." Episcopalian Bishop Thomas F. Gailor of Memphis, Tennessee, issued a similar statement: "The science of eugenics ... by devising methods for the prevention of the propagation of the feebleminded, criminal and unfit members of the community, is ... one of the most important and valuable contributions to civilization." Dr. Ada Comstock, president of Radcliffe College, declared publicly, "Eugenics is 'the greatest concern of the human race.' The development of civilization depends upon it." Dr. Albert Wiggam, an author and a leading member of the American Association for the Advancement of Science, pronounced his belief: "Had Jesus been among us, he would have been president of the First Eugenic Congress."[42]

While many of America's elite exalted eugenics, the original Galtonian eugenicists in Britain were horrified by the sham science they saw thriving in the United States and taking root in their own country. In a merciless 1913 scientific paper written on behalf of the Galton Laboratory, British scientist David Heron publicly excoriated the American eugenics of Davenport, Laughlin, and the Eugenics Record Office. Using the harshest possible language, Heron warned against "certain recent American work which has been welcomed in this country as of first-class importance, but the teaching of which we hold to be fallacious and indeed actually dangerous to social welfare." His accusations: "Careless presentation of data, inaccurate methods of analysis, irresponsible expression of conclusions, and rapid change of opinion."[43]

Heron lamented further, "Those of us who have the highest hopes for the new science of Eugenics in the future are not a little alarmed by many of the recent contributions to the subject which threaten to place Eugenics... entirely outside the pale of true science.... When we find such teaching—based on the flimsiest of theories and on the most superficial of inquiries—proclaimed in the name of Eugenics, and spoken of as 'entirely splendid work,' we feel that it is not possible to use criticism too harsh, nor words too strong in repudiation of advice which, if accepted, must mean the death of Eugenics as a science."[44]

Heron emphasized "that the material has been collected in a most unsatisfactory manner, that the data have been tabled in a most slipshod fashion, and that the Mendelian conclusions drawn have no justification whatever...." He went so far as to say the data had been deliberately skewed. As an example, he observed that "a family containing a large number of defectives is more likely to be recorded than a family containing a small number of defectives."[45] In sum, he called American eugenics rubbish.

Davenport exploded.

He marshaled all his academic and rhetorical resources and the propagandists of the ERO. Davenport and A. J. Rosanoff combined two defensive essays and a journal article denouncing Dr. Heron's criticism into a lengthy ERO Bulletin. The bulletin, entitled *Reply to the Criticism of Recent American Work by Dr. Heron of the Galton Laboratory*, was circulated to hundreds of public administrators, eugenic theorists and others whose minds needed to be swayed, assuaged or buttressed.[46]

As keeper of the eugenic flame and defender of its faithful, Davenport correctly portrayed Dr. Heron's assault to be against "my reputation [which] I regard as of infinitely less importance than the acquisition of truth; and if I resent these evil innuendoes it is not for myself at all, but only for the protection of the scientific interests which I am, for the time, custodian." In a rambling, point-by-point confutation, Davenport belittled Heron's attack as a vendetta by his Galtonian enemies in England. He explained away his faulty data as typographical. His rebuttal was rich with abstruse formulas in support of his subverted theses.[47]

In Davenport's mind, Mendel's laws hovered as the sacred oracle of American eugenics, the rigid determiner of everything tall and short, bright and dim, right and wrong, strong and weak. All that existed in the chaotic pool of life was subservient to Mendel's tenets as respun by Davenport. Indeed, Davenport cherished those tenets as if chiseled by the finger of God. Come what may, Davenport declared he would never "deny the truth of Mendelism." He defiantly proclaimed, "The principles of heredity are the same in man and hogs and sun-flowers."[48]

But the attacks did not stop. True, eugenics had ascended to a scientific standard throughout the nation's academic and intellectual circles, becoming almost enshrined in the leading medical journals and among the most progressive bureaucrats. The word itself had become a catchphrase of the intelligentsia. But soon the sweeping reality of the eugenics movement's agenda started filtering down to the masses. Average people slowly began to understand that the ruling classes were planning a future America, indeed a future world, that would leave many of them behind. Sensational articles began to appear in the press.

"14 million to be sterilized" was the warning from the Hearst syndicate of newspapers in late September of 1915. Alexander Graham Bell, long queasy about Davenport's obsession with defectives, reacted at once, contacting Cold Spring Harbor for some reassurance. Davenport wrote back on September 25: "I am very sorry that ripples of a very sensational fake article about the plans of the Eugenics Record Office to sterilize 14 million Americans has rippled"—he crossed out "has rippled"—". . . have disturbed the placid waters about [Bell's vacation home in] Beinn Bhreagh [Nova Scotia]." Davenport assured Bell he would warn others "against believing things . . . in the Hearst papers." Bell, only briefly comforted, wrote back, "Your note . . . is a great relief to me, as I was naturally disturbed over the newspaper notices—even though I didn't believe them."[49]

The articles did not stop, however. Crusading journalists and commentators began to expose American eugenics as a war of the wealthy against the poor. On October 14, 1915, the Hearst newspapers syndicated a series of powerful editorials pulling no punches. Typical was an editorial in the *San Francisco Daily News*:

WHERE TO BEGIN

The millions of Mrs. Harriman, relict of the great railroad "promoter," assisted by other millions of Rockefeller and Carnegie, are to be devoted to sterilization of several hundred thousands of American "defectives" annually, as a matter of eugenics.

It is true that we don't yet know all that the millions of our plutocracy can do to the common folks. We see that our moneyed plutocrats can own the governments of whole states, override constitutions, maintain private armies to shoot down men, women and children, and railroad innocent men to life imprisonment for murder, or lesser crimes. And IF WE SUBMIT TO SUCH THINGS, we ought not to be surprised if they undertake to sterilize all those who are obnoxious to them.

Of course, the proposition depends much on who are to be declared "defective."

The old Spartans, with war always in view, used to destroy, at birth, boys born with decided physical weakness. Some of our present-day eugenists go farther and damn children before their birth because of parents criminally inclined. Then we have eugenic "defectives" in the insane and the incurably diseased. The proposition is not wholly without justification. But isn't there another sort of "defective," who is quite as dangerous as any but whom discussion generally overlooks, especially discussion by the senile long-haired pathologists, and long-eared college professors involved in the Harriman-Rockefeller scheme to sterilize?

A boy is born to millions. He either doesn't work, isn't useful, doesn't contribute to human happiness, is altogether a parasite, or else he works to add to his millions, with the brutal, insane greed for more and more that caused the accumulation of the inherited millions. Why isn't such THE MOST DANGEROUS "DEFECTIVE" OF ALL? Why isn't the prevention of more such progeny THE FIRST DUTY OF EUGENICS? Such "defectives" directly attack the rights, liberties, happiness, and lives of millions.

Talk about inheriting criminal tendencies. Is there a ranker case of such than the inheritance of Standard Oil criminality as evidenced in the slaughter of mothers and their babes at Ludlow?

Sterilization of hundreds of thousands of the masses, by the Harrimans and Rockefellers? LET'S FIRST TRY OUT THE "DEFECTIVENESS" OF THE SONS OF BILLIONAIRES!

Let's first sterilize where sterilization will mean something immediate, far-reaching and thorough in the way of genuine eugenics![50]

More letters flew across the country as leading scholars began assessing the movement's image. Davenport worked on damage control. He began writing letters. Among the first was to Thomas D. Eliot, a major eugenic activist then living in San Francisco. "The article upon which the editorial in the *San Francisco Daily News* was based was entirely without any foundation in fact," Davenport assured Eliot. "The writer for the Hearst syndicate supplied them with an absolutely baseless and basely false article about imaginary plans of the Eugenics Record Office. As a matter of fact, the Eugenics Record Office exists only for the purpose of making studies primarily in human heredity and has nothing whatsoever to do with propaganda for sterilization. After the printing of this false article in scores of papers in this country my attention was called to it, and I wrote a letter to the *New York American* and requested them to publish the letter. This they refused to do...."[51]

Davenport scoffed, "We know the name of the unfortunate who wrote the article for the Hearst syndicate. To my protestation, he replies only that he proposes to publish a series of articles, intimating that he has worse ones in store [than] that already published. I tell you this so that you may be prepared for the future. It is quite within the range of possibility that he may state that the Rockefeller, Carnegie and Harriman millions are to be devoted to forcing the whites of the South to have children by the blacks in order to grade up the blacks. I can imagine even worse things." He dismissed Hearst readers as "paranoiacs and imbeciles," and urged his colleagues to stand fast.[52] But the press continued.

On February 17, 1916, a *New York American* reporter named Miss Hoffmann insisted on traveling up to New Haven, Connecticut, to interview the prominent Yale economist Irving Fisher about eugenics. Fisher, a leading raceologist, occupied a central role in the eugenics movement. The reporter had latched onto a sentence in a leading eugenic publication, which asserted, "Many women of the borderline type of feeblemindedness, where mental incapacity often passes for innocence, possess the qualities of charm felt in children, and are consequently quickly selected in marriage." Fisher did not know where the correct documentation was to support such a statement. "I should have turned her loose on you," he wrote to Davenport, "had I not known your sentiment on reporters especially of the Hearst journals!...Much as I dislike the tone of their articles...if we do not help them, they will do us positive injury... [and yet] in spite of their sensationalism, we can utilize them to create respect for the eugenics idea in the mind of the public."[53]

Fisher appended a typical progress report to his letter. "You will be glad to know," he wrote, "that I have interested the Dean here in trying to secure something in eugenics. You will doubtless hear from him.... I am delighted to see how other colleges have taken the matter up. Yale seems to be a little behind in this matter."[54]

Davenport was relieved that Fisher had steered the *New York American* reporter elsewhere, admitting, "I might have reacted in a way which I should subsequently have regretted."[55] Such scandals in the press prompted Alexander Graham Bell to distance himself from the eugenics movement.

Davenport surely sensed Bell's apprehension. When it came time to call the Spring 1916 scientific board meeting, Davenport struggled with the phrasing of his letter to Bell. "Do you authorize call for meeting here April Eighth." Vigorously scratched out. Slight variation: "Do you authorize *me* to call meeting here on April Eighth." Vigorously scratched out. Start again: "Do you...." Scratched out, starting once more: "Shall *I* issue call Director's meeting here on April Eighth."[56]

On the afternoon of April 8, 1916, too impatient for a letter to arrive, Bell telephoned a message to Cold Spring Harbor.

> Dr. Davenport: Greatly regret inability to attend meeting of Eugenics Board as I had intended. Detained at last moment by important matters, demanding my immediate attention. I believe I have now served for three years as chairman. I would be much obliged if you would kindly present my resignation on the Board and say that it would gratify me very much to have some member now appointed to the position.
>
> With best wishes for a successful meeting,
> Alexander Graham Bell[57]

Davenport was surely shaken. He sent off a note asking if Bell would at least stay on until the end of the year as chairman of the board of scientific directors; at the same time, he assured Bell that in the future more emphasis would be placed on positive human qualities. Bell reluctantly agreed, but his connection to the movement was now permanently frayed.

On April 20, 1916, Bell agreed to chair just one more meeting, the December 15 session, but with "the understanding that I will then resign as Chairman of the Board." He added, "I am very much pleased to know from your letter that more attention is now to be paid to the Eugenic positive side than heretofore."[58]

Just before the meeting, Bell once again reminded Davenport that he would participate in the year-end meeting, but "I hope that you do not forget that I am to be allowed to resign from the chairmanship at this meeting." After that December meeting, Bell severed his relations with the movement altogether. In a polite but curt letter, Bell informed Davenport, "I will no longer be associated with yourself and the other directors. With best wishes for the continuance of the work, and kind regards."[59]

By the end of 1917, Mrs. Harriman's privately funded Eugenics Record Office had merged with the Carnegie Institution's Experimental Station. Both entities were headed by Davenport. They existed virtually side-by-side at Cold Spring Harbor, and to a large extent functioned as extensions of one another. This created a consolidated eugenic enterprise at Cold Spring Harbor. To facilitate the legal merger of what everyone knew was an operational fact, Mrs. Harriman deeded the ERO's existing assets plus a new gift of $300,000 to the Carnegie Institution, thus providing for the ERO's continued operation. As part of the merger, the ERO transferred its collection of 51,851 pages of family documentation and index cards on 534,625 individuals. Each card offered lines for forty personal traits.[60]

The science of eugenics was now consolidated under the sterling international name of the Carnegie Institution. Eugenics was stronger than ever.

* * *

Eugenics did not reform despite its public pillorying. The movement continued to amass volumes of data on families and individuals by combining equal portions of gossip, race prejudice, sloppy methods and leaps of logic, all caulked together by elements of actual genetic knowledge to create the glitter of a genuine science.

A statistical study found that fewer than 12 percent of Negro songs were in a minor key. "It tends to justify the general impression that the negro is temperamentally sunny, cheerful, optimistic," reported *Eugenical News*. As such, the study purveyed as scientific evidence that while "slave songs...refer to 'hard trials and tribulations,'" the genetic constitution of Negroes under American apartheid nonetheless displayed a "dominant mood...of jubilation...."[61]

Eugenicists began compiling long lists of ship captains and their progeny to identify an invented genetic trait called "thalassophilia," that is, an inherited love of the sea. *Eugenical News* listed several captains who died or were injured in shipwrecks. "Such hardy mariners do not call for our sympathy," declared *Eugenical News*, "they were following their instinct."[62]

Behaviors, mannerisms, and personal attributes that we now understand to be shaped by environment were all deemed eugenic qualities. "When we look among our acquaintances," Davenport wrote, "we are struck by their diversity in physical, mental, and moral traits... they may be selfish or altruistic, conscientious or liable to shirk... for these characteristics are inheritable...."[63]

In painstakingly compiled family trait booklets, each numbered at the top right for tracking, the most personal and subjective measurements were recorded as scientific data. Family trait booklet #40688, of the Bohemian farmer Joseph Chloupek and his Irish wife Mary Sullivan, was typical. Question 12 asked for "special tastes, gifts or peculiarities of mind or body." For Chloupek, his traits were noted as "reading, affectionate, firm." His wife was noted as "very religious... broad minded in her religious attitude toward others." The rest of the family was similarly assessed, including Chloupek's mother, Eugenia, who was marked as a "good mother."[64]

Approximations were frequently entered as authentic scientific measurements. Question 13 called for the height either in inches, or, if preferred, with any of four notations: "very short, short, medium tall, very tall." Question 15 recorded hair color as "albino, flaxen, yellow-brown, light brown, medium brown, dark hair, black." Question 17 asked for the individual's skin to be described as "blond, intermediate, brunette, dark brown, black Negro, yellow, yellow-brown or reddish-brown." Question 26 asked for visual acuity, and the choices were "blind, imperfect, strong, or color blind"; in the case of the Chloupek family, the most common response was "good."[65]

A second genealogical tool, the family folder, recorded such eugenic "facts" as "participation in church activities" and "early moral environment." Special areas were set aside for notations as to whether the individual was known for "interest in world events or neighborhood gossip," or "modesty," or whether the person "holds a grudge." Question fifty-six asked for an evaluation of the individual's "optimism, patriotism, care for the good opinion of others."[66]

In ERO Bulletin #13, *How to Make a Eugenical Family Study*, coauthored by Davenport and Laughlin, field workers and information recorders were informed that eugenic authorities would explain the "eugenical meaning of the facts recorded."[67]

Even within the accepted parameters, the data was often only approximated. Heights for several dozen Jewish children were charted in one report with a special entry, "These weights recorded by nurses... are considered by Dr. Cohen as more accurate than those recorded on March 20." Physician Brett Ratner submitted extensive physical measurements of newborns, with a caveat. "The sheet... [includes] the length," he explained, "which is taken by the attending doctor by suspending the child by its legs, which is of course very inaccurate, and the chest was also done by the attending physician. Therefore, I cannot vouch for the chest and length measurement. The weights, however, are all absolutely accurate."[68]

Often, the science was filtered through personal animus, colored language and even name-calling. Character flaws were frequently accentuated in clinical eugenic descriptions, almost as if to pass the reader a cue. "James Dack was commonly known as 'Rotten Jimmy,'" read one typical description. "The epithet was given because of the diseased condition of his legs...although the term is said to have been equally applicable to his moral nature." No wonder Goddard admitted that in writing his revered eugenic text on the Kallikak family, "We have made rather dogmatic statements and have drawn conclusions that do not seem scientifically warranted by the data. We have done this because it seems necessary to make these statements and conclusions for the benefit of the lay reader...." In Vermont, a careful and methodical statewide survey condemned one man as eugenically unfit based on the genetic datum that he was "a big hopeless good for nothing."[69]

Davenport and Laughlin brashly predicted, "The day will yet come when among the first questions, asked by an employer of the applicant for a position, will be those relating to the occupations of his kin and the success they have had in such occupations."

Correcting the American ethic with a eugenic voice, they promulgated the stunning admonishment, "There are those who adhere to the obviously false doctrine that men are born equal and therefore it really doesn't matter who marries whom."[70]

The men and women of eugenics wielded the science. They were supported by the best universities in America, endorsed by the brightest thinkers, financed by the richest capitalists. They envisioned millions of America's unfit being rounded up and incarcerated in vast colonies, farms or camps. They would be prohibited from marrying and forcibly sterilized. Eventually—perhaps within several generation—only the white Nordics would remain. When their work was done at home, American eugenicists hoped to do the same for Europe, and indeed for every other continent, until the superior race of their Nordic dreams became a global reality.

Yet the very first sentence of the United States Constitution protected future generations. "We the People of the United States, in Order to form a more perfect Union, establish Justice...secure the Blessings of Liberty to ourselves and our Posterity, do ordain and establish this Constitution."[71] *Posterity* would be the monumental issue over which the forces of eugenics struggled. To eugenicists, the future of America and humanity itself was at stake.

In 1924, they would wage a pitched battle against a lone adversary. This adversary would not be a crusading journalist or an outspoken politician, but rather a helpless Virginia teenager named Carrie Buck. Declared feebleminded, she was actually a good student in a family of good students. Called a menace to society and to the future of mankind, she was actually just poor white trash from the back streets of Charlottesville, Virginia. This simple yet often eloquent girl would make the perfect test case. She was selected for exactly this reason.

* * *

Carrie Buck's mother, Emma, was one of Charlottesville's least respected citizens. Widowed and worthless, living on the margins of society, Emma was deemed a perfect candidate for feeblemindedness. After World War I, Virginia had a well-established policy of sweeping its social outcasts into homes for the feebleminded and epileptic. In Virginia, the two conditions, feeblemindedness and epilepsy, were virtually synonymous. They were also synonymous with another diagnosis, *shiftlessness*, that is, the genetic defect of being worthless and unattached in life.[72]

On April 1, 1920, Emma was hauled before a so-called Commission on Feeblemindedness. Justice of the Peace C. D. Shackleford convened the very brief hearing required. Physician J. S. Davis conducted the examination, referred to on the form as "an inquisition." The state's form enumerated sixty pointed questions. Question two, under Social History and Reaction, asked if Emma had ever been convicted of a crime. Emma's response: "Prostitution." In those days any woman might be charged with prostitution, whether for actually selling her body or simply for conducting herself in a fashion morally repugnant to the local authorities or even to the cop on the beat. Question eighteen, under Personal and Developmental History, asked if Emma had any diseases. She responded that she had syphilis. Question eight, under Physical Condition, asked specifically if Emma had ever had syphilis, to which her response was yes. Question nine, also under Physical Condition, asked if any venereal disease was present, and for the third time Emma confirmed that she had had syphilis. As to her moral character, the hearing officials wrote "notoriously untruthful." Indeed, question five, under Social History and Reaction, asked whether she had "conducted... herself in a proper conjugal manner." The examiners wrote "No."[73]

A few minutes later, Emma was officially deemed feebleminded. Shackleford signed the order of commitment, declaring she was "suspected of being feebleminded or epileptic." Five days later, Emma was driven to the Colony for Epileptics and Feebleminded. There she was consigned to Ward Five. She would remain at the colony for the rest of her life.[74]

Years before, in 1906, when Emma was still married, she had given birth to a daughter, Carrie. When Emma's husband died, the widow drifted into the social fringes of Charlottesville. At age three, Carrie was removed from Emma's custody and placed with another family. There were no formal adoption proceedings. Charlottesville peace officer J.T. Dobbs and his wife simply took the child into their Grove Street house. The Dobbses had a child of their own, approximately Carrie's age. Mrs. Dobbs needed extra help with the chores. Carrie was good at her chores, and also did well in school. School records show her performance was "very good—deportment and lessons." But when Carrie was in sixth grade, the Dobbses withdrew the girl from school so she could concentrate on the increasing load of housework—not only for their home on Grove Street, but for others in the neighborhood that Carrie was "loaned" to. Although Carrie never felt like she was a part of the Dobbs family, she was happy to be there. She recalled being obedient, and always considered herself "a good girl."[75]

One day in the summer of 1923, seventeen-year-old Carrie was discovered to be pregnant. She explained that she had been raped. "He forced himself on me," Carrie later recollected, "he was a boyfriend of mine and he promised to marry me." Years later, she would accuse a Dobbs nephew of being the rapist.[76]

The Dobbses would not listen to her explanations. They wanted Carrie—and her shame—out of the house at once. As Dobbs was the local peace officer, and familiar with the legal workings of the county, he knew just what to do. He filed commitment papers with Justice Shackleford. Dobbs claimed the girl was feebleminded, epileptic or both, and anyway, the family could no longer afford to board her. Shackleford scheduled a commitment proceeding.[77]

On January 23, 1924, Shackleford convened a brief hearing. Two doctors attended to render their expert opinions. The Dobbses testified that Carrie had experienced "hallucinations and...outbreaks of temper" and had engaged in "peculiar actions." Carrie was quickly declared "feebleminded" and transferred to the custody of the Colony for Epileptics and Feebleminded. For Shackleford, it was the second generation of Bucks he had sent to the colony—first the mother, Emma, and now her daughter, Carrie.[78]

It was not unusual for Virginia to use its Colony for Epileptics and Feebleminded as a dumping ground for those deemed morally unsuitable. Classifying promiscuous women as morons was commonplace. The colony's superintendent, Dr. Albert Priddy, admitted as much in a report: "The admission of female morons to this institution has consisted for the most part of those who would formerly have found their way into the red-light district and become dangerous to society...."[79]

But the numbers of morally condemned women were becoming economically daunting. "If the present tendency to place and keep under custodial care in State institutions all females who have become incorrigibly immoral [continues]," he argued, "it will soon become a burden much greater than the State can carry. These women are never reformed in heart and mind because they are defectives from the standpoint of intellect and moral conception and should always have the supervision by officers of the law and properly appointed custodians." Priddy's solution was the common eugenic remedy, sterilization.[80]

When Carrie was condemned, eugenical sterilizations were not yet legal in Virginia. Priddy's institution had certainly sterilized many women, but always as part of "therapeutic" treatment for unspecified types of "pelvic disease."[81] These therapeutic sterilizations on young, unsuspecting women were recorded as "voluntary," with informed consent transcripts to prove it. One such transcript read:

> Doctor: Do you like movies?
> Patient: Yes, sir.
> Doctor: Do you like cartoons?
> Patient: Yes, sir.
> Doctor: You don't mind being operated on, do you?
> Patient: No, sir.
> Doctor: Then you can go ahead.[82]

Priddy well understood how far outside the law such sterilizations were. In 1916, he had been taken to court for sterilizing several members of another Virginia family. On September 23, 1916, while the hardworking George Mallory was on shift at a nearby sawmill, his wife Willie and nine of their dozen children were at home in Richmond. Two family friends were visiting. Suddenly, two Richmond policemen burst in and declared the Mallory home "a disorderly house," that is, a brothel. It was later alleged that one of the policemen actually "made an indecent proposal" to one of the daughters.[83]

No matter, the younger children were turned over to the juvenile court, which, citing "vicious and immoral influences," transferred them to the Children's Home Society. Willie and her two eldest daughters, Jessie and Nannie, were confined at the City Detention Home, and then on October 14 referred to the Commission for the Feebleminded.[84]

Willie later recalled her experience. "A doctor examined my mind," she recounted, "and asked if I could tell whether salt was in the bread or not, and did I know how to tie my shoes. There was a picture hanging on the wall of a dog. He asked me if it was a dog or a lady. He asked me all sorts of foolish questions, which would take too long for me to tell you.... Then the doctor took his pencil and scratched his head and said, 'I can't get that woman in.'" But the attending juvenile probation officer, Mrs. Roller, was determined to have the family institutionalized. She told the doctor to write "unable to control her nerves," and added, "We can get her in for that."[85] He did so.

Mrs. Mallory, Jessie and Nannie were committed for lack of nervous control. Priddy had them now. Willie and Jessie were sterilized first. In late 1917, Priddy was getting ready to operate on the other daughter, Nannie, when he received another in a series of letters from George Mallory. Proud and strong-willed, Mallory expressed himself in powerful, if simple, terms. His English was lousy and his spelling atrocious. But his outrage was palpable. Grammar and form did not matter for Mallory. His family had been ripped from his home, and he wanted them back. On November 5, 1917, after several earlier letters were ignored, Mallory wrote an angry final demand.[86]

Dr Priddy

Dear sir one more time I am go write to you to ask you about my child I cannot here from her bye no means I have wrote three orfour times cant get hereing from her at all. We have sent her a box and I dont no wheather she recevied them or not. I want to know when can I get my child home again My family have been broked up on fake pertents same as white slavery. Dr what busneiss did you have opreatedeing on my wife and daughter with out my consent. I am a hard working man can take care of my family and can prove it and before I am finish you will find out that I am. I heard that some one told you lots of bad news but I have been living with her for twenty three years and cant no body prove nothing againts my wife they cant talk anything but cant prove nothing...just to think my wife is 43 years old and to be treated in that way, you ought to be a shamed of your selft of opreateding on her at that age just stop and think of how she have been treated what cause did you have opreateding her please let me no for there is no law for such treatment I have found that out I am a poor man but was smart anuf to find that out I had a good home as any man wanted nine sweet little children now to think it is all broke up for nothing I want to no what you are go do I earn 75$ a month I dont want my child on the state I did not put her on there. if you don't let me have her bye easy term I will get her by bad she is not feeble minded over there working for the state for nothing now let me no at once I am a humanbeen as well as you are I am tired of being treated this way for nothing I want my child that is good understanded let me know before farther notise. Now I want to know on return mail what are you go do wheather are go let my child come home let me here from her

Verly Truiley
Mr George Mallory

My last letter to you for my child with out trouble don't keep my child there I have told you not to opreated on my child if you do it will be more trouble.... [87]

Priddy was livid, and wrote Mallory back, threatening his own action. "Now, don't you dare write me another such letter or I will have you arrested in a few hours." Implying a threat of surgical consequences, he added, "If you dare to write me another such communication I will have you arrested and brought here too." Mallory's spelling was bad, but he retained an attorney who could spell quite correctly. He sued Priddy for sterilizing his wife and daughter Jessie. Mallory also filed a writ of *habeas corpus*, and by early 1918 his family was returned to him. Although Priddy's conduct was upheld on appeal, the judge warned Priddy not to sterilize any other patients until the law was changed. [88]

Enter Carrie Buck. She would be the test case.

Virginia's legislators had been reluctant to pass a eugenic sterilization law. "[We] were laughed at by the lawmakers who suggested they might fall victim to their own legislation," recalled Joseph DeJarnette, superintendent of the Western State Hospital in Staunton, Virginia. He added, "I really thought they ought to have been sterilized as unfit." [89]

In 1922, after numerous state laws had been vetoed or overturned by the courts on Constitutional grounds, Laughlin completed a massive 502-page compilation of state eugenical legislation. It was entitled *Eugenical Sterilization in the United States*. The dense volume, bristling with state-by-state legal analysis and precedent, included what lawyers and eugenicists unanimously declared to be a new "model sterilization law," updated since previous iterations of Laughlin's model legislation. It was indeed the complete legislator's guide. Laughlin was certain that a law that followed a rigid course of due process, proper notification to the patient, adversarial protection of the patient's rights, and a narrow, nonpunitive, health-based eugenical sterilization regimen could withstand a U.S. Supreme Court challenge. Burnishing the report's legal soundness was the fact that it was not issued by any of the Cold Spring Harbor entities, but was distributed as an official document of the Municipal Court of Chicago. Judge Olson, who headed Chicago's Municipal Court, concomitantly served as president of the Eugenics Research Association. Olson even wrote the introduction, saluting Laughlin, who "rendered the nation a signal service in the preparation of this work...."[90]

Laughlin personally sent a copy to Priddy. Now Priddy and his fellow Virginia eugenicists would carefully follow Laughlin's advice. In the fall of 1923, with a mandate from Virginia's State Hospital Board, Priddy and colony attorney Aubrey Strode authored comprehensive new legislation closely resembling the text and format of Laughlin's model statute. By March 30, 1924, Virginia's eugenics law, which now included numerous due process safeguards, was finally passed by both state houses and signed by the governor. It was to take effect on June 17, 1924.[91]

Although Carrie was condemned as feebleminded on January 23, 1924, she was not immediately admitted to the colony. Pregnant girls were not permitted in the facility. On March 28, Carrie gave birth to a daughter, Vivian. Since Carrie had been declared mentally incompetent, she could not keep the child. Ironically, the Dobbses took Vivian in.[92] Three generations of Bucks had intersected with J.T. Dobbs.

Carrie's arrival at the colony was delayed until June 4, just days before the new sterilization law took effect. A legal guardian, Robert Shelton, was properly appointed for her and properly paid $5 per day, just as the statute and due process required. On September 10, 1924, a colony review board properly met and ruled that Carrie "is feebleminded and by the laws of heredity is the probable potential parent of socially inadequate offspring, likewise afflicted...," and as such "she may be sexually sterilized...and that her welfare and that of society will be promoted by her sterilization...."[93]

Upon completion of the hearing, the board properly inquired if they could proceed. Colony attorney Strode properly advised that the Virginia act "had yet to stand the test of the Courts." Strode later recounted, "Whereupon, I was instructed to take to court a test case."[94]

Carrie's guardian, Shelton, was then asked by Strode to appeal the case "in order that we may test the constitutionality through our state courts, even to the Supreme Court of the United States." Shelton then secured ostensibly independent counsel to represent the eighteen-year-old in a legal challenge scheduled for November 18, 1924. Attorney Irving Whitehead was selected to represent Carrie. Whitehead was no stranger to the colony, however, and to many the arrangement seemed little more than a collusive defense. He was, after all, one of the original three directors appointed by the governor to manage the colony when it was established in 1910. Whitehead and his fellow trustees appointed Priddy as their first superintendent. Later, Whitehead had represented the institution on the State Board of Hospitals. In his official capacity, Whitehead had personally endorsed the sterilizations of some two dozen women, including the two Mallory women, and had even lobbied the Virginia legislature for broader legal authority. A building in the colony complex erected the year before was actually named after him. The Wednesday before the trial, Priddy recommended Whitehead for a government position.[95]

Yet it was Whitehead, a staunch eugenicist, founding father of the colony and an advocate of sterilization, who was to champion Carrie Buck's defense.

To bolster the argument that Carrie represented a biological menace, attention next fell on little Vivian. If the infant could somehow be deemed mentally defective, the Bucks would represent three generations of imbeciles—a clear threat to the state. Priddy asked a Red Cross social worker to send evidence certifying the infant as feebleminded, and was almost certainly startled to hear back from the social worker: "I do not recall and am unable to find any mention in our files of having said that Carrie Buck's baby was mentally defected."[96]

Priddy dispatched a note to eugenic activist Dr. Joseph DeJarnette, superintendent of the State Hospital at Staunton. DeJarnette would be called as a state expert witness. "A special term of the Court of Amherst will be held...November 18, 1924 to hear...the case of Carrie Buck's child, on which the constitutionality of the sterilization law depends. It is absolutely necessary that you be present and I would suggest you read up all you can on heredity like [the] jukes, callikaks [*sic*] and other noted families of that stripe." Priddy added, "I want you to help me in this matter by going over to Charlottesville…to get a mental test of Carrie Buck's baby.…The test you will make will be the usual one in line with the inclosed [*sic*] test sheet. We are leaving nothing undone in evidence to this case.…I am enclosing you a letter from Dr. Laughlin and think you will need it. Please return the inclosures [*sic*] as Col. Strode may want them for his files, he having had the correspondence with Dr. Laughlin."[97]

Priddy also assured DeJarnette that even though Vivian was only a few months old, she could still be deemed unfit. "We have an advantage," wrote Priddy, "in having both Carrie Buck and her mother, Emma, as inmates of this institution." Once more, the emphasis was on three generations.[98]

Shortly thereafter, Carrie's seven-month-old daughter Vivian was examined by a social worker. In a subsequent hearing the social worker was asked, "Have you any impression about the child?" Emphasizing the word *probabilities*, the social worker replied, "It is difficult to judge probabilities of a child as young as that, but it seems to me not quite a normal baby." In reply, she was led, "You don't regard her child as a normal baby?" The social worker cautiously responded, "In its appearance—I should say that perhaps my knowledge of the mother may prejudice me in that regard, but I saw the child at the same time as Mrs. Dobbs' daughter's baby, which is only three days older than this one, and there is a very decided difference in the development of the babies."[99]

Once more, the social worker was prompted, "You would not judge the child as a normal baby?" The social worker answered, "There is a look about it that is not quite normal, but just what it is, I can't tell." That was enough for the judge. Vivian was deemed defective, like her mother and grandmother before her.[100]

Priddy also requested expert eugenical testimony from Laughlin, who would not be able to travel to Virginia for the trial but agreed to file a deposition. He asked Priddy for Carrie's genealogy to help him prepare a proper eugenical verdict. Priddy had nothing. "As to our test case," Priddy wrote Laughlin, "I am very sorry I cannot make you out a genealogical tree such as you would like to have, but this girl comes from a shiftless, ignorant and moving class of people, and it is impossible to get intelligent and satisfactory data.…"[101]

Laughlin's deposition simply echoed Priddy's offhand words. "These people belong to the shiftless, ignorant and moving class of anti-social whites of the South," wrote Laughlin. His expert opinion went on: "Carrie Buck: Mental defectiveness evidenced by failure of mental development, having a chronological age of 18 years with a mental age of 9 years, according to Stanford Revision of Binet-Simon Test; and of social and economic inadequacy; has record during life of immorality, prostitution and untruthfulness; has never been self-sustaining; has had one illegitimate child, now about six months old and supposed to be mental defective."[102]

Laughlin's deposition then dispatched the mother, Emma Buck. "Mental defectiveness evidenced by failure of mental development," Laughlin averred, "having a chronological age of 52 years, with a mental age, according to Stanford Revision of Binet-Simon Test, of seven years and eleven months (7 yrs. 11 mos.); and of social and economic inadequacy. Has record during life of immorality, prostitution and untruthfulness; has never been self-sustaining, was maritally unworthy; having been divorced from her husband on account of infidelity; has had record of prostitution and syphilis...."[103]

Ultimately, Laughlin connected the dots, declaring that Carrie's "one illegitimate child, [was also] considered feeble-minded."[104] Three generations.

The judge took the case under advisement. While awaiting a decision, Priddy died of Hodgkin's disease, a cancer of the lymphatic system. Priddy's assistant, J. H. Bell, replaced him as defendant. Thereafter the case became known as *Buck v. Bell.*[105]

On April 13, 1925, the Amherst County Circuit Court upheld the original decision of the colony's special board. Carrie's attorney, Whitehead, immediately appealed the decision to the Virginia Court of Appeals. He petitioned on three Constitutional points: first, deprivation, without due process, of a citizen's rights to procreate; second, violation of the Fourteenth Amendment of the Constitution, providing for due process; and third, a violation of the Eighth Amendment of the Constitution, proscribing cruel and unusual punishment. Whitehead's brief was brief indeed, just five pages long. On the other hand, colony attorney Strode filed a forty-page brief carefully documenting the state's police powers and its need to protect public health and safety.[106]

Virginia's Court of Appeals upheld the colony's decision to sterilize Carrie, denying all claims of cruel and unusual punishment or lack of due process.[107] For Carrie, and the future of sterilization, there was nowhere to go but up. The circle of friends staging a collusive Constitutional challenge, papered wall to wall with documented safeguards and procedural rectitude, were now ready for their final step. Carrie's case was appealed to the highest court in America, the United States Supreme Court. The colony was confident. The board minutes for December 7, 1925, record: "Colonel Aubrey E. Strode and Mr. I. P. Whitehead appeared before the Board and outlined the present status of the sterilization test case and presented conclusive argument for its prosecution though the Supreme Court of the United States, their advice being that this particular case was in admirable shape to go to the court of last resort, and that we could not hope to have a more favorable situation than this one."[108]

If the Supreme Court would uphold Carrie Buck's sterilization, the floodgates of eugenic cleansing would be opened across the United States for thousands. Carrie's destiny, and indeed the destiny of eugenics, rested upon nine men—and most heavily on the one man who would ultimately write the court's opinion. That man was Justice Oliver Wendell Holmes Jr., considered by many to be America's clearest thinker and most important judicial authority.[109]

*　*　*

Oliver Wendell Holmes Jr. lived a life innervated by the great men of literature, propelled by his personal acts of courage, and eventually gilded by the judicial preeminence thrust upon him. He was the best America had to offer. Born in Massachusetts in 1841, his father was a famous physician, poet, and essayist. He had achieved literary esteem from his satirical columns in the *Atlantic Monthly*, later collected for the anthology *Autocrat of the Breakfast Table*. Young Oliver grew up in the company of his father's circle of literati, including Henry Wadsworth Longfellow, Ralph Waldo Emerson, and Nathaniel Hawthorne. Herman Melville was a neighbor at the Holmes' summerhouse.[110]

It was the law, however, that would capture the imagination of Oliver Wendell Holmes Jr. Judges and attorneys had peopled the Holmes family tree for three centuries. A maternal grandfather had sat on the Supreme Judicial Court of Massachusetts.[111]

Holmes was a Harvard scholar, but he had been brave enough to join the rush to war in 1861, even before taking the final exams needed for graduation. He joined the Twentieth Massachusetts Volunteers, known as the Harvard Regiment. He fought valiantly and was wounded three times, once in the chest at Ball's Bluff, once in the leg at Chancellorsville and once through the neck at Antietam during the single bloodiest day of the war. Some thought the scholar-turned-soldier fought to test his own manliness; others suggested it was for "duty and honor."[112] It was probably both.

118 || WAR AGAINST THE WEAK

Certainly, Holmes achieved hero status. One legend claims that when President Lincoln visited Fort Stevens, near Washington, D.C., Holmes had served as his escort. At some point the president stood up to get a better view of something, and a Confederate soldier promptly shot at his stovepipe hat. Holmes dragged the president down, admonishing, "Get down, you damn fool!" Far from insulted, a grateful Lincoln replied, "Goodbye, Captain Holmes. I'm glad to see you know how to talk to civilians."[113]

Even amid the wounds of war, Holmes never lost his fascination with the great thinkers. While recovering from injuries sustained at Chancellorsville, Holmes read the latest philosophical treatises. After the war, he returned to his beloved Harvard to earn a law degree and write legal theory.[114]

Soon, Holmes' rapier-like pronouncements on the purpose of American law as a champion of the people's will began to shape legal thought in the nation. He saw the law as a living, organic expression of the people, not just a sterile codex. "The life of the law has not been logic: it has been experience," Holmes lectured. "The felt necessities of the time, the prevalent moral and political theories, intuitions of public policy, avowed or unconscious, even the prejudices which judges share with their fellow men, have had a good deal more to do than the syllogism in determining the rules by which men should be governed. The law embodies the story of a nation's development through many centuries, and it cannot be dealt with as if it contained only the axioms and corollaries of a book of mathematics."[115]

His rise was rapid. In March of 1881, Holmes' provocative lectures on the nature of law were compiled into an anthology, *The Common Law*. It was an immediate success. Within ten months of the book's publication, in January of 1882, Holmes was elected a Harvard law professor by the university faculty. His reputation as an authority on jurisprudence widened. On December 8 of that same year, before serving his first full year as a professor, the governor of Massachusetts sent an urgent request for Holmes to leave Harvard and assume a seat as associate justice on the Massachusetts Supreme Court. So pressed was the governor that he implored Holmes to reply by 3:00 P.M. of the same day. Holmes replied on time and accepted the position. In 1899, Holmes was appointed chief justice of the Massachusetts Supreme Court.[116]

In 1902, President Theodore Roosevelt, impressed with Holmes' growing juridical prestige, appointed Holmes to the U.S. Supreme Court. There, Holmes assumed a legendary status as a defender of the Constitution and proud expositor of unpopular opinions that nonetheless upheld the rule of law. For more than a quarter century, his name was virtually synonymous with the finest principles of the legal system. During his tenure on the highest bench, he wrote nearly one thousand valued opinions.[117]

Holmes also became famous for powerful dissents, 173 in all. Many championed and clarified the most precious elements of free speech. In one such dissent, he argued "the ultimate good desired is better reached by free trade in ideas—that the best of truth is the power of the thought to get itself accepted in the competition of the market...." In 1928, he enunciated the lasting precept: "If there is any principle of the Constitution that more imperatively calls for attachment than any other it is the principle of free thought—not free thought for those who agree with us but freedom for the thought we hate." Yet Holmes was wise enough to assert that "the most stringent protection of free speech would not protect a man in falsely shouting fire in a theatre and causing a panic."[118]

Indeed, in 1931, his ninetieth birthday celebration would be an event for the nation, broadcast over the Columbia Radio System. Speeches lauded him as "America's most respected man of law."[119]

Into the hands of Oliver Wendell Holmes, defender of the noblest ideal of American jurisprudence, was Carrie Buck commended.

Buck v. Bell would be decided in May of 1927. But the eighty-six-year-old Holmes was in many ways defined by the Civil War and ethically shaped by the nineteenth century. While recovering from the wounds of Chancellorsville, his reading included Spencer's *Social Statics*, the turning-point tract that advocated social Darwinism and so significantly influenced Galtonian thought. Spencer argued the strong over the weak, and believed that human entitlements and charity itself were false and against nature. Indeed, Holmes' 1881 lecture series in *The Common Law* also asserted that the idea of inherent rights was "intrinsically absurd."[120]

Moreover, the warrior-scholar seemed to believe that "might makes right." In his essay entitled "Natural Law," Holmes defined truth. "Truth," he declared, "was the majority vote of that nation that could lick all others."[121] In a graduation speech to Harvard's class of 1895, Holmes declared the sanctity of blindly following orders. "I do not know what is true," he told the audience. "I do not know the meaning of the universe. But in the midst of doubt, in the collapse of creeds, there is one thing I do not doubt...that the faith is true and adorable which leads a soldier to throw away his life in obedience to a blindly accepted duty, in a cause he little understands, in a plan of a campaign of which he has no notion, under tactics of which he does not see the use."[122]

While Holmes' influential Supreme Court opinions and dissents exemplified and eloquently immortalized the highest virtues of American jurisprudence, his private exchanges reveal a different man. Holmes reviled "do-gooders" and in 1909 he quipped to a friend, "I doubt if a shudder would go through the spheres if the whole ant-heap were kerosened." In 1915, writing to John Wigmore, dean of Harvard Law School, Holmes sneered at "the squashy sentimentalism of a big minority" of people, who made him "puke." He was similarly nauseated by those "who believe in the upward and onward—who talk of uplift, who think... that the universe is no longer predatory. Oh, bring me a basin."[123]

In the years just prior to receiving *Buck v. Bell*, Holmes expressed his most candid opinions of mankind. In 1920, writing to English jurist Sir Frederick Pollack, Holmes confessed, "Man at present is a predatory animal. I think that the sacredness of human life is a purely municipal idea of no validity outside the jurisdiction. I believe that force, mitigated so far as it may be by good manners, is the *ultima ratio*, and between two groups that want to make inconsistent kinds of world I see no remedy except force."[124]

He was fond of a certain slogan, and in June of 1922 he repeated it to British scholar and future Labor Party Chairman Harold J. Laski. "As I have said, no doubt, often, it seems to me that all society rests on the death of men. If you don't kill 'em one way you kill 'em another—or prevent their being born." He added, "Is not the present time an illustration of Malthus?"[125]

In 1926, Holmes again confided to Laski, "In cases of difference between oneself and another there is nothing to do except in unimportant matters to think ill of him and in important ones to kill him."[126] Shortly thereafter, Holmes wrote Laski, "We look at our fellow men with sympathy but nature looks at them as she looks at flies...."[127]

The other men of the Supreme Court included Justice Louis Brandeis, the eminent Jewish human rights advocate. Another was the racist and anti-Semite James Clark McReynolds, who refused to even sit or stand next to Brandeis. The chief justice was former president William Howard Taft.[128]

On May 2, 1927, in the plain daylight of the Supreme Court, with only Justice Pierce Butler dissenting, Justice Holmes wrote the opinion for the majority.

> Carrie Buck is a feeble minded white woman who was committed to the State Colony above mentioned in due form. She is the daughter of a feeble minded mother in the same institution, and the mother of an illegitimate

THE UNITED STATES OF STERILIZATION || 121

feeble minded child. She was eighteen years old at the time of the trial of
her case in the circuit court, in the latter part of 1924. An Act of Virginia,
approved March 20, 1924, recites that the health of the patient and the welfare of society may be promoted in certain cases by the sterilization of mental defectives, under careful safeguard . . . without serious pain or substantial
danger to life; that the Commonwealth is supporting in various institutions
many defective persons who if now discharged would become a menace but
if incapable of procreating might be discharged with safety and become
self-supporting with benefit to themselves and to society; and that experience has shown that heredity plays an important part in the transmission of
insanity, imbecility, &c.[129]

Holmes' opinion summarized the extensive procedural safeguards
Virginia had applied, and concluded, "There is no doubt that in that
respect the plaintiff in error has had due process of law."[130] He continued,
and in many ways quoted Laughlin's model eugenical law verbatim.

The attack is not upon the procedure but upon the substantive law. It seems to
be contended that in no circumstances could such an order be justified. It certainly is contended that the order cannot be justified upon the existing
grounds. The judgment finds the facts that have been recited and that Carrie
Buck "is the probable potential parent of socially inadequate offspring, likewise afflicted, that she may be sexually sterilized without detriment to her
general health and that her welfare and that of society will be promoted by her
sterilization," and thereupon makes the order. . . . We have seen more than
once that the public welfare may call upon the best citizens for their lives. It
would be strange if it could not call upon those who already sap the strength of
the state for these lesser sacrifices, often not felt to be such by those concerned, in order to prevent our being swamped with incompetence.[131]

Then Holmes wrote the words that would reverberate forever.

It is better for all the world, if instead of waiting to execute degenerate offspring for crime, or to let them starve for their imbecility, society can prevent those who are manifestly unfit from continuing their kind. The
principle that sustains compulsory vaccination is broad enough to cover
cutting the Fallopian tubes.
    Three generations of imbeciles are enough.[132]

It was over. Carrie Buck was sterilized before noon on October 19, 1927. Her file was noted simply: "Patient sterilized this morning under authority of Act of Assembly...." Her mother Emma, residing elsewhere in the same institution, ultimately died some years later, and was ignominiously buried in a colony graveyard beneath tombstone marker #575. Little Vivian, the third generation to be declared an imbecile, was raised by the Dobbses, and enrolled in school, where she earned a place on the honor roll. In 1932, however, Vivian died of an infectious disease at the age of eight.[133]

Eugenical sterilization was now the law of the land. The floodgates opened wide.

*　*　*

In the two decades between Indiana's pioneering eugenical sterilization law and the Carrie Buck decision, state and local jurisdictions had steadily retreated from the irreversible path of human sterilization. Of the twenty-three states that had enacted legislation, Maine, Minnesota, Nevada, New Jersey, South Dakota and Utah had recorded no sterilizations at all. Idaho and Washington had performed only one procedure each, and Delaware just five. Even states with strong eugenics movements had only performed a small number: Kansas, for instance, had sterilized or castrated 335 men and women; Nebraska had sterilized 262 men and women; Oregon had sterilized 313; and Wisconsin had sterilized 144.[134]

Although some 6,244 state-sanctioned operations were logged from 1907 to July of 1925, three-fourths of these were in just one state: California. California, which boasted the country's most activist eugenic organizations and theorists, proudly performed 4,636 sterilizations and castrations in less than two decades. Under California's sweeping eugenics law, all feebleminded or other mental patients were sterilized before discharge, and any criminal found guilty of any crime three times could be asexualized upon the discretion of a consulting physician. But even California's record was considered by leading eugenicists to be "very limited when compared to the extent of the problem."[135]

Many state officials were simply waiting for the outcome of the Carrie Buck case. Once Holmes' ruling was handed down, it was cited everywhere as the law of the land. New laws were enacted, bringing the total number of states sanctioning sterilization to twenty-nine. Old laws were revised and replaced. Maine, which had not performed such operations before, was responsible for 190 in the next thirteen years. Utah, which had also abstained, performed 252 in the next thirteen years. South Dakota, which had performed none, recorded 577 in the next thirteen years. Minnesota, which had previously declined to act on its legislation, registered 1,880 in the next thirteen years.[136]

The totals from 1907 to 1940 now changed dramatically. North Carolina: 1,017. Michigan: 2,145. Virginia: 3,924. California's numbers soared to 14,568. Even New York State sterilized forty-one men and one woman. The grounds for sterilization fluctuated wildly. Most were adjudged feebleminded, insane, or criminal; many were guilty of the crime of being poor. Many were deemed "moral degenerates." Seven hundred were classed as "other." Some were adjudged medically unacceptable. All told, by the end of 1940, no fewer than 35,878 men and woman had been sterilized or castrated—almost 30,000 of them after *Buck v. Bell*.[137]

And the men and women of eugenics had more plans. They even had a song, created on the grounds of the Eugenics Record Office in the summer of 1910, which they chanted to the rambunctious popular melodies of the day. They sang their lyrics to the rollicking jubilation of ta-ra-ra-boom-de-ay.

> *We are Eu-ge-nists so gay,*
> *And we have no time for play,*
> *Serious we have to be*
> *Working for posterity.*

> Chorus:
> *Ta-ra-ra-boom-de-ay,*
> *We're so happy, we're so gay,*
> *We've been working all the day,*
> *That's the way Eu-gen-ists play*

> *Trips we have in plenty too,*
> *Where no merriment is due.*
> *We inspect with might and main,*
> *Habitats of the insane.*

> *Statisticians too are we,*
> *In the house of Carnegie.*
> *If to future good you list,*
> *You must be a Eu-ge-nist.*[138]

CHAPTER 7

Birth Control

The American masses were not rising up demanding to sterilize, institution-
alize and dehumanize their neighbors and kinfolk. Eugenics was a movement of the nation's elite thinkers and many of its most progressive
reformers. As its ideology spread among the intelligentsia, eugenics cross-
infected many completely separate social reform and health care movements, each worthwhile in its own right. The benevolent causes that became
polluted by eugenics included the movements for child welfare, prison
reform, better education, human hygiene, clinical psychology, medical treatment, world peace and immigrant rights, as well as charities and progressive
undertakings of all kinds. The most striking of these movements was also one
of the world's most overdue and needed campaigns: the birth control movement. The global effort to help women make independent choices about
their own pregnancies was dominated by one woman: Margaret Sanger.

Sanger was a controversial rabble-rouser from the moment she sprang
onto the world stage, fighting for a woman's most personal right in a completely male-dominated world order. In the early part of the twentieth century, when Sanger's birth control movement was in its formative stages,
women were second-class citizens in much of America. Even the most powerful women in America, such as Mrs. Harriman, could not vote in a federal
election, although the most uneducated coal miner or destitute pauper could.
Many husbands treated their wives like baby machines, without regard for
their health or the family's quality of life. Inevitably, in this state, many
women could not expect any role in the world beyond a life of childbearing
and childrearing. Sanger herself was the sixth of eleven children.[1]

Motherhood was to most civilizations a sacred role. Sanger, however,
wanted women to have a choice in that sacred role, specifically if, when and
how often to become pregnant. But under the strict morals laws of the day,
even disseminating birth control information was deemed a pornographic
endeavor.[2]

Sanger was not an armchair activist. She surrounded herself with the very misery she sought to alleviate. Working as a visiting nurse in New York City, Sanger encountered unwanted pregnancies and their consequences every day, especially in the teeming slums of lower Manhattan and Brooklyn. There, the oppressive reality of overpopulation and poverty cried out for relief. Without proper health care, poor women often died during pregnancy or in labor. Without proper prenatal care, children were often born malnourished, stunted or diseased, further straining family resources and subverting the quality of life for all. Infant mortality was high in the sooty slums of New York.[3]

In her autobiography, Sanger dramatized the moment that moved her to devote her life to the cause. It occurred one night in 1912 when she was called to the disheveled three-room flat of Jake and Sadie Sachs. The young couple already had three children and knew nothing about reproductive controls. Just months earlier, Sadie had lost consciousness after a self-induced abortion. Later, Sadie pleaded with Sanger for some information to help her avoid another pregnancy. Such information did exist, but it was not commonly available. One doctor advised that Sadie's husband "sleep on the roof." Now Sadie was pregnant again and in life-threatening physical distress. Sadie's frantic husband summoned nurse Sanger, who raced to the apartment and found the young woman comatose. Despite Sanger's efforts, Sadie died ten minutes later. Sanger pulled a sheet over the dead woman's face as her helpless, guilt-ridden husband shrieked, "My God! My God!"[4]

"I left him [Jake Sachs] pacing desperately back and forth," Sanger recounted in her autobiography, "and for hours I myself walked and walked and walked through the hushed streets. When I finally arrived home and let myself quietly in, all the household was sleeping. I looked out my window and down upon the dimly lighted city. Its pains and griefs crowded in upon me, a moving picture rolled before my eyes with photographic clearness: women writhing in travail to bring forth little babies; the babies themselves naked and hungry, wrapped in newspapers to keep them from the cold; six-year-old children with pinched, pale, wrinkled faces, old in concentrated wretchedness, pushed into gray and fetid cellars, crouching on stone floors, their small scrawny hands scuttling through rags, making lamp shades, artificial flowers; white coffins, black coffins, coffins, coffins interminably passing in never-ending succession. The scenes piled one upon another on another. I could bear it no longer."[5]

Sanger was never the same. A crusader at heart, she was thrust into a mission: to bring birth regulating information and options to all women. It was more than a health movement. It was women's liberation, intended to benefit all of society. Sanger and her circle of friends named the program "birth control." She traveled across the nation demanding the right to disseminate birth control information, which was still criminalized. She fought for access to contraception, and for the simple right of a woman to choose her own reproductive future. She herself became a worldwide *cause célèbre*. Her various advocacy organizations evolved into the worldwide federation known as Planned Parenthood. Sanger eventually assumed legendary status as a champion of personal freedoms and women's rights.[6]

Because Sanger challenged the moral as well as the legal order, and antagonized many religious groups that understandably held the right to life an inviolable principle, Sanger made many enemies. They dogged her everywhere she went, and in every endeavor.[7]

Sanger-hatred never receded. Decades after her death, discrediting Sanger was still a permanent fixture in a broad movement opposed to birth control and abortion. Their tactics frequently included the sloppy or deliberate misquoting, misattributing or misconstruing of single out-of-context sentences to falsely depict Sanger as a racist or anti-Semite.[8] Sanger was no racist. Nor was she anti-Semitic.

But Sanger was an ardent, self-confessed eugenicist, and she would turn her otherwise noble birth control organizations into a tool for eugenics, which advocated for mass sterilization of so-called defectives,[9] mass incarceration of the unfit[10] and draconian immigration restrictions.[11] Like other staunch eugenicists, Sanger vigorously opposed charitable efforts to uplift the downtrodden and deprived, and argued extensively that it was better that the cold and hungry be left without help, so that the eugenically superior strains could multiply without competition from "the unfit."[12] She repeatedly referred to the lower classes and the unfit as "human waste" not worthy of assistance, and proudly quoted the extreme eugenic view that human "weeds" should be "exterminated."[13] Moreover, for both political and genuine ideological reasons, Sanger associated closely with some of America's most fanatical eugenic racists.[14] Both through her publication, *Birth Control Review*, and her public oratory, Sanger helped legitimize and widen the appeal of eugenic pseudoscience.[15] Indeed, to many, birth control was just another form of eugenics.

But why?

The feminist movement, of which Sanger was a major exponent, always identified with eugenics. The idea appealed to women desiring to exercise sensible control over their own bodies. Human breeding was advocated by

American feminists long before Davenport respun Mendelian principles into twentieth century American eugenics. Feminist author Victoria Woodhull, for example, expressed the belief that encouraging positive and discouraging negative breeding were both indispensable for social improvement. In her 1891 pamphlet, *The Rapid Multiplication of the Unfit*, Woodhull insisted, "The best minds of to-day have accepted the fact that if superior people are desired, they must be bred; and if imbeciles, criminals, paupers and [the] otherwise unfit are undesirable citizens they must not be bred."[16]

Twenty years later, Sanger continued the feminist affinity for organized eugenics. Like many progressives, she applied eugenic principles to her pet passion, birth control, which she believed was required of any properly run eugenic society. Sanger saw the obstruction of birth control as a multitiered injustice. One of those tiers was the way it enlarged the overall menace of social defectives plaguing society.[17]

Sanger expressed her own sense of ancestral self-worth in the finest eugenic tradition. Her autobiography certified the quality of her mother's ancestors: "Her family had been Irish as far back as she could trace; the strain of the Norman conquerors had run true throughout the generations, and may have accounted for her unfaltering courage."[18] Sanger continued, "Mother's eleven children were all ten-pounders or more, and both she and father had a eugenic pride of race."[19]

Sanger always considered birth control a function of general population control and embraced the Malthusian notion that a world running out of food supplies should halt charitable works and allow the weak to die off. Malthus's ideals were predecessors to Galton's own pronouncements. Indeed, when Sanger first launched her movement she considered naming it "Neo-Malthusianism." She recounted the night the movement was named in these words: "A new movement was starting.... It did not belong to Socialism nor was it in the labor field, and it had much more to it than just the prevention of conception. As a few companions were sitting with me one evening, we debated in turn *voluntary parenthood, voluntary motherhood, the new motherhood, constructive generation*, and *new generation*. The terms already in use—*Neo-Malthusianism, Family Limitation*, and *Conscious Generation* seemed stuffy and lacked popular appeal.... We tried *population control, race control*, and *birth rate control*. Then someone suggested 'Drop the [word] *rate*.' *Birth control* was the answer...."[20]

Years later, Sanger still continued to see eugenics and birth control as adjuncts. In 1926, her organization sponsored the Sixth International Neo-Malthusian and Birth Control Conference. In a subsequent *Birth Control*

Review article referencing the conference, Jewish crusader Rabbi Stephen Wise, president of the American Jewish Congress, declared, "I think of Birth Control as an item...supremely important as an item in the eugenic program.... Birth control, I repeat, is the fundamental, primary element or item in the eugenic program."[21]

Indeed, Sanger saw birth control as the highest form of eugenics. "Birth control, which has been criticized as negative and destructive, is really the greatest and most truly eugenic method, and its adoption as part of the program of Eugenics would immediately give a concrete and realistic power to that science. As a matter of fact, Birth Control has been accepted by the most clear thinking and far seeing of the Eugenists themselves as the most constructive and necessary of the means to racial health."[22]

More than a Malthusian, Sanger became an outspoken social Darwinist, even looking beyond the ideas of Spencer. In her 1922 book, *Pivot of Civilization*, Sanger thoroughly condemned charitable action. She devoted a full chapter to a denigration of charity and a deprecation of the lower classes. Chapter 5, "The Cruelty of Charity," was prefaced by an epigraph from Spencer himself: "Fostering the good-for-nothing at the expense of the good is an extreme cruelty. It is a deliberate storing up of miseries for future generations. There is no greater curse to posterity than that of bequeathing them an increasing population of imbeciles."[23]

Not as an isolated comment, but on page after page, Sanger castigated charities and the people they hoped to assist. "Organized charity itself," she wrote, "is the symptom of a malignant social disease. Those vast, complex, interrelated organizations aiming to control and to diminish the spread of misery and destitution and all the menacing evils that spring out of this sinisterly fertile soil, are the surest sign that our civilization has bred, is breeding and is perpetuating constantly increasing numbers of defectives, delinquents and dependents. My criticism, therefore, is not directed at the 'failure' of philanthropy, but rather at its success."[24]

She condemned philanthropists and repeatedly referred to those needing help as little more than "human waste." "Such philanthropy...unwittingly promotes precisely the results most deprecated. It encourages the healthier and more normal sections of the world to shoulder the burden of unthinking and indiscriminate fecundity of others; which brings with it, as I think the reader must agree, a dead weight of human waste. Instead of decreasing and aiming to eliminate the stocks that are most detrimental to the future of the race and the world, it tends to render them to a menacing degree dominant."[25]

Sanger added, "[As] British eugenists so conclusively show, and as the infant mortality reports so thoroughly substantiate, a high rate of fecundity is always associated with the direst poverty, irresponsibility, mental defect, feeble-mindedness, and other transmissible taints. The effect of maternity endowments and maternity centers supported by private philanthropy would have, perhaps already have had, exactly the most dysgenic tendency. The new government program would facilitate the function of maternity among the very classes in which the absolute necessity is to discourage it."[26]

She continued, "The most serious charge that can be brought against modern 'benevolence' is that it encourages the perpetuation of defectives, delinquents and dependents. These are the most dangerous elements in the world community, the most devastating curse on human progress and expression. Philanthropy is a gesture characteristic of modern business lavishing upon the unfit the profits extorted from the community at large. Looked at impartially, this compensatory generosity is in its final effect probably more dangerous, more dysgenic, more blighting than the initial practice of profiteering and the social injustice which makes some too rich and others too poor."[27]

Like most eugenicists, she appealed to the financial instincts of the wealthy and middle class whose taxes and donations funded social assistance. "Insanity," she wrote, "annually drains from the state treasury no less than $11,985,695.55, and from private sources and endowments another twenty millions. When we learn further that the total number of inmates in public and private institutions in the State of New York—in alms-houses, reformatories, schools for the blind, deaf and mute, in insane asylums, in homes for the feeble-minded and epileptic—amounts practically to less than sixty-five thousand, an insignificant number compared to the total population, our eyes should be opened to the terrific cost to the community of this dead weight of human waste."[28]

She repeated eugenic notions of generation-to-generation hereditary pauperism as a genetic defect too expensive for society to defray. "The offspring of one feebleminded man named Jukes," she reminded, "has cost the public in one way or another $1,300,000 in seventy-five years. Do we want more such families?"[29]

Sanger's book, *Pivot of Civilization*, included an introduction by famous British novelist and eugenicist H. G. Wells, who said, "We want fewer and better children... and we cannot make the social life and the world-peace we are determined to make, with the ill-bred, ill-trained swarms of inferior citizens that you inflict upon us."[30]

Later, Sanger's magazine reprinted and lauded an editorial from the publication *American Medicine*, which tried to correct "the popular misapprehension that [birth control advocates] encourage small families. The truth is that they encourage small families where large ones would seem detrimental to society, but they advocate with just as great insistence large families where small ones are an injustice to society. They frown upon the ignorant poor whose numerous children, brought into the world often under the most unfavorable circumstances, are a burden to themselves, a menace to the health of the not infrequently unwilling mother, and an obstacle to social progress. But they frown with equal disapproval on the well-to-do, cultured parents who can offer their children all the advantages of the best care and education and who nevertheless selfishly withhold these benefits from society. More children from the fit, less from the unfit—that is the chief issue in Birth Control." But on this last point, however, Sanger disagreed with mainstream eugenicists—she encouraged intelligent birth control even for superior families.[31]

Sanger would return to the theme of more eugenically fit children (and fewer unfit) again and again. She preferred negative, coercive eugenics. "Eugenics seems to me to be valuable in its critical and diagnostic aspects, in emphasizing the danger of irresponsible and uncontrolled fertility of the 'unfit' and the feeble-minded establishing a progressive unbalance in human society and lowering the birth-rate among the 'fit.' But in its so-called 'constructive' aspect, in seeking to reestablish the dominance of [the] healthy strain over the unhealthy, by urging an increased birth-rate among the fit, the Eugenists really offer nothing more farsighted than a 'cradle competition' between the fit and the unfit."[32]

Sanger's solutions were mass sterilization and mass segregation of the defective classes, and these themes were repeated often in *Pivot of Civilization*. "The emergency problem of segregation and sterilization must be faced immediately. Every feeble-minded girl or woman of the hereditary type, especially of the moron class, should be segregated during the reproductive period. Otherwise, she is almost certain to bear imbecile children, who in turn are just as certain to breed other defectives. The male defectives are no less dangerous. Segregation carried out for one or two generations would give us only partial control of the problem. Moreover, when we realize that each feeble-minded person is a potential source of an endless progeny of defect, we prefer the policy of immediate sterilization, of making sure that parenthood is absolutely prohibited to the feeble-minded."[33]

Indeed, Sanger listed eight official aims for her new organization, the American Birth Control League. The fourth aim was "sterilization of the insane and feebleminded and the encouragement of this operation upon those afflicted with inherited or transmissible diseases...."[34]

For her statistics and definitions regarding the feebleminded, Sanger subscribed to Goddard's approach. "Just how many feebleminded there are in the United States, no one knows," wrote Sanger in another book, *Woman and the New Race*, "because no attempt has ever been made to give public care to all of them, and families are more inclined to conceal than to reveal the mental defects of their members. Estimates vary from 350,000 at the present time to nearly 400,000 as early as 1890, Henry H. Goddard, Ph.D., of the Vineland, N.J., Training School, being authority for the latter statement."[35]

Similarly, she accepted the view that most feebleminded children descended from immigrants. For instance, she cited one study that concluded, "An overwhelming proportion of the classified feebleminded children in New York schools came from large families in overcrowded slum conditions, and... only a small percentage were born of native parents."[36]

Steeped in eugenic science, Sanger frequently parroted the results of U.S. Army intelligence testing which asserted that as many as 70 percent of Americans were feebleminded. In January of 1932, the *Brooklyn Daily Eagle* sent Sanger a quote from a British publication asserting that one-tenth of England's population was feebleminded due to "random output of unrestricted breeding." In a letter, the *Eagle* editor asked Sanger, "Is that a fair estimate? What percentage of this country's population is deficient for the same reasons?" Sanger wrote her response on the letter: "70% below 15 year intellect." Her secretary then formally typed a response, "Mrs. Sanger believes that 70% of this country's population has an intellect of less than 15 years."[37] Her magazine, *Birth Control Review*, featured an article with a similar view. "The Purpose of Eugenics" stated, "Expert army investigators disclosed the startling fact that fully 70 per cent of the constituents of this huge army had a mental capacity below... fourteen years."[38]

When lobbying against the growing demographics of the defective, Sanger commonly cited eugenic theory as unimpeachable fact. For example, she followed one fusillade of population reduction rhetoric by assuring, "The opinions which I summarize here are not so much my own, originally, as those of medical authorities who have made deep and careful investigations."[39]

Sanger was willing to employ striking language to argue against the inherent misery and defect of large families. In her book, *Woman and the*

New Race, she bluntly declared, "Many, perhaps, will think it idle to go farther in demonstrating the immorality of large families, but since there is still an abundance of proof at hand, it may be offered for the sake of those who find difficulty in adjusting old-fashioned ideas to the facts. The most merciful thing that the large family does to one of its infant members is to kill it."[40]

At times, she publicly advocated extermination of so-called human weeds to bolster her own views. For example, her August 15, 1925, *Collier's* magazine guest editorial entitled "Is Race Suicide Probable?" argued the case for birth control by quoting eminent botanist and radical eugenicist Luther Burbank, "to whom American civilization is deeply indebted." Quoting Burbank, Sanger's opinion piece continued, "America ... is like a garden in which the gardener pays no attention to the weeds. Our criminals are our weeds, and weeds breed fast and are intensely hardy. They must be eliminated. Stop permitting criminals and weaklings to reproduce. All over the country to-day we have enormous insane asylums and similar institutions where we nourish the unfit and criminal instead of exterminating them. Nature eliminates the weeds, but we turn them into parasites and allow them to reproduce."[41]

Sanger surrounded herself with some of the eugenics movement's most outspoken racists and white supremacists. Chief among them was Lothrop Stoddard, author of *The Rising Tide of Color Against White World Supremacy*. Stoddard's book, devoted to the notion of a superior Nordic race, became a eugenic gospel. It warned: "'Finally perish!' That is the exact alternative which confronts the white race.... If white civilization goes down, the white race is irretrievably ruined. It will be swamped by the triumphant colored races, who will obliterate the white man by elimination or absorption.... Not to-day, nor yet to-morrow; perhaps not for generations; but surely in the end. If the present drift be not changed, we whites are all ultimately doomed."[42]

Stoddard added the eugenic maxim, "We now know that men are not, and never will be, equal. We know that environment and education can develop only what heredity brings." Stoddard's solution? "Just as we isolate bacterial invasions, and starve out the bacteria, by limiting the area and amount of their food supply, so we can compel an inferior race to remain in its native habitat ... [which will] as with all organisms, eventually limit ... its influence."[43]

Shortly after Stoddard's landmark book was published in 1920, Sanger invited him to join the board of directors of her American Birth Control League, a position he retained for years. Likewise, Stoddard retained a key position as a member of the conference committee of the First American Birth Control Conference.[44]

Another Sanger colleague was Yale economics professor Irving Fisher, a leader of the Eugenics Research Association. It was Fisher who had told the Second National Congress on Race Betterment, "Gentlemen and Ladies, you have not any idea unless you have studied this subject mathematically, how rapidly we could exterminate this contamination if we really got at it, or how rapidly the contamination goes on if we do not get at it."[45] Fisher also served on Sanger's Committee for the First American Birth Control Conference, and lectured at her birth control events. Some of these events were unofficial gatherings to discuss wider eugenic action. In a typical exchange before one such lecture in March of 1925, Laughlin wrote to Fisher, "I have received a letter from Mrs. Sanger verifying your date for the round-table discussion.... Dr. Davenport and I can meet you ... thirty minutes before Mrs. Sanger's conference opens ... so that we three can then confer on the business in hand in reference to our membership on the International Commission of Eugenics."[46]

Henry Pratt Fairchild served as one of Sanger's chief organizers and major correspondents.[47] Fairchild became renowned for his virulent anti-immigrant and anti-ethnic polemic, *The Melting Pot Mistake*. Fairchild argued, "Unrestricted immigration ... was slowly, insidiously, irresistibly eating away the very heart of the United States. What was being melted in the great Melting Pot, losing all form and symmetry, all beauty and character, all nobility and usefulness, was the American nationality itself." Like Stoddard, Fairchild compared ethnic minorities to a vile bacterium. "But in the case of a nationality," warned Fairchild, "the foreign particle does not become a part of the nationality until he has become assimilated to it. Previous to that time, he is an extraneous factor, like undigested, and possibly indigestible, matter in the body of a living organism. That being the case, the only way he can alter the nationality is by injuring it, by impeding its functions."[48] Like Fisher, Fairchild offered key speeches at Sanger's conferences, such as the 1925 Sixth International Neo-Malthusian and Birth Control Conference and the 1927 World Population Conference. In 1929, he became vice president and board member of Sanger's central lobbying group, the National Committee for Federal Legislation on Birth Control; in 1931 he served on the advisory board of Sanger's Birth Control Clinical Research Bureau, and later he served as vice president of the Birth Control Federation of America.[49]

Stoddard, Fairchild and Fisher were just three of the many eugenicists working in close association with Sanger and her birth control movement.

Therefore, even though Sanger was not a racist or an anti-Semite herself, she openly welcomed the worst elements of both into the birth control movement. This provided legitimacy and greater currency for a eugenics movement that thrived by subverting progressive platforms to achieve its goals of Nordic racial superiority and ethnic banishment for everyone else.

* * *

Because so many American eugenic leaders occupied key positions within the birth control movement,[50] and because so much of Sanger's rhetoric on suppressing defective immigration echoed standard eugenic vitriol on the topic,[51] and because the chief aims of both organizations included mass sterilization and sequestration, Sanger came to view eugenics and her movement as two sides of the same coin. She consistently courted leaders of the eugenics movement, seeking their acceptance, and periodically maneuvering for a merger of sorts.

The chief obstacle to this merger was Sanger's failure to embrace what was known as *constructive eugenics*. She argued for an aggressive program of negative eugenics, that is, the elimination of the unfit through mass sterilization and sequestration.[52] But she did not endorse constructive eugenics, that is, higher birth rates for those families the movement saw as superior.[53] Moreover, Sanger believed that until mass sterilization took hold, lower class women should practice intelligent birth control by planning families, employing contraception, and spacing their children. This notion split the eugenic leadership.

Some key eugenicists believed birth control was an admirable first step until more coercive measures could be imposed. However, other leaders felt Sanger's approach was a lamentable half-measure that sent the wrong message. A telling editorial in *Eugenical News* declared that the leaders of American eugenics would be willing to grant Sanger's crusade "hearty support" if only she would drop her opposition to larger families for the fit, and "advocate differential fecundity [reproductive rates] on the basis of natural worth."[54]

In other words, Sanger's insistence on birth control for all women, even women of so-called good families, made her movement unpalatable to the male-dominated eugenics establishment. But on this point she would not yield. In many ways this alienated her from eugenics' highest echelons. Even still, Sanger continued to drape herself in the flag of mainstream eugenics, keeping as many major eugenic leaders as close as possible, and pressing others to join her.

Typical was her attempt on October 6, 1921, to coax eugenicist Henry Osborn, president of the New York's Museum of Natural History, to join ranks with the First American Birth Control Conference. "We are most anxious to have you become affiliated with this group and to have your permission to add your name to the Conference Committee." When he did not reply, Sanger sent a duplicate letter five days later. Her answer came on October 21, not from Osborn, but from Davenport. Davenport, who vigorously opposed Sanger's efforts, replied that Osborn "believes that a certain amount of 'birth control' should properly be exercised by the white race, as it is by many of the so-called savage races. I imagine, however, that he is less interested in the statistical reduction in the size of the family than he is in bringing about a qualitative result by which the defective strains should have, on the average, very small families and the efficient strains, of different social levels, should have relatively larger families." Davenport declined on Osborn's behalf, adding, "Propaganda for birth control at this time may well do more harm than good and he is unwilling to associate himself with the forthcoming Birth Control Conference... [since] there is grave doubt whether it will work out the advancement of the race."[55]

Sanger kept trying. On February 11, 1925, she wrote directly to Davenport, inviting him to become a vice president of the Sixth International Neo-Malthusian and Birth Control Conference. Within forty-eight hours, America's cardinal eugenicist sharply declined. "As to any official connection on my part with the conference as vice president, or officially recognized participant or supporter, that is, for reasons which I have already expressed to you in early letters, not possible. For one thing, the confusion of eugenics (which in its application to humans is qualitative) with birth control (which as set forth by most of its propagandists, is quantitative) is, or was considerable and the association of the director of the Eugenics Record Office with the Birth Control Conference would only serve to confuse the distinction. I trust, therefore, you will appreciate my reasons for not wishing to appear as a supporter of the Birth Control League or of the conference."[56]

Not willing to take no for an answer, Sanger immediately wrote to Laughlin at Cold Spring Harbor, asking him to join a roundtable discussion at the conference. Among the conference topics devoted to eugenics was a daylong session entitled, "Sterilization, Crime, Eugenics, Biological Fertility and Sterility." Irving Fisher was considering participating, and by mentioning Fisher's name, Sanger hoped to entice Laughlin. When

Laughlin did not reply immediately, Sanger sent him a second letter at the Carnegie Institution in Washington on March 23, and then a third to Cold Spring Harbor on March 24. Fisher finally accepted and then wired as much to Laughlin, who then also accepted for the afternoon portion of the eugenic program.[57]

Ironically, during one of the conference's sparsely attended administrative sessions, when Sanger was undoubtedly absent, conservative eugenic theorist Roswell Johnson took the floor to quickly usher through a special "eugenic" resolution advocating larger families for the fit. It was exactly what Sanger opposed.[58]

Johnson, coauthor of the widely used textbook *Applied Eugenics*, introduced the resolution and marshaled a majority from the slight attendance while Sanger's main organizers were presumably out of earshot. It read: "*Resolved*, that this Conference believes that persons whose progeny give promise of being of decided value to the community should be encouraged to bear as large families, properly spaced, as they feel they feasibly can." Newspapers on both sides of the Atlantic energetically pounced on the resolution.[59]

Outraged, Sanger immediately repudiated the resolution—unconcerned with whether or not she alienated her allies in the mainstream eugenics movement. "It is my belief," she declared in the next available volume of *Birth Control Review*, "that the so-called 'eugenic' resolution, passed at the final session of the Sixth International Neo-Malthusian and Birth Control Conference, has created a lamentable confusion.... It was interpreted by the press as indicating that we believed we could actually increase the size of families among the 'superior' classes by passing resolutions recommending larger families."[60]

Despite the public row, Sanger continued to push for a merger with the Eugenics Research Association. The ERA had considered affiliation, but eventually declined. "For the time being... [the organization] would not seek formal affiliation with the Birth Control Conference."[61] Yet the overlap between Sanger's organizations and the most extreme eugenic bodies continued. The American Eugenics Society, founded in 1922, was the key advocacy and propaganda wing of the movement. Its board of directors, which included Davenport and Laughlin, also included two men who served on Sanger's organizational and conference boards, University of Michigan president Clarence C. Little and racist author Henry Pratt Fairchild. Moreover, the American Eugenics Society's advisory council included a number of men who also served in official capacities with

Sanger's various organizations, including Harvard sociologist Edward East, psychologist Adolf Meyer, and Rockefeller Foundation medical director William Welch.[62]

Therefore, it was only natural that the issue of merger continued to resurface, especially since Sanger's conferences and her publication, *Birth Control Review*, continued to trumpet the classic eugenic cause, often in the most caustic language. For example, a February 1924 birth control conference in Syracuse featured a paper entitled "Birth Control as Viewed by a Sociologist." The speech argued, "We need a eugenic program and by that I mean a program that seeks to improve the quality of our population, to make a stronger, brainier, and better race of men and women. This will require an effort to increase the number of superior and diminish that of the inferior and the weakling.... It is quite important that we cut down on the now large numbers of the unfit—the physical, mental and moral subnormals." This speech was quickly reprinted in the May 1924 issue of *Birth Control Review*, with the eugenic remarks highlighted in a special subsection headlined "Eugenics and Birth Control."[63]

In the December 1924 *Birth Control Review*, another typical article, this one by eugenicist John C. Duvall, was simply titled "The Purpose of Eugenics." In a section subtitled "Dangerous Human Pests," Duvall explained, "We therefore actually subsidize the propagation of the Jukes and thousands of others of their kind through the promiscuous dispensation of charitable relief, thereby allowing these classes of degenerates to poison society with their unbridled prolific scum, so that at the present time there are about one-half million of this type receiving attention in publicly maintained institutions, while thousands of others are at large to the detriment of our finer elements." The article added thoughts about eradicating such a problem. "It is interesting to note that there is no hesitation to interfere with the course of nature when we desire to eliminate or prevent a superfluity of rodents, insects or other pests; but when it comes to the elimination of the immeasurably more dangerous human pest, we blindly adhere to the inconsistent dogmatic doctrine that man has a perfect right to control all nature with the exception of himself."[64] It was the second time that year that Sanger's magazine had published virtually the same phrases declaring lower classes to be more dangerous than rats and bugs.[65] Such denunciations were commonplace in *Birth Control Review*.

No wonder then that in 1928, leaders of the American Eugenics Society began to suggest that its own monthly publication of eugenic proselytism, *Eugenics*, merge with Sanger's *Birth Control Review*. Leon Whitney, executive secretary of the American Eugenics Society and a Sanger ally, wrote Davenport on April 3, 1928, "It would be an excellent thing if both the American Birth Control League and the American Eugenics Society used the same magazine as their official organ, especially since they were both interested so much in the same problems." Whitney took the liberty of meeting with Sanger on the question, and reported to colleagues, "She felt very strongly about eugenics and seemed to see the whole problem of birth control as a eugenical problem." As to combining their publications, he added, "Mrs. Sanger took very kindly to the idea and seemed to be as enthusiastic about it as I was."[66]

But most of the eugenics movement's senior personalities recoiled at the notion. Furious letters began to fly across the eugenics community. On April 13, Paul Popenoe, who headed up California's Human Betterment Society, reviewed the Whitney letter with racial theorist Madison Grant, who happened to be traveling in Los Angeles. The next day, his agitation obvious, Popenoe wrote Grant a letter marked *"Confidential"* at the top. "I have been considerably disquieted by the letter you showed me yesterday, suggesting a working alliance between the American Eugenics Society and the American Birth Control League. In my judgment we have everything to lose and nothing to gain by such an arrangement. . . . The latter society . . . is controlled by a group that has been brought up on agitation and emotional appeal instead of on research and education. With this group, we would take on a large quantity of ready-made enemies which it has accumulated, and we would gain allies who, while believing that they are eugenists, really have no conception of what eugenics is. . . ."[67]

Popenoe reminded Grant that Sanger had personally repudiated the Johnson Resolution in favor of larger families. "If it is desirable for us to make a campaign in favor of contraception," stressed Popenoe in condescending terms, "we are abundantly able to do so on our own account, without enrolling a lot of sob sisters, grandstand players, and anarchists to help us. We had a lunatic fringe in the eugenics movement in the early days; we have been trying for 20 years to get rid of it and have finally done so. Let's not take on another fringe of any kind as an ornament. This letter is not for publication, but I have no objection to your showing it to Mr. Whitney or any other official of the American Eugenics Society. . . ."[68]

Grant dashed off an urgent missive to Whitney the next day, making clear, "I am definitely opposed to any connection with them. . . . When we organized the Eugenics Society, it was decided that we could keep clear of Birth Control, as it was a feminist movement and would bring a lot of unnecessary enemies. . . . I am pretty sure that Dr. Davenport and Prof. Osborn would agree with me that we had better go our own way indefinitely." Grant copied Davenport.[69]

Davenport was traveling when the letters started flying. On his return, he immediately began to rally the movement's leading figures against any "alliance with Mrs. Sanger." Davenport emphasized his feelings in a letter to Whitney. "Mrs. Sanger is a charming woman," he began, "and I have no doubt about the seriousness of her effort to do good. I have no doubt, also, that she may feel very strongly about eugenics. I have very grave doubts whether she has any clear idea of what eugenics is. . . . We have attached to the word, eugenics, the names of Mrs. E. H. Harriman and Andrew Carnegie—persons with an unsullied personal reputation, whose names connote good judgment and great means. Such valued associations have given to the word, eugenics, great social value and it is that which various organizations want to seize."[70]

He continued, "Now comes along Mrs. Sanger who feels that birth control does not taste in the mouth so well as eugenics and she thinks that birth control is the same as eugenics, and eugenics is birth control, and she would, naturally, seize with avidity a proposal that we should blend birth control and eugenics in some way, such as the proposed [joint] magazine. . . . The whole birth control movement seems to me a quagmire, out of which eugenics should keep."[71]

Davenport concluded with a clear threat to steer clear of any merger talk, or else. "I am interested in the work of the American Eugenics Society," he stated, "but I am more interested in preserving the connotation of eugenics unsullied and I should feel that if the Eugenics Society tied up with the birth control movement that it would be necessary for the Eugenics Record Office of the Carnegie Institution of Washington to withdraw its moral support."[72]

But the idea of a merger between eugenic and birth control groups never subsided. By the 1930s, both movements had fragmented into numerous competing and overlapping entities—many with similar names. Sanger herself had resigned from the American Birth Control League to spearhead other national birth control organizations. In 1933, when the Depression financially crippled many eugenics organizations, a union was again suggested. This time the idea was to merge the American Birth Control League and the American Eugenics Society precisely because the concept of a birth control organization now free of Sanger's strong will—but flush with funds—was attractive. But as all learned, no organization associated with birth control, whether or not Sanger was still associated with it, could be free from the presence of the birth control movement's founder.

On February 9, 1933, Fairchild wrote to Harry Perkins, president of the American Eugenics Society. "For two or three days I have been meaning to write to you, to report on recent developments. Things are moving pretty fast. Miss Topping has been asked by the Board of the A.B.C.L. [American Birth Control League] to spend two weeks or so interviewing various people, especially those not connected with any of the organizations involved, about the desirability of a merger.... Last Sunday I had a chance to talk with Margaret Sanger, and found her enthusiastic and entirely ready to cooperate. So about the only thing that remains to make it unanimous is an assurance that a working majority of the Board of the League is favorably inclined. There is every evidence that that requirement can be met. When that point is reached the main remaining question at issue will be that of finances. The Eugenics Society has none anyway, so that is easily disposed of. The main question is whether the supporters of the League, particularly the Rockefeller interests, will continue, or enlarge their contributions in case a merger is carried out."[73]

Within a month, the idea was again dead. "It looks as if the merger, after all, will not materialize in the immediate future," Fairchild informed Perkins. "It is the same old difficulty. The majority of the Board of the League seems to be in favor of a merger... a pet dream... cherished for years. However, they absolutely balk at the mention of Margaret Sanger. They all profess to love her dearly, and admit that she is one of the biggest women in the world, but they say that it is utterly impossible to work with her, and that any association which had her on its Board would go to pieces in a very short time, etc., etc., etc."[74]

Refusals by eugenic stalwarts carried their own organizational dangers. Fairchild and others actually feared Sanger would try to absorb large parts of the eugenics movement into her own. "As you may know," Fairchild warned Perkins, "I think the League is going to try to get a large number of the members of our Board and Advisory Council on to their Board. I shall not assist them in this effort, as I do not think the League is now, or ever can be as an independent organization, competent to function effectively in the field of Eugenics, although that is now their great objective. If, however, they do succeed in getting several of our members on their Board, it may make it possible for us to over-ride the objections to Mrs. Sanger by force of ballots if this ever seems desirable."[75]

The Great Depression continued to nudge the causes together. Still pending was the question of which movement would absorb the other. Perkins received yet another frank letter in mid May of 1933 from Popenoe. "Regarding amalgamation with The American Birth Control League," Popenoe wrote, "all of us out here were opposed to such a move when Whitney took it up five or six years ago and got in some premature and unfavorable publicity. Since then, conditions have changed a good deal. Mrs. Sanger's withdrawal from the League, followed by that of many of her admirers and of her husband's financial support, has crippled the League very badly in a financial way and it has also lost prestige scientifically for these and other reasons and because other agencies are now actively in the field.... The Birth Control League now has much less bargaining power than it had five or six years ago and if a coalition were worked out it could not expect to get such favorable terms as it would have asked for at that time. The same unfortunately applies in still greater measure to The American Eugenics Society because of its present depressed finances."[76]

Popenoe added candidly, "In effect, I should be perfectly willing to see the Eugenics Society swallow the Birth Control League.... I should not like to see the reverse situation in which the Birth Control League would swallow the Eugenics Society and tie us all up with its slogans and campaign practices.... If it comes to definite negotiations, the Birth Control people will naturally hold out for all they can get, but I think that a good poker player could get some big concessions from them."[77]

But no amount of maneuvering or economic desperation or organizational necessity would allow the equally doctrinaire movements to find a middle ground. The old men of eugenics would not permit it so long as Sanger would not compromise. Each side believed they possessed the more genuine eugenic truth. Both movements roamed the biological landscape in perpetual parallel, following the same lines but never uniting. Moreover, the thin space between the groups was mined. Once, on May 22, 1936, the executive secretary of the American Eugenics Society, George Reid Andrews, circulated to the directors a list of prestigious names to consider adding to the board. Sanger's name appeared on page one. Two weeks later Perkins received a handwritten note from another society officer: "Mr. Andrews has been dismissed ... with no opportunity to present his case."[78]

Sanger went on to lead numerous reform and women's advocacy organizations around the world. Her crusades evolved from birth control and contraception into sex education and world population control. She championed the cause of women on all continents and became an inspiring fig-

ure to successive generations. Her very name became enshrined as a beacon of goodwill and human rights.

But she never lost her eugenic *raison d'être*, nor her fiery determination to
eliminate the unfit. For instance, years after Sanger launched birth control,
she was honored at a luncheon in the Hotel Roosevelt in New York. Her
acceptance speech harkened back to the original nature of her devotion to
her cause. "Let us not forget," she urged, "that these billions, millions, thousands of people are increasing, expanding, exploding at a terrific rate every
year. Africa, Asia, South America are made up of more than a billion human
beings, miserable, poor, illiterate labor slaves, whether they are called that or
not; a billion hungry men and women always in the famine zone yet reproducing themselves in the blind struggle for survival and perpetuation. . . .[79]

"The brains, initiative, thrift and progress of the self supporting, creative human being are called upon to support the ever increasing and
numerous dependent, delinquent and unbalanced masses. . . . I wonder how
many of you realize that the population of the British Isles in Shakespeare's
time was scarcely more than six millions, yet out of these few millions came
the explorers, the pioneers, the poets, the Pilgrims and the courageous
founders of these United States of America. What is England producing
today with her hungry fifty million human beings struggling for survival?
She had then a race of quality, now it's merely quantity. One forgets that
the Italy of the Renaissance, of the painters, the sculptors, the architects,
was a loose collection of small towns—a tiny population that was yet the
nursery of geniuses. There again quality rises supreme above quantity.[80]

"This twentieth century of ours has seen the most rapid multiplication
of human beings in our history, quantity without quality, however. . . . Stress
quality as a prime essential in the birth and survival of our population. . . .[81]

"[The] suggestion I would offer as one worthy of national consideration is that of decreasing the progeny of those human beings afflicted with
transmissible diseases and dysgenic qualities of body and mind. While our
present Federal Governmental Santa Clauses have their hands in the taxpayer's pockets, why not in their generous giving mood be constructive and
provide for sterilizing as well as giving a pension, dole—call it what you
may—to the feebleminded and the victims of transmissible, congenital diseases? Such a program would be a sound future investment as well as a
kindness to the couples themselves by preventing the birth of dozens of
their progeny to become burdens, even criminals of another generation."[82]

Sanger did not deliver this speech in the heyday of Roaring Twenties
eugenics, nor in the clutches of Depression-era desperation, nor even in a

world torn apart by war. She was speaking at the Thirtieth Annual Meeting of the Planned Parenthood Federation on October 25, 1950. A transcript of her remarks was distributed to the worldwide press. A pamphlet was also distributed, entitled "Books on Planned Parenthood," which listed seven major topics, one of which was "Eugenics." The list of eugenic books and pamphlets included the familiar dogmatic publications from the 1930s covering such topics as "selective sterilization" and "the goal of eugenics."[83]

Almost three years later, on May 5, 1953, Sanger reviewed the goals of a new family planning organization—with no change of heart. Writing on International Planned Parenthood Federation letterhead, Sanger asserted to a London eugenic colleague, "I appreciate that there is a difference of opinion as what a Planned Parenthood Federation should want or aim to do, but I do not see how we could leave out of its aims some of the eugenic principles that are basically sound in constructing a decent civilization."[84]

Margaret Sanger gave hope to multitudes. For many, she redefined hope. In the process, she split a nation. But when the smoke cleared on the great biological torment of the twentieth century, Margaret Sanger's movement stands as a powerful example of American eugenics' ability to pervade, infect and distort the most dedicated causes and the most visionary reformers. None was untouchable. If one who loved humanity as much as Sanger could only love a small fraction of it, her story stands as one of the saddest chapters in the history of eugenics.

CHAPTER 8

Blinded

Why did blindness prevention rise to the top of the eugenic agenda in the 1920s?

Because mass sterilization, sequestration, birth control and scientific classifications of the mentally defective, socially unfit and racially inferior were just the leading edge of the war against the weak. Eugenic crusaders were keen to launch the next offensive: outlawing marriage to stymie procreation by those deemed inferior. To set a medicolegal precedent that could be broadly applied to all defectives, eugenicists rallied behind the obviously appealing issue of blindness. Who could argue with a campaign to prevent blindness?

Eugenicists, however, carefully added a key adjective to their cause: *hereditary*. Therefore, their drive was not to reduce blindness arising from accident or illness, but to prevent the far less common problem of "hereditary blindness." How? By banning marriage for individuals who were blind, or anyone with even a single case of blindness in his or her family. According to the plan, such individuals could also be forcibly sterilized and segregated—even if they were already married. If eugenicists could successfully lobby for legislation to prevent hereditary blindness by prohibiting suspect marriages, the concept of marriage restriction could then be broadened to include all categories of the unfit. Marriage could then be denied to a wide group of undesirables, from the feebleminded and epileptic to paupers and the socially inadequate.

Lucien Howe was a legendary champion in the cause of better vision. He is credited with helping preserve the eyesight of generations of Americans. A late nineteenth-century pioneer in ophthalmology, he had founded the Buffalo Eye and Ear Infirmary in 1876. He also aided thousands by insisting that newborns' eyes be bathed with silver nitrate drops to fight neonatal infection; in 1890, this practice became law in New York State under a statute sometimes dubbed "The Howe Law." His monumental two-volume study, *Muscles of the Eye* (1907), became a standard in the field. In 1918, Howe was elected president of the American Ophthalmologic Society, and he enjoyed prestige throughout American and European ocular medicine. For his accomplishments, he would be awarded a gold medal by the National Committee for the Prevention of Blindness. Later, he helped fund the Howe Laboratory of Ophthalmology at Harvard University. Indeed, so revered was the handlebar-mustachioed eye doctor that the American Ophthalmological Society would create the Lucien Howe Medal to recognize lifetime achievement in the field.[1]

Howe became a eugenic activist early on. He quickly rose to the executive committee of the Eugenics Research Association, then became a member of the International Eugenic Congress's Committee on Immigration, and ultimately became president of the Eugenics Research Association.[2] It was Howe who led the charge to segregate, sterilize and ban marriages of blind people and their relatives as a prelude to similar measures for people suspected of other illnesses and handicaps.

Eugenic leaders understood their campaign was never about blindness alone. Blindness was only the test case to usher in sweeping eugenic marriage restrictions. Eugenicists had sought such laws since the days of Galton, who had encouraged eugenically sound marriage and discouraged unsound unions. Of course marriage prohibitions for cultural, religious, economic and health reasons had flourished throughout history. In modern times, many such traditions continued in law throughout Europe. These mainly banned marriage to partners of certain ages, close familial relationships and serious health conditions. But the United States, with its numerous overlapping jurisdictions, led the world in marriage restriction laws, based on various factors of age, kinship, race and health. For example, marriage between whites and persons of African ancestry was criminalized in many states, including California, Maryland and North Dakota, plus the entire South. Montana outlawed marriage between whites and persons of Japanese or Chinese descent. Nevada forbade unions between whites and Malays. Several states legislated against intermarriage between whites and Native Americans.[3]

Eugenicists saw America's marriage laws as ways of halting procreation between defectives, because in addition to broad laws against race mixing, many states prohibited marriage for anyone deemed insane, epileptic, feebleminded or syphilitic. Delaware even criminalized marriage between paupers. No wonder radical British eugenicist Robert Rentoul proudly enumerated American state laws in his 1906 book *Race Culture; Or, Race*

Suicide?, commenting, "It is to these States we must look for guidance if we wish to . . . lessen the chances of children being degenerates."[4]

In preparing to instigate eugenic marriage legislation, Davenport circulated a state-by-state survey in 1913. It was part of an ERO bulletin entitled *State Laws Limiting Marriage Selection Examined in the Light of Eugenics*. In 1915, the *Journal of Heredity*, the renamed *American Breeders Magazine*, published an in-depth article by U.S. Assistant Surgeon General W. C. Rucker castigating the existing marriage laws as insufficient from a eugenic perspective. Rucker admitted that the movement preferred "permanent isolation of the defective classes," and continued, "neither the science of eugenics nor public sentiment is ready for [purely eugenic marriage] legislation." Hence, the only laws that would be viable, he suggested, would be "strictly . . . hygienic in intent."[5]

Enter the cause to prevent hereditary blindness.

In 1918, Howe began in earnest by compiling initial financial data from leading agencies serving the blind, tabulating an institution-by-institution cost per blind person. Cleveland's public school system spent $275 for each of its 153 blind pupils. The California School for the Deaf and Blind spent $396.90 per blind student. Maine's Workshop for the Blind topped the list, spending $865 for each of its forty individuals.[6]

Adding lost wages to custodial and medical care, Howe settled on the figure of $3.8 million as the national cost of blindness—a number he advertised to press his point. But how many people actually suffered from *hereditary* blindness? Howe knew from the outset that the number was small, estimated at about 7 percent of the existing blind population. No one knew for sure because so much blindness at birth was caused by problem pregnancies or poor delivery conditions. *Eugenical News* reported that the 1910 census initially counted 57,272 blind individuals in America, but then came to learn that nearly 4,500 of these cases were erroneously recorded. After further investigation, the Census Bureau reported that more than 90 percent of blind people had no blind relatives at all. Indeed, of 29,242 blind persons questioned, only thirty-one replied that both parents were also blind.[7]

Yet Howe and the eugenics movement seized upon hereditary blindness as their cause *du jour*. Howe and Laughlin contracted with a Pennsylvania printer to publish a fifty-two-page *Bibliography of Hereditary Eye Defects*, which included numerous European studies. The pages of *Eugenical News* became filled with articles on hereditary blindness. One issue contained four articles in a row on the topic. Howe became chairman of a Committee on Hereditary Blindness within the Section on Ophthalmology of the

American Medical Association. The AMA Section committee voted to add a geneticist—Laughlin was chosen—plus a practitioner "especially conversant with the good and also with the bad effects of sterilization." The sterilization expert chosen was Dr. David C. Peyton, of the Indiana Reformatory, who had succeeded eugenic sterilization pioneer Harry Clay Sharp.[8]

The AMA Section committee then began a joint program with the ERO to register family pedigrees of blind people. Four-page forms were printed. Each bore the distinct imprimatur of the "Carnegie Institution of Washington, Eugenics Record Office, founded by Mrs. E. H. Harriman," but at the top also declared official AMA cosponsorship. The subheadline read "in cooperation with the Committee on Hereditary Blindness, Section of Ophthalmology of the American Medical Association" and then credited Laughlin.[9]

Employing careful vagueness, the forms requested "any authentic family-record of what seem to be hereditary eye defects," and then explained how to "plot the family pedigree-chart." Ten thousand of these forms, entitled "Eye Defect Schedule," were printed at a cost of $91.76, half of which was defrayed by the ERO and half by the AMA Section. They were then mailed to America's leading institutions for the blind, as well as schools and help organizations, such as the Cleveland School for the Blind, the Blind Girls Home in Nashville, and the Illinois Industrial Home for the Blind.[10]

Even the ERO form admitted that delivering the family members' names could only hope to "lessen, to some extent at least, the frequency of hereditary blindness." But, cooperating with the request, many in the ophthalmologic community began handing over the names of those who were blind or related to blind people. "I am much interested in this investigation," Laughlin wrote to Howe, "and feel sure that under your leadership, the committee will be able to secure many interesting first-hand pedigrees which will not only throw light upon the manner of inheritance of the traits involved, but will as well provide first-hand information which may be used for practical eugenical purposes in cutting off the descent lines of individuals carrying the potentiality for offspring with seriously handicapping eye defects."[11]

The ERO now possessed yet another target list of unfit individuals.

By early 1921, ERO assistant director Howard Banker was able to brag to Ohio State University dean George Arps, "Records [have] already been collected of several hundred families, in which hereditary eye defects existed.... " Banker then confided, "In spite of evident reasons for drastic remedies, it does not seem advisable to recommend now any radical methods.... "[12]

Nonetheless, the outlines of anti-blind legislation were taking shape. Howe published a major article in the November 1919 edition of *Journal of Heredity*, entitled "The Relation of Hereditary Eye Defects to Genetics and Eugenics." The piece was not a clinical paper, but rather a call to legislative action. First, Howe guesstimated that the number of blind people in America had almost doubled to 100,000 since the 1910 census. (His own calculations of official reports from ten states, including the populous ones of New York, Massachusetts and Ohio, reported a total of only 23,630, indicating virtually no national increase.) Howe's article then addressed the entire blind population as though all of the exaggerated 100,000 suffered from a hereditary condition. Yet Howe knew that hereditary blindness constituted just a small percentage of the total, and even that fraction was falling fast. Because of medical and surgical advances, and as corrective lenses became more commonplace, estimates of hereditary blindness were constantly being reduced.[13]

As though his statistics and projections were authentic, Howe railed, "It is unjust to the blind to allow them to be brought into existence simply to lead miserable lives.... The longer we delay action to prevent this blindness, the more difficult the problem becomes." His plan? Give blind people and their families the option of being isolated or sterilized. "A large part, if not all, of this misery and expense," promised Howe, "could be gradually eradicated by sequestration or by sterilization, if the transmitter of the defect preferred the later." Howe suggested that authorities wait to discover a blind person, and then go back and get the rest of his family.[14]

Howe's article asked colleagues to carefully study sterilization laws applying to the feebleminded. "Where such eugenic laws have been enacted... [they] could be properly amended." Under Howe's plan, incarcerated blind people would be required to labor at jobs commensurate with their intelligence; such work would lessen their "sense of restraint." In a final flourish, Howe asked, "What are we going to do about it? That is the question at last forced on ophthalmologists...."[15]

By 1921, the ERO and AMA Section subcommittee had drafted sweeping legislation that pushed far beyond hereditary blindness or even general blindness. It targeted all people with imperfect vision. Under the proposal, any taxpayer could condemn such a person and his family as "defective." Such a measure would, of course, apply to anyone with blurry vision or even glasses, or any family that included someone with imperfect vision. According to the plan, one ophthalmologist and one eugenic practitioner, such as Laughlin, would render the official assessment. The ERO and

AMA Section subcommittee's draft law was entitled, "An Act for the Partial Prevention of Hereditary Blindness."[16]

The draft law read: "When a man and woman contemplate marriage, if a visual defect exists in one or both of the contracting parties, or in the family of either, so apparent that any taxpayer fears that the children of such a union are liable to become public charges, for which that taxpayer would probably be assessed, then such taxpayer... may apply to the County Judge for an injunction against such a marriage." The judge would then "appoint at least two experts to advise him concerning the probabilities of the further transmission of the eye defect." The experts were specified as a qualified ophthalmologist and "a person especially well versed in distinguishing family traits which are apt to reappear...." Upon the advice of the two experts, the judge could then decide to prohibit any planned marriage, which might yield "at least one child who might have more or less imperfect vision...."[17]

On January 6, 1921, the ERO distributed the draft law for review by several dozen of its core coterie. The mailing list of names was then marked with a plus next to those who approved, and a minus for those opposed. The people consulted included the leading psychologists of the day, such as Goddard, Terman, Yerkes, and Meyer. Apparently, not a few of the respondents either wore glasses or had a family member who did. The vote was divided. Many, such as psychologists Terman and Arps, voted in favor. Several were undecided, but at least half of those polled were opposed.[18]

Eugenicist Raymond Pearl, of Johns Hopkins University, promptly wrote back with his objections. "It makes the primary initiatory force any taxpayer," complained Pearl. "This opens the way at once for all sorts of busybodies to work out personal spite by holding up peoples' marriages pending an investigation.... Anyone who wore glasses contemplating getting married might under the terms of the law stated easily have their progress held up by some neighbor who wanted to make trouble.... Only busybodies would be likely to interest themselves in taking any action under it."[19]

Nonetheless, the ERO leadership sent the draft language to every fellow of the AMA's Ophthalmology Section. The nine-page list of ophthalmologists was similarly annotated with a plus or minus sign. Most of the doctors did not respond. But among those who did, not surprisingly, the yeas outpaced the nays. Dr. James Bach of Milwaukee was marked plus. Dr. Olin Barker of Johnstown, Pennsylvania, was marked plus, and was also noted for sending in a patient's family tree. Dr. David Dennis of Erie,

Pennsylvania, was marked plus and noted for sending in three family trees. The ophthalmologist mailing list's adjusted tally: 88 yes, 40 no.[20] That level of support was enough for the ERO.

On April 5, 1921, a New York State senator sympathetic to the eugenic cause introduced Bill #1597. It would amend the state's Domestic Relations Law with Howe's measure. It required "the town clerk upon the application for a marriage license to ascertain as to any visual defects in either of such applicants, or in a blood relative of either party.... " The clerk or any taxpayer could then apply to the local county judge who would then appoint either two physicians, one an ophthalmologist and the other a eugenic doctor, or one person who could fulfill both roles. Based on their testimony, the clerk was then empowered to prohibit the marriage.[21]

To lobby for the bill, Howe and other eugenicists created a special advisory committee to the Committee to Prevent Hereditary Blindness. Howe was hardly alone within the ophthalmologic community. His advisory committee included some of the leading doctors in the field. The long list included Dr. Clarence Loeb of Chicago, associate editor of the *Journal of Ophthalmology*; Dr. Frank Allport of Chicago, former chairman of the AMA's Committee on Conservation of Blindness; Dr. G. F. Libby of Denver, author of the "Hereditary Blindness" entry in the *Encyclopedia of Ophthalmology*; William Morgan of New York, president of the National Committee for the Prevention of Blindness; Professor Victor Vaughan of Ann Arbor, former president of the AMA's Committee of Preventive Medicine; as well as many other vision experts.[22]

In September of 1921, Howe and the ERO tried to extend the advisory committee beyond the field of ophthalmology. They sent personalized form letters to prominent New York State doctors, judges and elected officials. The invitations requested permission to add their names to the advisory committee, couching membership as an honorary function. The goal was to create the appearance of a groundswell of informed support among the state's administrative and medical establishment for the marriage restriction measure.[23]

Usually, the prominent individuals solicited were only too happy to see their names added to prestigious letterhead advancing a good cause. Few had any understanding of hereditary blindness or the specifics of Howe's legislative proposal. Often, respondents stated that they knew little about the subject, but were only too happy to join the committee. Only rarely did an individual decline. One who did decline was Dr. H. S. Birkett, an ear, nose and throat doctor with no knowledge of ophthalmologic health; he wrote back, "As this seems to be associated largely with an Ophthalmologic Committee, I would feel myself rather out of place. . . . I hardly think that my name would be an appropriate one on such a Committee." ERO organizers routinely kept track of how many eminent people joined or refused. It was all for appearances. At one point, an ERO notation asked for "more judges."[24]

The ERO's sweeping anti-blindness measure did not succeed in 1921.[25] But Howe refused to give up. On January 12, 1922, Howe reminded Laughlin that the intent was to target a broad spectrum of defectives, but beginning with known medical diseases was still the best idea. "We tried to legislate against too many hereditary defects," Howe recounted, "It would be better to limit the legislation to hereditary blindness, insanity, epilepsy and possibly hereditary syphilis." Crafting such legislation required care. Howe conceded, "The phraseology as concocted by doctors and scientists is quite different from that which Constitutional lawyers would have recommended."[26]

Howe was relentless in keeping the idea alive. Lawyers associated with Columbia University were called upon to refine the text to pass Constitutional muster. In one reminder letter, Howe asked Laughlin, "Have you heard anything from our friends connected with the Law Department of Columbia, as to what progress they have made in their attempt to formulate that law for the prevention of hereditary blindness? . . . When members of a committee are supposedly resting, that is the time to get work out of them."[27]

On July 22, 1922, Howe wrote to Laughlin from his New York estate, aptly named "Mendel Farm." Howe expressed his undying devotion to the Mendelian cause and his still-burning determination to "hunt" those with vision problems and subject them to eugenic countermeasures. "As today is . . . the centenary of the birth of our 'Saint' Gregor," wrote Howe with some gaiety, "I feel like sending a word to you, to Drs. Davenport, Little— indeed to every one of the earnest workers at Cold Spring Harbor. . . . If our good old Father Mendel is still counting peas grown in the celestial garden, he probably takes time on this anniversary, to lean over the golden bars, and as he rubs his glasses to look down on what is being done at Cold Spring Harbor and several other institutions like it, his mouth must stretch into a very broad grin when he thinks how little attention was paid to him on earth and what a big man he is now."[28]

Returning to the idea of hunting down the families of the visually impaired, Howe wrote, "Can you suggest any appeal which could be made to the State Board of Health so as to induce them to set one or two of their field workers to hunting up other defective members of certain families whose names appear so frequently among the pupils of schools for the blind? . . . With remembrances to Mrs. Laughlin and best wishes always."[29]

Laughlin replied that he too wanted to "hunt" for those with imperfect vision. "A state survey hunting hereditary eye defects and other degeneracy, but laying principal emphasis on eye disorders, would constitute a splendid piece of work." Howe responded with a letter, eager as ever, declaring that the schools could easily provide the family trees. "Probably the director of almost every school of the blind can remember two or three pupils from branches of the same family who are there because of albinism, cataract, optic atrophy or some similar condition. . . . But," he cautioned, "superintendents have not been trained as field workers [to trace the extended families]."[30]

Therefore, Howe again pushed for the New York State Board of Health to undertake such a statewide hunt. Fortunately, New York State Commissioner of Public Health Hermann M. Biggs was already a member of Howe's advisory committee to prevent hereditary blindness. "I will ask one or two doctors in New York or elsewhere to send letters to you for Dr. Biggs advocating such an investigation," wrote Howe. He also offered to personally train the state's field workers.[31]

An official New York State hunt for the visually impaired never occurred. But Howe continued his pursuit of the names. In 1922, twenty of forty-two state institutions for the blind filled out forms on a total of 2,388 individuals in their care, constituting approximately half of America's institutionalized total. The numbers only further infuriated Howe. By his calculations, institutionalized blind people cost taxpayers $28 to $39 per inmate per month, higher than the feebleminded at $15.21 per month and prison inmates at $18.93 per month. No wonder that on February 10, 1923, Howe sent a letter jointly addressed to Davenport and Laughlin suggesting that any blindness-prevention law include a provision to imprison the visually impaired. In a list, Howe's second point read: "If the hereditary blind whose intended marriage has been adjudged to be dangerous, prefer to go to prison at the expense of the taxpayer that would probably be cheapest for the community and kindest to possible children . . . and a better protection against future defectives." Howe repeated the idea twice more in that letter.[32]

In the same long February 10 letter, Howe promised to send a report to the secretary of the AMA's Section on Ophthalmology. But he was waiting for additional names of blind people to come in so he could forward the latest tally. Howe also assured that he was working closely with Columbia University law professor J. P. Chamberlain to revise the hoped-for legislation.[33]

Several months later, in July of 1923, Professor Chamberlain wrote an article for the *American Bar Association Journal* advocating what he called "repressive legislation" to restrict marriages. "The effect of the modern doctrine of eugenics is being felt in state legislative halls," Chamberlain began. "There is a growing tendency to segregate them [defective persons] in colonies for their own well being and to protect society... and along with this repressive legislation is another trend... legislation limiting the rights of certain classes of persons to marry and requiring preliminary evidence of the fitness of the parties to the ceremony." Professor Chamberlain assured the nation's attorneys that protecting future generations was sound public policy and within any state's police powers. Once a proper "standard of deficiency" could be written into the statutes, marriage restriction could be enforced against the defective as well. "The past record makes it appear probable that the law will not lag behind medical science."[34]

Howe floated another attempt at legislation to prevent hereditary blindness when on February 1, 1926, Bill #605 was introduced to the New York State Assembly. This time, the proposal required a sworn statement from any marriage applicants averring, "Neither myself nor, to the best of my knowledge and belief, any of my blood relatives within the second degree have been affected with blindness...." No definition of blindness was offered. Once again, the bill empowered the town clerk to prohibit the marriage, and even made initial consultation with experts optional. Ironically, even Howe could not craft a definition for blindness. In a letter to another ophthalmologist, he confessed that in a conversation with a federal official, Howe had been called upon to define the condition; both had been at a loss for words. "He was as much in doubt as I," wrote Howe, adding, "Please tell me what better measure you can suggest." Bill #605 was never enacted.[35]

But Howe continued his crusade. Even as he was pushing his anti-blindness legislation, Howe was also orchestrating a second marriage restriction against not just the visually impaired, but anyone judged unfit. His idea was to require a large cash bond from any marriage applicant suspected of being "unfit." Again, no definitions or standards were set. The couple applying for a marriage license would be required to post a significant cash bond against the possibility that their defective children might be a cost to the state. Howe suggested bonds of as much as $14,000, equivalent to over $130,000 today.[36] In other words, marriage by those declared eugenically inferior would be made economically impossible by state law.

Howe had come up with his idea for a general marriage bond as early as 1921. At the time, Laughlin had praised Howe's concept. "Your plan for offering bond is, I believe, a practical one," Laughlin wrote Howe on March 30, 1921. He continued, "For one thing, it presents in very clear and clean cut manner to the average tax-payer the problem of paying for social inadequates from the purse of the tax payer. There is nothing like touching the purse of the tax payer in order to arouse his interest. . . . " Laughlin was pleased with the larger implications because Howe's idea represented a "feature in future eugenical control, not only of hereditary blindness but of hereditary defects of all sorts." Howe's bonding plan, wrote Laughlin, would "place the responsibility for the reproduction of defectives upon the possible parents of such." Moreover, Laughlin wrote, cash bonding would be most useful in "border-line" cases where no one could be sure.[37]

Within a year, Howe was asking Columbia University's Professor Chamberlain to draft legislative language to enforce bonding. In May of 1922, Laughlin sent yet another letter of encouragement to Howe, asserting that should any law to "bond parents against the production of defective children" withstand court challenge, "a great practical eugenical principle will have been established."[38]

In late December of 1922, in a letter inviting Mr. and Mrs. Howe to join the Laughlins for lunch at Cold Spring Harbor, Laughlin could not hide his continuing enthusiasm. "The bonding principle," wrote Laughlin, ". . . securing the state against the production of defectives has, I think, great possibilities. Perhaps the greatest single amendment which can be made to the present marriage laws for the prevention of the production of degenerates. If you can develop the principle and secure its adoption, you will have deserved the honor of the eugenical world."[39]

Eventually, the marriage bond proposal was introduced to the New York State Assembly as a part of Bill #605, Howe's amended anti-blindness effort. Under the proposal, any town clerk, depending on the severity of the suspected defectiveness, could set the bond, up to $14,000. The amount of $14,000 represented Howe's estimate for supporting and educating a blind child. The bond could be released once the wife turned forty-five years of age. Eugenicists were hopeful and even published the entire text of Bill #605 in *Eugenical News*. Marriage bonding legislation, however, died in New York when Bill #605 was voted down.[40]

Even still, the Eugenics Record Office wove the notion into the model eugenics legislation it distributed to the various states. In a memo, Laughlin asserted that the principle should be viewed "in reference not only to the blind, but also to all other types of social inadequacy (and this is the goal sought)." He added, "If this principle were firmly established it would doubtless become the most powerful force directed against the production of defectives and inadequates."[41]

During the 1920s, while Howe was trying to establish marriage prevention and marriage bonding, he and Laughlin were also working on a third concept. It was known by several names and was ultimately called "interstate deportation." Under this scheme, once a family was identified as unfit, family members could be uprooted and deported back to the state or town of their origin—presumably at the expense of the original locale. This would create a financial liability for any town or state, forcing them to view any suspected defective citizens as an intolerable expense. The plan held open the possibility of mass interstate deportations to jurisdictions that would simply refuse the deportees, leading to holding pens of a sort. Some eugenicists called for "colonies." Margaret Sanger advocated "wide open spaces" for the unfit. After all, the United States government had already set the precedent by creating a system of reservations for Native Americans.

It was Howe's initiative for marriage prevention and bonding that opened the door. In a review of Howe's marriage restrictions, Laughlin wrote in the spring of 1921, "It is easy for the eugenicist to plan a step further and to urge further development of our deportation services which means only that the community which produces a non-supporting defective must maintain him ... it means more inter-state deportation and finally, within the state, deportation to counties in which defectives are born or have citizenship or long residence."[42]

By late 1922, Howe and other sympathetic ophthalmologic colleagues, along with Laughlin and the Carnegie Institution, were formulating deportation specifics. Howe was developing a eugenic "debit and credit" system to rank individuals. Towns, counties and states would then be charged when their defectives moved elsewhere in the nation. "Of course our national deportation system is based upon this theory," Laughlin acknowledged to Howe in a December 5, 1922, letter. A few weeks later, Laughlin again lauded a system of bonding "each state, community and family ... for its own degenerates." He adding that "the matter of deportation ... [is] only one other phase in the application of this greater principle."[43]

Once more, bonding marriages against hereditary blindness was to be the precedent for national deportation. "You have done a splendid service," Laughlin wrote Howe in March of 1925, "in directing the work of the Committee on Prevention of Hereditary Blindness. The whole thing appeals so strongly to me because I believe it is a step in the direction of working out . . . the matter of placing responsibility for the production of hereditary inadequates upon families, towns, states and nations which produce them."[44]

Eventually, the eugenics movement developed a constellation of bonding, financial responsibility and deportation principles which it tried to implement based on precedents set by Howe's hereditary blindness countermeasures. The program's goal was to create enclaves of eugenically preferred citizens, which would be achieved when the unfit were systematically expelled from an area. It was defective cleansing. An outline of the measure was published as a lead essay in *Eugenical News*. The section headlined "Interstate Deportation" declared, "There is now, however, a substantial and growing movement for the inter-state and inter-town return of charity cases and ne'er-do-wells from the host communities to the communities which produced them."[45]

Setting up an argument for property confiscation, the *Eugenical News* outline explained that the cost of relocation and maintenance would be borne first by the community the family had come from, but then ultimately by the defective family itself. "In many communities the town or the county or the state has a legal claim upon any property of the producing family, particularly the parents. . . . "[46] The government would have the power to turn any family deemed unfit into a family of paupers.

The *Eugenical News* essay also challenged the concept of free movement within the United States. "It remains to be seen whether an individual inadequate can simply move in on a community and claim legal residence." *Eugenical News* asked, "Is there a legal recourse, for example, in the case of 'dumping' the undesirables of one community on another, of 'exiling' or 'driving out of town' undesirable persons? Perhaps the time will come when there will be no place where such undesirables can go, in which case the logical place for them is the community and family where they were produced." But in the end, after describing a thorough program of dislocation and deportation, the article made the final result clear: "Compulsory segregation or sterilization of potential parents of certain inadequates."[47]

Throughout the essay outlining the new set of eugenic responsibilities and countermeasures, Howe was credited for his tireless efforts. One article declared, "He threw the weight of his professional experience, as an ophthalmologist, into this particular field...." [48]

But most of Howe's most radical plans never took root, in large part because the famed ophthalmologist died before he could complete his work. He died on December 17, 1928, at age eighty, in his Belmont, Massachusetts, home. The next month *Eugenical News* eulogized the man who had served as president of the Eugenics Research Association until shortly before his death. "Lucien Howe was a true gentleman, a broad scholar, and he loved his fellow men." This statement echoed the tribute of the American Ophthalmological Society, which adopted the following resolution: "A student of quality, an author of distinction, a scholar in the house of scientific interpretation and original research, Dr. Howe, a former president of this Society, has added to its reputation and has maintained its tradition." For eight decades, the American Ophthalmological Society has awarded the Lucien Howe Medal for service to the profession and mankind. [49]

CHAPTER 9

Mongrelization

The U.S. Census Bureau would not cooperate with eugenics. No agency collected and compiled more information on individuals than the bureau. Its mission was clear: to count Americans and create a demographic portrait for policymakers. A fundamental principle of census taking is the confidentiality and sanctity of individual records. In the early twentieth century, American eugenics coveted this information.

For years, eugenic leaders tried—with little result—to convince the Census Bureau to change its ways. They targeted the 1920 census. In 1916, Alexander Graham Bell, representing the Eugenics Record Office, was among the first to formally suggest that the bureau add the father's name and the mother's maiden name to the data gathered on each individual.[1] The Census Bureau declined to make the addition.

But shortly after Bell's first entreaty, Laughlin proposed a survey of all those in state custodial and charitable facilities, as well as jails. The Census Bureau agreed, and soon thereafter its director of statistical research, Joseph A. Hill, granted Laughlin the assignment. Laughlin was credentialed as a "special agent of the Bureau of the Census."[2] This first joint program, however, would not lead to an alliance with the Census Bureau, but to a bureaucratic war.

Since the 1880s, the Census Bureau had compiled statistics on what it called "the defective, dependent and delinquent" population, referring to the insane as *defective*, the elderly and infirm as *dependent*, and prisoners as *delinquent*. Laughlin insisted on changing the Census Bureau's terminology to "the socially inadequate" and adding to its rolls large, stratified contingents of the unfit, especially along racial lines. Laughlin's concept of social inadequacy would encompass those who "entail a drag upon those members of the community who have sufficient insight, initiative, competence, physical strength and social instincts to enable them to live effective lives...."[3]

The Census Bureau refused. It stubbornly claimed that Laughlin's newly concocted term, *socially inadequate*, if used publicly, would surely "call forth criticism and protest." Nor would it accept any of Laughlin's substitute categories, such as "submerged tenth" or "the sub-social classes." To adhere to the legal descriptions of the project—and follow the most conservative line—the Census Bureau insisted on its traditional appellations, "defective, dependent and delinquent."[4]

A war of nomenclature erupted, one Laughlin described as a "tempest in a teapot." It raged for more than two years. First, the Census Bureau polled its own stable of social science experts, who reacted with "caustic criticism." Unwilling to back down, Laughlin consulted his own bevy of experts, and then, disregarding any direction from the Census Bureau, employed the term *socially inadequate* anyway when he requested information from 576 state and federal institutions. To rub his point in the Census Bureau's face, Laughlin asked the institutions not only for data, but also for their opinions about his choice of terminology. All but three of the institutions endorsed his new term, and he eventually swayed those three as well, achieving unanimity. Laughlin saw this as more than vindication for his position.[5]

The Census Bureau did not. Although the outbreak of World War I interrupted the project, in May of 1919 the bureau finalized and then published Laughlin's work under the title it chose, *Statistical Directory of State Institutions for the Defective, Dependent and Delinquent Classes*. Determined to have the last word, Laughlin published a vituperative article in the *Journal of Sociology*, recounting the quarrel in detail. Quoting page after page of support for his position from prominent sociologists and officials he had worked with, Laughlin publicly castigated the Census Bureau for lack of leadership and scientific timidity.[6]

Following the irksome, years-long experience, the Census Bureau refused all but cosmetic cooperation with eugenicists. Laughlin, in his capacity as secretary of the Eugenics Research Association, wrote to Samuel Rogers, director of the census, in 1918, asking if the bureau planned to identify nine classes of "socially inadequate." Rogers formally replied that no such data would be gathered, except the names and addresses of the deaf and blind, as previously collected.[7]

At a 1919 conference, the ERA Executive Committee decided to try to convince the Census Bureau to conduct "an experimental genealogical survey of a selected community." Three days later, the ERA formally petitioned Census Bureau Director Rogers to add two additional columns titled *Ancestry* to the paper questionnaires or enumeration sheets. "In the interest of race betterment," the two new columns, to be situated between the existing columns eleven and twelve, would identify the mother, by maiden name, and the father. "Family ties would be established," explained the ERA request, "and thus all census enumeration records would become available for genealogical and family pedigree-studies." The ERA predicted that these records would "constitute the greatest and most valuable genealogical source in the world." Writing in the *Journal of Heredity*, Laughlin advocated the two additional columns so that any "individual could be located from census to census and generation to generation.... Such investigations would be of the greatest social and political value."[8]

The proposals became more and more grandiose as the government's capacity for data retrieval and analysis increased. But any cooperation between the Census Bureau and American eugenics was for all practical purposes destroyed by Laughlin's dogmatic insistence on employing charged terminology more pejorative than the Census Bureau was willing to adopt.[9]

Despite a year-to-year cascade of petitions, letters, scientific articles and eugenic rationales urging the agency to create a massive registry of American citizens that could be marked as fit or unfit, the Census Bureau stands out as one federal organization that simply refused to join the movement.[10]

Rebuffed by the Census Bureau, Laughlin turned his attention to other government agencies, using his official bureau contacts with hundreds of state and federal institutions. His goal was to create further classifications that other bureaus and agencies of the federal government could adopt. An official 1922 booklet distributed by the U.S. House of Representatives to administrators of state institutions was entitled "Classification Standards to be Followed in Preparing Data for the Schedule 'Racial and Diagnostic Records of Inmates of State Institutions', prepared by Harry Laughlin." It listed sixty-five racial classifications. Classification #15 was German Jew, #16 was Polish Jew, #17 was Russian Jew, #25 was North Italian, #26 was South Italian, #30 was Polish ("Polack"), #61 was Mountain White, #62 was American Yankee, #63 was American Southerner and #64 was Middle West American.[11] If the Census Bureau would not adopt his eugenic classifications, Laughlin hoped the states would.

Virginia was eager, thanks to its registrar of vital statistics, Walter Ashby Plecker. Plecker considered himself a product of the Civil War, even though he was born in Virginia in 1861, just as the conflict began.

Memories of his youth in Augusta County, Virginia, during the turbulent Reconstruction years, were influenced greatly by a beloved Negro family servant called Delia. In many ways, Delia represented the emotional strength of the whole family. As was common, she essentially raised Plecker as a young boy, exercising "extensive control" over his activities and earning his lasting gratitude. Plecker's sister sobbed at Delia's wedding at the thought of losing the connection, and Delia broke down as well. When Plecker's mother fell ill for the last time, she sent for Delia to nurse her back to health if possible. In his mother's final hour, it was Delia who comforted her at her deathbed, and when the moment came, it was Delia who tenderly placed her fingers on the woman's eyelids and shut them for the last time. No wonder Delia was remembered in the mother's will. No wonder that Plecker, as executor of his mother's estate, warmly wrote the first bequest check to Delia. From Plecker's point of view, Delia was family.[12]

Fond memories of Delia did not prevent Walter A. Plecker from becoming a fervent raceologist and eugenicist, however. He detested the notion of racial and social mixing in any form. His obsession with white racial purity would turn him into America's preeminent demographic hunter of Blacks, American Indians and other people of color. In the process, Plecker fortified Virginia as the nation's bastion of eugenic racial salvation. Plecker's fanaticism propelled him into a lifelong crusade to codify the existence of just two races: white and everything else.

Plecker began his career in medicine, receiving a degree from the University of Maryland at Baltimore, and then continuing in obstetrics at the New York Polyclinic. He opened a practice in Virginia and quickly became involved with family records, at one point serving as a pension examiner. Plecker moved his practice to Birmingham, Alabama, for several years, but soon returned to his beloved Virginia. He settled in Elizabeth City County, one of the eight original Virginia shires created in 1634. Elizabeth City County was intensely proud of its genealogical heritage. The historic county's citizens included many so-called First Families of Virginia, that is, Colonial settlers. Meticulous family records had been kept, but were in large part destroyed during the numerous battles and town burnings of the Revolutionary War, the War of 1812, and the Civil War. After the Civil War, Elizabeth City County meticulously restored and reorganized its population records.[13]

In 1900, Elizabeth City County created a health department, along with a section of vital statistics to document births and death. A few years later, Plecker was hired as a county health officer, where he fastidiously recorded life cycle events. One triracial Hampton, Virginia, family that he first encountered in 1905 made quite an impact on him. After delivering their baby boy, Plecker at bedside registered the mother as "Indian and colored," and the husband as "Indian and white." Later, the woman's daughter ran off with a white man, marrying in another state. The young couple then returned to Hampton as a second-generation racially mixed marriage.[14] Plecker was appalled by the racial permissiveness of Virginia's system.

Later, when Plecker observed a local Negro death rate twice that of whites, he began to investigate, pursuing a goal of "near 100% registration of births and deaths." Population statistics and registration became more than a fascination; they became his mission. His proficiency at registering citizens made Plecker a natural pick in 1912 to help draft the state's new law creating the Bureau of Vital Statistics. At age fifty-one, Plecker was invited to head the new agency as registrar and to set his own salary. He was so dedicated to population registration that he magnanimously asked "for little more than subsistence." Virginia's 1912 statute established registration of the state's citizens by race—without clear definitions. Yet for three hundred years Virginia had produced racially mixed citizens by virtue of the state's original Colonial settlement, its indigenous Indian population, a thriving slave system, and waves of European immigration.[15]

But a desire for general population registration was not what drove Plecker. He was hardly devoted to the statistical sciences or demographics. He was simply a racist. Plecker's passion was for keeping the white race pure from any possible mixture with Black, American Indian or Asian blood. The only real goal of bureaucratic registration was to prevent racially mixed marriages and social mixing—to biologically barricade the white race in Virginia.

In an official Virginia State Health Bureau pamphlet, Plecker declared: "The white race in this land, is the foundation upon which rests its civilization, and is responsible for the leading position which we occupy amongst the nations of the world. Is it not, therefore, just and right that this race decide for itself what its composition shall be, and attempt, as Virginia has, to maintain its purity?"[16]

Plecker was no authority on eugenics, however. He was a proud member of the American Eugenics Society, but that required no real scientific expertise for membership. Nor did Plecker really comprehend the tenets of Mendelian genetics or heredity. Years after he became a leading exponent of eugenic raceology, Plecker wrote to Laughlin for advice on race mixing formulas, and confided, "I am not satisfied with the accuracy of my own knowledge as to the result of racial intermixture with repeated white crossings." He added that he just didn't understand Davenport's complex protoplasmic discussion of skin color, explaining, "I have never felt justified in believing that...children of mulattoes are really white under Mendel's Law."[17]

Although he cloaked his crusade under the mantle of eugenic science, Plecker did not mind confessing his real motive to Laughlin. "While we are interested in the eugenical records of our citizens," Plecker wrote the ERO, "we are attempting to list only the mixed breeds who are endeavoring to pass into the white race."[18] In other words, Plecker could not be distracted with complex formulas and eugenic charts tracing a spectrum of racial and subracial lineages. In Virginia, you were either ancestrally white or you weren't.

Plecker introduced new techniques in registering births and deaths. In July of 1921, for instance, the Bureau of Vital Statistics mailed a special warning to each of Virginia's 2,500 undertakers. Plecker reminded them that under the law, death certificates could not simply be mailed, but must be delivered in person for verity's sake. Nor could a body be removed or buried without a proper burial permit. An extra permit was needed to ship a body. Moreover, Plecker demanded that coffin dealers provide monthly reports of "all sales of which there is any doubt, giving the address of purchaser, or head of the family, and name of deceased with place and date."[19] Under Plecker's rule, no one was permitted to die in Virginia without leaving a long racial paper trail.

Plecker would enforce similar regimens with midwives and obstetricians, town clerks and church clerics—anyone who could attest to the racial makeup of those who lived and died in Virginia. Over the next several years, he created a cross-indexed system that recorded more than a million Virginia births and deaths since 1912. He also catalogued thousands of annual marriages, each filed under both husband's and wife's name. The data quickly became too voluminous for index cards. Plecker created a complicated but unique system to store the massive troves of information. Clerks would type all the names "on to sheets of the best linen paper, using unfading carbon ribbons," Plecker once explained in a flourish of braggadocio, adding, "We make these in triplicate and bind them in books. These [names] can be quickly referred to as easily as you can find a word in the dictionary." Eventually, Plecker hoped to secure state funding to reconstruct as many records as possible going back to 1630 and then "indexing these by our system."[20]

Plecker planned to add the names of all epileptics, insane, feebleminded and criminals, which would be gathered from the state's hospitals, prisons, city bureaus and county clerks, bestowing on Virginia a massive eugenical database that would reach back to the first white footfalls on Virginia soil. "The purpose will be to list degenerates and criminals," he assured.[21] Of course the ERO was also assembling hundreds of thousands of names, but its extensive rolls only amounted to a patchwork of lineages from counties speckled around the country. Plecker's vision would deliver America's first statewide eugenic registry—a real one.

It is important to understand that while carrying the banner of eugenics, Plecker's true passion never varied. It was always about preserving the purity of the white race. Millions of inscribed linen pages and thousands of leather-bound volumes could be filled, but Plecker would never achieve his real goal without dramatic legislative changes. Existing state laws outlawing mixed-race marriages, including Virginia's, were simply too permissive. In the first place, most states varied on what exactly constituted a Negro or colored person. At least six states forbade whites from marrying half-Negroes or mulat-toes. Nearly a dozen states prohibited whites from marrying those of one-quarter or even one-eighth Negro ancestry. Others were simply vague. Virginia's own blurred statutes had allowed extensive intermarriage through the generations: between whites and light-skinned Negroes, White-Indian-Negro triracials, mulattoes, and others. Plecker and the ERO called this process the "mongrelization" of Virginia's white race.[22]

To halt mongrelization, a coalition of Virginia's most powerful whites organized a campaign to create the nation's stiffest marriage restriction law. It would ban marriage between a certified white person and anyone with even "one drop" of non-Caucasian blood. The key would be mandatory statewide registration of all persons, under Plecker's purview as registrar of the Bureau of Vital Statistics. Leading the charge for the new legislation were Plecker and two friends, the musician John Powell and the journalist Earnest S. Cox.[23]

Powell was one of Virginia's most esteemed composers and concert pianists. Ironically, he built his musical reputation on performing his *Rhapsodie Negre*, which wove Negro themes and spirituals into a popular sonata form. Later, as Powell became more race conscious, he claimed that Negroes had stolen their music from the "compositions of white men." Powell decried the American melting pot as a "witch's cauldron."[24]

Cox led the White America Society, and authored the popular racist tome, *White America* (1923), which warned of the mongrelization of the nation. "[The] real problems when dealing with colored races," trumpeted Cox, "[is] the sub-normal whites who transgress the color line in practice and the super-normal whites who [only] oppose the color line in theory." *Eugenical News* effusively reviewed Cox's book, stating, "America is still worth saving for the white race and it can be done. If Mr. E. S. Cox can bring it about, he will be a greater savior of his country than George Washington. We wish him, his book and his 'White America Society' godspeed." Plecker, Cox and Powell created a small but potent white supremacist league known as the Anglo-Saxon Clubs, which would become pivotal in the registration crusade.[25]

Despite their virulent racism, the Anglo-Saxon Clubs claimed they harbored no ill will toward Negroes. Why? Because now it was just science—eugenic science. The Anglo-Saxon Clubs could boast, "'One drop of negro blood makes the negro' is no longer a theory based on race pride or color prejudice, but a logically induced, scientific fact." As such, even the group's constitution proclaimed its desire "for the supremacy of the white race in the United States of America, without racial prejudice or hatred."[26] This was the powerful redefining nature of eugenics—in action.

The Anglo-Saxon Clubs and their loose confederation of local branches successfully petitioned the Virginia General Assembly and quickly brought about Senate Bill #219 and House Bill #311, each captioned "An Act to Preserve Racial Integrity." The legislation would require all Virginians to register their race and defined whites as those with "no trace whatsoever of any blood other than Caucasian." As one Norfolk editorialist described the proposal, "Each person, not already booked in the Vital Statistics Bureau will be required to take out a sort of passport correctly setting forth his racial composition...." This passport or certificate would be required before any marriage license could be granted. Pure whites could only marry pure whites. All other race combinations would be allowed to intermarry freely.[27]

The Anglo-Saxon Clubs found a powerful ally in their campaign. The state's leading newspaper, the *Richmond Times-Dispatch*, allowed its pages to become a megaphone for the legislation. In July of 1923, for example, Cox and Powell published side-by-side articles entitled "Is White America to Become a Negroid Nation?" The men claimed their proposed legislation was based on sound Mendelian eugenics that now conclusively proved that when two human varieties mixed, "the more primitive...always dominates in the hybrid offspring." The *Richmond Times-Dispatch* supported the idea in an editorial.[28]

On February 12, 1924, Powell enthralled a packed Virginia House of Delegates with his call to stop Negro blood from further mongrelizing the state's white population. "POWELL ASKS LAW GUARDING RACIAL PURITY" proclaimed the *Richmond Times-Dispatch*'s page one headline. Subheads read "Rigid Registration System is Needed" and "Bill Would Cut Short Marriage of Whites with Non-Whites." The newspaper's lead paragraph called the address "historic." Leaving little to doubt, the article made clear that a "rigid system of registration" would halt the race mixing and mongrelization arising from centuries of procreation by whites with Negro slaves and their descendants. Such preeminent eugenic raceologists as Madison Grant were quoted extensively to reaffirm the scientific necessity underpinning the legislative effort. Lothrop Stoddard, a member of Margaret Sanger's board of advisors, was also quoted, declaring, "I consider such legislation . . . to be of the highest value and greatest necessity in order that the purity of the white race be safeguarded from possibility of contamination with nonwhite blood. . . . This is a matter of both national and racial life and death."[29]

Virginia's legislature, in Richmond, was soon scheduled to debate what was now dubbed the "Racial Integrity Act." It was the same 1924 session of the legislature that had enacted the law for mandatory sterilization of mental defectives that was successfully applied to Carrie Buck. On February 18, 1924, with the forthcoming debate in mind, the *Richmond Times-Dispatch* published a rousing editorial page endorsement that legislators were sure to read. Employing eugenic catchphrases, the newspaper reminded readers that when "amalgamation" between races occurred, "one race will absorb the other. And history shows that the more highly developed strain always is the one to go. America is headed toward mongrelism; only . . . measures to retain racial integrity can stop the country from becoming negroid in population. . . . Thousands of men and women who pass for white persons in this state have in their veins negro blood . . . it will sound the death knell of the white man. Once a drop of inferior blood gets in his veins, he descends lower and lower in the mongrel scale."[30]

Despite the bill's popular appeal, legislators were unwilling to ratify the measure without two adjustments. First, the notion of mandatory registration was considered an "insult to the white people of the state," as one irritated senator phrased it. Plecker confided to a minister, "The legislature was about to vote the whole measure down when we offered it making registration optional." Mandatory registration was deleted from the bill. Second, a racial loophole was permitted (over Plecker's objection), this to accommodate the oldest and most revered Virginia families who proudly boasted of descending from pre-Colonial Indians, including Pocahontas. Plecker's original proposal only allowed those with one-sixty-fourth Indian blood or less to be registered as white. This was broadened by the senators to one-sixteenth Indian blood, with the understanding that many of Virginia's finest lineages included eighteenth- and nineteenth-century Indian ancestors.[31]

Virginia's Racial Integrity Act was ratified on March 8, 1924, and became effective on June 15. Falsely registering one's race was defined as a felony, punishable by a year in prison.[32]

As soon as the law was enacted, Plecker began circulating special bulletins. The first went out in March of 1924, even before the effective date of the law. Under the insignia of the Virginia Department of Health, a special "Health Bulletin," labeled "Extra #1" and entitled "To Preserve Racial Integrity," laid out strict instructions to all local registrars and other government officials throughout the state. "As color is the most important feature of this form of registration," the instructions read, "the local registrar must be sure that there is no trace of colored blood in anyone offering to register as a white person. The penalty for willfully making a false claim as to color is one year in the penitentiary.... The Clerk must also decide the question of color before he can issue a marriage license.... You should warn any person of mixed or doubtful color as to the risk of making a claim as to his color, if it is afterwards found to be false." Health Bulletin Extra #1 defined various levels of white-Negro mixtures, such as mulatto, quadroon, octoroon, colored and mixed. Along with the bulletin, Plecker distributed the first 65,000 copies of State Form 59, printed on March 17, "Registration of Birth and Color—Virginia."[33]

Health Bulletin #2 was mailed several days later and warned, "It is estimated that there are in the state from 10,000 to 20,000, possibly more, near white people, who are known to possess an intermixture of colored blood, in some cases to a slight extent, it is true, but still enough to prevent them from being white. In the past, it has been possible for these people to declare themselves as white.... Then they have demanded the admittance of their children into the white schools, and in not a few cases have intermarried with white people.... Our Bureau has kept a watchful eye upon the situation." Bulletin #2 reminded everyone that a year of jail time awaited anyone who violated the act.[34]

Plecker quickly began using his office, letterhead and the public's uncertainty about the implications of the new law to his advantage. His letters and bulletins informed and sometimes hounded new parents, newly-weds, midwives, physicians, funeral directors, ministers, and anyone else the Bureau of Vital Statistics suspected of being or abetting the unwhite.[35]

April 30, 1924
Mrs. Robert H. Cheatham
Lynchburg, Virginia

We have a report of the birth of your child, July 30th, 1923, signed by Mary Gildon, midwife. She says that you are white and that the father of the child is white. We have a correction to this certificate sent to us from the City Health Department at Lynchburg, in which they say that the father of this child is a negro. This is to give you warning that this is a mulatto child and you cannot pass it off as white. A new law passed by the last legislature says that if a child has one drop of negro blood in it, it cannot be counted as white. You will have to do something about this matter and see that this child is not allowed to mix with white children. It cannot go to white schools and can never marry a white person in Virginia.
      It is an awful thing.

                                        Yours very truly,
                                        W. A. Plecker
                                        STATE REGISTRAR[36]

Plecker followed this with a short note to the midwife, Mary Gildon.

This is to notify you that it is a penitentiary offense to willfully state that a child is white when it is colored. You have made yourself liable to very serious trouble by doing this thing. What have you got to say about it?
                                        Yours very truly,
                                        W. A. Plecker
                                        STATE REGISTRAR[37]

      Plecker's friend Powell of the Anglo-Saxon Clubs was copied on both letters. A small handwritten notation at the top left read, "Dear Mr. Powell: This is a specimen of our daily troubles and how we are handling them."[38]
      Plecker acted on rumor, consulted arcane tax and real estate documents, and of course whatever records were available from various eugenic sources. On July 29, 1924, Plecker wrote to W. H. Clark, who lived at Irish Creek in Rockbridge County. "I do not know you personally and have no

positive assurance as to your racial standing, but I do know that an investigation made some time ago by the Carnegie Foundation of the people of mixed descent in Amherst County found the Clark family one of those known to be thus mixed. We learned also that members of this family and of other mixed families have crossed over from Amherst County and are now living on Irish Creek." After informing Clark that his ancestors included "three Indians who mixed with white and negro people," Plecker asserted that the man was now one of five hundred individuals who would be removed from the list of white people.[39]

Adding a threat of prosecution, Plecker warned, "We do not expect to be easy upon anyone who makes a misstatement and we expect soon to be in possession of facts which we can take into court if necessary." Plecker seemed to enjoy taunting the racially suspect. He sardonically added that he looked forward to tarring even more of Clark's extended family. "I will be glad to hear what you have to say," quipped Plecker, "and particularly to have the dates and places of the births and marriages of yourself, your parents and grandparents."[40]

Plecker was equally ruthless with his own registrars. One was Pal S. Beverly, a registrar in Pera, Virginia. Beverly had bitterly complained that registration of his own family as white had been overruled by Plecker. Records unearthed by Plecker showed Beverly to be a so-called "Free Issue" Negro, that is, a class of freed slave. "Because of your constant agitation," Plecker wrote him on October 12, 1929, "of the question that you are a white man and not a member of the 'Free Issue' group of Amherst, as you and your ancestors have been rated, we wrote to you recently asking for the names of your father and of his father and your grandfather's mother."[41]

Plecker had probed Beverly's family tree for generations. The registrar laid it out for him in stunning and damning detail. "The certificate of death of your mother Leeanna (or Leander) Francis Beverly, Nov. 5, 1923, states that she was the wife of Adolphus Beverly," informed Plecker. "This certificate was signed by you when you were our local registrar." Plecker then checked Adolphus Beverly's 1881 marriage license and discovered that Beverly's father was listed as colored. Plecker then investigated Adolphus Beverly's father, Frederick. In the Personal Property Tax Book for the years 1846 through 1851, Frederick was listed as a freed slave. Frederick was born in 1805 and was recorded in the census along with his older brother, Samuel—and on and on.[42]

"I am notifying you finally," Plecker informed Beverly, "that you can have no other rating in our office under the Act of 1924 than that of a mulatto or colored man, regardless of your personal appearance, voting list, or statements which any persons may make to petitions in your behalf.... I want to notify you further that any effort that you make to register yourself or your family in our office as white is, under the Racial Integrity Act of 1924, a felony making you liable to a penalty of one year in the penitentiary." For extra measure, he added that the bureau had identified numerous other mixed-race individuals in the county named Beverly.[43]

As promised, Plecker began decertifying the extended family members of Pal Beverly. Among them was Mascott Hamilton of Glasgow, in Rockbridge County, Virginia. After Plecker's ruling, Hamilton's children were thrown out of the white school they attended. When Hamilton threatened to sue, Plecker gleefully replied, "I am glad to learn from you the fact that your children are kept out of the white schools...." He presented the point-by-point documentation: "You and your wife belong to the group of people known as 'free issues' who are classed in Amherst County where they started as of free Negro stock, the name they were called by before the War Between the States to distinguish them from slave Negroes.... Your wife's mother married Price Beverly, a grandson of Frederick Beverly, who was a son of Bettie Buck or (Beverly) who was a slave and set free and sent to Amherst by her owner Peter Rose of Buckingham County, together with her sons Frederick and Samuel. Your wife's grandmother, Aurora Wood married Richard, a son of the freed negro, Frederick Beverly."[44]

The litany continued. "The children which you refer to were probably your wife's by her divorced husband Sam Roberts, who is shown to be an illegitimate son of Jennie Roberts. You did not marry Dora till 1925. The Roberts family is also of true 'free issue' stock. Your wife gave birth to one child two months after she was married to Sam Roberts. Does she say that the father was a white man and not her husband? What a mess—trying to be white!!"[45]

Plecker scoffed, "Your wife's history shows a complete line of illegitimacy and she claims this as the ground upon which she hopes to be classified as white. It would be difficult to find a white family except of feebleminded people in the state with such a record." Ending with his standard threat, Plecker warned, "It is a penitentiary offense to try to register as white a child with any ascertainable trace of negro blood, and that when you go into court you will have this charge to face."[46]

Similarly denigrating correspondence was mailed across the state. In May of 1930, Plecker notified the wife of Frank C. Clark, of rural Alleghany

County, that her protestations of a white appearance and years of living as white were meaningless. "The question of whether or not there is any trace of negro blood present is determined by the record of ancestors and not by the appearance of an individual at the present day after securing crossings of white blood. Neither does the securing of marriage licenses, and registering children falsely as white establish the racial origin." Her father-in-law's colored marriage license, and the state's pre–Civil War tax records, "establishes the colored ancestry of your husband Frank C. Clark."[47]

Plecker then enumerated the genealogical details of Mrs. Clark's mother, Elena, her grandmother, Ella, and even her great-great-grandmother, Creasy, "who was said to have been 'a little brown-skinned Negro who lived to be nearly one hundred years old.'" In closing, Plecker admonished, "All descendants of the people referred to above are colored and will be so considered in our office. They cannot legally marry into the white race nor attend white schools. Anyone who registers the births of descendants of the above as white ... makes himself or herself liable to one year in the penitentiary."[48]

In one case, four mulattoes from one family married white spouses, two in Washington, D.C., one in a distant Virginia town, and one in an undetermined location. When they returned to their hometown, Plecker tracked them all down and called the police. The couples "fled before the warrants issued for their arrest were served," Plecker recounted to a friend.[49]

In another case, Plecker investigated a Grayson County couple married five years earlier. The couple had just given birth to a son. After a review of the birth certificate and other records, the man was found to be white, but Plecker determined his wife to be of Negro descent. Plecker essentially unmarried the couple. He ruled, "They were married illegally and under the laws of Virginia, they are not legally married. Both are liable to the State Penitentiary." That ruling and any attendant information was forwarded to the Commonwealth Attorney for prosecution.[50]

Plecker's relentless crusade continued for years. His typical workday began at 8:30 in the morning and ended at 5:00, and he usually put in a half-day on Saturdays. Two assistants, Miss Marks and Miss Kelly, helped him manage his constant correspondence as he probed for clues about individuals' racial composition and then consummated his investigations with elaborate, combative missives.[51]

More than just prohibiting marriage and school admittance, he also tried to keep everyone but certified whites from riding in the white railroad coaches. He even pressured white cemeteries. When Riverview Cemetery in Charlottesville tried to bury someone of suspected Negro bloodline, Plecker protested, "This man is of negro ancestry.... To the white owner of a lot, it might prove embarrassing to meet with negroes visiting at one of their graves on the adjoining lot."[52]

When he didn't possess actual documentation, the registrar was more than willing to fake it. In 1940, fifteen citizens in Pittsylvania County had petitioned Plecker to bar the five children of the King Family from attending white school "on account of being of negroid mixture." Plecker contacted the chairman of the Pittsylvania County school board seeking information on the five students admitting "we have no information in regard to them . . . [and] no way of proving facts from the record." Plecker explained, "We are particularly desirous of knowing whether a negro man is the reputed father of these children, and if possible, his name." Until that time, Plecker assured the school official, "We will preserve [the petitions of the fifteen people] in our files as evidence . . . and upon that information we will designate any of these children found in our records as colored—regardless."[53]

In one episode, the Bedford County clerk, Mr. Nichols, contacted Plecker to confirm the racial status of a young man seeking to marry a white girl. The young man's complexion was one of mixed parentage. Plecker wrote back, "We do not know whether we can establish his racial descent until we have had further information as to his family.... [But] if this young man has the appearance of being mulatto and cannot prove the contrary, I would suggest that no license be granted to him." Two days later, the young couple went to the next county, Roanoke County, and successfully secured their marriage license. Plecker discovered it after the fact, haranguing the issuing clerk, "We have no positive information as to the man's pedigree, we can only surmise it from Mr. Nichols' observation as to his appearance. [But] shall this man . . . be turned loose upon the community to raise more mulatto children?"[54]

Plecker proselytized and chastised anyone who would listen. His Bureau of Vital Statistics regularly published radical racist and eugenic literature, which was distributed to thousands of doctors, ministers, teachers, morticians and racial integrity advocates. One series of official tracts, entitled the "New Family Series," was aimed at youngsters to heighten their awareness of "dangers threatening the integrity and supremacy of the white race." The bureau's 1925 annual report to the governor was itself widely disseminated as a special health bulletin. In that report, Plecker lamented, "Not a few white women are giving birth to mulatto children. These

174 || WAR AGAINST THE WEAK

women are usually feebleminded, but in some cases they are simply depraved. The segregation or sterilization of feebleminded females is the only solution to the problem."[55]

The 1924 state publication, *Eugenics and the New Family*, insisted, "The variation in races is not simply a matter of color of skin, eyes, and hair and facial and bodily contour, but goes through every cell of the body. The mental and moral characteristics of a black man cannot even under the best environments and educational advantages become the same as those of a white man. But even if the negro's attainments should be considerable, these could not be transmitted to his offspring since personally acquired qualities are not inheritable. Neither can the descendents of the union of the two races if left to their own resources, be expected to develop or maintain the highest type of civilization."[56]

When Virginia's Racial Integrity Act was passed in 1924, Plecker became an immediate hero among raceologists and eugenicists across America. He addressed major eugenic conferences and authored special articles on the topic for *Eugenical News*, the American Eugenics Society's *Eugenics*, and various eugenic research anthologies. Laughlin was so impressed that he cited Plecker's work in the 1929 edition of the *American Year Book* "for leadership in establishing new racial integrity laws in the American states."[57]

Plecker's audience expanded beyond eugenic circles. The American Public Health Association invited him to read a paper before its fifty-third Annual Meeting in October of 1924, in Detroit. At the event, Plecker preached to the nation's most important public health officials that whites and nonwhites could not "live in close contact without injury to the higher [whites], amounting in many cases to absolute ruin. The lower [nonwhites] never has been and never can be raised to the level of the higher." The association was so taken with Plecker's advocacy that it reprinted much of his speech in the *American Journal of Public Health*. The journal praised Virginia's law as "the most perfect expression of the white ideal, and the most important eugenical effort that has been made in the past 4,000 years." Such platforms only served to legitimize Plecker's views.[58]

Soon Plecker was pushing for similar "one drop" racial integrity laws in other states. Exporting such legislation was essential to his strategy since Virginians of any complexion could easily cross state lines to marry. In one article Plecker complained, "White and coloreds...quietly move to Washington or northern States and become legally married. In some instances, they even return to their home State and live in marriage relations...."[59]

To help make Virginia's race law a national standard, Virginia Governor E. Lee Trinkle proudly distributed copies of the Racial Integrity Act to every governor in America, with a personal letter requesting that they propose similar legislation in their own states. John Powell reported to one interested Midwestern legislator, "He [Trinkle] received thirty-one replies. Nineteen of these, most of them from southern governors, were noncommittal; eleven, the majority from the north and west, strongly approved; the only disapproval came from the governor of Minnesota."[60]

Powell added, "Of course, laws against intermarriage cannot solve the negro problem in any of its aspects—industrial, economic, political, social, biological or eugenical. They can, however, delay the evil day and give time for the evolvement of an effective solution . . . a real and final solution."[61]

Even if some governors were hesitant, legislators and activists across the nation were eager to replicate the law. Ohio senator Harry Davis requested more information, which Plecker provided along with a detailed briefing on the difficulties of lobbying such a bill. A Maryland lawmaker, John R. Blake, asked for a copy of the law plus a recommendation for a speaker to address the legislature. When the race-minded Reverend Wendell White of South Carolina wrote for more information on such a law, Plecker gladly sent it, bemoaning the vague response of that state's governor. Plecker encouraged the clergyman, "If such men as you and others will get behind him [the governor of South Carolina] and the legislature, you can get this or a better law across."[62]

To help, Plecker's Bureau of Vital Statistics mailed literature to legislators in "all of the States, appealing to them to join Virginia in a united move to preserve America as a White Nation." The first two states to emulate Virginia's statute were Alabama and Georgia. Wisconsin attempted to follow suit. Other states were slow to approve "one drop" measures, in part because of increasing civil rights activism. With methodical lobbying, however, the eugenics movement hoped to spur more such laws. To that end, Laughlin asked Plecker to compile a special chart for *Eugenical News* entitled "Amount of Negro Blood Allowed in Various States for Marriage to Whites."[63]

Plecker's bureaucratic ire did not confine itself to white and Negro unions. Asians were also barred from marrying whites. For instance, on February 28, 1940, Spotsylvania Circuit Court Clerk A. H. Crismond issued a marriage license to a local couple, Philip N. Saure and Elsie M. Thomas. Upon checking, Plecker discovered that the groom was a native of the Philippines and the bride an Italian-American born in Pittsburgh.

Assuming the woman was white, Plecker chided, "You as Clerk were not authorized to issue a marriage license to a person of any of the colored races, including Filipinos." He lectured the clerk parenthetically in typical eugenic prose, "The Italians from the Island of Sicily are badly mixed with former negro slaves, and if this woman is from there, it is ... [possible] she herself would have a trace of negro blood."[64]

At about the same time, Plecker informed a California researcher that Virginia was also disallowing marriages between whites and Hindus because they were "of the colored races ... who are considered either Mongolian or Malay, I am not sure which." He told a South Boston, Virginia, contact that Portuguese were admixed with Negroes, and equally disqualified. His eugenic tracts bemoaned the presence of 500,000 to 750,000 Mexicans in Texas and called for their expulsion south of the border.[65]

But Plecker harbored a special animus toward one ethnic group. He despised Native Americans. Because he believed that American Indian tribes had intermixed for generations with whites and some Negroes, Plecker was satisfied that pure Indians no longer existed. To him, they were all mongrels. Worse, because Virginia's Racial Integrity Act contained a historic loophole for those with no more than one-sixteenth Indian ancestry, Plecker saw the exemption as a demographic escape tunnel for those of mixed Negro lineage. From the outset, Plecker embarked upon a furious campaign to eradicate American Indian identity.

Virginia's fabled history of settlement began with Indians. Years before any European landed in America, the Algonquin ruled the wooded lands which later became known as Virginia. Dashing Stream and his wife, Scent Flower, gave birth to Powhatan, who rose to become a noble chief ruling a federation of Algonquin tribes. Powhatan's daughter was the beautiful Pocahontas, who in legend and perhaps in fact saved Captain John Smith by persuading her father to spare Smith's life when he was Powhatan's captive. Ultimately, in a well-documented saga, she married John Rolfe and sailed for England, where in 1617 she died of smallpox at the age of twenty-two. Their Virginia descendants included the Randolphs, the Bollingses, the Rolfes, the Pendletons, the Smiths, the Wynnes, the Yateses, and many others who helped build Virginia during the earliest Colonial times and eventually constituted Virginia's aristocracy.[66]

But three hundred years of population admixture, genocide and oppressive living conditions for those who remained had reduced the continent's many once proud tribes to a decimated remnant. The U.S. Census Bureau counted Indians in varying ways at various times, employing an array of definitions, all subject to local discretion, throughout the nineteenth and early twentieth centuries. Partially as a result of these inconsistencies, Indian demographic statistics ebbed and flowed in American population records, and their legal status was complex and troubled. But on June 2, 1924, Congress finally granted citizenship to all Indians not already naturalized under its Indian Citizenship Act. This law was ratified less than two weeks before the effective date of Virginia's own Racial Integrity Act. The new federal Indian law, together with Virginia's one-sixteenth Indian exemption, outraged Plecker.[67]

He embarked on a systematic effort to identify the lower class descendants of American Indians who had intermarried with whites and Negroes, and to reclassify them from Indian or white to mongrel. Among his main targets were the Monacan Indians, mainly of Amherst County, who descended from the Monacan Confederacy and dated back to Pocahontas's day. Others he pursued included the Rappahannock, Chickahominy and Pamunkey tribes. These Indian communities were small and often cloistered. Some two hundred dwelled in Rockbridge County. In King William County there were probably fewer than 250. In another county, there were just forty individuals who called themselves Indians but whom Plecker claimed derived instead from the illegitimate daughter of a Negro and a white.[68] All were targets for the registrar.

American Indians throughout the state vigorously objected when the Bureau of Vital Statistics attempted to reclassify them as Negro, or mongrel, or even nonwhite. "We had considerable trouble," Plecker admitted in a correspondence, "in establishing the position of the American Indian, and admitted those with one-sixteenth or less of Indian blood, to accommodate our Pocahontas descendants and one or two other cases known to us in the State. That clause, however, has given us much trouble, as a number of groups who have but a trace of Indian blood, the rest being negro and white, are claiming exemption under that clause. In at least one county, some who are descendent of antebellum 'free negroes' with a considerable admixture of illegitimate white blood, are claiming themselves Indians and seem to have been meeting with success."[69]

Most of Virginia's Indians were rural poor, living in modest cabins near mission churches. It was easy to marginalize them as unfit. Physically, most of them bore only the strong, classically handsome features of American Indians, including high cheekbones, thick black hair and their traditional complexion. Some, however, did possess blond hair, reflecting clear Anglo-Saxon parentage. A few, presumably descended from intermarriages with free Negroes in the prior century, possessed darker skin.[70]

Virginia's registrar, however, only allowed for two classifications, white and nonwhite. All 1,300 of Virginia's local registrars were under orders to watch for Indians with any trace of Negro ancestry registering as white. In at least one case, the local registrar consulted a hair comb hanging inside a Monacan church. "If it passes through the hair of an applicant," explained Plecker, "he is an Indian. If not, he is a negro." In a private letter, Plecker described the hair comb as being "about as reliable as some of their [the Indians'] other tests." In *Eugenical News*, he bragged that his "systematic effort to combat" what he called "near-whites" included utilizing "living informants" as well as the state's oldest tax and registration records.[71] If he couldn't get them one way he would get them another way.

Plecker employed his usual pejorative tactics in erasing "Indian" as a racial category from the state's records. He sarcastically accused one Indian family in Rockbridge County of having a bloodline that included several Indians who had intermarried with some whites and Negroes. He instructed local registrar Aileen Goodman to change their classification to "colored" and brashly notified the accused individual that, "In the future, no clerk in Virginia is permitted to issue a marriage license . . . [to] persons of mixed descent with white people and our Bureau expects to make it very plain to clerks that this law must be absolutely enforced." The Rockbridge family members were no longer Indians.[72]

Even when no Negro bloodline was apparent, Plecker was adamant. He identified one man in Lexington, Virginia, as "one-fourth Indian, three-fourths white, who cannot be distinguished from a white man. He attended one of the colleges of Virginia, studied law, and married into a good family in Rockbridge County. There are several similar cases in Southwest Virginia where Indians . . . have married white women and their children are passing as white." He informed the local registrar, "You see [to it] that the mixed people of your territory are registered either as colored or 'free issue.'" Disallowing even the category "mixed Indian," Plecker instructed, "the term 'mixed' without the word 'Indian' after it might be acceptable but we would prefer one of the other terms." The Lexington, Virginia, family members were no longer Indians.[73]

At one point Plecker visited an Indian church following its Sunday service, and after two hours sternly informed the assembled that no matter how they protested, they would be registered as "colored and would continue to be so and that none of them would be considered anything else." Some years later, when the clerk of Charles City tried to issue a marriage license to a member of the church, Reable Adkins, and even included the birth certificate attesting to the man's white lineage, Plecker simply changed the records. "We received this certificate for this birth with both parents given as white," he acknowledged. "Of course we will not accept the certificate in that way.... All of the Adkins group and others associated with them under their Chickahominy Charter are classed in our office as colored and never as white or Indian. In reply to your inquiry as to whether a marriage license should be issued to them other than colored, when they present birth certificates stating that they are Indian, I wish to state emphatically that this should not be done.... They are negroes and should always be classed as negroes, regardless of any birth certificate they present.... When the certificates come in to us we index and classify them as negroes." A special form was usually attached to the back of the certificate nullifying the category. Adkins family members were no longer Indians.[74]

Plecker's interference even extended beyond Virginia. For example, Plecker wrote to William Bradby of Detroit, Michigan, advising that his birth certificate claiming to be of "half-breed Indian" parentage would be disallowed. Leaving no room for argument, Plecker declared simply, "We do not recognize any native-born Indian as of pure Indian descent unmixed with negro blood." Bradby's family members were no longer Indians. [75]

To bolster his assertion that Indians simply no longer existed, only mongrel mixtures, Plecker turned for scientific support to the Carnegie Institution and its Eugenics Record Office. For years prior to the passage of Virginia's Racial Integrity Act, the ERO had focused on the Indians of Virginia as examples of the unfit. In 1926, the Carnegie Institution financed and published the results of extensive fieldwork by two of its Virginia researchers who had examined some five hundred tribal members in one area. The Carnegie Institution's book, printed under its own imprimatur with Davenport's close supervision, was entitled *Mongrel Virginians*.[76]

Mongrel Virginians was heralded for its academic completeness. It asserted that all living descendants of the several hundred Indians in question "have been visited time and again by one or both of the authors. In addition every known white, colored or Indian person in the county, state or nation who could furnish information concerning the deceased or living has been consulted and asked to give any material of value to the investigation." The Carnegie report lumped all of these Indians into one new group, which they called the "Win Tribe." Indeed, the subtitle of

Mongrel Virginians was *The Win Tribe*. No one had ever heard of a Win Tribe prior to this volume. The book explained "Win" stood for "White-Indian-Negro."[77]

"The Wins themselves claim to be of Indian descent," the book asserted. "They are described variously as 'low down' yellow negroes, as Indians, [and] as 'mixed.' No one, however, speaks of them as white. The Wins themselves in general claim the Indian descent although most of them realize they are 'mixed,' preferring to speak of the 'Indian' rather than of a possibility of a negro mixture in them."[78]

The Carnegie report assessed their usefulness to society as follows: "It is evident from this study that the intellectual levels of the negro and the Indian race as now found is below the average for the white race. In the Wins, the early white stock was probably at least of normal ability, i.e. for the white.... [Today, however,] the whole Win tribe is below the average, mentally and socially. They are lacking in academic ability, industrious to a very limited degree and capable of taking little training. Some of them do rather well the few things they know, such as raising tobacco or corn—a few as carpenters or bricklayers, but this has been the result of years of persistent supervision by the white landlords. Less than a dozen men work even reasonably well without a foreman.... Very few could tell the value of either twenty-five or seventy-five cents."[79]

Nor did the Carnegie report find redeeming qualities in the Indian culture it described. "There is practically no music among them," the study reported, "and they have no sense of rhythm even in the lighter mulatto mixtures. As is well known, the negro is 'full' of music. Some of them [the mulattoes] have been given special training in music, but no Win has ever shown any semblance of ability in this line."[80] No mention was made of the Indians' legendary rhythmic dances or songs and their many drums and other musical instruments.

Mongrel Virginians was accorded credibility because of its prestigious authorship, and its touted academic rigor. "Amidst the furor of newspaper and pamphlet publicity on miscegenation which has appeared since the passage of the Virginia Racial Integrity Law of 1924," the report assured, "this study is presented not as a theory or as representing a prejudiced point of view, but as a careful summary of the facts of history."[81]

Plecker seized on *Mongrel Virginians* to prove his point and help him reclassify Indians. He helped popularize the book around the state with his own enthusiastic reviews. *Eugenical News* extolled the study to the movement at large.[82]

Despite *Mongrel Virginians*, Indians and others fought back. Several sued Plecker from the beginning and made substantial progress in the courts. Plaintiffs' attorneys were often unyielding in their objections. One such attorney, J. R. Tucker, demanded that Plecker stop interfering with a birth certificate and threatened, "I find nowhere in the law any provision which authorizes the Registrar to constitute himself judge and jury for the purpose of determining the race of a child born and authorizing him to alter the record.... I desire and demand a correct copy of the record... without comment from you and without additions or subtractions, and I hereby notify you that unless I obtain a prompt compliance... I shall apply to a proper court for a mandamus to compel you."[83]

In a candid note, Plecker admitted to his cohort Powell that his bureau's strategy was based in no small way on simple intimidation. Tucker's ultimatum had rattled Plecker. "In reality," he conceded, "I have been doing a good deal of bluffing, knowing all the while that it could not be legally sustained. This is the first time my hand has absolutely been called."[84]

As early as November of 1924, one judge by the name of Henry Holt ruled against Plecker, setting the stage for a test case. "In twenty-five generations," wrote the judge in an incisive opinion, "one has thirty-two millions of grandfathers, not to speak of grandmothers, assuming there is no intermarriage. Half of the men who fought at Hastings were my grandfathers. Some of them were probably hanged, and some knighted. Who can tell? Certainly in some instances there was an alien strain. Beyond peradventure, I cannot prove that there was not." Nor could the judge find any two ethnologic authorities who could agree on the definition of pure Caucasian.[85]

Powell and Plecker worried about the judge's ruling. The commonwealth attorney was willing to pursue an appeal as a test case, but he also warned that the entire Racial Integrity Act might be struck down. They decided not to pursue the appeal. Plecker in turn assisted efforts to get the legislature to reduce the Pocahontas exemption, causing raucous debate within the state house and in the newspapers of Virginia.[86]

Plecker continued his crusade even after retiring in 1946 at the age of eighty-four. To the last day he was publishing racist pamphlets decrying mongrelization, defending the purity of the white race, decreeing demographic status family-by-family in a state and in an era when demographic status defined one's existence. In a final flourish, Plecker submitted his resignation with the declaration, "I am laying down this, my chief life work, with mingled feeling of pleasure and regret." He hoped to be dubbed "Registrar Emeritus."[87]

During his tenure, Walter A. Plecker dictated the nature of existence for millions of Americans, the living, the dead and the never born. His verdicts, often just his suspicions, in many ways defined the lives of an entire generation of Virginians—who could live where, who could attend what school and obtain what education, who could marry whom, and even who could rest in peace in what graveyard. It was not achieved with an army of soldiers, but rather with a legion of registrars and millions of registration forms. He was able to succeed because his campaign was not about racism, nor mere prejudice, nor even white supremacy. It was about science.

Now that science was ready to spread across the seas.

PART TWO

Eugenicide

CHAPTER 10

Origins

One morning in June of 1923, John C. Merriam, the Carnegie Institution's newly installed president, telephoned Charles Davenport at Cold Spring Harbor. Anticipation was in the air. A long-awaited eugenic countermeasure, loosely called "the plan," finally seemed within reach. "The plan" would create an American eugenic presence throughout the world even as inferior strains were eliminated in the United States. It was now important to be politically careful. Merriam, however, was worried about the behavior of Harry H. Laughlin.[1]

Merriam's hopeful phone call to Davenport had been years in the making. American eugenics had always sought a global solution. From the beginning, ERO leaders understood all too well that America was a nation of immigrants. But American eugenicists considered most of the immigrants arriving after 1890 to be genetically undesirable. This was because the 1890s witnessed the onset of the great Eastern and Southern European exodus to the United States, with throngs of non-English-speaking families crowding into the festering slums of New York and other Atlantic seaboard cities.[2]

Eugenicists viewed continued immigration as an unending source of debasement of America's biological quality. Sterilizing thousands of the nation's socially inadequate was seen as a mere exercise, that is, fighting "against a rising tide," unless eugenicists could also erect an international barrier to stop continuing waves of the unfit. Therefore the campaign to keep defective immigrants out of the country was considered equally important to the crusade to cleanse America of its genetic undesirables. This meant injecting eugenic principles into the immigration process itself—both in the U.S. and abroad.

Immigration had always been a complex, emotionally-charged concept in the United States. A thousand valid arguments encompassing economics, health conditions, overcrowding, demographics and humanitarianism

185

perpetually fed competing passions to either increase or decrease immigration. Moreover, the public and political mood twisted and turned as conditions in the country changed. Between 1880 and 1920, more than twenty million immigrants had flooded into the United States, mainly fleeing Europe's upheaval. More than eight million of that number arrived between 1900 and 1909.[3]

America's turn-of-the-century welcome was once poetically immortalized with the injunction: "Give me your tired, your poor, your huddled masses yearning to breathe free, the wretched refuse of your teeming shore."[4] But after World War I, American society was in ethnic, economic and demographic turmoil. Now-curtailed war industries laid off millions. Returning "dough boys" needed work as well, only adding to widespread joblessness. Inflation ate into wages. African-Americans who had gone to war now expected employment as well; they had fought for their country, and now they wanted their sliver of the American dream. Dislocation bred discontent. Massive labor strikes paralyzed much of America during 1919, with some 22 percent of the workforce joining a job action at some point during that year.[5]

Moreover, demographic upheaval was reweaving the very fabric of American social structure. Boy soldiers raised on the farm suddenly turned into hardened men during trench warfare; upon returning they often moved to cities, ready for a new life. Postwar immigration boomed—again, concentrated in the urban centers. The 1920 census revealed that for the first time in American history, the population majority had shifted from rural to urban areas. America was becoming urbanized, and mainly by immigrants. The 1920 census meant wrenching Congressional reapportionment, that is, a redrawing of district lines for seats in the House of Representatives. Eleven rural states were set to lose seats to more urbanized states. The House had expanded its available seats to 435 to preserve as much district *status quo* as possible.[6] But immigration remained the focal point of a political maelstrom.

To further inflame the day, race riots and ethnic strife ripped through the cities. African-Americans, back from soldiering, were tired of racism; they wanted a semblance of rights. At the same time, the Ku Klux Klan rose to never before seen prominence. The threat of Bolshevism worried the government and the average man. The Red Scare in the summer of 1919 pitted one *ism* against another. Marxism, communism, Bolshevism, and socialism sprang into the American consciousness, contending with capitalism. Race riots against African-Americans and mob violence against

anarchistic Italians and perceived political rabble-rousers ignited throughout the nation. A man named J. Edgar Hoover was installed to investigate
subversives, mainly foreign-born.[7]

As the twenties roared, they also growled and groaned about immigration. Along with the most recent huddled masses came widespread vexation
about the future of American society. Legitimate social fears, ethnic combat and economic turmoil stimulated a plethora of restrictive reforms,
some sensible, some extreme.

The best and worst of the nation's feelings about immigration were
exploited by the eugenicists. They capitalized on the country's immigration
stresses, as well as America's entrenched racism and pervasive postwar racial
anxiety. Seizing the moment, the men of the Carnegie Institution injected a
biological means test into the very center of the immigration morass, drag-
ging yet another field of social policy into the sphere of eugenics.

As early as 1912, the eugenics movement's chief immigration strategist,
Harvard professor Robert DeCourcy Ward, advocated eugenic screening
of immigrant candidates before they even reached U.S. shores. Davenport
enthusiastically wrote a colleague, "I thoroughly approve of the plan which
Ward urges of inspection of immigrants on the other side."[8] Bolstered by
other eugenic immigration activists, such as ophthalmologist Lucien
Howe, Laughlin became the point man in the movement's efforts. Their
goals were to rewrite immigration laws to turn on eugenic terminology,
and to install an overseas genetic surveillance network.

Key to any success was Albert Johnson. Johnson was an ambitious
small-town personage who would eventually acquire international potency.
Born in 1869 in Springfield, Illinois, on the northern edge of the Mason-
Dixon Line, Johnson grew up during the tempestuous Reconstruction
years. His high school days were spent in provincial Kansas communities,
including the newly created village of Hiawatha, and later Atchison, the
state's river and railroad center. But Johnson was an urban newspaperman
at heart, working first as a reporter on the *Herald* in St. Joseph, Missouri,
and then the *St. Louis Globe-Democrat*. Within a few years he had joined the
ranks of east coast journalists, becoming managing editor of Connecticut's
New Haven Register in 1896, and two years later serving as a news editor of
the *Washington Post*. After his stint with the *Post*, Johnson moved to
Tacoma, Washington, where he worked as editor of the *Tacoma News*.
Johnson then returned to his small-town roots as editor and publisher of
the local newspaper in Hoquiam, Washington. In 1912, while publisher, he
successfully ran for Congress. Johnson chaired the House Committee on

Immigration and Naturalization for twelve years, beginning in 1919. In that pivotal position, Johnson would shape American immigration policy for decades to come.[9] During his tenure, Johnson acted not only as a legislator, but also as a fanatic raceologist and eugenicist.

Even before Johnson rose to chair the Immigration Committee, Congress had enacted numerous immigration restrictions that were reactive, not eugenic, in nature, even if the legislation employed much of the same terminology. For example, a 1917 statute barred immigration for "all idiots, imbeciles, feebleminded persons, epileptics, insane persons... [and] persons of constitutional psychopathic inferiority." Laughlin and his colleagues wanted to rewrite these classifications along strictly biological and racial lines. His idea? New legislation to create a corps of eugenic "immigration attachés" stationed at American consulates across Europe and eventually the entire world. These consuls would exclude "all persons sexually fertile... who cannot... demonstrate their eugenical fitness... mental, physical and moral." Laughlin's proposed law was of paramount importance to eugenic stalwarts. As a leading immigration activist told Davenport in an October 1, 1920, letter, any new system would need to "heavily favor the Nordics" and ensure that "Asiatics, Alpines and Meds... [are] diminished."[10]

The *Journal of Heredity*, formerly the *American Breeders Magazine*, trumpeted one of the movement's rationales for overseas screening in an article entitled "Immigration Restriction and World Eugenics." The article declared, "Just as we isolate bacterial invasions, and starve out the bacteria by limiting the area and amount of their food supply, so we can compel an inferior race to remain in its native habitat... [which will] as with all organisms, eventually limit... its influence."[11]

Premier racial theorist Madison Grant, president of the Eugenics Research Association and vice president of the Immigration Restriction League, was a close ally and confidant of Johnson's. Grant's influence with Congress on immigration was a recognized asset for the eugenics movement, and was well utilized. Davenport would periodically send him materials, including confidential reports done by social workers on individual New York immigrants deemed defective, "which you may be able to use with Congress." As far as Johnson was concerned, any immigration was too much immigration. In fact, Johnson had already introduced without success an emergency measure to suspend all immigration for two years.[12]

It wasn't long before Laughlin became the designated eugenic authority for Johnson's committee. Laughlin began in 1920 by offering Johnson the same definition of the "socially inadequate" previously rejected by the Census Bureau, together with the same flawed data. Unlike the Census Bureau, however, Johnson readily accepted these notions. He invited Laughlin to testify before a full House committee to formally espouse his raceology and lobby for the new legislation.[13]

Laughlin enthusiastically testified for two mornings, on April 16 and 17, 1920, invoking a gamut of eugenic arguments, from the history of the Jukes to the Tribe of Ishmael to the high cost of institutionalizing defective stock. At one point, when Laughlin was explaining one of his new terms for mental incompetence, a committee member interrupted and asked him how to spell it. Laughlin replied: "M-O-R-O-N. It is a Greek word meaning a foolish person."[14]

To stem the supply of morons and stymie further degeneracy, Laughlin asked Johnson to allow him to enable "testing the worth of immigrants... in their home towns, because that is the only place where one can get eugenical facts.... For example, whether he comes from an industrious or shiftless family." But just as the terms *feeblemindedness* and *blindness* were vague and fundamentally undefined, the exact nature of *shiftlessness* was also unclear. Laughlin assured Johnson that this could be remedied. "General shiftlessness could easily be made into a technical term," he explained, "by a little definition in the law. It could be made a technical term by describing it by a 50-word paragraph...."[15]

Laughlin emphasized that the quality and character of the individual candidate for immigration were not as important as his ancestral pedigree. "If the prospective immigrant is a potential parent, that is, a sexually fertile person," testified Laughlin, "then his or her admission should be dependent not merely upon present literacy, social qualifications and economic status, but also upon the possession in the prospective immigrant and in his family stock of such physical, mental, and moral qualities as the American people desire.... The lesson," he emphasized, "is that... the family stock should be investigated, lest we admit more degenerate 'blood.'"[16]

Johnson, a proud champion of immigration quotas, was greatly impressed with Laughlin's expertise and saw its usefulness in drafting any restrictive legislation. The chairman promised to invite Laughlin back as an expert to help the committee deliberate on his proposal for eugenic attachés. Laughlin's two-day testimony and proposed law were published by the House under the title "The Biological Aspects of Immigration."[17]

When Laughlin came back to consult, an encouraged Johnson created a new title for him: "Expert Eugenics Agent." Laughlin was now empowered to conduct wide-ranging racial and immigration studies, and to present them as reliable Congressional data. His new authority included the power to print and circulate official committee correspondence and question-naires, and mail them *en masse* at House expense. The first of these was a survey entitled "Racial and Diagnostic Record of State Institutions." It was printed on official House letterhead, with the committee members' names routinely listed at the top, but now with Laughlin's name added as "Expert Eugenics Agent." The form asked 370 state institutions—hospitals, pris-ons, asylums—in the forty-eight states to report the nationalities, races and problematic natures of their residents. Perhaps intentionally, private insti-tutions were not queried, limiting the survey and its resulting data to the most needy and troubled within immigrant groups.[18]

Laughlin's target for the survey data was the 1924 legislative session. This was when temporary immigration quotas, enacted under Johnson's baton in 1921, were scheduled to be revised. Those restrictive quotas had calculated the percentages of the foreign born nation-by-nation, as enumerated by the 1910 census, and then limited each nation's new annual immigration to only 3 percent of that number. This had the effect of turning America's demo-graphic clock back to 1910. But to eugenicists, this restrictive quota was not restrictive enough. Laughlin and his colleagues wanted to turn the clock back to 1890, before mass influxes from Eastern and Southern Europe had begun. Laughlin's study of "Racial and Diagnostic Records of State Institutions" would statistically prove that certain racial and national types were criminalistic and amoral by genetic nature.[19]

But the hundreds of state hospitals, prisons and other institutions spread across the United States all saw their residents' ancestries through different eyes using different terminology. To guide institutions in stan-dardizing their responses, Laughlin circulated a supplemental Congres-sional publication entitled "Classification Standards to be Followed in Preparing Data for the Schedule 'Racial and Diagnostic Records of Inmates of State Institutions.'" His title, "Expert Eugenics Agent," was printed on the cover. The booklet listed sixty-five racial classifications to be employed. Classification #15 was German Jew, #16 was Polish Jew, #17 was Russian Jew, #18 was Spanish-American (Indian), #19 was Spanish-American (White), #25 was North Italian, #26 was South Italian, #29 was Russian, #30 was Polish (Polack), #61 was Mountain White, #62 was American Yankee, #63 was American Southerner, and #64 was Middle West American. Crimes to be classified for genetic purposes included sev-eral dozen categories ranging from homicide and arson to driving recklessly, disorderly conduct, and conducting business under an assumed name. The data collected would all go into one mammoth Mendelian database to help set race-based immigration quotas.[20]

The Carnegie Institution was no bystander to Laughlin's operation. Laughlin regularly kept Carnegie president John Merriam briefed on the special Congressional privileges and testing regimens placed at the disposal of the eugenics movement. Merriam authorized Carnegie statistician J. Arthur Harris to validate the reliability of the data Laughlin would offer Congress. However, Laughlin's derogatory raceological assertions were now becoming more public, and Merriam feared that his views would not be popular with America's vocal minorities.[21]

In November of 1922, Laughlin's statistics-filled presentation to Congress was published as "Analysis of America's Modern Melting Pot." It contained copious racial and ethnic denigrations. Johnson declared that the entire session would be published officially with the pejorative subtitle "Analysis of the Metal and Dross in America's Modern Melting Pot." The dross was the human waste in American society. Laughlin's testimony insisted, "Particularly in the field of insanity, the statistics indicate that America, during the last few years, has been a dumping ground for the mentally unstable inhabitants of other countries."[22]

During his testimony about the melting pot, Laughlin told the House, "The logical conclusion is that the differences in institutional ratios, by races and nativity groups... represents real differences in social values, which represent, in turn, real differences in the inborn values of the family stocks from which the particular inmates have sprung. These degeneracies and hereditary handicaps are inherent in the blood." Laughlin asked for authority to conduct additional racial studies of "Japanese and Chinese... Indians... [and] Negroes." He appended a special statistical qualification for Jews, explaining, "The Jews are not treated as a separate nation, but are accredited to their respective countries of birth." As such, he urged a separate "study of the Jew as immigrant with special reference to numbers and assimilation."[23]

Laughlin's constant racial and ethnic derogations were no longer confined to scholarly journals, but were now echoing in Congressional hearing rooms. Indeed, a graphic raceological immigration exhibit from a recent eugenics conference had been installed for public examination in the Immigration Committee's hearing rooms. All these ethnic and racial revilements in turn opened Carnegie and the movement to increasingly vituperative attacks from the large immigrant groups that were becoming ever more entrenched in the country. But Laughlin was unbending. "If immigration is to be made a biological or racial asset to the American people," he railed, "radical statutory laws must be enforced." At one point he authored an immigration treatise under the Carnegie Institution's credential, which concluded that America was being infested by defective immigrants; as its prime illustration, the treatise offered "The Parallel Case of the House Rat," which traced rodent infestation from Europe to the rats' ability "to travel in sailing ships."[24]

Incendiary or not, Laughlin's rhetoric and eugenic data were producing results with Congress. It was exactly the scientific justification Johnson and other government figures needed to implement greater quotas and deploy the overseas network they wanted. Johnson was increasingly becoming not just a congressman favoring racial immigration quotas, but a eugenic organizational leader in his own right. In 1923, while chairing Congress's House Immigration and Naturalization Committee, Johnson also joined an elite new private entity with a Congressional-sounding name. The new seven-man *ad hoc* panel was called the "Committee on Selective Immigration." Chaired by Johnson's friend, raceologist Madison Grant, and vice-chaired by immigration specialist Robert DeCourcy Ward, the body also included Laughlin as secretary and eugenic ophthalmologist Lucien Howe.[25]

The Committee on Selective Immigration's first report concluded that America needed the Nordic race to thrive. "Immigrants from northwestern Europe furnish us the best material for American citizenship and for the future upbuilding of the American race. They have higher living standards than the bulk of southeastern Europeans; are of higher grade of intelligence; better educated; more skilled; better able to understand, appreciate and support our form of government." In contrast, the committee concluded, "Southern and eastern Europe . . . have been sending large numbers of peddlers, sweatshop workers, fruit-stand keepers [and] bootblacks. . . ."[26]

Citing the research on "inferiors" produced by Laughlin and other experts, the eugenic committee assured, "Had mental tests been in operation [years ago] . . . over 6 million aliens now living in this country, free to vote, and to become the fathers and mothers of future Americans, would have never been admitted." Relying on Laughlin and other commonly accepted eugenic principles, the *ad hoc* committee advocated passage of Laughlin's overseas surveillance laws and declared that racial quotas "based on the 1890 census [are] sound American policy. . . ."[27] Because Johnson functioned as both a member of the elite eugenic panel and as chairman of the House Immigration Committee, eugenic immigration quotas based on 1890 demographics now seemed assured.

Suddenly, in June of 1923, Johnson was thrust into new importance within the eugenics movement. On June 16, he was elected president of the Eugenics Research Association. Prior to this he hadn't even been a member of the organization. Nonetheless, this now positioned Johnson, with all his governmental powers, at the narrow pinnacle of eugenic organizational leadership. At the same time, Secretary of Labor James J. Davis, whose department was responsible for the domestic aspects of immigration, had signaled his willingness to cooperate in creating the overseas eugenic network to investigate immigrant families. The battle for negative eugenics—prevention—could now be waged at its source.[28]

No wonder that four days later, on June 20, Merriam anxiously telephoned Davenport. Secretary Davis had just sent a letter to President Warren Harding supporting the eugenic immigration legislation, and Davis was eager to secure any scientific underpinnings to justify it. Davis was due to sail to Europe on July 4, and now he contacted Merriam to ask if Laughlin might accompany him. Merriam answered that the Carnegie Institution would of course cooperate. That was the exciting part of Merriam's telephone conversation with Davenport. But then Merriam expressed his concerns about Laughlin.[29]

Laughlin was unpracticed in politics and was now expostulating scientific conclusions that were provoking reproach. Merriam told Davenport that the Carnegie Institution was quite aware of Laughlin's shortcomings and wanted to ensure that nothing stood in the way of a quiet success for "the plan" and its incorporation into the expected 1924 immigration reforms. Laughlin did not merely verbalize extremist views; many saw him as a eugenic zealot who would do anything to accomplish his goals. Yet in this situation, some political caution was necessary. "It is understood," Merriam repeated to Davenport moments later, "that the desire to have Dr. Laughlin associated with the Secretary is not for the purpose of changing our plans but is rather due to the fact that the Secretary recognizes that our work...can be useful to him.... It is not expected that there will be any modification of our plan, but rather that the Secretary will help to carry out the plans which you and Dr. Laughlin have worked out."[30]

Minutes later, Merriam went to the unusual extreme of dictating a letter to Davenport explicitly reiterating his concerns. "In order that there may be no misunderstanding...regarding Dr. Laughlin's work," Merriam wrote, "I wish to be frank and say that I have heard a number of quite different criticisms"—he scratched out the word *different* and penned in the word *frank*—". . . quite frank criticisms of Dr. Laughlin's conclusions drawn from his recent studies. . . . Because the genetics and eugenics work is so important it is necessary that we be exceedingly guarded, lest conclusions go beyond the limits warranted by the facts and therefore ultimately diminish the effectiveness of our scientific work." Merriam closed with a warning, "I am sure that neither you nor Dr. Laughlin will underestimate my interest in this problem or my recognition of its very great importance."[31]

Davenport in turn spoke to Laughlin, advising him that Secretary Davis had invited Laughlin to join him in sailing to Europe. Davenport also verbalized Merriam's concerns about Laughlin. When Merriam's letter arrived in Cold Spring Harbor a few days later, Davenport issued a pointed memorandum to Laughlin driving home Merriam's censure by quoting verbatim: "In order that there may be no misunderstanding. . . regarding Dr. Laughlin's work I wish to be frank and say that I have heard a number of quite frank criticisms of Dr. Laughlin's conclusions drawn from his recent studies. . . . Because the genetics and eugenics work is so important, it is necessary that we be exceedingly guarded lest conclusions go beyond the limits warranted by the facts and therefore ultimately diminish the effectiveness of our scientific work. . . . I am sure that neither you nor Dr. Laughlin will underestimate my interest in this problem or my recognition of its very great importance."[32]

The next Monday, Davis appointed Laughlin "Special Immigration Agent to Europe," making it official with a certificate. Laughlin had a penchant for titles that used the word *agent*. First he was retained as a "Special Agent of the Bureau of the Census." Then Johnson dubbed him the House's "Expert Eugenics Agent."[33] Now in his latest agent capacity he would tour Europe for six months, quietly investigating the family trees of aspiring immigrant families.

If he could establish the scientific numbers necessary to pronounce certain ethnic and racial groups as either eugenically superior or inferior, America's whole system of immigration could change. Laughlin wanted all potential immigrants to be ranked in one of three classes. "Class 1: Not sexually fertile, now or potentially, and not debarred on account of cacogenesis [genetic dysfunction]. Class 2: Sexually fertile, now or potentially, and not debarred on account of cacogenesis. Class 3: Sexually fertile, now or potentially, and debarred on account of cacogenesis."[34] Laughlin now found himself the syndic of America's genetic future.

Despite the urgings of the Carnegie Institution, Laughlin was unwilling to sail with Davis in July. He needed more time. Instead, he and his wife departed aboard the *S.S. Belgoland* about a month later, in time to attend an international eugenics meeting in Lund, Sweden. For the next six months, Laughlin would travel throughout Europe, setting up shop at American consulates and rallying logistical support from like-minded European eugenics groups.[35]

Scandinavia was first. In Sweden, he contacted the American embassy in Stockholm, as well as consular officials in Uppsala and Göteborg. In Denmark, he visited the consulate in Copenhagen. Laughlin concluded that Sweden was actually hoarding its superior strains by discouraging emigration through such groups as the Society for the Prevention of Emigration and an investigation undertaken by the government's Emigration Commission. Working with Sweden's official State Institute of Race-Biology, Laughlin launched ancestral verifications of four immigrant candidates, all young men, one from Kalmartan, one from Valhallavägen, and two from Stockholm. The American consul was to provide a social worker to undertake the field work along the lines of an earlier Laughlin study that was being translated into Swedish.[36]

He was sure his work in Sweden would yield scientific proof that Nordics were superior human beings. Writing from Europe, he expressed his elation to Judge Harry Olson of Chicago. "It seems that the Swedish stock has been selected for generations by a very hard set of national conditions—severe climate, relatively poor soil. The strenuous struggle for existence seems to have eliminated the weaklings.... Of course, the original Nordic stock was sound, else it would have died out entirely... [and] could not have made a good stock." Indeed, Laughlin thought that Swedish emigrants "must be considered her finest product in international commerce."[37]

His optimism faded as he traveled south. In Belgium, Laughlin contacted the American consul in Brussels to initiate investigations of four applicants whose visas had not yet been approved—two men and a woman from Brabant, and a woman from Brussels. His fellow eugenic activist Dr. Albert Govaerts, who had studied the previous year in Cold Spring Harbor, helped Laughlin get organized and performed the physical examinations. The Solvay Institute, with the consent of Vrije University, provided desk space.[38]

In Italy, he liaised with that country's Commissioner General of Emigration who agreed to help prepare field studies of four Italians seeking to emigrate to the U.S. Laughlin was convinced Italy had "an excess of population" and that the Italian government was "desirous of finding an outlet for their 'unemployed.'" With this in mind, he began investigating the four Italians.[39]

In England, an office was set up for Laughlin in the Eugenics Education Society headquarters outside London. Four Britons who had applied to emigrate were selected for familial examination. They included two Middlesex Jews (a teenage man named Morris and a woman in her twenties), plus a young woman from Devonshire and a young man from Hampshire. U.S. Public Health Service officers stationed in England were to perform the medical examinations.[40]

Laughlin reported back to Davenport that the various investigations "were made by a field worker...in much the same fashion as similar individual and family histories are made by eugenical field workers in the United States." The help of U.S. consuls was indispensable to "securing the most intimate individual and family histories of would-be emigrants to America...awaiting visas." Indeed, the individuals themselves were actually selected by the consuls, "who are giving their full cooperation in the work," Laughlin added. He hoped consular officials would go further and glean confidential family character information from local priests. If immigrant candidates felt the questions were too intrusive or offensive, Laughlin explained, field workers would "simply withdraw to the American Consulate, and announce that if the would-be immigrant desires to have his passport vised [issued a visa], he must provide the information concerning his own 'case history' and 'family pedigree.'" Laughlin boasted that the consuls would "smooth the way for perfecting these field studies."[41]

Mental tests to identify feeblemindedness were of course part of the investigation, although Laughlin did not indicate what language was being used in the various non-English-speaking countries. Where U.S. Public Health staff was not available for medical examinations, Laughlin proposed contract nurses or physicians. Secretaries and stenographers stationed around the Continent would be employed to type up the results.[42]

The purpose of Laughlin's family probes was not to help the United States properly ascertain the intellectual, economic, political or social caliber of individual immigrants, which fell well within any government's prerogative, but rather to determine how much tainted blood an applicant had received from his forebears. Ancestral blood, not individual worth, would be Laughlin's sole determinant.

He was receiving excellent cooperation until he arrived in Paris in late November of 1923. There he set up a mailing account at the local American Express office at 11 Rue Scribe, and was then ready to begin work. But when he contacted American Consul General A. M. Thackera to begin his local probes, the embassy balked. Someone at the embassy checked Regulation 124, dating back to 1896. It was against regulations for American consuls to correspond with officials of other American departments. Laughlin, as Special Immigration Agent to Europe, was officially a representative of the Department of Labor. Obviously, the rule would not allow them to collaborate with Laughlin.[43]

To resolve the problem, a conference was held in Paris on Sunday, December 2, 1923, attended not only by Consul General Thackera, but also by his British counterpart, Consul General Robert Skinner, as well as Consul General-at-Large Robert Frazer. They could find no way around the regulations. So they cabled Washington for instructions. By the end of the week, the State Department sent notice that the rule had been waived, so long as the diplomats "confined themselves to facts and did not render opinion or try to outline policy," as Laughlin reported it. The project proceeded unimpeded, mainly because the consuls were eager to cooperate.[44]

Before he was done, Laughlin had visited twenty-five U.S. consular offices in ten countries: Sweden, Denmark, Belgium, Italy, Holland, Germany, Switzerland, England, Spain and France, as well as the French colony of Algiers. Not only did Laughlin proudly establish eugenic testing procedures and precedents wherever he went, he created a network of friendly American consuls throughout the Continent, a feat he bragged about to the ERO. In fact, going beyond on-site work with the twenty-five consulates, Laughlin also mass-mailed every American consulate in Europe and the Near East—128 consulates in all—advising them of his project and seeking detailed local demographic data. Within months, two consulates had already provided partial reports directly to Laughlin, and more than two dozen others had sent the requested information to the State Department to be forwarded to Laughlin, who was still traveling. Eventually eighty-seven consulates supplied the requested population and ethnic information directly to Laughlin. Only eleven did not respond.[45]

During his whirlwind tour, Laughlin found little time for sightseeing. Moreover, as he traveled from city to city and incurred mounting expenses for stenographers, field investigators, report printing and other general living expenses, he was advancing his own money. He was still collecting a salary as ERO assistant director, but he complained more than once, "I am bearing my own expense." He was uncertain if he would ever be reimbursed. In late 1923, Laughlin petitioned Davenport, "If these studies prove profitable, and I am permitted to continue them beyond the first of January [1924], I respectfully request that provision be made for my expenses."[46]

Assistant Secretary of Labor Henning had promised a $500 stipend, and Laughlin had applied to receive it, but Henning's secretary then notified Laughlin that the department had "no means of sending you cash in advance...." Laughlin confided to Davenport, "I am a little uneasy about the 500 Dollars. The Department of Labor promised, but did not deliver."[47]

Carnegie and the ERO were not helpful, still apprehensive about Laughlin's growing reputation for outlandish race science. Even the prestigious scientific journal *Nature* had publicly castigated Laughlin in a review of his 1922 study on eugenic sterilization. For Laughlin, the tension with his own organization was palpable. To counter the bad reviews, he began sending a disenchanted Merriam as many complimentary European reviews of his work as he could. He also dispatched frequent optimistic reports back home justifying his investment of time, but noted that, in return, "I have not heard very many times from Cold Spring Harbor."[48]

At one point in late November of 1923, an almost desperate Laughlin admitted that the British and Belgian family case studies had already exhausted the anticipated $500 Labor Department reimbursement, and "the Swedish and Italian studies will need additional funds." He asked for financial assistance from the Carnegie Institution, and also mentioned this request to Davenport, so formally as to almost be provocative. "I ... do not feel like going into the matter any further without authorization for expenses from the director of the Eugenics Record Office," Laughlin wrote to Davenport, who was, of course, the director of the ERO. He added, "I should also like the assurance that in case the Department of Labor does not supply the money which I have actually spent for field assistance, I should be reimbursed [by the ERO]."[49]

Finally, on December 21, the Carnegie Institution decided to be more forthcoming with support for Laughlin's European endeavors. Davenport dispatched a letter to Laughlin in Belgium assuring him that the Department of Labor would reimburse all legitimate expenses. At the end of the letter he casually appended exactly what he knew Laughlin most wanted to hear: "Did I tell you that $300 has been appropriated for your traveling expenses in the budget of this Department [at Carnegie], and a check will be made out to you for it January first?"[50]

In mid-February of 1924, Laughlin sailed into New York Harbor after an exhausting six-month eugenic mission to Europe. Now it was time for the special immigration agent to compile his ideas and data into a scientific report to Congress. His government allies were more than ready. Several weeks before Laughlin sailed home, the seven-man *ad hoc* Committee on Selective Immigration published a detailed endorsement of his conclusions and proposed legislation, including overseas eugenic screening. Signing on to that report was House Immigration Committee Chairman Johnson, acting in his alter ego as member of the seven-man committee. The published report noted that although Laughlin was still in Europe, they knew he would agree with its contents.[51]

On February 17, 1924, just after Laughlin returned, Davis in his capacity as secretary of labor also advocated Laughlin's ideas in a special editorial in the *New York Times*. Davis declared that the program suggested by Laughlin must be enacted "so that America may not be a conglomeration of racial groups ... but a homogenous race striving for the fulfillment of the ideals upon which this Government was founded."[52]

On March 8, Laughlin again testified before Johnson's immigration committee, this time presenting a massive table- and chart-bedecked report bearing the charged title "Europe as An Emigrant-Exporting Continent and the United States as an Immigrant-Receiving Nation." True to form, Laughlin declared the existence of an "American Race." He admitted that America was created by "a transplanted people," but that the "nation was established by its founders. The pioneers 'got in on the ground floor.'" As such, this new American race "is a race of white people." Therefore, he summarized, the nation's racial character "is being modified to some degree by the changed racial character of the immigration of the last two generations."[53]

His voluminous charts and reports displayed samplings of the twelve family pedigrees he had assembled in Europe, as well as abundant columns of immigrant data and U.S. population trends. In exhibit after exhibit, Laughlin piled racial ratio upon racial ratio and population percentage upon population percentage, offering copious scientific reinforcement of his conclusions. The majority of Johnson's committee expressed complete support for both Laughlin and his research. At one point a congressman asked Laughlin to respond to denunciations of his work. "I decline to get into controversy with any heckler-critics," he retorted, "...I shall answer criticisms by supplying more first-hand facts." Johnson piped in, "Don't worry about criticism, Dr. Laughlin, you have developed a valuable research and demonstrated a most startling state of affairs."[54]

Johnson's committee was also willing to lobby within other government agencies in support of Laughlin's work. For example, when it became obvious that the State Department itself was now balking at releasing the confidential information that twenty-five consulates had submitted for Laughlin, immigration committee members bristled. "I think we ought to have a show-down on this," snapped one congressman.[55]

The issue was finally decided some weeks later in a private meeting. On June 17, Carnegie president Merriam and Laughlin met at Washington's elite Cosmos Club with Assistant Secretary of State Wilbur Carr, who headed up the consular service. Carnegie officials correctly believed that Carr had become "very favorably inclined toward cooperation with the Institution in this matter." At their meeting, Merriam explained the ERO's interest and Carr agreed to share the information, so long as Laughlin abided by a working understanding. Inasmuch as Laughlin held multiple government positions, any Carnegie Institution activities on the topic inside the United States would continue under the purview of the Department of Labor, the House Immigration Committee or any other domestic agency. But any overseas activity would need both general State Department approval and prior agreement by the ranking diplomat in the foreign locale. As part of the arrangement, Laughlin also agreed that any future demographic publications gleaned from consular data would be submitted in advance to the State Department "to prevent any possible embarrassment of the Federal Government."[56]

Two days later, with the arrangement sealed, Secretary of Labor Davis delivered a formal, interdepartmental request directly to Secretary of State Charles Evans Hughes asking that the confidential consular data be made available to Laughlin. Laughlin was prepared to assemble a detailed, highly personal, multifolder case study of immigrant candidates and their ancestry. Folder D, section 2b, for example, catalogued the family's "moral qualities." With the new information, Laughlin could offer vivid examples of his new system of human "filtering."[57]

The State Department sought to "prevent any possible embarrassment of the Federal Government" by Laughlin for the same reason the Carnegie Institution and Merriam expressed jitters. By this time Laughlin was more than a controversial pseudoscientist increasingly challenged by immigrant groups and others; he was in some quarters a complete laughingstock. And when Laughlin was excoriated in the popular press, all of eugenics and the Carnegie Institution itself were also opened to ridicule.[58]

Perhaps no better example of the ridicule directed at Laughlin at the time was a forty-seven-page lampoon written under the pseudonym Ezekiel Cheever, who in reality was probably either the irreverent *Baltimore Sun* commentator H. L. Mencken or one of his associates. Cheever's booklet, a special edition of his *School Issues*, was billed on its cover as a "Special Extra Eugenics Number" in which Cheever "wickedly squeals on Doctor Harry H. Laughlin of the Carnegie Institution and other Members of the

Eugenics Committee of the United States of America for feeding scientifically and biologically impure data to Honorable Members of the House of
Representatives concerning the Immigration Problem." In page after page
of satirical jabs, Laughlin's statistics were cited verbatim and then dismem-
bered for their preposterousness.[59]

For example, Cheever deprecated Laughlin's reliance on IQ testing to
gauge feeblemindedness. "Undoubtedly, one of the greatest blunders made
by scientific men in America the past fifty years," he wrote, "was the premature publication of the results of the Army [intelligence] tests." Mocking
Laughlin's scientific racism, Cheever titled one section "Nigger in the
Wood-Pile," which charged, "If the opinions advanced by Doctor Laughlin and based upon this same unscientific rubbish, are as unreliable as they
appear when the rubbish is revealed in a true light, then it would seem that
the Carnegie Institution of Washington must either disclaim any part of
the job or confess that the job, despite Carnegie Institution's part is a rotten
one, provided Carnegie Institution does not wish to be regarded as on a par
with the Palmer Institute of Chiropractic."[60]

Cheever scolded "Honorable Albert Johnson, Chairman of the House's
Committee on Immigration and Naturalization and a member of the
Eugenics Committee, [who] announced at the hearings: 'I have examined
Doctor Laughlin's data and charts and find that they are both biologically
and statistically thorough, and apparently sound.' It is now in order for
Congress to examine Honorable Albert Johnson and ascertain if as much
can be said about him."[61]

In a section titled "Naughty Germ Plasms," referring to Laughlin's
race-based state institution surveys, Cheever jeered, "If the reader will
examine the schedules sent out to cooperating institutions he will get a new
and somewhat startling view as to what constitutes 'the more serious crimes
or felonies.' Under adult types of crime there were listed: Drunkenness,
Conducting business under an assumed name, Peddling without license,
Begging, and Reckless driving. Among the serious crimes or felonies of the
juvenile type he will find: Trespass, Unlawful use of automobiles, Begging,
Truancy, Running away, Being a stubborn and disobedient child. If Doctor
Laughlin can devise a means for locating germ plasms that are responsible
for such heinous crimes, his fame will overshadow that of Pasteur."[62]

Often, the booklet used Laughlin's own words against him. Cheever
quoted from one passage in Laughlin's testimony that confessed, "At the
beginning of this investigation there were in existence no careful or
extended studies of this particular subject; the figures that were generally

202 || WAR AGAINST THE WEAK

given were either guesswork or based upon very small samples of the population."[63]

"Either Doctor Laughlin is exceedingly stupid," scorned Cheever, "or else he is merely a statistical legerdemain [sleight of hand artist]."[64]

Extracts from Cheever's booklet were syndicated in the *Baltimore Sun*. Other attacks followed. One severe assessment of his work by a reviewer named Jennings, writing in *Science Magazine*, caused eugenic circles particular distress because it appeared in a scholarly publication. "Can't you get out some sort of reply to Jennings," immigration guru Robert DeCourcy Ward wrote Laughlin. "He has been making a lot of trouble about your Melting Pot Report.... I hate to have that man talk and write without getting any real come-back from you." Impervious as always, Laughlin shrugged off Jennings, and also dismissed Cheever as "more of a political attack trying to answer scientific data."[65]

Davenport had no choice but to also deflect complaints arising from the steady stream of critical articles. Not a few of these were sent directly to the Carnegie Institution. Writing on Carnegie Institution letterhead, Davenport defensively replied to one man who had read Cheever's pieces in the *Baltimore Sun*, asserting that Laughlin had been unduly libeled. Indeed, Davenport's rebuttal likened the Cheever articles to the ridicule launched against Davenport himself years earlier by Galtonian eugenicists in England. He closed by saying that Cheever was so "out for blood" that he should be imprisoned.[66]

But no amount of public rebuke would dissuade Johnson, and that was all Laughlin cared about. Johnson continued to publish Laughlin's testimony as though it were solid scientific truth. Using Laughlin's biological data as a rationale, he pressed for new immigration quotas keyed to the national ancestral makeup reflected in the 1890 census. During April and May of 1924, the House and Senate passed the Immigration Act of 1924, and President Calvin Coolidge signed the sweeping measure into law on May 26. This legislation would radically reduce non-Nordic immigration, since the representation of Eastern and Southern Europeans was radically less in 1890 than it had been in 1910. The Italian quota, for example, would be slashed from 42,000 per year to just 4,000. Many called the new legislation the "National Origins Act" because it limited new immigration to a quota of just 2 percent of the "national origins" present in America according to the 1890 census.[67]

But tempestuous debate still surrounded the statistical validity of the 1890 census, and no one knew how reliable its reporting had been.

Statisticians quarreled over just who was Irish or German or Italian, and/or whose name sounded sufficiently Irish or German or Italian to be counted as such. Quotas could not be established until the disputed 1890 percentages were settled. So the 1924 law charged the Census Bureau with the duty of studying the numbers and reporting their conclusions to a so-called "Quota Board," which would be comprised of the three relevant cabinet secretaries: Davis of Labor, Herbert Hoover of Commerce, and Frank Kellogg of State. Quotas were to be announced by the president himself in 1927.[68]

Eugenicists tried mightily to influence the Quota Board's deliberations. Just how the quotas were set would dictate the success or failure of this latest eugenic legislative crusade. A common rallying cry was expressed in A. P. Schultz's raceological tome, *Race or Mongrel*, which proclaimed, "The principle that 'all men are created equal' is still considered the chief pillar of strength of the United States.... Only one objection can be raised against it, that it does not contain one iota of truth."[69]

Constant permutations and reevaluations of the demographic data were bandied back and forth throughout 1926. Politically-spun rhetoric masked true feelings. One senator, for example, staunchly announced he would not permit the new quotas to discriminate against Jews, Italians or Poles, but he concluded with the traditional eugenic view that any quota system must stop discriminating against Northwestern Europeans, that is, Nordics. As ethnic groups ramped up their pressure, however, some of the most stalwart quota crusaders began to falter.[70]

In the second half of 1926, the quota champion himself, Albert Johnson, came up for reelection. By now the immigrants in his district had come together in opposition to further restrictions. He began to equivocate. In August of 1926, Johnson gave a campaign speech opposing the "national origins" provisions because too many foreign elements would vote for repeal anyway. At one point he publicly declared in a conciliatory tone, "If the national origins amendment... is going to breed bad feeling in the United States... and result in friction at home, you may rest assured it will not be put into effect." He added that his own "inside information" was that the quotas would never be instituted.[71] Disheartened eugenicists sadly concluded that Johnson and his allies had completely succumbed to the influence of foreign groups.

Johnson's inside information proved somewhat prophetic. On January 3, 1927, Secretaries Davis, Hoover and Kellogg delivered to President Coolidge country-by-country quota recommendations, accompanied by a carefully crafted cover letter declaring that they could come to no reliable consensus about the true percentages of national origins in 1890. "It may be stated," the joint letter cautioned, "that the statistical and historical information available from which these computations were made is not entirely satisfactory." On January 6, Congress requested the official letter and its recommendations. The White House delivered them the next day. Eugenicists assumed that although there was room for argument, some form of quotas would be enacted at once.[72]

But before the sun set that day, the White House delivered a replacement cover letter to the Senate. This one was similar, bearing the same January 3 date, again addressed to President Calvin Coolidge and again signed by all three cabinet secretaries. But the key phrase warned the President more forcefully: "Although this is the best information we have been able to secure, we wish to call attention to the reservations made by the committee and to state that, in our opinion, the statistical and historical information available raises grave doubts as to the whole value of these computations as a basis for the purposes intended. We therefore cannot assume responsibility for such conclusions under these circumstances."[73]

In other words, within hours the demographic information went from merely problematic to absolutely worthless. Quotas could not be reliably ordained under the circumstances. On the last day of the 1927 session, Congress passed Senate Joint Resolution 152 postponing implementation of the new quotas for one year. House debate on the question ran less than thirty minutes. A year later, in 1928, quotas were once more postponed, again after a protracted statistical and political standoff replete with Congressional letter-writing campaigns and fractious newspaper editorials. Eugenicists were outraged and saw it as a triumph by organized foreign elements.[74]

Even before the first postponement, Laughlin began investigating the heritage of the individual senators themselves. "We are working on the racial origin study of present senators," Laughlin reported to a eugenic immigration activist, "and will line the study up with the data which you sent on members of the [original] Constitutional Convention. It will make an exceedingly interesting comparison," he added, "showing the drift of composition in the racial make-up of the American people, or at least of their leaders."[75]

Finally, in 1929, after indecisive demographic scuffles between census scholars and eugenic activists trying to preserve Nordic preference, compromise quotas were agreed upon by scholars formally and informally advising Congress and the president. Admitting that the numbers were "tainted" and "far from final," binding quotas were nonetheless created.

The new president, Herbert Hoover, promulgated the radical reductions based on the accepted analysis of the 1890 census. Even those quotas did not last long. Two years later they succumbed to redistricting pressures, political concerns and the momentum of the coming 1930 census. Finally, the quotas were revised based on national percentages from the 1920 census.[76]

Laughlin's quest for an overseas network of eugenic investigators achieved only brief success. The system was installed in Belgium, Great Britain, the Irish Free State, Norway, Sweden, Denmark, Germany, Czechoslovakia, Italy, Holland and Poland, and for a time the system eugenically inspected some 80 percent of the would-be emigrants from those countries. On average, 88 of every 1,000 applicants were found to be mentally or physically defective. Laughlin aimed to have one eugenicist stationed in each capital. But overseas examination was short-lived for lack of the extraordinary funding and complicated bilateral agreements required. Moreover, too many foreign governments ultimately objected to such examinations of their citizens.[77] Long after the examinations ceased, however, America's consuls remained eugenically aware of future immigrants and refugees as never before. Their biological preferences and prejudices would become insurmountable barriers to many fleeing oppression in the world of the 1930s.

Quotas and the National Origins Act ruled immigration until 1952. Only the 1952 McCarran-Walter Immigration and Naturalization Act amended almost a century of racial and eugenic American law to finally declare: "The right of a person to become a naturalized citizen . . . shall not be denied or abridged because of race or sex or because such person is married."[78]

American eugenics felt it had secured far less than half a loaf. For this reason, it was important that inferior blood be wiped away worldwide by analogous groups in other countries. An international movement would soon emerge. During the twenties, the well-funded eugenics of Laughlin, Davenport and so many other American raceologists would spawn, nurture and inspire like-minded individuals and organizations across Europe.

CHAPTER 11

———

Britain's Crusade

By the time four hundred delegates crowded into an auditorium at the University of London to witness the opening gavel of the First International Congress of Eugenics in 1912, Galton had died and Galtonian eugenics had already been successfully dethroned. America had appropriated the epicenter of the worldwide movement. Eugenic imperialism was vital to the followers of Davenport, as they envisioned not just a better United States, but a totally reshaped human species everywhere on earth.

Nowhere was American influence more apparent than in the cradle of eugenics itself, England. The same centuries of social consternation that had shaped Galton also shaped the new generation of eugenicists who supplanted him. Several storm fronts of historic population anxieties collided over England at the turn of the century. Urban overcrowding, overflowing immigration, and rampant poverty disrupted the British Empire's elegant Victorian era. After the Boer War, the obvious demographic effects of Britain's far-flung imperialism and fears over a declining birth rate and future manpower further inflamed British intellectuals, who were reexamining the inherent quality and quantity of their citizens.[1]

English eugenicists did what they did for Britain in a British context, with no instructions or coordination from abroad and precious little organizational assistance from anyone in America. While Britain's movement possessed its own great thinkers, however, British eugenic science and doctrine were almost completely imported from the United States. With few exceptions, American eugenicists provided the scientific roadmaps and the pseudoscientific data to draw them. During the early years, the few British attempts at family tracing and eugenic research were isolated and unsuccessful. Hence, while the population problems and chronic class conflicts were quite British, the proposed solutions were entirely American.

Galton died in 1911, more than a year before the First International Congress, but his marginalization had begun when Mendel's work was rediscovered in the United States. Quaint theories of felicitous marriages among the better classes, yielding incrementally superior offspring, were discarded in favor of wholesale reproductive prohibition for the inferior classes. Eugenic thought may have originated in Britain, but eugenic action began in America.

In the first decade of the twentieth century, while Galton and his circle were still publishing thin pamphlets, positing revolutionary positions at elite intellectual get-togethers and establishing a modest biometric laboratory, America was busy building a continent-wide political and scientific infrastructure. In that first decade, no government agency in Britain officially supported eugenics as a movement. But in America, the U.S. Department of Agriculture and its network of state college agricultural stations lent its support as early as 1903. Galton in London did not enjoy the backing of billionaires. But on Long Island, the vast fortunes of Carnegie, Rockefeller and Harriman financed unprecedented eugenic research and lobbying organizations that developed international reach. By 1904, when Galton and his colleagues were still moderating their theories, Charles Davenport was already creating the foundations of a movement that he would soon commandeer from his British predecessors. Before 1912, the Eugenics Record Office would begin extensive family-by-family lineage investigations in prisons, hospitals and poor communities. In England the one major attempt at tracing family pedigrees was a lone, protracted effort that took more than a decade to complete and another decade to publish.[2]

Americanized eugenics began to take root in England in the twentieth century under the pen of a Liverpool surgeon named Robert Reid Rentoul. In many ways, Rentoul helped lay the philosophical groundwork for British eugenics, and he would become a leading voice in the movement. A distinguished member of the Royal College of Surgeons, Rentoul worked with the feebleminded and had undertaken intense studies of America's eugenic activities. In 1903, he published a twenty-six-page pamphlet, *Proposed Sterilization of Certain Mental and Physical Degenerates: An Appeal to Asylum Managers and Others*. He urged both voluntary and compulsory sterilization to prevent reproduction by the unfit. As precedents, Rentoul devoted several pages to the legislative efforts in Minnesota, Colorado, Wisconsin and other U.S. states. The pamphlet's appendix included an abstract of Minnesota's early marriage restriction law. Rentoul lobbied for similar legislation in the United Kingdom. In one speech before the influential

Medico-Legal Society in London, he proposed that all physicians and lawyers join the call to legalize forced sterilization.[3]

Rentoul's ideas quickly ignited the passions of new eugenic thinkers, including those who gathered at a meeting of the London Sociological Society on the afternoon of May 16, 1904. Galton delivered an important address entitled "Eugenics, its Definition, Scope and Aims," stressing actuarial progress, marriage preferences and general education. "Over-zeal leading to hasty action," he cautioned, "would do harm, by holding out expectations of a near golden age, which will certainly be falsified and cause the science to be discredited." He added, "The first and main point is to secure the general intellectual acceptance of *Eugenics* as a hopeful and most important study." But the famous novelist and eugenic extremist H. G. Wells then rose to publicly rebuke Galton, bluntly declaring, "It is in the sterilization of failures, and not in the selection of successes for breeding, that the possibility of an improvement of the human stock lies." On that afternoon in Britain the lines were clearly drawn—it was positive eugenics versus negative eugenics.[4]

Rentoul continued his study of American eugenics throughout 1905, specifically fixing on the emerging notion of "race suicide" as espoused by the likes of American raceologist E. A. Ross and President Theodore Roosevelt. In 1906, Rentoul published his own in-depth eugenic polemic entitled *Race Culture; Or, Race Suicide?*, which became a veritable blueprint for the British eugenic activism to come. In page after page, Rentoul mounted statistics and percentages to document Great Britain's mental and physical social deterioration. But as remedies, Rentoul held up America's marriage restriction laws, advocacy by American physicians for sterilization, and recent state statutes. He explained the fine points of the latest legislative action in New Jersey, Delaware, Minnesota, Ohio, Indiana, North Dakota and other U.S. jurisdictions. "I cannot express too high an appreciation," Rentoul wrote, "of the many kindnesses of the U.S.A. officials to me in supplying information."[5]

Rentoul declared that he vastly preferred Indiana's vasectomies and salpingectomies to the castrations performed in Kansas and Massachusetts. But he added that the Kansas physician's pioneering efforts at asexualization were enough to justify "erecting a memorial to his memory." In one chapter, Rentoul cited an incident involving Dr. Oliver Wendell Holmes, the father of the future Supreme Court justice. When called to attend to a mentally unstable child, Dr. Holmes complained that to be effective, "the consultation should have been held some fifty years ago!" Rentoul also quoted

Alexander Graham Bell's eugenic denigration of charity: "Philanthropy in this country is doing everything possible to encourage marriage among deaf mutes." Rentoul urged his countrymen to duplicate American-style surveys of foreigners housed in its mental institutions and other asylums.[6]

Rentoul summarized his vision for Britain's eugenic future with these words: "It is to these States we must look for guidance if we wish to... lessen the chances of children being degenerates."[7]

Of course Rentoul's scientific treatise also addressed America's race problem in a eugenic context. In a passage immediately following references to such strictly local curses as Jack the Ripper, Rentoul asserted, "The negro is seldom content with sexual intercourse with the white woman, but culminates his sexual *furor* by killing the woman, sometimes taking out her womb and eating it. If the United States of America people would cease to prostitute their high mental qualities and recognize this negro as a sexual pervert, it would reflect greater credit upon them; and if they would sterilize this mentally afflicted creature instead of torturing him, they would have a better right to pose as sound thinkers and social reformers."[8]

The next year a few dozen eugenic activists formed a provisional committee, which a year later, in 1908, constituted itself as the Eugenics Education Society. Many of its founders were previously members of the Moral Education League, concerned with alcoholism and the proper application of charity. David Starr Jordan, president of the Eugenics Section of the American Breeders Association, was made a vice president of the Eugenics Education Society. The new group's biological agenda was to cut off the bloodlines of British degenerates, mainly paupers, employing the techniques pioneered in the United States. The two approved methods were sterilization—both voluntary and compulsory— and forcible detention, a concept euphemized under the umbrella term "segregation." Sympathetic government and social service officers were intrigued but ultimately unconvinced, because England, although steeped in centuries of class prejudice, was nonetheless not yet ready for American-style coercive eugenics.[9]

True, some in government explored eugenic ideas early on. For example, in August of 1906 the Lancashire Asylums Board unanimously resolved: "In view of the alarming increase of the insane portion of our population, immediate steps [should] be taken to inquire into the best means for preventing the propagation of those mentally afflicted...." But that resolution only called for an inquiry. Then the office of the secretary of state considered establishing a penal work settlement for convicts, vagrants

and the weak-minded on the Island of Lundy, thus setting the stage for segregating defectives. But this proposal floundered as well.[10]

It wasn't that England lacked the legal or sociological precedents for a
eugenics program. Pauperism was thought to be hereditary and had long
been judged criminal. Class conflict was centuries old. But America's solutions simply did not translate. Marriage restriction and compulsory segregation were anathema to British notions of liberty and freedom. Even
Galton believed that regulated marriages were an unrealistic proposition in
a democratic society. He knew that "human nature would never brook
interference with the freedom of marriage," and admitted as much publicly.
In his published memoir, he recounted his original error in even suggesting
such utopian marriages. "I was too much disposed to think of marriage
under some regulation," he conceded.[11]

As for sterilization, officials and physicians alike understood that the
use of a surgeon's knife for either sterilization or castration, even with the
consent of the family or a court-appointed guardian, was plainly criminal.
This was no abstruse legal interpretation. Reviewers commonly concluded
that such actions would be an "unlawful wounding," in violation of Section
Twenty of the 1861 Offense against the Person Act. Thus fears of imprisonment haunted every discussion of the topic. Ministry of Health officials
understood that in the event of unexpected death arising from the procedure, guardians or parents and physicians alike could be prosecuted for
manslaughter. Such warnings were regularly repeated in the correspondence of the Eugenics Education Society, in memorandums from the
Ministry of Health, and in British medical journals. Even the *Journal of the
American Medical Association* and *Eugenical News* made the point clear.[12]

America enjoyed a global monopoly on eugenic sterilization for the
first decades of the twentieth century. What was strictly illegal in the
United Kingdom was merely extralegal—a gray area—in America.
Therefore Indiana prison physician Harry Clay Sharp was able to sterilize
scores of inmates long before his state passed enabling legislation in 1907.
Moreover, while American states maintained control over their own medical laws, in Britain only Parliament could pass such legislation. British
eugenicists understood what they did about sterilization by observing the
American experience.

Nor did organized British eugenics immediately launch any field studies to trace the ancestries of suspected degenerates. Indeed, the whole idea
of family investigation caused discomfort to many in Britain, especially
members of the peerage, who cherished their lineages and genealogies.

Eugenicists believed that the firstborn in any family was more likely to suffer crippling diseases and insanity than later children, and this undermined the inheritance concepts attached to primogeniture, by which the eldest often inherited everything. Essentially, they thought the peerage itself had become unsound. In fact, Galton and his chief disciple, Karl Pearson, described the House of Lords as being occupied by men "who have not taken the pains necessary to found or preserve an able stock."[13]

Only a sea change in British popular sentiment from top to bottom, and an overhaul of legal restraints, would enable eugenical activity in England. Hence the Eugenics Education Society well understood that *education* would indeed have to be its middle name. That mission never changed. Almost twenty years later, when the organization shortened its name to the Eugenics Society, its chief organizers admitted, "It was believed that the object of the Society being primarily education was so universally established as to make the word *education* in the title redundant."[14] In reality, of course, "education" meant little more than constant propagandizing, lobbying, letter writing, pamphleteering, and petitioning from the intellectual and scientific sidelines, where British eugenics dwelled.

From its inception in 1908, the Eugenics Education Society had adopted American attitudes on negative eugenics. But with a movement devoid of any firsthand research in English society, the newly born EES was reduced to appropriating American theory from Davenport and company, and then trying to force it into the British sociological context. Although an aging Galton agreed to become the society's first "honorary president," by 1910 Galton and Pearson both understood that their ideas were not really welcome in the society. The Galton Laboratory and the simple biometric ancestral outlines recorded at various collaborating institutions by Pearson were seen as innocuous vestiges of the current movement. The society's main function was suasion, not science.[15]

Throughout late 1909, parlor lectures were given to inquisitive audiences in Derby, Manchester, Leeds and Birmingham. Groups in Liverpool, Glasgow, Cardiff and London scheduled talks as well. Such propagandizing was repugnant to Galton and Pearson, who saw themselves as scientists. Moreover, while monies were being raised for a Lecture Fund to defray the society's travel expenses, much of Pearson's research remained unpublished. In a January 3, 1910, interview with *The Standard* of London, Pearson complained about "four or five memoirs [scientific reports] on social questions of which the publication is delayed from lack of funds ... the problem of funds is becoming so difficult that the question of handing it over to be pub-

lished outside this country has already arisen." Almost derisively, he clarified, "The object of the Galton Laboratory is scientific investigation, and as
scientific investigators, the staff do not attempt any form of propaganda.
That must be left to outside agencies and associations."[16]

By 1912, America's negative eugenics had been purveyed to like-
minded social engineers throughout Europe, especially in Germany and
the Scandinavian nations, where theories of Nordic superiority were well
received. Hence the First International Congress of Eugenics attracted
several hundred delegates and speakers from the United States, Belgium,
England, France, Germany, Italy, Japan, Spain and Norway.[17]

Major Leonard Darwin, son of Charles Darwin and head of the EES,
was appointed congress president. But the working vice presidents
included several key Americans, including race theorist David Starr Jordan,
ERO scientific director Alexander Graham Bell, and Bleeker van
Wagenen, a trustee of New Jersey's Vineland Training School for Feebleminded Girls and Boys and secretary of the ABA's sterilization committee.
Of course Charles Davenport also served as a working vice president.[18]

Five days of lectures and research papers were dominated by the U.S.
contingent and their theories of racial eugenics and compulsory sterilization. The report from what was dubbed the "American Committee on
Sterilization" was heralded as a highlight of the meeting. One prominent
British eugenicist, writing in a London newspaper, identified Davenport as
an American "to whom all of us in this country are immensely indebted, for
the work of his office has far outstripped anything of ours...."[19]

Although Galton had died by this point, a young Scottish physician and
eugenic activist by the name of Caleb Saleeby informed his colleagues that
if Galton were still alive, he would agree that eugenics was now an
American science. If Galton could "read the recent reports of the American
Eugenics Record Office," wrote Saleeby, "which have added more to our
knowledge of human heredity in the last three years than all former work
on that subject put together, [Galton] would quickly seek to set our own
work in this country upon the same sure basis."[20]

By the final gavel of the First International Congress of Eugenics,
Galton's hope of finding the measurable physical qualities of man had
become officially passé among British eugenicists. Saleeby cheerfully
reported, "'Biometry'...might have never existed so far as the Congress
was concerned." Indeed, Pearson declined to even attend the congress. In
newspaper articles, Saleeby denounced biometrics as a mere "pseudoscience."[21]

The society had by now successfully purveyed the notion that defective individuals needed to be segregated. Whenever social legislation arose, the society's several dozen members would implore legislators and key decision makers to consider the eugenic agenda. For example, when the Poor Laws were being revised in 1909, a typical form letter went out. "The legislation for the reform of the Poor Law will be prominently before parliament. It is most essential that, when the reforms are made, they should include provisions for the segregation of the most defective portion of the community; it will be the business of the Society, during the coming year, to appeal to the country on this ground...." [22]

But the crusade to mass incarcerate and segregate the unfit did not achieve real impetus until England considered a Mental Deficiency Act in 1913. Like so many freestanding social issues invaded by eugenics, mental illness, feeblemindedness and pauperism had long been the subject of legendary argument in England. From 1886 to 1899, Britain passed an Idiots Act, a Lunacy Act, and a Defective and Epileptic Children Act. With the arrival of the twentieth century, the nation sought an updated approach. [23]

From 1904 to 1908, a Royal Commission on the Care and Control of the Feebleminded had deliberated the question of segregating and sterilizing the mentally unfit. The commission's ranks included several British eugenicists who had formed other private associations ostensibly devoted to the welfare of the feebleminded, but which were actually devoted to promoting eugenic-style confinement and surgical measures. The associations sounded charitable and benevolent. But such groups as The National Association for the Care and Protection of the Feebleminded and The Lancashire and Cheshire Association for the Permanent Care of the Feebleminded really wanted to ensure that the "feebleminded"—whatever that meant—did not reproduce more of their kind. [24]

The ambitious British eugenic plans encompassed not just those who seemed mentally inferior, but also criminals, debtors, paupers, alcoholics, recipients of charity and "other parasites." Despite passionate protestations from British eugenicists, however, the commission declined to recommend either widespread segregation or any form of sterilization. [25]

But eugenicists continued their crusade. In 1909 and 1910, other so-called welfare societies for the feebleminded, such as the Cambridge Association for the Care of the Feebleminded, contacted the Eugenics Education Society to urge more joint lobbying of the government to sanction forced sterilization. Mass letter-writing campaigns began. Every candidate for Parliament was sent a letter demanding they "support measures...

that tend to discourage parenthood on the part of the feebleminded and other degenerate types." As in America, sterilization advocacy focused first and foremost on the most obviously impaired, in this case, the feebleminded, but then escalated to include "other degenerate types." Seeking support for the Mental Deficiency Act, society members mailed letters to every sitting member of Parliament, long lists of social welfare officials, and virtually every education committee in England. When preliminary governmental committees shrank from support, the society simply redoubled its letter-writing campaign.[26]

Finally the government agreed to consider the legislation. Home Secretary Winston Churchill, an enthusiastic supporter of eugenics, reassured one group of eugenicists that Britain's 120,000 feebleminded persons "should, if possible, be segregated under proper conditions so that their curse died with them and was not transmitted to future generations." The plan called for the creation of vast colonies. Thousands of Britain's unfit would be moved into these colonies to live out their days.[27]

But while on its surface the proposed Mental Deficiency Act seemed confined to the feebleminded, many of whom already resided in institutions, the bill was actually a stalking-horse for more draconian measures. The society planned to slip in language that could snare millions of unwanted, pauperized and other eugenically unsound families. EES president Major Leonard Darwin revealed his true feelings in a speech to the adjunct Cambridge University Eugenics Society.

"The first step to be taken," he explained, "ought to be to establish some system by which all children at school reported by their instructors to be specially stupid, all juvenile offenders awaiting trial, all ins-and-outs at workhouses, and all convicted prisoners should be examined by trained experts in mental defects in order to place on a register the names of all those thus ascertained to be definitely abnormal." Like his colleagues in America, Darwin wanted to identify not just the so-called unfit, but their entire families as well.[28]

Darwin emphasized, "From the Eugenic standpoint this method would no doubt be insufficient, for the defects of relatives are only second in importance to the defects of the individuals themselves—indeed, in some cases [the defects of relatives] are of far greater importance." British eugenicists were convinced that just seeming normal was not enough—the unfit were ancestrally flawed. Even if an individual appeared normal and begat normal children, he or she could still be a "carrier" who needed to be sterilized. One society leader, Lord Riddell, explained, "Mendelian theory has disclosed that human characteristics are transmitted through carriers in a weird fashion. Mental-deficients may have one normal child who procreates normal children; another deficient child who procreates deficients and another apparently normal child who procreates some deficients and some normals. Mathematically, this description may not be quite accurate, but it will serve the purpose."[29]

More than a decade after Rentoul first proposed mimicking U.S. laws, British eugenicists now lobbied to install American-style marriage restrictions. Once again, it was the seemingly "normal" people that British eugenicists feared. Saleeby explained, "The importance ... will become apparent when we consider the real meaning of the American demonstration that many serious defects are Mendelian recessives. It is that there are many persons in the community, personally normal, who are nevertheless 'impure dominants' in the Mendelian sense, and half of whose germ cells accordingly carry a defect. According to a recent calculation, made in one of the bulletins of the Eugenics Record Office, about one-third of the population in the United States is thus capable of conveying mental deficiency, the 'insane tendency,' epilepsy, or some other defect.... Their number would be increased ... [unless] Dr. Davenport's advice as to the mating of defectives with normal persons were followed, for all their offspring would then belong to this category."[30]

Leonard Darwin and his colleagues hoped "a system will also be established for the examination of the family history of all those placed on the register as being unquestionably mentally abnormal, especially as regards the criminality, insanity, ill-health and pauperism of their relatives, and not omitting to note cases of marked ability." Their near kin were to be shipped off to facilities, and marriages would be prohibited or annulled.[31]

But once the plan to incarcerate entire families became known, revolted critics declared that the eugenic aspects of the Mental Deficiency Act would "sentence innocent people to imprisonment for life." In a newspaper article, Saleeby strongly denied such segregation need always be permanent. In a section subheadlined "No Life Sentences," Saleeby suggested, "All decisions to segregate these people must be subject to continual revision...."[32] Under the society's actual plan, however, incarcerations of ordinary people would occur not because of any observable illness or abnormality—but simply because of a suspect lineage.

Leonard Darwin authored a revealing article on the proposed law in February of 1912 for the society's publication, *Eugenics Review*. He confessed to the membership, "It is quite certain that no existing democratic government would go as far as we Eugenists think right in the direction of limiting the liberty of the subject for the sake of the racial qualities of future generations. It is here we find the practical limitation to the possibility of immediate reform: for it is unwise to endeavor to push legislation beyond the bounds set by public opinion because of the dangerous reaction which would probably result from neglecting to pay attention to the prejudices of the electorate."[33]

The First International Congress of Eugenics convened in London in July of 1912, at the height of the Parliamentary debate about the Mental Deficiency Act. Saleeby hoped the American contingent could offer their latest science on feeblemindedness as grist to sway lawmakers. But while the American delegation had spent over a year preparing a report on methods to terminate defective family lines, they were focused on sterilization of the unfit, not segregation. On the eve of the congress, Saleeby bemoaned the lost opportunity in a newspaper editorial. "It so chances, most unfortunately," he wrote, "that though the American Committee on Sterilization will present a preliminary report on the practicability of surgical measures for the prevention of parenthood on the part of defectives, no paper is being read on Mental Deficiency, of all subjects that which we should most have desired to hear discussed and reported widely at the present time."[34]

Saleeby added, "Dr. Davenport, the director of the American Eugenics Office ... is to read a paper, but unfortunately he will not deal with the feebleminded." Nonetheless, Saleeby saw progress. "Four years after a Report [by the Royal Commission on the Care and Control of the Feebleminded] which the American Students altogether superseded in 1909, thanks to their introduction of the Mendelian method, we have at last got a Mental Deficiency Bill through its second reading in the House of Commons."[35]

Parliament, however, could not endorse the wholesale segregation into colonies envisioned by the society. Political parties clashed on the issue. Catholics, laborites and libertarians staunchly attacked the legislation. At the end of 1912, *Eugenics Review* informed its members, "It is with the deepest regret that we have had to relinquish all hope of seeking this much-needed measure become law this Session." The clauses most important to the society were stricken. Clause 50, for example, had mandated an American-style marriage restriction—it was rejected. But eugenics' supporters in the House of Commons promised to revive the bill for the next session. "Our efforts to secure this result," *Eugenics Review* continued, "must not, however, be in the slightest degree relaxed...." Speaking to its several branches and affiliates throughout the nation, the publication urged: "Members of Eugenic societies should continue to urge on their representatives in Parliament by every available means ... and should unsparingly condemn their abandonment on account of the mere demands of party."[36]

Throughout 1913, the society continued to press for eugenic action along American lines. One eugenically-minded doctor reintroduced the marriage restriction clause, asking that existing marriages to so-called defectives be declared "null and void." This clause was refused. So were sweeping efforts to round up entire families. But in August of 1913, much of the bill was passed, partly for eugenic reasons and partly for social policy reasons. Britain's Mental Deficiency Act took effect in April of 1914. The act defined four classes: idiot, imbecile, feebleminded and moral defective. People so identified could be institutionalized in special colonies, sanitariums or hospitals established for the purpose. A Board of Control, essentially replacing the old Lunacy Commission, was established in each area to take custody of defectives and transport them to the colonies or homes. A significant budget was allocated to fund the new national policy.[37]

In many ways, this measure was simply an attempt to provide care and treatment for the needy. Colonies for epileptics, the insane, the feebleminded and those suffering from other maladies were already a part of Britain's national medical landscape. But to eugenicists, institutionalization was the same as incarceration. In a journal article, Saleeby explained to British readers, "The permanent care for which the Act provides is, under another name, the segregation which the principles of negative eugenics requires. ... In the United States, public opinion and understanding appear to be so far advanced that the American reader need not be appealed to."[38]

But as the law was finally rendered, the families of identified individuals were in no danger of being rounded up. Marriage restrictions were also rejected. The society admitted that the watered-down act "does not go as far as some of its promoters may have wished." In a review, one of its members conceded that legislators could not in good conscience enact profound new policies "where so much is debatable, so much untried, or still in experimental stages." Quickly, however, twenty-four Poor Law unions—charitable organizations—in the north of England purchased land to create colonies. Others proceeded much more slowly. It was all complicated because standards for certifying mental defectives varied widely from place to place.[39]

The eugenicists intended to press on, but several months later they were interrupted by the outbreak of World War I.

* * *

American eugenicists enjoyed a gargantuan research establishment, well funded and well staffed. The list of official and quasi-official bodies supporting or engaged in eugenical activities was long: the Carnegie Institution's Experimental Station, the Eugenics Record Office, the Eugenics Section of the American Breeders Association (which had by now changed its name to the American Genetic Association), the U.S. Army, the Department of Agriculture, the Labor Department, agencies of the State Department, and a Committee of Congress. Moreover, scores of state, county and municipal agencies and institutions added their contributions, as did a network of biology, zoology, genetic and eugenic departments at some of the country's most respected private and state universities. Buttressing all of it was a network of organizations, such as the Eugenics Research Association in New York, the Human Betterment Foundation in California, the Race Betterment Foundation in Michigan, as well as professional organizations throughout the medical and scientific fields. A labyrinth of American laws, enough to fill a five hundred–page guide to sterilization legislation, innervated the sterilization enterprise.[40]

At any given time there were hundreds of field workers, clinicians, physicians, social workers, bureaucrats and raceologists fanning out across America, pulling files from dimly-lit county record halls, traipsing through bucolic foothills and remote rural locations, measuring skulls and chest sizes in prisons, asylums and health sanitariums, and scribbling notes in the clinics and schools of urban slums. They produced a prodigious flow of books, journal articles, reports, columns, tables, charts, facts and figures where tallies, ratios and percentages danced freely, bowed and curtsied to make the best possible impression, and could be relied upon for encores as required. Little of it made sense, and even less of it was based on genuine science. But there was so much of it that policymakers were often cowed by the sheer volume of it.

British eugenic groups were merely eager end users.

But the Eugenics Education Society understood that it would be nearly impossible to apply American eugenic principles to the British social context without native research. Certainly, Galton and Pearson had been devoted to statistics from the beginning. Galton was the one who came up with the idea of family pedigree. His first efforts at organized human measurement, self-financed, were launched in the 1880s. Galton even created his own short-lived Eugenics Record Office in 1904, which was soon merged with Pearson's Biometric Laboratory. But lack of funds, lack of manpower and lack of momentum made these slow and careful pursuits far too tentative for the new breed of British eugenicists. Although pedigrees were faithfully published in the Galton Laboratory's multivolume *Treasury of Human Inheritance*, this was done not so much to show transmissible flaws as a prelude to sterilization, but rather to track the incidence of disease and defect, demonstrating the need to carefully control one's progeny.[41]

After a few years, Pearson and his circle of biometricians became bitter and isolated from the movement at large. At one point the Carnegie Institution routinely dispatched a staff scientist from its Department of Physiological Psychology, Professor Walter Miles, to tour European eugenic and biological laboratories. Miles made a proper appointment at Pearson's laboratory with the receptionist. But when Miles arrived, he was rudely refused entry. Nor was Miles even allowed to announce his presence or leave a message. Miles complained in a confidential memo, "She said that Dr. Pearson was an extremely busy man and could not be interrupted." The Carnegie representative was also denied a courtesy tour in the computational section of the lab away from Pearson. "The porter," continued Miles, "would not even take my card with a written statement on it that I had called and was exceedingly sorry... not to have been able to visit the Laboratory." An irritated Carnegie lab director in Boston later demanded an explanation of Pearson. An antagonistic exchange of letters culminated in a blunt message from the Boston director to Pearson declaring that the Carnegie Institution "will have to forgo the privilege of having personal contact with you or your associates.... It is more than obvious that visitors are not wanted."[42]

Galtonian biometrics and sample pedigrees remained handy relics within the British eugenics establishment, but the Eugenics Education Society was convinced it needed more substantial homegrown research to advance its legislative agenda. It tried to utilize ERO-style pedigrees in 1910 when a Poor Law reform committee asked for information. From the society's point of view, the "conclusion that pauperism is due to inherent defects which are hereditarily transmitted" was inescapable. In some cases pauper pedigrees reached back four generations, enabling society lobbyists to declare, "There is no doubt that there exists a hereditary class of persons who will not make any attempt to work."[43]

Yet the Royal Commission on the Poor Law—in both its minority and majority reports—found the few cases unconvincing. The eugenic viewpoint "was almost wholly neglected," as the society's liaison committee

BRITAIN'S CRUSADE || 221

bemoaned. "It soon appeared," a 1910–1911 society annual report admitted, "that before anything could be ascertained concerning the existence of a biological cause of pauperism, research must be made into a number of pauper family histories."[44]

Ernest J. Lidbetter stepped forward to emulate the American model. He would lead the society's charge toward a semblance of convincing research. But it took him twenty-two years to complete his work and publish his results. When he eventually did so, it was amid accusations and acrimony by and among his colleagues.[45]

Lidbetter was neither a physician nor a scientist. Since 1898, he had been a case investigator with the Poor Law Authority in London. He was eventually assigned to Bethnal Green, one the East End's most poverty-wracked districts. It had been a zone of impoverishment for decades. Once the society began probing pauper heritage, the eugenic match was made. In about 1910, Lidbetter became a proponent of the society's hereditarian view of pauperism, speaking to his fellow relief officers through the Metropolitan Relieving Officer's Association, university circles and at willing venues. The EES thanked Lidbetter for his help when several workhouses contributed family tree data to the society.[46]

Lidbetter's outlook was expressed perfectly in his lecture to a few dozen colleagues one Wednesday night in 1913, at a board meeting of the Metropolitan Relieving Officer's Association. Research into hereditary pauperism, far advanced in America and accepted in many official circles, was just starting in England. Eugenic notions were completely new to his audience. Lidbetter displayed heredity diagrams and insisted that England was plagued by a biologically distinct "race of chronic pauper stocks." He insisted that doubters "had to be answered, not in the light of their opinion, but by a series of cases checked, tested and confirmed over and over again." Hence he urged their cooperation in assembling pauper pedigrees from amongst their poverty cases.[47]

Attempts to create more than token samples of degenerate family trees were interrupted by the Great War, which began in the fall of 1914. British eugenics understandably slid into the background. In 1918, after shell-shocked soldiers climbed out of Europe's muddy trenches, British eugenics slowly regrouped. Lidbetter did not resume his examination of degenerate families until March of 1923, more than a decade after he had begun. By this time the Eugenics Education Society had been infused with other scientists, including the esteemed agronomist and statistician Ronald A. Fisher. Fisher had calculated the Mendelian and genetic secrets of various strains of potatoes and wheat, and he had used this information to create more effective manures at an experimental agriculture station north of London. He and others were now applying the coefficients and correlations so successful in mixing fertilizer and spawning stronger crops to complex hereditary formulas for humans. Fisher tacked the essence of Pearson's biometric measurements and agrarian science onto American Mendelism to create his own strain of eugenics.[48]

Lidbetter finally resumed his simple work in March of 1923, with a survey of all the indigents of Bethnal Green's workhouses and welfare clinics. He counted 1,174 people. But the society, especially its so-called Research Committee, which now included Fisher, insisted on proper statistical "control groups." Lidbetter, a welfare worker, was lost. Control groups? Should he compare streets, or maybe homes, perhaps families, or would one school against another be a better idea? In any event there was no money to finance such as effort. Eventually someone donated a token £20, which allowed a student to begin field work in the summer of 1923. But as the project sputtered on, it made little progress.[49]

The society shopped around for a few hundred pounds here and there, with little luck. In September of 1923, Laughlin showed up. He was in the middle of his Congressional immigration mission. The society provided him office space for three weeks so he could undertake American-style pedigree research on eugenically suspect immigration applicants. The society's difficulties were instantly apparent to him. England was helping too many of its indigent citizens. Laughlin wrote to his colleague Judge Harry Olson in Chicago. "England has a particular hard eugenic problem before her, because her Poor Law system has worked anti-eugenic, although from the standpoint of pure charity, it has saved much individual suffering."[50]

Eugenicists from Laughlin to Lidbetter were staunchly opposed to charitable works as a dysgenic force, that is, a factor that promoted eugenically unacceptable results. Lidbetter, a Poor Law officer charged with helping the disadvantaged, regularly lectured his fellow relief officers that charity only "created an environment in which the worst could survive as well as the best." He believed that poor people were "parasites" and that "public and private charity tended to encourage the increase of this class."[51]

Disdain for charity dramatically increased during and after World War I, especially among eugenic theorists such as David Starr Jordan, Laughlin and indeed many Britons. They postulated that in war, only the strong and brave killed each other. In other words, in war, the finest eugenic specimens of every nation would die off *en masse*, leaving the cowards, the infirm, the physically incapable and the biologically weak to survive and multiply.[52]

In articles, speeches and booklets, eugenicists lamented the loss of life. In his 1915 booklet, *War and the Breed*, David Starr Jordan wrote as a concerned American, years before the U.S. entered the conflict. Jordan mourned the dead young men of Scotland, Oxford and Cambridge. He quoted one war dispatch: "Ypres cost England 50,000 out of 120,000 men engaged. The French and Belgian loss [is estimated] at 70,000 killed and wounded, that of the Germans at 375,000. In that one long battle, Europe lost as many men as the North lost in the whole Civil War."[53] More then seven million would ultimately die in the Great War.

Yet eugenicists seemed more distressed that the strong were dying on the battlefield while the inferior remained. Jordan railed in his volume, "Father a weed, mother a weed, do you expect the daughter to be a saffron root?" The Eugenics Education Society published another typical article entitled "Skimming the Cream, Eugenics and the Lost Generation." War was denounced as dysgenic because "the cream of the race will be taken and the skimmed milk will be left."[54]

Lidbetter's research efforts were still unable, however, to attract the financial or investigative resources needed to convince British policymakers to do away with their unfit by a widespread American-style program of sterilization. By 1926, the quest for financing had compelled the society to plead with a Harvard eugenic psychologist, "English finances are indescribable, and we greatly fear our work will be brought to a standstill for want of the small sum needed, namely £300–£500 per year."[55]

An internal struggle developed within the society as skilled statisticians, such as Fisher, tried to oust Lidbetter from the Research Committee leadership in an attempt to improve the appearance of studies. The minutes of acrimonious meetings were doctored to conceal the degree of organizational strife. Financial resources dwindled. Lidbetter's meagerly paid assistant quit over money. At one point the society was unable to acquire the family index cards Lidbetter had accumulated. The society's general secretary, Cora Hodson, wrote to the new assistant, "I am trying to persuade Mr. Lidbetter to let us duplicate his index...keeping cards here....I may not succeed...."[56]

But Lidbetter's new assistant also quit within a year, again for lack of money. On September 15, 1927, Hodson revealed to a member, "I am rather seriously troubled about Mr. Lidbetter's research work. Funds have dropped tragically off....We are now faced with the loss of [an assistant]...simply for want of an adequate salary."[57]

Years of solitary and unfinanced effort had produced precious little data to support the society's vituperative rhetoric against so-called defectives. When the issue of publishable "results" came up, the society was forced to inform its membership, "It is impossible to speak of the 'result' of an investigation such as this after so short a period of work. The sum of money available was enough to provide an investigator for only a few months.... Much useful work has been recorded and the outline of seven promising pedigrees prepared. In none of these however was it possible in the time available to prepare the work in such detail as to warrant publication."[58]

Eventually, in 1932, after many society squabbles and a cascade of attempted committee coups, Lidbetter arranged to publish his results. He planned a multivolume set. "There is good hope of funds for the publication of a first volume to be contributed from the U.S.A.," a society official wrote. But that funding fell through. The first book in his series was finally released in England, but it was also the last; the other volumes were dropped. During the first three decades of the twentieth century, British eugenicists were forced to rely mainly on American research because it was the only other English-language science available to them, except for materials from Scandinavia and Germany—and these too had generally been translated by American sources. In February of 1926, the society secretary had sent off a note to a member, "Do you read German? The most thoughtful articles on the new methods are in a Swiss medical journal."[59]

At one point Saleeby bragged that he had accumulated a eugenic bibliography 514 pages long. But this bibliography was in fact the work of University of California zoology professor Samuel J. Holmes, and it was published by the university's academic press.[60]

As late as mid-1925, EES secretary Hodson was still seeking elementary information on heredity. On June 17, 1925, she dispatched a letter to Yale University's Irving Fisher, who headed the Eugenics Research Association. "My Council is considering the question of trying to extend the knowledge of heredity by liaison with our Breeders Associations. They are eager to get as much information as possible about the very successful work in Eugenics done by the American Breeders' Association, and I shall be most grateful if you will... forward any particulars that you think will be useful, or to tell me with whom I should communicate on the matter." She was referred to the Eugenics Record Office at Cold Spring Harbor.[61]

When Hodson tried to interest British high schools in adding eugenics to their curriculums, she wrote to the American Eugenics Society for information. "We are just making a beginning over here," she wrote, "with definite eugenic teaching in schools and it will be most helpful to me to be able to say that something concrete is being done in the United States, even if I cannot give chapter and verse for statistics."[62]

When British officials needed information on sterilization, they often wrote to America, bypassing the Eugenics Education Society—which had in 1926 changed its name to the Eugenics Society. In the spring of 1928, for example, when the medical officer for the County Council of Middlesex sought preliminary information on "sterilization of mental defectives," he wrote a letter directly to the American Social Hygiene Association, a Rockefeller-endowed organization in New York. In his response, the acting director of ASHA's Division of Legal and Protective Measures took the liberty of mentioning to the Middlesex medical officer Laughlin's vast legislative guide, *Eugenical Sterilization in the United States*. ASHA contacted Laughlin and asked him to send anything additional "which might be of aid to him. We are sure he would appreciate anything you may be able to send."[63]

By the late twenties, thousands of Americans had been forcibly sterilized. British eugenicists believed that America was lighting the way while Britain cowered in the shadows. British eugenicists were steadfast in their determination to introduce similar legislation in England. This meant a continued reliance on the science of Laughlin and Davenport.

The tradition already existed. On January 29, 1924, Laughlin had lectured at a society meeting. He described the American approach. "Then we go down still further and include the great mass of people, about nine-tenths of humanity. Then there is the submerged tenth, the socially inadequate persons who must be prevented from reproducing. If we try to classify them by types, we must call them the insane, the feebleminded, the paupers, the epileptic, the criminals, and so on. These people, and the family stocks that produce them . . . must be cut off and prevented from reproducing at all."[64]

Laughlin emphasized that it was not enough to sterilize an individual; his entire extended family needed to be sterilized as well. "I do not believe that humanity would ever make . . . eugenical progress if it simply prevented these individuals from reproducing. In order to prevent the reproduction of such individuals, we have to go up higher into the upper strata, and find out which families are reproducing these degenerates. The remedy lies in drying up the source. It is the pedigree rather than the individual basis of selection that counts in racial fortunes." This mandate was published more than a year later in the April 1925 *Eugenics Review* as a reminder. The society was determined to follow the American lead and sterilize all suspects, not just the obvious ones.[65]

In 1927, still desperate for research, Hodson circulated a draft letter endorsing eugenics in Britain. Members of the society were to sign these letters and mail them *en masse* to the editors of the *Times*—without disclosing their affiliations. "Two distinguished American authors," the proposed letter began, "have recently calculated that 1,000 college graduates will have scarcely 200 grown up great-grandsons, whilst 1,000 miners will have 3,700. We have no reason to doubt these figures, though unfortunately British statistics give us no means of checking them accurately. . . . We have nothing based on past experiences to guide us. . . . "[66] The nation was still reeling from a devastating coal miners' strike and Hodson's letter was surely designed to inflame.

The society was sending strategic letters to newspaper editors because it intended to make its strongest push to legalize sterilization. The first step in the British game plan, segregation, was faltering. Sterilization was needed. Medical, welfare and eugenic circles had been debating the subject for years. The British Medical Association's section on medical sociology had examined the subject extensively in 1923; Hodson appeared before the group and proclaimed that at least 10 percent of the nation must be forcibly sterilized at once—or many more would need to be sterilized within one or two generations. This warning became a popular slogan for society advocates.[67]

By 1926, British intelligence testers were surprised to discover that the number of mental defectives had vastly increased and maintenance costs were running as high as £4 million annually. Within three years, government investigators, employing mental tests designed by the Americans Goddard, Terman and Yerkes, claimed that the numbers of the mentally deficient had almost doubled in two decades, from 156,000 in 1909 when numbers were being gathered during the first Royal Commission to some 300,000 in 1929. The rate of mental deficiency had nearly doubled as well, they claimed, from 4.6 per thousand to 8.56 per thousand.[68] There was no way to know if the numbers had genuinely doubled or were merely a result of Terman and Goddard's questionable methodology—which had recently deemed 70 percent of American military recruits feebleminded.

The alarming new intelligence statistics were produced by the government's Mental Deficiency Committee, established to investigate mental defectives under the leadership of Sir Arthur Wood. Wood was a former assistant secretary of the medical branch of the Board of Education. Several eugenic advocates were associated with the Mental Deficiency Committee, and the resulting 1929 three-volume *Wood Report* closely resembled eugenic thinking on the deterioration of British intelligence levels. The committee used a new category, the "Social Problem Group," to describe the subnormal tenth of the nation. The Social Problem Group was comprised not only mental deficients, but also criminals, epileptics, paupers, alcoholics and the insane. Wood speculated that Britain was afflicted by a large number of problem types who although not certifiable, were nevertheless "carriers." The committee thanked the eugenics movement for its service in addressing the problem, but declined to endorse sterilization.[69] It was a significant setback.

To the additional outrage of eugenic activists, government policymakers now recommended that the many colonies and custodial institutions governed under the Mental Deficiency Act stop operating as mere long-term warehouses of people. Instead, these facilities "should be used for the purpose of stabilizing, training and equipping defectives for life in the community, [rather] than providing permanent homes," as one society memo glumly reported. The society complained that these colonies would soon be "turned into 'flowing lakes' rather than remain as 'stagnant pools.'" Deinstitutionalization would reverse all the society had sought to achieve.[70]

Sterilization was now more imperative than ever. By early 1929, the society mounted a fresh campaign to pass a national sterilization act. In mid-February of 1929, they sent a petition to Minister of Health Neville Chamberlain, a future prime minister. "Segregation as a remedy is failing," the resolution advised, "principally owing to the increasing number of deficients and the enormous costs."[71]

Within sixty days, a preliminary sterilization bill was drafted and circulated. It proposed coercive sterilization for those certified as feebleminded or about to be released from an institution; it also mandated broad marriage prohibitions, gave the state the power to unmarry couples, and criminalized the concealment of sterilization from a spouse. A postscripted suggestion declared, "If ever we have a proper system of registration, each person would have a card (or some equivalent), and on this card [eugenic] events, such as cancellation of marriage should be entered." Sir Frederick Willis had assembled the draft law almost two years earlier and passed it along to the society with one condition. "Should you care to use this draft, I should prefer that it should not be known that I have had anything to do with it; it does not necessarily represent my view."[72]

Eugenic stalwarts began propagandizing in earnest. Lord Riddell created a position paper for the Medico-Legal Society, a copy of which was duly forwarded to Chamberlain. Citing the many billions devoted to caring for the unfit, Riddell cautioned, "Unless we are careful, we shall be eaten out of house and home by lunatics and mental deficients." Riddell then quoted Harvard eugenicist Edward East. "Professor East says 'We are getting a larger and larger quantity of human dregs at the bottom of our national vats.'" Assuring that vasectomy did not reduce sex drive, Riddell asserted, "This is confirmed by replies sent to questionnaires put to 75 normal, intelligent, mostly professional American men who had undergone voluntary sterilization.... The dangers for men are negligible, and for women, in light of the Californian experience, not very serious."[73]

Indeed, Riddell emphasized that the proposed British law was efficacious because, "In California, where the law is similar to that now advocated, the results have been highly satisfactory."[74]

A Committee for Legalising Sterilization was formed in about 1930, and it began proffering intellectual position papers and suggestions for a draft law fused with layers of standard eugenic dogma. The phrase "voluntary sterilization" was employed to make it more palatable to the British public. The bill also provided so-called "safeguards" that would allow court-appointed guardians to make the decision for the individual—which technically constituted a voluntary decision. One report from the Committee for Legalising Sterilization repeatedly pointed to the 8,515 compulsory sterilizations performed throughout America, and especially California, as precedents. The CLS explained that California had performed 5,820 surgeries up until January 1, 1928, and had increased that number to 6,255 by January 1, 1929. These procedures were largely recorded as "voluntary." The committee's report explained, "In the California institutions, the defectives have been made to feel that by asking for sterilization, they are behaving in a laudable and socially useful manner."[75]

Eugenicists also capitalized on legitimate economic fears arising from years of crippling domestic strikes and the worldwide depression. Lord Riddell had challenged both the Medico-Legal Society and the Ministry of Health with visceral economic rhetoric. He calculated that the annual cost of caring for a growing population of the unfit could skyrocket to well above £16 million. "One is appalled by the prospect of multiplying these vast colonies of the lost, and... the injustice... of erecting splendid new buildings to house lunatics and mental defectives, when thousands of sound citizens are unable to secure decent dwellings at a moderate rent." He hammered, "As it is, the abnormal citizen receives far more care and attention than the normal one.... Consider an alternative solution—namely sterilization."⁷⁶

In 1930 the society launched another attempt to create a consensus of sorts among welfare organizations, the medical establishment and the British populace. A sudden endowment helped enormously. The society's financial problems disappeared when a wealthy Australian sheep rancher who periodically visited England (but spent most of his time at his villa in Nice, France) endowed the society. His name was Henry Twitchen. A bizarre and diseased man whom society elders called a "queer being," Twitchen had become enamored with eugenics in the early twenties and had promised to bequeath his fortune to the society. He died in 1929. Although his fortune had shrunk by that time, the £70,000 he donated changed everything for the organization now known as the Eugenics Society. One society official happily remembered that the money suddenly made the organization "rich."⁷⁷ Money meant travel expenses, pamphlet printing, better orchestrated letter-writing campaigns and the other essentials of political crusades.

Lidbetter's study, for whatever it was worth, was still unpublished. To compensate for their total lack of scientific evidence other than the American offerings, which even then were becoming increasingly discredited, in mid-1930 the society reached out to Germany, where expanding eugenic research was producing prodigious volumes of literature. German eugenicists were only too happy to forward packets of materials, including a five-page explication of the existing German literature on feeblemindedness along with four reprints. One of these essays, "Psychiatric Indications for Sterilization," was translated by the society and published as a pamphlet. Most of all, the German studies reflected the control groups that the statisticians demanded. One essay explained, "My procedure is to ascertain the number of psychopaths a) in an affected family, b) in families carefully selected ... [and] a sample of the average population."⁷⁸

Packets of documentation from Germany did not prevent Hodson from expressing her continuing admiration for American eugenics. On June 11, 1930, Hodson wrote to her counterpart at the American Eugenics Society that her recent review of "the wide and far-seeing development of the task in the United States" only reinforced her belief in the primacy of America's movement. "I used to say, when asked," Hodson added, "that I thought probably Germany was taking Eugenics most seriously, but I am quite sure that now the American Eugenics Society leads the world."

British efforts, Hodson admitted, "are not covering even one-third of the field of your committees."[79]

Hodson's continuing appreciation for American eugenics was understandable. Throughout the first half of 1930, Hodson had corresponded with Davenport in preparation for a gathering of international eugenic scientists in September. Davenport would serve as president of the conference. In February of 1930, Hodson wrote him for approval of conference dates and discussion topics, and then asked if she could print the program in both French and English for distribution. Hodson hoped that Davenport's latest views on race mixing would "wake up our Government people...." She added, "There is another point of importance for England in this connection—our anthropologists are not working much in unison.... [The conference's work] might be a focus in getting their activities combined...."[80]

In March of 1930 she wrote Davenport asking if any good films could be brought over from the ERO to screen at the conference. "Our English films I should offer only in the last resort as we are not really proud of them." A few days later, Davenport wrote back answering Hodson's cascade of questions, approving or rejecting detail after detail. In April, Hodson sent a letter to colleagues explaining, "Dr Davenport hopes that this year, the American interest in standardisation of human measurements may be linked up with the work proceeding in that direction in England...."[81]

In May, Davenport mailed Hodson another long list of approvals and declinations of her ideas. Typical was his review of her draft letters, which Davenport had to approve. "I think the draft of Letter #2 is to be preferred to #1. Of course, it is much weaker than #1 but may serve as a penultimate. Something like your draft #1 might serve as an ultimate and then we can prepare an ultissimum, if that has no effect."[82] Davenport was accustomed to treating Hodson like a secretary, not a general secretary.

A month later, however, Davenport cancelled his trip altogether, saying he was suddenly in poor health and in need of a long rest. It was after this unexpected cancellation that Hodson finally turned to the Germans for information, in July of 1930, since German eugenicists would now be running the conference in Davenport's absence.[83]

That summer Britain first confronted American-style eugenics. Dr. Lionel L. Westrope was the doctor at the High Teams institution located in London's Gateshead district. He impressed Ministry of Health officials as "an enthusiast on the question of the sterilisation of the unfit and was inclined to mix up the therapeutic and sociological aspects of these cases."

Around June of 1930, supervisors discovered that Westrope was castrating young men. He admitted to having performed two in May of 1930, and a third on an unknown date.[84]

William George Wilson had been admitted as a diagnosed imbecile to the Gateshead mental ward about a decade earlier. Later, Wilson was described as "thoroughly degenerate ... extremely dirty and absolutely indifferent as to his personal appearance." Wilson also masturbated excessively, so much so "that there was actually hemorrhage from the penis." His mother reportedly caught the boy masturbating once and asked for help. Westrope castrated Wilson, then twenty-two years old, and reported, "the improvement was wonderful. Not only did the patient cease to masturbate, but, three months after the operation, he began to take some interest in his appearance...." But a year later Wilson died, supposedly of pneumonia.[85]

Nonetheless, Westrope was encouraged. In February of 1930, an eight-year-old boy named Henry Lawton was brought to Gateshead for being an "epileptic imbecile, unable to talk" and for suffering what Westrope called "fits." After admission, Henry was discovered writhing on his stomach, as though in a "sexual connection." When staffers rolled him over they found his penis to be erect. No determination was made as to whether the writhing was a "fit," an epileptic seizure or just ordinary prepubescent activity. On May 7, 1930, the boy was castrated.[86]

Five days later, fifteen-year-old Richard Pegram was arrested for allegedly sexually assaulting a woman. The record stated that Pegram "pushed up against her and said that he was 'horny.'" When asked to explain, Pegram flippantly replied, "Well, I had the 'horn.'" Police immediately brought the young man to Gateshead. Within days, he too was castrated.[87]

When the Ministry of Health learned of Westrope's illegal surgeries, a flurry of anxious memos and reports were exchanged as astonished officials tried to find some way to justify what they themselves knew was criminal castration. Westrope claimed he had parental consent. Officials bluntly rejected this assertion. One wrote, "Consent or no consent, the surgeon is guilty of unlawful wounding ... and in the case of [the] death, manslaughter." As officials passed the reports back and forth, some of them scribbled in the margins that two of the boys had not even been certified as mentally defective. One wrote, "This was NOT a case of certified mental defect." Another penned in the margin, "Not a certified case." Hence there was no possibility of arguing therapeutic necessity.[88]

Westrope himself simply claimed that it had not occurred to him that the procedure might be illegal. But in fact anyone associated with the surgeries might have been held civilly or criminally responsible, including Board of Control officials themselves. The Board of Control had custody over the boys. On August 1, 1930, facing the prospect of criminal prosecution, Board of Control Chairman Sir Lawrence Brock wrote a letter to a Ministry of Health attorney providing all the details and admitting that the boys had been castrated "as the result of sexual misbehavior." Brock then added, "If sterilization is to be carried out by Medical Officers of Poor Law Institutions it would in any case seem to be preferable to adopt the American method [of vasectomy] and not resort to the extremer course of actual castration."[89]

The matter was hushed up as some sort of therapeutic necessity or medical oversight. Westrope was not prosecuted and remained at his post at Gateshead. He was, however, required to submit an immediate letter of apology, and to promise not to do it again. On October 14, Westrope, writing on Gateshead Borough letterhead, penned a short note to Ministry of Health officials: "I now hereby give an undertaking, that I will not perform the operation again, until such time as the operation may be legalized." Two days later, a supervising doctor came by and asked Westrope to sign the note, which he did. Nine years later, Westrope was still presiding at Gateshead, and even sat as a merit judge in awarding gold medals to ambulance crews who distinguished themselves by promptly delivering patients to the institutions.[90]

The campaign to legalize sterilization continued in 1930, Westrope's misconduct notwithstanding. However, despite efforts to convince policymakers, the British people simply could not stomach the notion. Labor was convinced that the plan was aimed almost exclusively at the poor. Catholics believed that eugenics, breeding and sterilization were all offenses against God and the Church, and indeed in some cases a form of murder.[91]

With a sense that eugenic marriage restrictions and annulments, as well as sterilization, would soon be enacted in Britain, the Vatican spoke out. On December 31, 1930, Pope Pius XI issued a wide-ranging encyclical on marriage; in it he condemned eugenics and its fraudulent science. "That pernicious practice must be condemned," he wrote, "which closely touches upon the natural right of man to enter matrimony but affects also in a real way the welfare of the offspring. For there are some who over solicitous for the cause of eugenics . . . put eugenics before aims of a higher order, and by public authority wish to prevent from marrying all those whom, even though naturally fit for marriage, they consider, according to the norms and conjectures of their investigations, would, through hereditary transmission, bring forth defective offspring. And more, they wish to legislate to deprive these of that natural faculty by medical action [sterilization] despite their unwillingness. . . . [92]

"Public magistrates have no direct power over the bodies of their subjects; therefore, where no crime has taken place and there is no cause present for grave punishment, they can never directly harm, or tamper with the integrity of the body, either for the reasons of eugenics or for any other reason."[93]

Making clear that the destruction of a child for any "eugenic 'indication'" was nothing less than murder, the encyclical went on to quote Exodus: "Thou shalt not kill."[94]

Disregarding religious and popular sentiment, the society pressed on. Articles that they promoted continued to warn British readers of the dangers posed by family lines such as America's Jukes; readers were also reminded of the success California was having with sterilization. But Labor and Catholics would not budge. Nor would their representatives in Parliament.[95]

Two more papal decrees, issued in March of 1931, denounced both positive and negative eugenics. On July 21, 1931, A. G. Church exercised his right under the House of Commons' Ten Minute Rule to put the issue to a test. Under the Ten Minute Rule, debate would be massively curtailed. Church was a member of the Eugenics Society's Committee on Voluntary Sterilization, and in his ten minutes he stressed the strictly "voluntary" nature of his measure. But then he let it slip. He admitted that, indeed, the voluntary proposal offered that day was only the beginning. Ultimately, eugenicists favored compulsory sterilization.[96]

Sterilization opponents in the House of Commons "crushed" Church, as it was later characterized. In the defeat that followed, Church was voted down 167 to 89. He was not permitted to introduce his legislation. Society leaders were forced to admit that it was Labor's opposition and the Church's encyclicals that finally defeated their efforts.[97]

Still unwilling to give up, within a few weeks the society began inviting more experts to form yet another special commission. Constantly trumpeting the successes in California and other American states, the society convinced Minister of Health Chamberlain to convene a special inquiry to investigate the Social Problem Group and how to stop its proliferation. The man selected to lead the commission was Board of Control Chairman Brock, the same man who had presided over the Gateshead debacle.[98]

The Brock Commission convened in June of 1932. One of its first acts was to ask the British Embassy in Washington and its consulates throughout the nation to compile state-by-state figures on the numbers of men and women sterilized in America. British consular officials launched a nation-wide fact-finding mission to compile America's legislation precedents and justifications. Numerous state officials, from Virginia to California, assisted consular officials. Reams of interlocutory reports produced by the Brock Commission advocated using American eugenic sterilization as a model, and in 1934 the commission formally recommended that Britain adopt similar policies. Section 86 of the recommendations, entitled "The Problem of the Carrier," endorsed the idea that the greatest eugenic threat to society was the person who seemed "normal" but was actually a carrier of mental defect. "It is clear that the carrier is the crux of the problem," the Brock Report concluded, bemoaning that science had not yet found a means of identifying such people with certainty.[99]

But for opponents, the Brock Report only served to confirm their rejection of sterilization in Britain. The Trades Union Congress condemned the idea, insisting that protracted unemployment might itself be justification for being classed "unfit." In plain words, Labor argued that such applications of eugenics could lead to "extermination." The labor congress's resolution declared: "It is quite within the bounds of human possibility that those who want the modern industrial evils under the capitalist system to continue, may see in sterilization an expedient, degrading though it may be, to exterminate the victims of the capitalist system."[100]

No action was ever taken on Brock's recommendations. By this time it was 1934, and the Nazis had implemented their own eugenic sterilization regime. In Germany, the weak, political dissidents, and Jews were being sterilized by the tens of thousands.[101] The similarities were obvious to the British public.

CHAPTER 12

Eugenic Imperialism

American eugenicists saw mankind as a biological cesspool. After purifying America from within, and preventing defective strains from reaching U.S. shores, they planned to eliminate undesirables from the rest of the planet. In 1911, the Eugenics Section of the American Breeders Association, in conjunction with the Carnegie Institution, began work upon its *Report of the Committee to Study and to Report on the Best Practical Means for Cutting Off the Defective Germ-Plasm in the Human Population.* The last of eighteen points was entitled "International Co-operation." Its intent was unmistakable: the ERO would undertake studies "looking toward the possible application of the sterilization of defectives in foreign countries, together with records of any such operations...." The American eugenics movement intended to turn its sights on "the extent and nature of the problem of the socially inadequate in foreign countries."[1] This would be accomplished by incessant international congresses, federations and scientific exchanges.

Global eugenics began in 1912 with the First International Congress of Eugenics in London. At that conference, the dominant American contingent presented its report on eliminating all social inadequates worldwide. Their blueprint for world eugenic action was overwhelmingly accepted, so much so that after the congress the Carnegie Institution published the study as a special two-part bulletin.[2]

International cooperation soon began to coalesce. That first congress welcomed delegations from many countries, but five in particular sent major consultative committees: the United States, Germany, Belgium, Italy and France. During the congress, these few leaders constituted themselves as a so-called International Eugenics Committee. This new body first met a year later. On August 4, 1913, prominent eugenic leaders from the United States, England, Belgium, Denmark, France, Germany, Italy and Norway converged on Paris. This new international eugenics oversight

235

committee would function under various names and in various member configurations as the supreme international eugenics agency, deciding when and where congresses would be held, which national committees and institutions would be recognized, and which eugenic policies would be pursued. The dozen or so men scheduled a second planning session for one year later, August 15, 1914, in Belgium. They also scheduled the Second International Congress of Eugenics, which would be open to delegates from all nations and held two years later, in 1915, in New York.[3]

But in August of 1914, Germany invaded Belgium.

A continent-wide war ignited before Europe's eyes. The Belgian planning session was cancelled, and the Second International Congress of Eugenics was postponed. While Europe fought, and indeed even after the United States entered the war, America continued its domestic eugenic program and held its place as the world leader in eugenic research, theory and activism.[4]

When the war ended four years later, international eugenics reorganized, with America retaining its leadership. The Second International Congress of Eugenics was rescheduled for September 1921, still in New York, under the auspices of the Washington-based National Research Council, the administrative arm of America's prestigious, Congressionally-chartered National Academy of Sciences. The National Academy of Sciences functioned as a way of uniting America's disparate scientific establishments. As it had for the first congress, the State Department mailed the invitations around the world. Although the National Research Council was the official authorizing body, Davenport wrote his colleagues that it was "up to the New York group to put this Congress through."[5]

The "New York group" was led by Laughlin, Mrs. Harriman and Madison Grant, author of *The Passing of the Great Race*. In addition to being among the world's leading raceologists, Grant was a trustee for the American Museum of Natural History. The museum became the titular sponsor of the second congress. The museum's premises were used for the congress's meetings and exhibits, its staff helped with the details, and its president, Henry Osborn, a eugenicist himself, was named president of the international gathering. The museum's name was prominently displayed on the published proceedings, as though the congress were just another museum function.[6] All of this imbued the event with a distinctly evolutionary and anthropological quality. This was exactly the intent of congress organizers. They wanted the event to be seen as a milestone in the natural history of the human species.

The second congress was rich with typical raceological dogma and dominated by American biological precepts. Alexander Graham Bell assumed the honorary presidency. The proceedings were divided into four sections: comparative heredity, the human family, racial differences and "Eugenics and the State." Delegates from every continent attended to share eugenic principles and to form legislative game plans they could take back home. Osborn's opening address represented a challenge from America. "In certain parts of Europe," he set forth, "the worst elements of society have gained the ascendancy and threaten the destruction of the best." He recognized that "To each of the countries of the world, racial betterment presents a different aspect.... Let each... consider its own problems...." But in the final analysis it came down to one mandate: "As science has enlightened government in the prevention and spread of disease, it must also enlighten government in the prevention of the spread and multiplication of worthless members of society...."[7]

Osborn also repeated the standard eugenic idea: "The true spirit of American democracy that *all men are born with equal rights and duties* has been confused with the political sophistry that *all men are born with equal character and ability to govern themselves....*"[8]

Not only was the rhetoric American, but so was the science. Out of fifty-three scientific papers, all but twelve were produced by American eugenicists on American issues, all conforming to the Carnegie Institution's sociopolitical strategies. Topics included Indiana's Tribe of Ishmael, Kentucky's mountain people and Lucien Howe's proposals on hereditary blindness.[9]

Some European eugenicists complained about America's domination of the global congress. Sweden's Hermann Lundborg, for example, railed to Davenport in a rambling handwritten missive that America was trying to hijack the worldwide movement. "I have been hoodwinked.... By what right do you in America usurp the words Second International, when the Congress is not international. It is an injustice which not only I, but I believe the majority of my [Swedish] section do not approve of."[10]

Such protests did not deter Davenport and his colleagues. Indeed, in a special presentation on the essence of eugenic research, Davenport explained his dedication. "Why do we investigate?" he asked. "Alas! We have now too little precise knowledge in any field of eugenics. We can command respect for our eugenic conclusions only as our findings are based on rigid proof...." Davenport reminded the delegates that wealthy American benefactors had made the critical difference between mere ideas and hard data. "It is largely due to the extraordinary vision of Mrs. E. H. Harriman, the founder of the Eugenics Record Office, that in this country, eugenics is more a subject of research than [mere] propaganda."[11]

Money made the difference for the international convention as well. Mrs. Harriman donated an extra $2,500 to fund the more than 120 exhibits erected throughout the museum. These included a prominent exhibit on sterilization statutes in the United States. The Carnegie Institution extended a special grant of $2,000 to defray travel expenses for several of the key European speakers, and to cover general expenses for the delegates. Other wealthy eugenicists contributed significant sums and were named patrons of the gathering. They included sanitarium owner John Kellogg, working through his Race Betterment Foundation, and YMCA benefactor and prominent political contributor Cleveland H. Dodge.[12]

In recalling the congress some weeks later for the Indiana Academy of Science, Carnegie researcher Arthur Estabrook quoted Osborn: "That all men are born with equal rights and duties has been confused with the political sophistry that all men are born with equal character and ability to govern themselves...."[13]

During the congress Davenport orchestrated the renaming and broadening of the International Eugenics Committee into a Permanent International Commission on Eugenics. This renamed entity would sanction all eugenic organizations in "cooperating" member countries, which now included Belgium, Czechoslovakia, Denmark, France, Great Britain, Italy, the Netherlands, Norway, Sweden, Argentina, Brazil, Canada, Colombia, Cuba, Mexico, Venezuela, Australia, New Zealand and the United States. Germany was not included because it refused to sit on the same panel with its World War I enemies Belgium and France. Germany was also struggling under the punitive terms of the Treaty of Versailles, which made international eugenic cooperation difficult.[14]

Multinational eugenics gathered momentum during the next two years. In October of 1922, the Permanent International Commission assembled in Brussels. The meeting was once again steered by Davenport and his circle. Representatives from Belgium, Denmark, Great Britain, France, the Netherlands and Norway began coordinating their efforts. The commission resolved to learn more about eugenic campaigns in India and Japan, and also voted unanimously to invite Germany back into its ranks.[15]

In September of 1923, Laughlin kicked off his first European immigration tour by attending the Permanent International Commission meeting in Lund, Sweden. Preparations for this meeting prevented Laughlin from sailing to Europe in July with Secretary of Labor James Davis. At the Lund meeting, Laughlin advanced most of the motions that the commission adopted.[16]

The 1923 meeting proved a watershed event for the movement. The group ratified the four-point "Ultimate Program" devised by the American Eugenics Society, calling for each nation to undertake research, education, administrative measures and "conservative legislation" within its borders. And although it welcomed news of their efforts, the commission stopped short of extending membership to Japan and India.[17]

To keep the eugenic directorate truly elite, commission rules permitted no more than three representatives of each cooperating country to be empanelled. Davenport and Laughlin sat at the apex of this group. All commission members were dedicated to the American-espoused belief in Nordic supremacy, a sentiment which was also growing in Germany. Yet Germany was still not a full participant on the commission. Although Germany was willing to rejoin the group, German race scientists told commissioners that Germany still "could not cooperate with representatives of certain nations." In personal correspondence, German eugenicists specified whom they meant: the French.[18] Commission leaders said they would wait.

During the next two years, with Germany still in the periphery, Davenport and Laughlin were able to extend U.S. domination of the commission's scope, science, and political agenda. Resolutions were binding on the dozen or so members, committing them to pursue the agreed-upon legislative and scientific strategies. Because of this, policy developed on Long Island leapt across the ocean directly into the capitals of other nations.[19]

For example, in 1925 Davenport introduced a resolution based on Laughlin's strategy of investigating immigrant families and screening them for eugenical fitness. Likening human beings to farm animals, Davenport's resolution read: "Whereas every nation has a right to select those who shall be included in its body politic, and whereas some knowledge of both family history and past personal performance are as essential a part of the information about a human immigrant and potential parent, as about an imported horse or cow, therefore [be it] resolved that each immigrant-receiving country may properly enquire into the family and personal history of each immigrant."[20] Commission members, working through scientific and intellectual societies back home, then pressured for changes in immigration regulations along these lines.

Worldwide uniformity was important to Davenport. To push usage of the ERO's standard family pedigree form in all countries, Davenport issued a message: "Members are reminded that a standardized form of pedigree was worked out by the Federation and has been widely published in most countries." He also asked all cooperating national societies to lobby for national registration and census schemes similar to models already developed by his colleagues in Norway and Holland. Davenport tempered his worldwide eugenic mandates by assuring he would "avoid anything which might savour of interference in national affairs," adding, "nevertheless, it is clear that in certain directions, such work might be usefully undertaken."[21]

By 1925, the commission was comprised not just of individuals, but also of constituent eugenic societies and institutions. Hence it was time to adopt another new name, the International Federation of Eugenic Organizations (IFEO). The new name was meant to further extend the organization's scope, and also reflected Davenport and Laughlin's desire to energize and standardize the movements in many countries. Ultimately, uniformity of eugenic action was written into IFEO membership rules. As president of the IFEO, Davenport issued a memorandum to member societies restating the federation's goals: "To endeavor to secure some measure of uniformity in the methods of research, and also sufficient uniformity in the form of presentation of results to make international work of worldwide use. To endeavor to promote measures tending to eugenic progress, whether international or national, on comparable lines."[22]

Even though Davenport was an influential steering force, federation members were independent thinkers. They advanced their own substantial legislative and scientific contributions for consideration by the federation. The Nordic countries of Scandinavia were especially active in this regard. Indeed, Europe's northwestern nations were the most receptive to eugenics. Predominantly Catholic countries were the most resistant. Whether resistant or receptive, however, each country's eugenics movement developed its own literature in its own language, its own racial and genetic societies, its own raceological personalities and its own homegrown agenda. Nonetheless, the movement's fundamental principles were American and shepherded by Americans. Many foreign eugenicists traveled to America for training at Cold Spring Harbor and to attend meetings, congresses and conferences. As the epicenter of eugenics, and by virtue of its domination of the IFEO, American eugenic imperialism was able to take root throughout Europe and indeed the world.[23]

Belgium's Société Belge d'Eugénique was organized in 1919. The Belgian Eugenics Society announced in *Eugenical News* that it was "fully awake to the needs of the time in connection with preservation of the race.

Its leaders realize that the safeguarding of public health through hygienic measures is not sufficient, but that due attention must be paid to the prevention of the transmission of hereditary traits that would be injurious to the race." The new society's nine sections included ones for social hygiene, documentation and legislation. Within two years, the Belgian Eugenics Society launched its own journal, which the ERO at Cold Spring Harbor quickly declared to be of "high order."[24]

Dr. Albert Govaerts led the Belgian movement. He was allied with Laughlin from the beginning. After the second international congress in New York in 1921, Govaerts stayed on and traveled to Cold Spring Harbor for a term of study, which was funded by a fellowship from America's postwar Commission for Relief in Belgium Educational Foundation.[25]

Govaerts's work at the ERO concentrated on hereditary tuberculosis studies, and his research was published in the *American Review of Tuberculosis* in 1922. After Govaerts returned to Belgium, his original tables and calculations remained on file at the ERO. By early 1922, Govaerts's Belgian Eugenics Society had installed eugenic lectures and courses at the University of Brussels. They also succeeded in garnering recognition of the budding science from the Belgian government. Later in 1922, a government-supported National Office of Eugenics opened in Brussels at the distinguished Solvay Institute. The National Office of Eugenics trained eugenic field workers and operated as a Belgian version of the ERO.[26]

Laughlin and Govaerts often worked as a team. Laughlin used Govaerts's office as a headquarters during his 1923 sojourn throughout Europe as a Congressional immigration agent, and he even stayed in his home when visiting Brussels. Eugenicists never secured sterilization laws in Belgium, but Govaerts boasted of his lobbying efforts for a "eugenical prenuptial examination" to be required of all marriage applicants. *Eugenical News* reported that Govaerts "very graciously states that Belgian eugenicists are deeply indebted to the Eugenics Record Office for the service rendered in aiding the Belgian society to establish its new office."[27]

In Canada, eugenic passions became inflamed over many issues, including the birth rate of French Canadians. But perhaps no debate was more heated than the one prompted by problems associated with immigrant groups. Hard-working Asian and European immigrants flowed into Canada throughout the 1890s as the country's infrastructure expanded. In 1905, Ontario carried out its first census of the feebleminded. Shortly after Indiana passed its 1907 sterilization law, Ontario's Provincial Inspector of Hospitals and Public Charities argued that Rentoul's concepts could end the hereditary production of tramps, prostitutes and other immoral characters. Another Canadian physician pointed to the example of a Chicago doctor who advocated asexualization.[28]

By 1910, Canada's British-American Medical Association was studying the sterilization laws in California and Indiana. Similar legislation proposed in Ontario and Manitoba did not succeed. But the movement for human breeding and sterilization of the unfit continued. The first Canadian sterilization law was passed by Alberta's legislature in 1928. Alberta's Sexual Sterilization Act targeted mental defectives who "risk... multiplication of [their] evil by transmission of [their] disability to progeny." Alberta's Eugenics Board authorized the sterilization of four hundred people in its first nine years. In 1937, certain safeguards were eliminated by the new Social Credit government, and the door was opened to forced sterilization. Until the law was repealed in 1972, of some 4,700 applications, 2,822 surgeries were actually authorized. The majority of Alberta's sterilized were young women under the age of twenty-five, many under the age of sixteen. Following the example of America's hunt for mongrels, Alberta disproportionately sterilized French-Canadian Catholics, Indians and Métis (individuals of mixed French-Canadian and Indian descent). Indians and Métis constituted just 2.5 percent of Canada's population, but in later years represented 25 percent of Alberta's sterilized.[29]

British Columbia passed its own law in 1933, creating a three-person Eugenics Board comprised of a judge, a psychiatrist and a social worker. Because records were lost or destroyed, no one will ever know exactly how many were sterilized in British Columbia, although one study discussed the fates of over fifty women who had undergone the operation.[30]

In Switzerland, the eminent psychiatrist and sexologist Dr. Auguste Forel was a leading disciple of eugenics beginning in 1910. He was also a proponent of U.S.-style sterilization laws. The wealthy industrialist Julius Klaus was another early advocate, endorsing eugenic registers to identify Switzerland's unfit. When he died in 1920, Klaus bequeathed more than a million Swiss francs, or about $4.4 million in modern money, to establish a fund for Swiss eugenic investigations and related advocacy. Klaus's will specifically forbade using the fund for charitable works to "ameliorate the condition of physical and mental defectives."[31]

Swiss eugenic scientists were suddenly endowed. The anthropologist Otto Schlaginhaufen became director of the Zurich-based Julius Klaus Foundation for Heredity Research, Social Anthropology and Racial Hygiene as well as the Institution for Race Biology. These organizations were dedicated to "the promotion of all scientifically based efforts, whose ultimate goal is . . . to improve the white race." In 1923, Schlaginhaufen and Forel, now fully funded, ascended to the Permanent International Eugenics Commission.[32]

Swiss eugenics focused on the exclusion of certain ethnic groups, as well as Forel's notion of sexology, that is, the study of sexual behavior, especially as it related to women. Forel believed women wished to be and should be "conquered, mastered and subjugated" to fulfill their national reproductive duty. In 1928, Switzerland's first sterilization law was passed in Canton Vaud, where Forel practiced. It targeted a vaguely-defined "unfit." Only Vaud passed such a law, but physicians across the country performed sterilizations for both medical and eugenical reasons. Although the extent of Swiss sterilizations remains unknown, one scholar ascertained that some 90 percent of the operations were conducted on women.[33]

In Denmark, eugenics was organized by two of Davenport's earliest confederates, August Wimmer and Sören Hansen. Wimmer was a psychiatrist at the University of Copenhagen, and Hansen was president of the Danish Anthropological Committee. As Nordic raceologists seeking to stamp out defective strains within an already eugenically elite country, their affiliation with Davenport was natural. One Danish physician even traveled to the Vineland Training School in New Jersey to study under H. H. Goddard, whose texts on the Kallikaks and revision of the Binet-Simon test became standard in Danish eugenical publications. Although resistant at first, in 1912 the government launched a massive eugenical registration of deaf-mutes, the feebleminded and other defectives. It was not until a decade later that the first eugenic marriage restriction law was adopted. So-called "therapeutic sterilization" was common, but compulsory sterilization would not be legalized until 1929.[34]

A government commission reexamined the sterilization issue in 1926, looking to America for guidance. In November of 1927, Laughlin arranged for his lengthy legislative guide on sterilization to be sent by Chicago judge Harry Olson directly to a member of the Danish sterilization commission. In 1929, Hansen proudly reported to *Eugenical News* that his country had finally adopted what he termed, "the first 'modern' eugenical sterilization law to be enacted in Europe."[35]

Shortly after the passage of Denmark's legislation, the Rockefeller Foundation began supporting eugenic research in that country. Denmark's leading eugenic scientist, Dr. Tage Kemp, received much of the financial support. The first grants were awarded in 1930 for blood group research.

The next year Kemp received a special Rockefeller fellowship to continue his research. In 1932, Kemp traveled to Cold Spring Harbor for further study. He wanted eugenic and genetic research to achieve greater scientific and medical exactitude. "I was notably impressed by the importance of the careful execution of the several observations," he wrote Rockefeller officials, adding, "these ought as far as possible to be carried out and re-examined (after-examined) by an investigator with medical education." Rockefeller officials agreed, granting Kemp a second fellowship in 1934. They would continue to fund race biology and human genetics in Denmark throughout the 1930s.[36]

Kemp was among the new breed of eugenic geneticists the Rockefeller Foundation was cultivating to lift eugenics out of mere racial rhetoric and into the realm of unemotional science. A Rockefeller report explained their confidence in Kemp. "Race biology today suffers immensely from its mixture with political dogmas and drives. Dr. Kemp, through his personality and training, is as free from these as possible."[37]

In Norway, the raceologist Jon Alfred Mjøen endorsed American eugenics from the outset. He propounded his theories from a well-equipped animal and human measurement lab as well as a grand personal library, crammed floor to ceiling with books and files. At the second congress in New York, Mjøen suggested the resolution that ultimately led to the formation of the American Eugenics Society. In his opening address to the convention, Osborn singled out Mjøen and Lundborg. "It is largely through the active efforts of leaders like Mjøen and Lundborg," he acknowledged, "that there is a new appreciation of the spiritual, moral and physical value of the Nordic race."[38]

Davenport toured eugenic facilities in Norway, and Mjøen visited New York on several occasions. Mjøen was also a frequent contributor to, and topic of, *Eugenical News*. The dapper Norwegian was often pictured arm-in-arm with leading American eugenicists, such as Leon Whitney. Norway passed its sterilization law in 1934, and in 1977 amended it to become a mostly voluntary measure. Some 41,000 operations were performed, about 75 percent of them on women.[39]

The Swedish government's State Institute of Race-Biology opened its doors in 1922. It was an entire school dedicated to eugenic thought, and it would leave a multilayered movement in its wake. Sweden alternately shared and coordinated its programs with the IFEO. Sweden's first sterilization law was passed in 1934. It began by sterilizing those who had "mental illness, feeble-mindedness, or other mental defects" and eventually widened its scope to include those with "an anti-social way of life." Eventually, some 63,000 government-approved sterilizations were undertaken on a range of "unfit" individuals, mainly women. In some years women represented a mere 63 percent of those sterilized, but in most years the percentage who were women exceeded 90 percent.[40]

American influence rolled across the Continent. Finland, Hungary, France, Romania, Italy and other European nations developed American-style eugenic movements that echoed the agenda and methodology of the font at Cold Spring Harbor. Soon the European movements learned to cloak their work in more medically and scientifically refined approaches, and many were eventually funded by such philanthropic sponsors as the Rockefeller Foundation and the Carnegie Institution. In the late twenties and thirties, these foundations liberally granted money to studies that adhered to a more polished clinical regimen.[41]

Throughout the twenties and thirties, America's views were celebrated at the numerous international gatherings held in America, such as the Third International Congress of Eugenics, which in 1932 was hosted once again at New York City's American Museum of Natural History. Theory became doctrine when proliferated in the many eugenic newsletters, books, and journal articles published by the American movement. America's most venerable universities and academic authorities also reinforced the view that eugenic science was legitimate.[42]

Some nations, such as France and Italy, rejected their native eugenic movements. Some, such as Holland, only enacted broadly-based registration laws. Some, such as Lithuania and Brazil, enacted eugenic marriage laws. Some, such as Finland, went as far as forced sterilization.[43]

One nation, Germany, would go further than anyone could imagine.

CHAPTER 13

Eugenicide

Murder was always an option.

Point eight of the *Preliminary Report of the Committee of the Eugenic Section of the American Breeders Association to Study and to Report on the Best Practical Means for Cutting Off the Defective Germ-Plasm in the Human Population* specified euthanasia as a possibility to be considered.[1] Of course euthanasia was merely a euphemism—actually a misnomer. Eugenicists did not see euthanasia as a "merciful killing" of those in pain, but rather a "painless killing" of people deemed unworthy of life. The method most whispered about, and publicly denied, but never out of mind, was a "lethal chamber."

The lethal chamber first emerged in Britain during the Victorian era as a humane means of killing stray dogs and cats. Dr. Benjamin Ward Richardson patented a "Lethal Chamber for the Painless Extinction of Lower Animal Life" in the 1880s. Richardson's original blueprints show a large wood- and glass-paneled chamber big enough for a Saint Bernard or several smaller dogs, serviced by a tall slender tank for carbonic acid gas, and a heating apparatus. In 1884 the Battersea Dogs Home in London became one of the first institutions to install the device, and used it continuously with "perfect success" according to a sales proposal at the time. By the turn of the century other charitable animal institutions in England and other European countries were also using the chamber.[2]

This solution for unwanted pets was almost immediately contemplated as a solution for unwanted humans—criminals, the feebleminded and other misfits. The concept of the lethal chamber was common vernacular by the turn of the century. When mentioned, it needed no explanation; everyone understood what it meant.

In 1895, the British novelist Robert Chambers penned his vision of a horrifying world twenty-five years into the future. He wrote of a New York where the elevated trains were dismantled and "the first Government Lethal Chamber was opened on Washington Square." No explanation of

"Government Lethal Chamber" was offered—or necessary. Indeed, the idea of gassing the unwanted became a topic of contemporary chitchat. In 1901, the British author Arnold White, writing in *Efficiency and Empire*, chastised "flippant people of lazy mind [who] talk lightly of the 'lethal chamber'.... "[3]

In 1905, the British eugenicist and birth control advocate H. G. Wells published *A Modern Utopia*. "There would be no killing, no lethal chambers," he wrote. Another birth control advocate, the socialist writer Eden Paul, differed with Wells and declared that society must protect itself from "begetters of anti-social stocks which would injure generations to come. If it [society] reject the lethal chamber, what other alternative can the socialist state devise?"[4]

The British eugenicist Robert Rentoul's 1906 book, *Race Culture; Or, Race Suicide?*, included a long section entitled "The Murder of Degenerates." In it he routinely referred to Dr. D. F. Smith's earlier suggestion that those found guilty of homicide be executed in a "lethal chamber" rather than by hanging. He then cited a new novel whose character "advocate[d] the doctrine of 'euthanasia' for those suffering from incurable physical diseases." Rentoul admitted he had received many letters in support of killing the unfit, but he rejected them as too cruel, explaining, "These [suggestions] seem to fail to recognize that the killing off of few hundreds of lunatics, idiots, etc., would not tend to effect a cure."[5]

The debate raged among British eugenicists, provoking damnation in the press. In 1910, the eugenic extremist George Bernard Shaw lectured at London's Eugenics Education Society about mass murder in lethal chambers. Shaw proclaimed, "A part of eugenic politics would finally land us in an extensive use of the lethal chamber. A great many people would have to be put out of existence, simply because it wastes other people's time to look after them." Several British newspapers excoriated Shaw and eugenics under such headlines as "Lethal Chamber Essential to Eugenics."[6]

One opponent of eugenics condemned "much wild and absurd talk about lethal chambers.... " But in another article a eugenicist writing under the pseudonym of Vanoc argued that eugenics was needed precisely because systematic use of lethal chambers was unlikely. "I admit the word 'Eugenics' is repellent, but the thing is essential to our existence.... It is also an error to believe than the plans and specifications for County Council lethal-chambers have yet been prepared."[7]

The Eugenics Education Society in London tried to dispel all "dark mutterings regarding 'lethal chambers.'" Its key activist Saleeby insisted,

"We need mention, only to condemn, suggestions for 'painless extinction,' lethal chambers of carbonic acid, and so forth. As I incessantly have to repeat, eugenics has nothing to do with killing...." Saleeby returned to this time and again. When lecturing in Battle Creek, Michigan, at the First National Conference on Race Betterment in 1914, he emphasized a vigorous rejection of "the lethal chamber, the permission of infant mortality, interference with [pre]-natal life, and all other synonyms for murder."[8]

But many British eugenicists clung to the idea. Arthur F. Tredgold was a leading expert on mental deficiency and one of the earliest members of the Eugenics Education Society; his academic credentials eventually won him a seat on the Brock Commission on Mental Deficiency. Tredgold's landmark *Textbook on Mental Deficiency*, first published in 1908, completely avoided discussion of the lethal chamber. But three subsequent editions published over the next fourteen years did discuss it, with each revision displaying greater acceptance of the idea. In those editions Tredgold equivocated: "We may dismiss the suggestion of a 'lethal chamber.' I do not say that society, in self-defense, would be unjustified in adopting such a method of ridding itself of its anti-social constituents. There is much to be said for and against the proposal...." By the sixth edition, Tredgold had modified the paragraph to read: "The suggestion [of the lethal chamber] is a logical one.... It is probable that the community will eventually, in self-defense, have to consider this question seriously." The next two editions edged into outright, if limited, endorsement. While qualifying that morons need not be put to death, Tredgold concluded that for some 80,000 imbeciles and idiots in Britain, "it would be an economical and humane procedure were their existence to be painlessly terminated.... The time has come when euthanasia should be permitted...."[9]

Leaders of the American eugenic establishment also debated lethal chambers and other means of euthanasia. But in America, while the debate began as an argument about death with dignity for the terminally ill or those in excruciating pain, it soon became a palatable eugenic solution. In 1900, the physician W. Duncan McKim published *Heredity and Human Progress*, asserting, "Heredity is the fundamental cause of human wretchedness.... The surest, the simplest, the kindest, and most humane means for preventing reproduction among those whom we deem unworthy of this high privilege [reproduction], is a gentle, painless death." He added, "In carbonic acid gas, we have an agent which would instantaneously fulfill the need."[10]

By 1903, a committee of the National Conference on Charities and Correction conceded that it was as yet undecided whether "science may conquer sentiment" and ultimately elect to systematically kill the unfit. In 1904, the superintendent of New Jersey's Vineland Training School, E. R. Johnstone, raised the issue during his presidential address to the Association of Medical Officers of American Institutions for Idiotic and Feebleminded Persons. "Many plans for the elimination [of the feeble-minded] have been proposed," he said, referred to numerous recently published suggestions of a "painless death." That same year, the notion of executing habitual criminals and the incurably insane was offered to the National Prison Association.[11]

Some U.S. lawmakers considered similar ideas. Two years later, the Ohio legislature considered a bill empowering physicians to chloroform permanently diseased and mentally incapacitated persons. In reporting this, Rentoul told his British colleagues that it was Ohio's attempt to "murder certain persons suffering from incurable disease." Iowa considered a similar measure.[12]

By 1910, the idea of sending the unfit into lethal chambers was regularly bandied about in American sociological and eugenic circles, causing a debate no less strident than the one in England. In 1911, E. B. Sherlock's book, *The Feebleminded: a guide to study and practice*, acknowledged that "glib suggestions of the erection of lethal chambers are common enough...." Like others, he rejected execution in favor of eugenic termination of bloodlines. "Apart from the difficulty that the provision of lethal chambers is impracticable in the existing state law...," he continued, "the removal of them [the feebleminded] would do practically nothing toward solving the chief problem with the mentally defective set..., the persistence of the obnoxious stock."[13]

But other eugenicists were more amenable to the idea. The psychologist and eugenicist Henry H. Goddard seemed to almost express regret that such proposals had not already been implemented. In his famous study, *The Kallikak Family*, Goddard commented, "For the low-grade idiot, the loathsome unfortunate that may be seen in our institutions, some have proposed the lethal chamber. But humanity is steadily tending away from the possibility of that method, and there is no probability that it will ever be practiced." Goddard pointed to familywide castration, sterilization and segregation as better solutions because they would address the genetic source.[14]

In 1912, Laughlin and others at the Eugenics Section of the American Breeders Association considered euthanasia as the eighth of nine options. Their final report, published by the Carnegie Institution as a two-volume bulletin, enumerated the "Suggested Remedies" and equivocated on euthanasia. Point eight cited the example of ancient Sparta, fabled for drowning its weak young boys in a river or letting them die of exposure to ensure a race of warriors. Mixing condemnation with admiration, the Carnegie report declared, "However much we deprecate Spartan ideals and her means of advancing them, we must admire her courage in so rigorously applying so practical a system of selection.... Sparta left but little besides tales of personal valor to enhance the world's culture. With euthanasia, as in the case of polygamy, an effective eugenical agency would be purchased at altogether too dear a moral price."[15]

William Robinson, a New York urologist, published widely on the topic of birth control and eugenics. In Robinson's book, *Eugenics, Marriage and Birth Control (Practical Eugenics)*, he advocated gassing the children of the unfit. In plain words, Robinson insisted: "The best thing would be to gently chloroform these children or to give them a dose of potassium cyanide." Margaret Sanger was well aware that her fellow birth control advocates were promoting lethal chambers, but she herself rejected the idea completely. "Nor do we believe," wrote Sanger in *Pivot of Civilization*, "that the community could or should send to the lethal chamber the defective progeny resulting from irresponsible and unintelligent breeding."[16]

Still, American eugenicists never relinquished the notion that America could bring itself to mass murder. At the First National Conference on Race Betterment, University of Wisconsin eugenicist Leon J. Cole lectured on the dysgenic effects of charity and medicine on eugenic progress. He made a clear distinction between Darwin's concept of natural selection and the newer idea of simple "selection." The difference, Cole explained, "is that instead of being *natural* selection it is now *conscious* selection on the part of the breeder.... Death is the normal process of elimination in the social organism, and we might carry the figure a step further and say that in prolonging the lives of defectives we are tampering with the functioning of the social kidneys!"[17]

Paul Popenoe, leader of California's eugenics movement and coauthor of the widely-used textbook *Applied Eugenics*, agreed that the easiest way to counteract feeblemindedness was simple execution. "From an historical point of view," he wrote, "the first method which presents itself is execution.... Its value in keeping up the standard of the race should not be underestimated."[18]

Madison Grant, who functioned as president of the Eugenics Research Association and the American Eugenics Society, made the point clear in *The Passing of the Great Race*. "Mistaken regard for what are believed to be divine laws and a sentimental belief in the sanctity of human life tend to prevent both the elimination of defective infants and the sterilization of such adults as are themselves of no value to the community. The laws of nature require the obliteration of the unfit and human life is valuable only when it is of use to the community or race."[19]

On November 12, 1915, the issue of eugenic euthanasia sprang out of the shadows and into the national headlines. It began as an unrelated medical decision on Chicago's Near North Side. At 4 A.M. that day, a woman named Anna Bollinger gave birth at German-American Hospital. The baby was somewhat deformed and suffered from extreme intestinal and rectal abnormalities, as well as other complications. The delivering physicians awakened Dr. Harry Haiselden, the hospital's chief of staff. Haiselden came in at once. He consulted with colleagues. There was great disagreement over whether the child could be saved. But Haiselden decided the baby was too afflicted and fundamentally not worth saving. It would be killed. The method—denial of treatment.[20]

Catherine Walsh, probably a friend of Anna Bollinger's, heard the news and sped to the hospital to help. She found the baby, who had been named Allan, alone in a bare room. He was naked and appeared to have been lying in one position unattended. Walsh urgently called for Haiselden, "to beg that the child be taken to its mother," and dramatically recalled, "It was condemned to death, and I knew its mother would be its most merciful judge."[21]

Walsh pleaded with Haiselden not to kill the baby by withholding treatment. "It was not a monster—that child," Walsh later told an inquest. "It was a beautiful baby. I saw no deformities." Walsh had patted the infant lightly. Allan's eyes were open, and he waved his tiny fists at her. She kissed his forehead. "I knew," she recalled, "if its mother got her eyes on it she would love it and never permit it to be left to die." Begging the doctor once more, Walsh tried an appeal to his humanity. "If the poor little darling has one chance in a thousand," she pleaded, "won't you operate and save it?"[22]

Haiselden laughed at Walsh, retorting, "I'm afraid it might get well." He was a skilled and experienced surgeon, trained by the best doctors in Chicago, and now chief of the hospital's medical staff. He was also an ardent eugenicist.[23]

Chicago's health commissioner, Dr. John Dill Robertson, learned of the deliberate euthanasia. He went to the hospital and told Haiselden he did not agree that "the child would grow up a mental defective." He later recollected, "I thought the child was in a dying condition, and I had doubts that an operation then would save it. Yet I believed it had one chance in

100,000, and I advised Dr. Haiselden to give it this one chance." But Haiselden refused.[24]

Quiet euthanasia of newborns was not uncommon in Chicago. Haiselden, however, publicly defended his decision to withhold treatment as a kind of eugenic expedient, throwing the city and the nation into moral turmoil amid blaring newspaper headlines. An inquest was convened a few days later. Some of Haiselden's most trusted colleagues were impaneled on the coroner's jury. Health Commissioner Robertson testified, "I think it very wrong not to save life, let that life be what it may. That is the function of a physician. I believe this baby might have grown up to be an average man.... I would have operated and saved this baby's life...."[25]

At one point Haiselden angrily interrupted the health commissioner's testimony to question why he was being singled out when doctors throughout Chicago were routinely killing, on average, one baby every day, under similar circumstances. Haiselden defiantly declared, "I should have been guilty of a graver crime if I had saved this child's life. My crime would have been keeping in existence one of nature's cruelest blunders." A juror shot back, "What do you mean by that?" Haiselden responded, "Exactly that. I do not *think* this child would have grown up to be a mental defective. I know it."[26]

After tempestuous proceedings, the inquest ruled, "We believe that a prompt operation would have prolonged and perhaps saved the life of the child. We find no evidence from the physical defects that the child would have become mentally or morally defective." The doctor jurors concluded that the child had at least a one-in-three chance—some thought an "even chance"—of surviving. But they also decided that Haiselden was within his professional rights to decline treatment. No law compelled him to operate on the child. The doctor was released unpunished, and efforts by the Illinois attorney general to indict him for murder were blocked by the local prosecutor.[27]

The medical establishment in Chicago and throughout the nation was rocked. The *Chicago Tribune* ran a giant banner headline across the width of its front page: "Baby Dies; Physician Upheld." One reader in Washington, D.C., wrote a letter to the editor asking, "Is it not strange that the whole country should be so shaken, almost hysterical, over the death of a babe never consciously alive...?" But the nation was momentarily transfixed.[28]

Haiselden considered his legal vindication a powerful victory for eugenics. "Eugenics? Of course it's eugenics," he told one reporter. On another occasion he remarked, "Which do you prefer—six days of Baby Bollinger or seventy years of Jukes?"[29]

Emboldened, Haiselden proudly revealed that he had euthanized other such newborns in the past. He began granting high-profile media interviews to advertise his determination to continue passively euthanizing infants. Within two weeks, he had ordered his staff to withhold treatment from several more deformed or birth-defected infants. Haiselden would sometimes send instructions via cross-country telegraph while on the lecture tour that arose from his eugenic celebrity. Other times he would handle it personally, like the time he left a newly delivered infant's umbilical cord untied and let it bleed to death. Sometimes he took a more direct approach and simply injected newborns with opiates.[30]

The euthanasia of Allan Bollinger may have begun as one doctor's controversial professional decision, but it immediately swirled into a national eugenic spectacle. Days after the inquest ruling, *The Independent*, a Hearst weekly devoted to pressing issues of the day, ran an editorial asking "Was the Doctor Right?" *The Independent* invited readers to sound off. In a special section, *The Independent* published supportive letters from prominent eugenicists, including Davenport himself. "If the progress of surgery," wrote Davenport, "is to be used to the detriment of the race . . . it may conceivably destroy the race. Shortsighted they who would unduly restrict the operation of what is one of Nature's greatest racial blessings—death."[31]

Haiselden continued to rally for eugenic euthanasia with a six-week series in the *Chicago American*. He justified his killings by claiming that public institutions for the feebleminded, epileptic and tubercular were functioning as lethal chambers of a sort. After clandestinely visiting the Illinois Institution for the Feebleminded at Lincoln, Illinois, Haiselden claimed that windows were deliberately left open and unscreened, allowing drafts and infecting flies to swarm over patients. He charged that Lincoln consciously permitted "flies from the toilets, garbage and from the eruptions of patients suffering from acute and chronic troubles to go at will over the entire institution. Worse still," he proclaimed, "I found that inmates were fed with the milk from a herd of cattle reeking with tuberculosis."[32]

At the time, milk from cattle with tuberculosis was a well-known cause of infection and death from the disease.[33] Lincoln maintained its own herd of seventy-two cows, which produced about 50,000 gallons of milk a year for its own consumption. Ten diseased cows had died within the previous two years. State officials admitted that their own examinations had determined that as many as half of the cows were tubercular, but there was no way to know which ones were infected because "a tubercular cow may be the fattest cow in the herd." Lincoln officials claimed that their normal pasteurization "by an experienced employee" killed the tuberculosis bacteria. They were silent on the continuous handling of the milk by infected residents.[34]

Medical watchdogs had often speculated that institutions for the feeble-minded were really nothing more than slow-acting lethal chambers. But Haiselden never resorted to the term *lethal chamber*. He called such institutions "slaughterhouses."[35]

In tuberculosis colonies, residents continuously infected and reinfected each other, often receiving minimal or no treatment. At Lincoln, the recently established tuberculosis unit housed just forty beds for an estimated tubercular population of hundreds. Lincoln officials asserted that only the most severely infected children were placed in that ward. They stressed that other institutions for the feebleminded recorded much higher mortality rates, some as high as 40 percent.[36]

Eugenicists believed that when tuberculosis was fatal, the real culprit was not bacteria, but defective genes. The ERO kept special files on mortality rates resulting from hereditary tuberculosis, compiled by the Belgian eugenicist Govaerts and others.[37]

Tuberculosis was an omnipresent topic in textbooks on eugenics. Typical was a chapter in Davenport's *Heredity in Relation to Eugenics* (1911). He claimed that only the submerged tenth was vulnerable. "The germs are ubiquitous.... Why do only 10 percent die from the attacks of this parasite? ... It seems perfectly plain that death from tuberculosis is the result of infection added to natural and acquired non-resistance. It is then highly undesirable that two persons with weak resistance should marry...." Popenoe and Johnson's textbook, *Applied Eugenics*, devoted a chapter to "Lethal Selection," which operated "through the destruction of the individual by some adverse feature of the environment, such as excessive cold, or bacteria, or by bodily deficiency."[38]

Some years earlier, the president of the National Conference on Charities and Correction had told his institutional superintendents caring for the feebleminded, "We wish the parasitic strain ... to die out." Even an article in *Institution Quarterly*, Illinois's own journal, admitted, "it would be an act of kindness to them, and a protection to the state, if they could be killed."[39]

No wonder that at one international conference on eugenics, Davenport proclaimed without explanation from the podium, "One may even view with satisfaction the high death rate in an institution for low grade feeble-minded, while one regards as a national disaster the loss of ... the infant child of exceptional parents."[40]

Haiselden himself quipped, "Death is the Great and Lasting Disinfectant."[41]

Haiselden's accusations of deliberate passive euthanasia by neglect and abuse could neither be verified nor dismissed. Lincoln's understaffed, overcrowded and decrepit facility consistently reported staggering death rates, often as high as 12 percent per year. In 1904, for example, 109 of its epileptic children died, constituting at least 10 percent and probably far more of its youth population; cause of death was usually listed as "exhaustion due to epileptic seizures." Between 1914 and 1915, a bout of dysentery claimed eight patients; "heat exhaustion" was listed as the cause. During the same period, four individuals died shortly after admission before any preliminary examination at all; their deaths were categorized as "undetermined."[42]

For some of its most vulnerable groups, Lincoln's death rate was particularly high. As many as 30 percent of newly admitted epileptic children died within eighteen months of admission. Moreover, in 1915, the overall death rate among patients in their first two years of residence jumped from 4.2 percent to 10 percent.[43]

Tuberculosis was a major factor. In 1915, Lincoln reported that nearly all of its incoming patients were designated feebleminded; roughly 20 percent were classified as epileptics; and some 27 percent of its overall population were "in the various stages of tubercular involvement." No isolation was provided for infected patients until the forty-bed tuberculosis unit opened. Lincoln officials worried that the statistics were "likely to leave the impression that the institution is a 'hot-bed' for the spread of tuberculosis." Officials denied this, explaining that many of the children came from filthy environments, and "the fact that feebleminded children have less resistance, account[s] for the high percentage of tuberculosis found among them."[44]

Lincoln officials clearly accepted the eugenic approach to feeblemindedness as gospel. Their reports and explanations were laced with scientific quotations on mental deficiency from Tredgold, who advocated euthanasia for severe cases, and Barr, who extolled the wisdom of the Kansas castrations. Lincoln officials also made clear that they received many of their patients as court-ordered institutionalizations from the Municipal Court of Chicago; as such, they received regular guidance from the court's supervising judge, Harry Olson. *Eugenical News* praised Olson for operating the court's psychopathic laboratory, which employed Laughlin as a special consultant on sterilization. Olson was vital to the movement and hailed by

Eugenical News as "one of its most advanced representatives." In 1922, Olson became president of the Eugenics Research Association.[45]

Moreover, staff members at Lincoln were some of the leading eugenicists in Illinois. Lincoln psychologist Clara Town chaired the Eugenics Committee of the Illinois State Commission of Charities and Corrections. Town had helped compile a series of articles on eugenics and feeblemindedness, including one by her friend Henry H. Goddard, who had invented the original classifications of feeblemindedness. One reviewer described Town's articles as arguments that there was little use in caring for the institutionalized feebleminded, who would die anyway if left in the community; caring for them was little more than "unnatural selection."[46]

For decades, medical investigators would question how the death rates at asylums, including the one in Lincoln, Illinois, could be so high. In the 1990s, the average life expectancy for individuals with mental retardation was 66.2 years. In the 1930s, the average life expectancy for those classified as feebleminded was approximately 18.5 years. Records suggest that a disproportionate percentage of the feebleminded at Lincoln died before the age of ten.[47]

Haiselden became an overnight eugenic celebrity, known to the average person because of his many newspaper articles, speaking tours, and his outrageous diatribes. In 1917, Hollywood came calling. The film was called *The Black Stork*. Written by *Chicago American* reporter Jack Lait, it was produced in Hollywood and given a massive national distribution and promotion campaign. Haiselden played himself in a fictionalized account of a eugenically mismatched couple who are counseled by Haiselden against having children because they are likely to be defective. Eventually the woman does give birth to a defective child, whom she then allows to die. The dead child levitates into the waiting arms of Jesus Christ. It was unbridled cinematic propaganda for the eugenics movement.[48]

In many theaters, such as the LaSalle in Chicago, the movie played continuously from 9 A.M. until 11 P.M. National publicity advertised it as a "eugenic love story." Sensational movie posters called it a "eugenic photoplay." One advertisement quoted Swiss eugenicist Auguste Forel's warning: "The law of heredity winds like a red thread through the family history of every criminal, of every epileptic, eccentric and insane person. Shall we sit still...without applying the remedy?" Another poster depicted Haiselden's office door with a notice: "BABIES NOT TREATED." In 1917, a display advertisement for the film encouraged: "Kill Defectives, Save the Nation and See 'The Black Stork.'"[49]

The Black Stork played at movie theaters around the nation for more than a decade.[50]

Gassing the unwanted, the lethal chamber and other methods of euthanasia became a part of everyday American parlance and ethical debate some two decades before President Woodrow Wilson, in General Order 62, directed that the "Gas Service" become the "Chemical Warfare Service," instructing them to develop toxic gas weapons for world war. The lethal chamber was a eugenic concept more than two decades before Nevada approved the first such chamber for criminal executions in 1921, and then gassed with cyanide a Chinese-born murderer, the first such execution in the world. Davenport declared that capital punishment was a eugenic necessity. Popenoe's textbook, *Applied Eugenics*, listed execution as one of nine suggested remedies for defectives—without specifying criminals.[51]

In the first decades of the twentieth century, America's eugenics movement inspired and spawned a world of look-alikes, act-alikes and think-alikes. The U.S. movement also rendered scientific aid and comfort to undisguised racists everywhere, from Walter Plecker in Virginia right across Europe. American theory, practice and legislation were the models. In France, Belgium, Sweden, England and elsewhere in Europe, each clique of raceological eugenicists did their best to introduce eugenic principles into national life; perhaps more importantly, they could always point to recent precedents established in the United States.

Germany was no exception. German eugenicists had formed academic and personal relationships with Davenport and the American eugenic establishment from the turn of the century. Even after World War I, when Germany would not cooperate with the International Federation of Eugenic Organizations because of French, English and Belgian involvement, its bonds with Davenport and the rest of the U.S. movement remained strong. American foundations such as the Carnegie Institution and the Rockefeller Foundation generously funded German race biology with hundreds of thousands of dollars, even as Americans stood in breadlines.[52]

Germany had certainly developed its own body of eugenic knowledge and library of publications. Yet German readers still closely followed American eugenic accomplishments as the model: biological courts, forced sterilization, detention for the socially inadequate, debates on euthanasia. As America's elite were describing the socially worthless and the ancestrally unfit as "bacteria," "vermin," "mongrels" and "subhuman," a superior race of Nordics was increasingly seen as the final solution to the globe's eugenic problems.[53]

America had established the value of race and blood. In Germany, the concept was known as *Rasse und Blut*.

U.S. proposals, laws, eugenic investigations and ideology were not undertaken quietly out of sight of German activists. They became inspirational blueprints for Germany's rising tide of race biologists and race-based hatemongers, be they white-coated doctors studying *Eugenical News* and attending congresses in New York, or brown-shirted agitators waving banners and screaming for social upheaval in the streets of Munich.

One such agitator was a disgruntled corporal in the German army. He was an extreme nationalist who also considered himself a race biologist and an advocate of a master race. He was willing to use force to achieve his nationalist racial goals. His inner circle included Germany's most prominent eugenic publisher. In 1924, he was serving time in prison for mob action.[54] While in prison, he spent his time poring over eugenic textbooks, which extensively quoted Davenport, Popenoe and other American raceological stalwarts.[55] Moreover, he closely followed the writings of Leon Whitney, president of the American Eugenics Society, and Madison Grant, who extolled the Nordic race and bemoaned its corruption by Jews, Negroes, Slavs and others who did not possess blond hair and blue eyes. The young German corporal even wrote one of them fan mail.[56]

In *The Passing of the Great Race*, Madison Grant wrote: "Mistaken regard for what are believed to be divine laws and a sentimental belief in the sanctity of human life tend to prevent both the elimination of defective infants and the sterilization of such adults as are themselves of no value to the community. The laws of nature require the obliteration of the unfit and human life is valuable only when it is of use to the community or race."[57]

One day in the early 1930s, AES president Whitney visited the home of Grant, who was at the time chairing a eugenic immigration committee. Whitney wanted to show off a letter he had just received from Germany, written by the corporal, now out of prison and rising in the German political scene. Grant could only smile. He pulled out his own letter. It was from the same German, thanking Grant for writing *The Passing of the Great Race*. The fan letter stated that Grant's book was "his Bible."[58]

The man writing both letters to the American eugenic leaders would soon burn and gas his name into the blackest corner of history. He would duplicate the American eugenic program—both that which was legislated and that which was only brashly advocated—and his group would consistently point to the United States as setting the precedents for Germany's actions. And then this man would go further than any American eugenicist ever dreamed, further than the world would ever tolerate, further than humanity will ever forget.

The man who sent those letters was Adolf Hitler.[59]

TOP
Francis Galton's original
scrap of paper inventing the
term *eugenics*. UNIVERSITY
COLLEGE LONDON ARCHIVES

BOTTOM LEFT
Gregor Mendel, discoverer
of the principles of heredity.
AMERICAN PHILOSOPHICAL
SOCIETY

BOTTOM RIGHT
Francis Galton, father
of eugenics. UNIVERSITY
COLLEGE LONDON ARCHIVES

First Race Betterment Conference Banquet, 1914. AMERICAN PHILOSOPHICAL SOCIETY

Chicago Tribune report of Dr. Harry Haiselden's infant euthanasia, November 1915.
COURTESY OF MARTIN PERNICK

The eugenic photoplay—

"The Black Stork"

is being sold on a state rights basis by the

Sheriott Pictures Corporation
218 West 42nd Street
New York City

TOP
Movie industry ad for
The Black Stork, April 1917.
COURTESY OF MARTIN
PERNICK

BOTTOM
Image from *The Black Stork*
depicting a euthanized baby
floating to Jesus. "ARE YOU
FIT TO MARRY" PRESERVATION
FUNDED BY THE UNIVERSITY
OF MICHIGAN HISTORICAL
FILM COLLECTION; COURTESY
OF MARTIN PERNICK AND
JOHN ALLEN

LEFT AND ABOVE
Charles B. Davenport
AMERICAN PHILOSOPHICAL SOCIETY

RIGHT
Harry H. Laughlin
PICKLER MEMORIAL LIBRARY

Charles B. Davenport leads a training session with field workers at the ERO, 1913.

COLD SPRING HARBOR LABORATORY ARCHIVE

Eugenics Training Class 1914

Harry H. Laughlin and Charles B. Davenport pose with ERO field workers, 1914.

COLD SPRING HARBOR LABORATORY ARCHIVE

Eugenics Record Office files. COLD SPRING HARBOR LABORATORY ARCHIVE

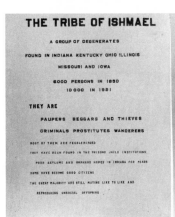

Exhibit poster showing dwellings of the so-called Tribe of Ishmael.

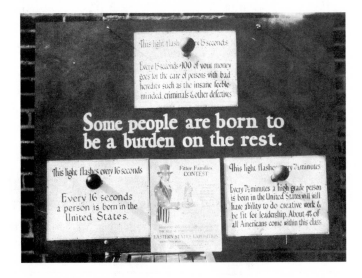

Exhibit: "Some people are born to be a burden on the rest," circa 1926.

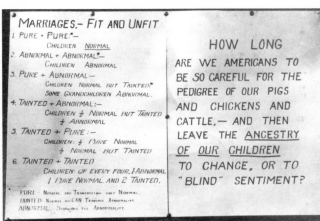

Exhibit poster: "Marriages, Fit and Unfit."

ERO copy of the September 1910 edition of *Archiv für Rassen- und Gesellschafts-Biologie*, featuring articles by German eugenics founding father Alfred Ploetz, Ernst Rüdin (who later became president of the International Federation of Eugenic Organizations), and Roderick Plate (who would become a demographic and statistical expert for Nazi killer Adolf Eichmann). AMERICAN PHILOSOPHICAL SOCIETY

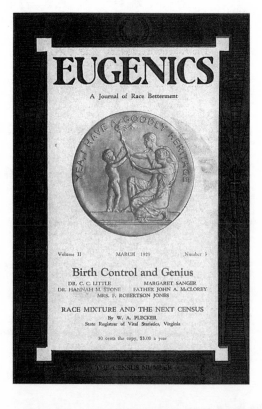

Eugenics, March 1929 edition, featuring articles by Virginia racist Walter Plecker and birth control advocate Margaret Sanger.
VERMONT STATE PUBLIC RECORDS DIVISION

Carrie Buck standing in a park, date unknown.
AUTHOR'S COLLECTION

Code list for IBM
Hollerith punch card
system used in the
Jamaica Race Crossing
Study, 1928. AMERICAN
PHILOSOPHICAL SOCIETY

Column   List of data for Columns of Hollerith Cards.
1.   Sex and Race.
2.   Age.
3,4.   Vertex height (Omitting hundreds).
5,6.   Weight   (Omitting hundreds).
7.   span / V.H.   Classes 96 -7, 98 -9 &c.
8.   Acrom. -Sizlion / V.H.
9,10. Horizontal distances,gnath-acnion: V.H. / chen girth ; cephalic index.
11.   Sitting height ÷ V.H.
12.   Kneeling ht ÷ V.H.
       Nasal breadth
13.   Nasion - Subnasale
14.   Interpupillary distance ÷ Head breadth.
       Nose breadth
15.   Pinna length
16.   Head height ÷ head breadth.
17.   "   "   ÷ head length.
18.   Cranial capacity.
19.   Face length (nas.-gnath.) / Zizygomatic
       Chest transv.
20.   Bit. ÷ prsternale   ♂ only.
21 -22. Chest girth.
23.   Cristal breadth ÷ acromion breadth.
24.   Foot length.
25.   Foot index. Foot breadth / Foot length breadth
26.   Hand length / breadth
27.   Hair color; eye color.
28.   Hair form.
29.   Teeth.
30.   Outline . figure.
31.   Drawing of man.
32.   Army beta.
33.   Army alpha.
34.   Form substitution.
35 -40. Seashore Music Test.
41 -43. Form discrimination.

TOP LEFT
Margaret Sanger
PLANNED PARENT-
HOOD FEDERATION

TOP RIGHT
**ERO's copy of
German eugenicist
Erwin Baur
photograph.**
AMERICAN
PHILOSOPHICAL
SOCIETY

BOTTOM
**Ernst Rüdin,
director of the
Kaiser Wilhelm
Institute for
Anthropology,
Human Heredity
and Eugenics.** MAX
PLANCK INSTITUT
FÜR PYSCHIATRIE,
HISTORISCHES
ARCHIV DER KLINIK

Edwin Katzen-Ellenbogen at Buchenwald with a warm hat he claimed he never wore.

EUGENICAL NEWS

CURRENT RECORD OF HUMAN GENETICS AND RACE HYGIENE

GERMAN POPULATION AND RACE POLITICS.*

AN ADDRESS BY DR. FRICK, REICHSMINISTER FOR THE INTERIOR, BEFORE THE FIRST MEETING OF THE EXPERT COUNCIL FOR POPULATION- AND RACE-POLITICS HELD IN BERLIN, JUNE 28, 1933.

German Men and Women:

While thanking you for your ready coöperation I take the liberty of giving you today a survey of the work which we propose to do, as well as an outline of the goal which we want to reach.

The National Socialist movement under the leadership of Adolf Hitler deserves the merit of having preserved the German nation from utter political disruption and the Reich from disorganization. It would be a great mistake to believe that now the main problem has been solved. Any one with perspicacity knows that the most difficult task has yet to be accomplished, namely to stop the national and cultural ruin. Germany is one of the countries which not only had to bear the main burden of the World War with tremendous losses of its best men and racial constituents, but it is also the one country which has been threatened (during as well as after the war) with a decreasing birth-rate. While at the beginning of this century we still had approximately two million births per year, today these only amount to 975,000. In 1900 there were

*Translated from the German for the EUGENICAL NEWS by A. Hellmer.

36 living births to 1000 inhabitants—this number dwindled down to about 15 in 1932. The number of children, then, is alarmingly decreasing—the post-war two-child system has been superseded; the German nation has adopted the one-child and even the no-child system.

Notwithstanding the successful results achieved in reducing mortality and lengthening the span of life by means of general hygiene, control of infectious diseases, social hygiene and medicine, the decrease in mortality is not sufficient to preserve the population as a biologic national entity. Today rough figures of births and deaths give only an inadequate estimate; in order to recognize the true situation from the standpoint of population-politics, we must carefully sift the vital debits and credits by considering the age distribution of the various classes. According to the figures of the Statistical Reichsmat, the German nation with its present low birth-rate is no longer able to maintain itself by its own strength. With 15 births to 1000 inhabitants the figure falls about 30% below the increase needed to maintain the population number in the future. With the present birth-rate neither Berlin nor other large German cities, not even the medium and small towns are able to maintain their population number. Only the rural communities still have a minimal excess of births, but this is no longer sufficient to offset the loss in the German

and twins; and in this work there will not be investigated alone interesting twins, but all twins and families of definite geographical origin must be considered. It is desirable to determine what traits of bodily and

interior is given in Fig. 2.

The EUGENICAL NEWS extends best wishes to Dr. O. Freiherr von Verschuer for the success of his work in his new and favorable environment.

FIG. 1. Haus der Volksgesundheit on south bank of the Main. Verschuer's Institute is located on the second floor.

mental sort, what diseases and anomalies in mankind are hereditary; according to what laws they are transmitted from one generation to the other; how far external influences that may act in inhibiting or accelerating fashion must be taken into consideration, and many other points.

The plan of the Universität Institut für Erbbiologie und Rassenhygiene, which occupies the second floor of the People's Hygiene House, shows a larger and a smaller lecture hall and a series of rooms for investigation, for the director and assistants, for library, for archives, for special investigators and for clinical rooms, about sixty-two rooms in all, mostly, however, of small size. Some idea of the Institute.

FIG. 2. Corridor through the clinical part of the Institute.

ON THE BREEDING OF ARYANS

And Other Genetic Problems of War-time Germany

TAGE U. H. ELLINGER

DURING a visit to Germany in the winter of 1939-40, I had an opportunity to meet some of my fellow geneticists, who seemed to be working undisturbed by the campaign and the "mopping up" in Poland, and by the hectic preparations for the assaults on a great many peaceful countries such as Denmark, Norway, Holland, and Belgium. The following unpretentious notes, written for laymen, may perhaps interest some of their many American friends.

Quite a few of them were busy treating or rather mistreating the sex cells of animals and plants in order to produce new varieties. I was introduced to all kinds of extraordinary creatures produced in that way, mice without toes or with corkscrew tails, flies that violated the very definition of a fly by having four wings instead of two, funny-looking moths, and strange plants.

Radiation, especially with X-rays, is the principal means of producing such new kinds, or rather monsters, of animals and plants, and the wizard in this business was a Russian, Dr. Timofeeff-Ressovsky, who has found an asylum at the Kaiser Wilhelm Institute for Brain Research. An industrial concern has presented him with the enormous machines with which he radiates the minute sex cells of tiny little Drosophila flies.

Timofeeff is a fanatic and an enthusiast. I was really spellbound while he gave me a three-hour lecture on his work, incessantly gesticulating as he walked up and down the floor. The German staff of the Institute looked at this strange and temperamental Russian with amusement and sincere admiration. They even granted him a freedom of speech and opinion they would deny any other human being.

Genuine German thoroughness characterized Professor Nachtsheim's elaborate experiments on the heredity of disease. Since one can not very well make human patients mate and produce numerous babies to suit the analysis of a pathological problem, he had resorted to rabbits. These obliging and fertile animals suffer from a great many troubles like our own. It was a pathetic sight to look at the hundreds of incurables in the rabbit houses. But at that, our own institutions housing people afflicted with hereditary diseases are no less disconsolate.

When I went to see the famous old "Geheimrat" Fischer at the Kaiser Wilhelm Institute for Anthropology, an S.A. man in black uniform was present during the conversation. He was introduced as Dr. Abel, and the director afterwards asked him to show me around. The result was that we spent several days together.

Twins in the "New Order"

Twins have, of course, for a long time been a favorite material for the study of the relative importance of heredity and environment, of nature and nurture. It does, however, take a dictatorship to oblige some ten thousand pairs of twins, as well as triplets and even quadruplets, to report to a scientific institute at regular intervals for all kinds of recordings and tests.

I was particularly interested in their laboratory for the study of the inheritance of behavior and mental capacities. For this purpose, the twins were placed in two identical rooms, separated by a narrow corridor for the observer, who had a free view over both rooms through big windows in the walls. These, however, were fitted with that remarkable kind of glass through which you can see in one direction

TOP LEFT

Nazi Interior Minister Wilhelm Frick propagandizing for forced sterilization in *Eugenical News*, March–April 1934.

AUTHOR'S COLLECTION

TOP RIGHT

Lavish praise for and photos of "Verschuer's Institute" in *Eugenical News*, June 1936.

AUTHOR'S COLLECTION

BOTTOM

Enthusiastic essay about Nazi breeding experiments in *Journal of Heredity*, 1942.

AUTHOR'S COLLECTION

THE HUMAN BETTERMENT FOUNDATION
(A NON-PROFIT CORPORATION)

EXECUTIVE OFFICES:
SUITE 321, PACIFIC SOUTHWEST BUILDING
PHONE WAKEFIELD 6374

PASADENA, CALIFORNIA

REPORT TO THE BOARD OF DIRECTORS OF THE HUMAN BETTERMENT
FOUNDATION, FOR THE YEAR ENDING FEBRUARY 12, 1935.

During the past year a large part of the resources of the
Foundation has been devoted to a second survey of sterilization in
California, in collaboration with the California Bureau of Juvenile
Research. The task of collecting case histories from the state institutions was completed by the Foundation workers, and all of these
were tabulated and coded. Hollerith cards were then punched for each
individual, 67 facts being thus recorded for each of about 8,000 persons.
These cards are now being sorted to make 1500 scatter diagrams, which
will reveal the correlations of all of the facts. Males and females are
tabulated separately, as are also the insane and feeble-minded. In this
immense but highly important task, the staff of the Foundation has been
helped by six SERA workers furnished by the Bureau of Juvenile Research.
The study should be completed within the next year, and will not merely
bring up to date the previous study which this Foundation made (beginning in 1926), but will extend it and result in a great deal of new
and valuable information.

The Foundation had a poster exhibit in connection with the
annual meeting of the American Public Health Association in Pasadena in
September. Frank C. Reid of the Foundation staff spent most of the
time during the week in charge of this exhibit, contacting leaders in
the field of public health from all over the United States. An outstanding feature of the meeting was also the eugenics exhibit of the German
government, which was devoted largely to sterilization. This is re-
~~garded as~~ ~~prolonged~~ ~~whole~~ ~~the Germany and other European countries,~~
Goethe of Sacramento, one of the members of the Foundation, wrote as
ws to Mr· Gosney:

"You will be interested to know that your work has
played a powerful part in shaping the opinions of the group
of intellectuals who are behind Hitler in this epoch-making
program. Everywhere I sensed that their opinions have been
tremendously stimulated by American thought, and particularly
by the work of the Human Betterment Foundation. I want you,
my dear friend, to carry this thought with you for the rest
of your life, that you have really jolted into action a great government of 60,000,000 people."

The offices of the Foundation are frequently visited by distin-
ed people from out of town. Among the most recent callers who were in-
ted in eugenic sterilization were: Dr. Harry H. Benjamin of New York
Dr. Wilhelm Krauss of the Institute of Race Biology, Sweden; and

Human Betterment Foundation Annual Report for 1935 citing a letter from
board member C. M. Goethe to racist eugenicist E. S. Gosney, bragging: "You
will be interested to know that your work has played a powerful part in shaping the
opinions of the group of intellectuals who are behind Hitler in this epoch-making
program. Everywhere I sensed that their opinions have been tremendously stimulated
by American thought, and particularly by the work of the Human Betterment
Foundation. I want you, my dear friend, to carry this thought with you for the
rest of your life, that you have really jolted into action a great government of 60
million people." VERMONT STATE PUBLIC RECORDS DIVISION

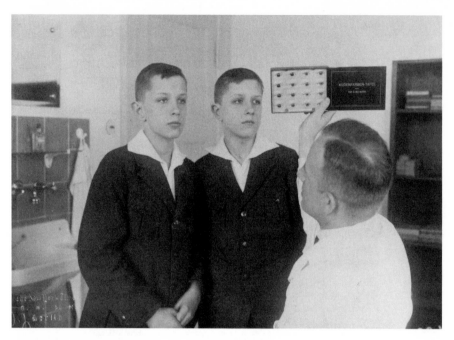

ABOVE AND BELOW, LEFT

Nazi eugenicist Dr. Otmar Freiherr von Verschuer examining twins; his assistant, Josef Mengele, continued the experiments at Auschwitz. MAX PLANCK GESELLSCHAFT ARCHIV

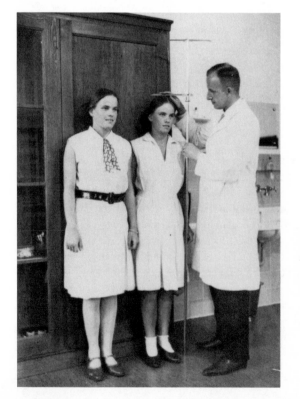

Auschwitz's murderous doctor, Josef Mengele.
AUSCHWITZ ARCHIV

CHAPTER 14

Rasse und Blut

Negative eugenic solutions appeared in Germany at the end of the nine-teenth century.

From 1895 to 1900, German physician Gustav Boeters worked as a ship's doctor in the United States and traveled throughout the country. He learned of America's castrations, sterilizations and numerous marriage restriction laws. When Boeters returned to Germany, he spent the next three decades writing newspaper articles, drafting proposed legislation and clamoring to anyone who would listen to inaugurate eugenic sterilization. Constantly citing American precedents, from its state marriage restriction statutes to sterilization laws from Iowa to Oregon, Boeters passionately argued for Germany to follow suit. "In a cultured nation of the first order—the United States of America—that which we strive toward [sterili-zation legislation], was introduced and tested long ago. It is all so clear and simple." Eventually, Boeters became so fixated on the topic that he was considered delusional and was forced to retire from his post as a medical officer in Saxony—but not before prompting German authorities to seri-ously consider eugenic laws.[1]

While Boeters was touring America, so was German physician Alfred Ploetz. A socialist thinker, Ploetz had traveled to America in the mid-1880s to investigate utopian societies. He became caught up in the post–Civil War American quest to breed better human beings. In Chicago, in 1884, he studied the writings of leading American utopians. He also spent several months working at the Icarian Colony, an obscure utopian community in Iowa. Ploetz was disappointed to find the Icarians socially disorganized, and he began to believe that racial makeup was the key to social success.[2]

Ploetz also opened a medical practice in Springfield, Massachusetts, and began to breed chickens. Later, he moved to Meriden, Connecticut, where he graduated to human breeding projects. By 1892, Ploetz had already compiled 325 genealogies of local families and hoped to gather even more from a nearby secret German lodge. A colleague recalled that Ploetz was convinced "the Anglo-Saxons of America would be left behind, unless they adopted a policy that would change the relative proportions of the population."[3]

Like his medical and utopian colleagues, Ploetz was undoubtedly a devotee of the late nineteenth century's hygiene and sanitary movement that sought to eradicate germs and disease. One of the leading exponents of this movement was Benjamin Ward Richardson, inventor of the lethal chamber and author of *Hygeia, A City of Health*. The same conflicts that perplexed late-nineteenth-century British and American social Darwinists, from Spencer to New York's human breeding advocates, also confronted German hereditarians. By the mid-1880s, Ploetz had propounded a eugenic racial theory. Galton's term *eugenics* had not yet been translated, and Ploetz coined the term *Rassenhygiene (racial hygiene)*. He articulated his notions of racial and social health in a multivolume 1895 work, *The Foundations of Racial Hygiene*. Volume one was entitled *Fitness of Our Race and the Protection of the Weak*. His colleagues later argued that the term *Rassenhygiene* should not be translated into English as *race hygiene*, but as *eugenics*. The two were one and the same.[4]

Ploetz believed that a better understanding of heredity could help the state identify and encourage the best specimens of the German race. Ironically, while Ploetz believed in German national eugenics and harbored strong anti-Semitic sentiments,[5] he included the Jews among Germany's most valuable biological assets. After returning to Germany, Ploetz in 1904 helped found the journal *Archiv für Rassen- und Gesellschaftsbiologie (Archives of Race Science and Social Biology)*, and the next year he organized the Society for Racial Hygiene (*Gesellschaft für Rassenhygiene*) to promote eugenic research. Both entities functioned as the principal clearinghouses for German eugenics for years to come. Understandably, Ploetz emerged as Germany's leading race theorist and was often described as "the founder of eugenics as a science in Germany."[6]

Even as Boeters and Ploetz were formulating their American-influenced ideas, German social theorist Alfred Jost argued in his 1895 booklet, *The Right to Death*, that the state possessed the inherent right to kill the unfit and useless. The individual's "right to die" was not at issue; rather, Jost postulated, it was the state's inherent "rights to [inflict] death [that are] the key to the fitness of life."[7] The seeds of German negative eugenics were planted.

With Nordic superiority as the centerpiece of American eugenics, Davenport quickly established good personal and professional relations with German race hygienists. As director of the Carnegie Institution's Station for Experimental Evolution, Davenport was more than happy to correspond frequently with German eugenic thinkers on matters major and mundane. In the first decade of the twentieth century, typed and hand-written letters sailed back and forth across the Atlantic, encompassing requests for copies of the latest German research to replies to German appeals for Carnegie donations for a memorial to Mendel.[8]

Quickly, Davenport and the Carnegie Institution became the center of the eugenic world for German researchers. America was enacting a grow-ing body of eugenic laws and governmental practices, and the movement enjoyed wealthy backers and the active support of U.S. officials. While a small group of German social thinkers merely expounded theory, America was taking action. At the same time, by virtue of their blond and blue-eyed Nordic nature as well as their stellar scientific reputation, Germany's bud-ding eugenicists became desirable allies for the Americans. A clear partner-ship emerged in the years before World War I. In this relationship, however, America was far and away the senior partner. In eugenics, the United States led and Germany followed.

One of Davenport's earliest German allies was the anthropologist and eugenicist Eugen Fischer. Fischer was among the first "corresponding scien-tists" recruited by Davenport when the Cold Spring Harbor facility opened in 1904. Before long Davenport and Fischer were exchanging their latest research, including studies on eye color and hair quality. In 1908, Fischer expanded into research on race mixing between whites and Hottentots in Africa, focusing on the children known as "Rehoboth bastards." Miscegen-ation fascinated Davenport. He and his colleagues, both German and American, jointly pursued studies on race mixing for years to come.[9]

When Davenport elevated eugenics into a global movement, he chose German eugenicists for a major role, and British leaders went along. Indeed, the First International Congress of Eugenics in London was sched-uled for July of 1912 to coincide with summer visits to Great Britain by leading German and American eugenicists. At the time, these two groups were seen as the giants of eugenic science.[10] But in fact there was only room for one giant in the post-Galtonian world—and that would be America.

When Ploetz founded the Society for Racial Hygiene in Berlin in 1905, it was little more than an outgrowth of his own social circle and his publica-tion, *Archiv für Rassen- und Gesellschaftsbiologie.* By the end of 1905, the Society for Racial Hygiene had just eighteen German and two non-German members. Even when so-called "branches" opened in other

German cities, these chapters usually claimed only a handful of members. The society was less a national organization devoted to Germany's territorial borders than it was a *Germanic* society devoted to the Nordic roots and Germanic language innervating much of northwestern Europe. Ploetz himself maintained Swiss citizenship, as did some of his key colleagues. Thinking beyond Germany's borders, Ploetz expanded the group within a few years into the International Society for Race Hygiene. So-called branches were established in Norway and Sweden, but again, these branches were comprised of just a handful of eugenic compatriots.[11]

As society members traveled through other traditionally Germanic and Nordic lands, however, they recruited more fellow travelers. By 1909, Ploetz's growing international organization numbered more than 120 members, although most were German nationals. In the summer of that year, the organization gained prestige when Galton agreed to become its honorary president, just as he had for the budding Eugenics Education Society.[12]

Two years later, in 1911, Ploetz raised his group's profile again, this time by participating in the International Hygiene Exhibition in Dresden. But the Anglo-American bloc was clearly reluctant to see the German wing rise on the world eugenic stage. After a series of negotiations, the Anglo-American group for all intents and purposes absorbed Ploetz's budding international network into their larger and better-financed movement.[13]

Ploetz was brought in as a lead vice president of the First International Congress of Eugenics in London in 1912. He was one of about fifteen individuals invited back to Paris the next year to create the Permanent International Eugenics Committee. This new and elite panel evolved into the International Eugenics Commission and later became the International Federation of Eugenic Organizations, which governed the entire worldwide movement. After some failed attempts to regain leadership, Ploetz and his societies finally bowed to American eugenicists and their international eugenics agencies.[14]

After 1913, the United States continued to dominate by virtue of its widespread legislative and bureaucratic progress as well as its diverse research programs. These American developments were closely followed and popularized within the German scientific and eugenic establishment by Géza von Hoffmann, an Austro-Hungarian vice consul who traveled throughout the United States studying eugenic practices. Von Hoffmann's 1913 book, *Racial Hygiene in the United States* (*Die Rassenhygiene in den Vereinigten Staaten von Nordamerika*), exhaustively detailed American laws on sterilization and marriage restrictions, as well as methods of field investigation and data collection. With equal thoroughness, he delineated America's eugenic organizational structure—from the Rockefeller Foundation to the institutions at Cold Spring Harbor. Then, in alphabetical order, he summarized each state's eugenic legislation. A comprehensive eighty-four-page bibliography was appended, with special subsections for such topics as "euthanasia" and "sterilization."[15]

Most importantly, von Hoffmann's comprehensive volume held up American eugenic theory and practice as the ideal for Germany to emulate. "Galton's dream," he wrote, "that racial hygiene should become the religion of the future, is being realized in America.... America wants to breed a new superior race." Von Hoffmann repeatedly chided Germany for allowing mental defectives to roam freely when in America such people were safely in institutions. Moreover, he urged Germany to follow America's example in erecting race-based immigration barriers. For years after *Racial Hygiene in the United States* was published, leading German eugenicists would credit von Hoffmann's book on America's race science as a seminal reference for German biology students.[16]

Laughlin and the Eugenics Record Office were the leading conduits of information for von Hoffmann. The ERO sent von Hoffmann its special bulletins and other informational summaries. In turn, von Hoffmann hoped to impress Laughlin with updates of his own. He faithfully reported the latest developments in Germany and Austria, such as the formation of a new eugenic research society in Leipzig, a nascent eugenic sexology study group in Vienna, and genetic conference planning in Berlin.[17]

But it was the American developments that captivated von Hoffmann. Continually impressed with Laughlin's ideas, he frequently reported the latest American news in German medical and eugenic literature. "I thank you sincerely," von Hoffmann wrote Laughlin in a typical letter dated May 26, 1914, "for the transmission of your exhaustive and interesting reports. The far-reaching proposal of sterilizing one tenth of the population impressed me very much. I wrote a review of [the] report... in the *Archiv für Rassen- und Gesellschaftsbiologie* [Ploetz's journal]."[18]

Eager to be a voice for German eugenics in America, von Hoffmann also contributed articles about German developments to leading U.S. publications. In October of 1914, his article "Eugenics in Germany" appeared in the *Journal of Heredity*, explaining that while sterilization was being debated, "the time has not yet come for such a measure in Germany." In the same issue, the *Journal of Heredity* published an extensive review of Fischer's book about race crossing between Dutch and Hottentots in

Africa, and the resulting "Rehoboth bastard" hybrids. Indeed, German eugenic philosophy and progress were popular in the *Journal of Heredity*. In 1914, for example, they published an article tracing the heredity of Bismarck, an article outlining plans for a new experimental genetics lab in Berlin, an announcement for the next international genetics conference in Berlin, and reviews of the latest German books.[19]

In the fall of 1914, the Great War erupted. During the war, "the eugenics movement in Germany stood entirely still," as one of Germany's top eugenic leaders later remembered in *Journal of Heredity*. Ploetz withdrew to his estate. Sensational headlines in American newspapers reported and denounced German atrocities against civilians, such as bayoneting babies and mutilating women's breasts. Many of these stories were later found to be utterly unfounded. But despite the headlines, the American eugenics movement strengthened ties with its German scientific counterparts. In 1916, Madison Grant's *The Passing of the Great Race* declared that the white Nordic race was destined to rule the world, and confirmed the Aryan people's role in it. German nationalists were heartened by America's recognition of Nordic and Aryan racial superiority. Reviews of the book inspired a spectrum of German scientists and nationalists to think eugenically even before the work was translated into German.[20]

American fascination with the struggling German eugenics movement continued right up until the United States entered the war in April of 1917. In fact, the April issue of *Eugenical News* summarized in detail von Hoffmann's latest article in *Journal of Heredity*. It outlined Germany's broad plans to breed its own eugenically superior race after the war to replace German men lost on the battlefield. The article proposed special apartment buildings for desirable single Aryan women and cash payments for having babies.[21]

America entered the war on April 6, 1917. Millions died in battle. At the eleventh hour of the eleventh day of the eleventh month of 1918, a defeated Germany finally agreed to an armistice, ending the bloody conflict. The Weimar Republic was created. A peace treaty was signed in June of 1919. American eugenics' partnership with the German movement resumed.[22]

Laughlin prepared a detailed pro-German speech for the Ninth Annual Meeting of the Eugenics Research Association, held at Cold Spring Harbor in June of 1920. In the text, Laughlin analyzed Germany's newly imposed democratic constitution point by point, identifying the clauses that authorized eugenic and racial laws. These included a range of state powers, from "Article 7 ... [allowing] protection of plants from disease and pests" to

"Articles 119 to 134 inclusive [which] prescribe the fundamental law of Germany in reference to the social life." Declaring that "modern civilization" itself depended on German and Teutonic conquest, Laughlin closed by assuring his colleagues, "From what the world knows of Germanic traits, we logically concede that she will live up to her instincts of race conservation...." Laughlin never actually delivered the speech, probably because of time constraints, so *Eugenical News* published it in their next issue, as did a subsequent edition of the official British organ, *Eugenics Review*. Reprints of the *Eugenics Review* version were then circulated by the ERO.[23]

Scientific correspondence also resumed. Shortly after Laughlin's enthusiastic appraisal, a eugenicist at the Institute for Heredity Research in Potsdam requested ERO documentation for his advisory committee's presentation to the local government. Davenport dispatched materials and supporting statements "that will be of use to you in your capacity as advisor to the Government in matters of race hygiene." ERO staffers had missed their exchanges with German colleagues, and Davenport assured his Potsdam friend, "I read your letter to our staff at its meeting on Monday and they were interested to hear from you." Information about the new advisory committee was published in the very next issue of *Eugenical News*. German race scientists reciprocated by sending their own research papers for Davenport's review, covering a gamut of topics from inherited human traits to mammalian attributes.[24]

But efforts by German eugenicists to join America's international movement were still hampered by the aftershocks of the war. Under the Treaty of Versailles, Germany agreed to pay the Allies massive war reparations, 132 billion marks or 33 billion dollars. This crippled the finances of all of Germany, including its raceologists. Meanwhile, German nationalists were enraged because France and Belgium now occupied the Rhineland. France's army had long included African soldiers from its colonies—such as Senegal, Mali and North Africa—who were now mingling with German women and would ultimately father several hundred children of mixed race in Germany.[25]

Infuriated Germans refused to cooperate with international committees that included Belgian or French scientists. Nor did they have the money to travel, even within Europe. The International Congress of Hygiene, for instance, originally scheduled for May of 1921 in Geneva, was cancelled because "the low value of the currency of many countries and the high value of the Swiss franc make it impossible for many countries to send delegates," as one published notice explained.[26]

Hence German scientists were unable and unwilling to attend the Second International Congress of Eugenics in New York in September of 1921. Instead, they sent bitter protest letters to Cold Spring Harbor, denouncing the French and Belgian occupation of their land and seeking moral support from colleagues in America. Indeed, even though invitations to the congress were mailed to eugenicists around the world by the State Department, the Germans were excluded due to escalating postwar diplomatic and military tensions. Three weeks before the Second Congress, Davenport wrote to one prominent Berlin colleague, Agnes Bluhm, "profound regrets that international complications have prevented formal invitations to the International Eugenics Congress in New York City." He added his "hope that by the time of the following Congress such complications will have been long removed." So once again American science took center stage in international eugenics. Alienated from much of the European movement, Germany's involvement in the field was now mainly limited to correspondence with Cold Spring Harbor.[27]

In 1922, Germany defaulted on its second annual reparations payment. France and Belgium invaded Germany's rich industrial Ruhr region on January 11, 1923, to seize coal and other assets. During the height of the harsh Ruhr occupation, the Weimar government began printing money day and night to support striking German workers. This shortsighted move made Germany's currency worthless nearly overnight, leading to unprecedented hyperinflation.[28]

All of these factors contributed to Germany's isolation from organized eugenics. Efforts by Davenport in 1920 and 1921 to include German scientists in the International Eugenics Commission were rebuffed. None of the players wanted to sit together. Determined to bring German eugenicists back into the worldwide movement, Davenport traveled to Europe in 1922. He selected Lund, Sweden, as the site of the 1923 conference, because, as he confided to a German colleague, "it would be convenient to Berlin." It also circumvented Allied nations such as Belgium, England and France. Davenport then arranged for his colleagues on the IEC to take the first step and formally invite German representatives to join the commission. But tensions over the Rhineland and reparations were still too explosive for the Germans to agree. By the spring of 1923, Davenport had to concede in frustration, "German delegates would not meet in intimate association with the French."[29]

Davenport wrote to one key German eugenicist, "I implore you, that you will use your influence to prevent such a backward step. The only way we can heal the wounds caused by the late war is to repress these sad memories from our scientific activities. It will do a lot to restore international science and to set an example for other scientific organizations to follow if a delegate is sent to the meeting of the Commission to be held in Lund next autumn."[30]

But the occupation of the Ruhr by French and Belgian forces further inflamed angry German eugenicists. "Cooperative work between Germans and French seems to be impossible so long as the Ruhr invasion lasts," one embittered German eugenic leader wrote Davenport. "If in America a foreign power had entered and held in its grasp the chief industrial area surely no American man of science would sit with a representative of that other nation at a table. Therefore, one should correspondingly not expect Germans to do this."[31]

Weimar continued to print money around the clock, creating hour-to-hour hyperinflation. Fabulous stories abounded of money being carted around in wheelbarrows and being used to stoke furnaces. One famous story centered on a Freiburg University student who ordered a cup of coffee listed on the menu for 5,000 marks; by the time he ordered a refill, the second cup cost 9,000 marks. Another told of an insurance policy redeemed to buy a single loaf of bread. The American dollar, which had traded for 1,500 marks in 1922, was worth 4.2 trillion marks by the end of 1923.[32]

German extremists tried to exploit the hyperinflation crisis to start a political revolution to abrogate the Treaty of Versailles. Among the agitators was Adolf Hitler. In November of 1923, Hitler organized the Beer Hall Putsch in Munich. He hoped to seize power in Bavaria and march all the way to Berlin. His rebellion was quickly put down. Hitler was sentenced to five years in prison, to be served at Landsberg Fortress. Referring to his jail cell as his "university," Hitler read voraciously. It was during these prison years that Hitler solidified his fanatical eugenic views and learned to shape that fanaticism into a eugenic mold.[33]

Where did Hitler develop his racist and anti-Semitic views? Certainly not from anything he read or heard from America. Hitler became a mad racist dictator based solely on his own inner monstrosity, with no assistance from anything written or spoken in English. But like many rabid racists, from Plecker in Virginia to Rentoul in England, Hitler preferred to legitimize his race hatred by medicalizing it, and wrapping it in a more palatable pseudoscientific facade—eugenics. Indeed, Hitler was able to recruit more followers among reasonable Germans by claiming that science was on his side.

The intellectual outlines of the eugenics Hitler adopted in 1924 were strictly American. He merely compounded all the virulence of long-established American race science with his fanatic anti-Jewish rage. Hitler's extremist eugenic science, which in many ways seemed like the logical extension of America's own entrenched programs and advocacy, eventually helped shape the institutions and even the machinery of the Third Reich's genocide. By the time Hitler's concept of Aryan superiority emerged, his politics had completely fused into a biological and eugenic mindset.

When Hitler used the term *master race*, he meant just that, a biological "master race." America crusaded for a biologically superior race, which would gradually wipe away the existence of all inferior strains. Hitler would crusade for a master race to quickly dominate all others. In Hitler's view, eugenically inferior groups, such as Poles and Russians, would be permitted to exist but were destined to serve Germany's master race. Hitler demonized the Jewish community as social, political and racial poison, that is, a biological menace. He vowed that the Jewish community would be neutralized, dismantled and removed from Europe.[34]

Nazi eugenics would ultimately dictate who would be persecuted, how people would live, and how they would die. Nazi doctors would become the unseen generals in Hitler's war against the Jews and other Europeans deemed inferior. Doctors would create the science, devise the eugenic formulas, write the legislation, and even hand-select the victims for sterilization, euthanasia and mass extermination.

Hitler's deputy, Rudolf Hess, coined a popular adage in the Reich, "National Socialism is nothing but applied biology."[35]

While in prison, at his "university," Hitler codified his madness in the book *Mein Kampf*, which he dictated to Hess. He also read the second edition of the first great German eugenic text, *Foundation of Human Heredity and Race Hygiene* (*Grundriss der menschlichen Erblichkeitslehre und Rassenhygiene*), which had been published in 1921. Germany's three leading race eugenicists, Erwin Baur, Fritz Lenz and Eugen Fischer, authored the two-volume set.[36] All three of the book's authors were closely allied to American eugenic science and Davenport personally. Their eugenics originated at Cold Spring Harbor.

Baur, an intense racist, closely studied American eugenic science and formulated his ideas accordingly. He was comfortable confiding to his dear friend Davenport just how those ideas fused with nationalism. For example, in November of 1920, about a year before *Foundation of Human Heredity and Race Hygiene* went to press, Baur wrote to Davenport in almost perfect English, "The Medical Division of the Prussian Government has asked me to prepare a review of the eugenical laws and *Vorschriften* [regulations] which have already been introduced into the differed States of your country." He emphasized, "Of especial interest are the marriage certificates (*Ehebestimmung*)—certificates of health required for marriage, laws forbidding marriage of hereditarily burdened persons among others—[and] further the experiments made in different states with castration of criminals and insane.[37]

"It is at present extraordinarily difficult [here in Germany] to gather together the desired material [about U.S. legislation]," Baur continued. "I am thinking, however, that perhaps in your institute [Carnegie Institution] all this material has been already gathered. That, perhaps, there may be some recent printed report on the matter. If my idea is correct I would be exceedingly thankful to you if you could help me with a collection of the material."[38]

Baur then bitterly complained about confiscatory war reparations under the Treaty of Versailles, and the presence of French and Belgian-African troops as enforcers. "The entire work of eugenics is very difficult with us," Baur reported, "all children in the cities are entirely insufficiently nourished. Everywhere milk and fat are lacking, and this matter will become yet greater if we now shall give up to France and Belgium the *milch* [milk] cows which they have requisitioned [for war reparations]. The entirely unnecessary huge army of occupation eats us poor, but eugenically the worst is what we call the Black Shame, the French negro regiments, which are placed all over Germany and which in the most shameful fashion give free rein to their impulses toward women and children. By force and by money they secure their victims—each French negro soldier has, at our expense, a greater income than a German professor—and the consequence is a frightful increase of syphilis and a mass of mulatto children. Even if all French-Belgian tales of mishandling by German soldiers were true, they have been ten times exceeded by what now—in peace!—happens on German soil.[39]

"But I have wandered far from my theme," Baur continued. "We have under the new government an advisory commission for race hygiene... [which] will in the future pass upon all new bills from the eugenical standpoint. It is for this commission that I wish to prepare the *Referate* [reports] on American eugenic laws." Baur added that the Carnegie researcher Alfred Blakeslee's "paper is in press [for publication in Germany], the plate is at the lithographers."[40]

Baur was one of the principal German scientists Davenport had implored to join the International Eugenics Commission.[41]

Baur's coauthor, Fritz Lenz, like many German eugenicists, was long an aficionado of American sterilization. He lectured German audiences that they were lagging far behind America. Like Baur, Lenz was among the German eugenic leaders Davenport beckoned to join him at the helm of world eugenics. Lenz reluctantly refused Davenport's entreaties to attend either an international commission or congress, and in 1923 candidly declared to Davenport, "Europe goes with rapid steps toward a new frightful war, in which Germany will chiefly participate.... That is the position in Europe and, therefore, I do not believe the time for international congresses has arrived so long as France occupies the Ruhr, that is, not before the second World War. I do not wish this certainly; I know that our race in it would suffer more heavily than in the past World War but it cannot be avoided."[42]

Lenz suggested to Davenport that while he could not participate in international gatherings, German and American eugenics could and should continue to advance eugenic science between them, mainly by corresponding. California eugenic leader Popenoe had already established a vigorous exchange with Lenz. Lenz wanted such bilateral contact extended to the ERO as well. "I would be thankful," he wrote Davenport, "if I also could secure the publications of the Eugenics Record Office in order to notice them [report on them] in the *Archiv für Rassen- und Gesellschaftsbiologie* [*Archives of Race Science and Social Biology*]. I have much missed the bulletins of these last years." Lenz closed his letter with "the hope of a work of mutual service."[43] Lenz later predicted, "The next round in the thousand year fight for the life of the Nordic race will probably be fought in America."[44]

The third coauthor of *Foundation of Human Heredity and Race Hygiene* was Eugen Fischer, a Carnegie Institution "corresponding scientist" since 1904. Fischer was a close colleague of Davenport's, and they would form an international eugenic partnership that would last years.[45]

The two-volume *Foundation of Human Heredity and Race Hygiene* that Hitler studied focused heavily on American eugenic principles and examples. The book's short bibliography and footnotes listed an abundance of American writers and publications, including the *Journal of Heredity*, various *Bulletins* of the Eugenics Record Office, Popenoe's *Applied Eugenics*, Dugdale's *The Jukes*, Goddard's *The Kallikak Family* and Davenport's own three books, *Heredity in Relation to Eugenics*, *The Hill Folk* and *The Nams*. Of course, the Baur-Fischer-Lenz work also featured themes and references from von Hoffmann's *Racial Hygiene in the United States* and Hitler's favorite, Madison Grant's *The Passing of the Great Race*.[46]

The Baur-Fischer-Lenz volumes also included repeated explorations and reiterations of American eugenic issues. World War I U.S. Army testing had revealed that "the high percentage of blue eyes [among recruits] is remarkable." The authors then noted the decline of blue-eyed men since the trait was measured in Civil War recruits. The anthropological fine points of American immigration were probed. For example, Fischer wrote, "In the children of Jews who have emigrated from eastern or central Europe to the United States, the skulls are narrower than those of their broad-skulled parents, and this comparative narrowness is more marked in proportion to the number of years that have elapsed since the migration.... Sicilians acquire somewhat broader heads in the United States." Repeated references to American Indian, Negro, and Jewish characteristics were liberally sprinkled throughout the volumes. They also included information on the Eugenics Record Office and Indiana's pioneering sterilization doctor, Harry Clay Sharp.[47]

The Baur-Fischer-Lenz volumes were well received in Cold Spring Harbor. Davenport promised he would write a review for *Eugenical News*. Both *Eugenical News* and *Journal of Heredity* ran favorable reviews of each subsequent revised edition. One of Popenoe's reviews in *Journal of Heredity*, this one in 1923, lauded the work as "worthy of the best traditions of German scholarship, and . . . to be warmly recommended." Popenoe especially praised Lenz's sixteen-point program, which outlined plans to cut off defective lines of descent and the "protection of the Nordic race."[48]

It was no accident that Hitler read *Foundation of Human Heredity and Race Hygiene*. It was published by Julius Lehmann of Lehmanns Verlag, Germany's foremost eugenic publishing house. Someone at Lehmanns happily reported to Lenz that Hitler had read his book. Lehmanns Verlag also published Ploetz's *Archiv für Rassen- und Gesellschaftsbiologie*, the *Monatsschrift für Kriminalbiologie (Monthly Journal of Criminal Biology)*, and von Hoffmann's *Racial Hygiene in the United States*. The year after Hitler was imprisoned, Lehmanns published the German translation of Grant's *The Passing of the Great Race*.[49]

Julius Lehmann was not just a publisher with a proclivity for race biology. He was a shoulder-to-shoulder coconspirator with Hitler during the 1923 Beer Hall Putsch, and was at Hitler's side on November 8, 1923, when the National Socialists launched their abortive coup against the Bavarian government. After the beer hall ruckus, Bavarian officials were held hostage at Lehmann's ornate villa until the uprising was suppressed. As the revolt collapsed, Lehmann, a financial supporter as well as a friend, convinced the Nazi guards to allow their captives to escape rather than execute them. Lehmann was the connection between the theory of the Society for Racial Hygiene and the action of militants such as the Nazis.[50]

Hitler openly displayed his eugenic orientation and thorough knowledge of American eugenics in much of his writing and conversation. For example, in *Mein Kampf* he declared: "The demand that defective people be prevented from propagating equally defective offspring is a demand of the clearest reason and, if systematically executed, represents the most humane act of mankind. It will spare millions of unfortunates undeserved sufferings, and consequently will lead to a rising improvement of health as a whole."[51]

Hitler mandated in *Mein Kampf* that "The Peoples' State must set race in the center of all life. It must take care to keep it pure.... It must see to it that only the healthy beget children; that there is only one disgrace: despite one's own sickness and deficiencies, to bring children into the world.... It must put the most modern medical means in the service of this knowledge. It must declare unfit for propagation all who are in any way visibly sick or who have inherited a disease and can therefore pass it on, and put this into actual practice."[52]

Hitler railed against "this... bourgeois-national society [to whom] the prevention of the procreative faculty in sufferers from syphilis, tuberculosis, hereditary diseases, cripples, and cretins is a crime.... A prevention of the faculty and opportunity to procreate on the part of the physically degenerate and mentally sick, over a period of only six hundred years, would not only free humanity from an immeasurable misfortune, but would lead to a recovery which today seems scarcely conceivable.... The result will be a race which at least will have eliminated the germs of our present physical and hence spiritual decay."[53]

Repeating standard American eugenic notions on hybridization, Hitler observed, "Any crossing of two beings not at exactly the same level produces a medium between the level of the two parents. This means: the offspring will probably stand higher than the racially lower parent, but not as high as the higher one.... Such mating is contrary to the will of Nature for a higher breeding of all life."[54]

In some cases, Hitler's eugenic writings resembled passages from Grant's *The Passing of the Great Race*. Grant wrote, "Speaking English, wearing good clothes and going to school and to church do not transform a Negro into a white man. Nor was a Syrian or Egyptian freedman transformed into a Roman by wearing a toga and applauding his favorite gladiator in the amphitheater."⁵⁵

In a similar vein, Hitler wrote, "But it is a scarcely conceivable fallacy of thought to believe that a Negro or a Chinese, let us say, will turn into a German because he learns German and is willing to speak the German language in the future and perhaps even give his vote to a German political party." He also noted, "Surely no one will call the purely external fact that most of this lice-ridden [Jewish] migration from the East speaks German a proof of their German origin and nationality."⁵⁶

Grant wrote, "What the Melting Pot actually does in practice can be seen in Mexico, where the absorption of the blood of the original Spanish conquerors by the native Indian population has produced the racial mixture which we call Mexican and which is now engaged in demonstrating its incapacity for self-government. The world has seen many such mixtures and the character of a mongrel race is only just beginning to be understood at its true value."⁵⁷

In a similar vein, Hitler wrote, "North America, whose population consists in by far the largest part of Germanic elements who mixed but little with the lower colored peoples, shows a different humanity and culture from Central and South America, where the predominantly Latin immigrants often mixed with the aborigines on a large scale."⁵⁸

Mein Kampf also displayed a keen familiarity with the recently-passed U.S. National Origins Act, which called for eugenic quotas. "There is today one state in which at least weak beginnings toward a better conception [of immigration] are noticeable. Of course, it is not our model German Republic, but the [United States], in which an effort is made to consult reason at least partially. By refusing immigrants on principle to elements in poor health, by simply excluding certain races from naturalization, it professes in slow beginnings a view which is peculiar to the Peoples' State."⁵⁹

In page after page of *Mein Kampf*'s rantings, Hitler recited social Darwinian imperatives, condemned the concept of charity, and praised the policies of the United States and its quest for Nordic purity. Perhaps no passage better summarized Hitler's views than this from chapter 11: "The Germanic inhabitant of the American continent, who has remained racially pure and unmixed, rose to be master of the continent; he will remain the master as long as he does not fall a victim to defilement of the blood."⁶⁰

Hitler proudly told his comrades just how closely he followed American eugenic legislation. "Now that we know the laws of heredity," he told a fellow Nazi, "it is possible to a large extent to prevent unhealthy and severely handicapped beings from coming into the world. I have studied with great interest the laws of several American states concerning prevention of reproduction by people whose progeny would, in all probability, be of no value or be injurious to the racial stock.... But the possibility of excess and error is still no proof of the incorrectness of these laws. It only exhorts us to the greatest possible conscientiousness.... It seems to me the ultimate in hypocrisy and inner untruth if these same people [social critics]—and it is them, in the main—call the sterilization of those who are severely handicapped physically and morally and of those who are genuinely criminal a sin against God. I despise this sanctimoniousness...."[61]

Reflecting upon the race mixing caused by occupying French-African troops and his hope for Nordic supremacy, Hitler later told one reporter, "One eventually reaches the conclusions that masses of men are mere biological plasticine [clay]. We will not allow ourselves to be turned into niggers, as the French tried to do after 1918. The nordic blood available in England, northern France and North America will eventually go with us to reorganize the world."[62]

Moreover, as Hitler's knowledge of American pedigree techniques broadened, he came to realize that even he might have been eugenically excluded. In later years, he conceded at a dinner engagement, "I was shown a questionnaire drawn up by the Ministry of the Interior, which it was proposed to put to people whom it was deemed desirable to sterilize. At least three-quarters of the questions asked would have defeated my own good mother. If this system had been introduced before my birth, I am pretty sure I should never have been born at all!"[63]

Nor did Hitler fail to grasp the eugenic potential of gas and the lethal chamber. Four years before *Mein Kampf* was written, a psychiatrist and a judge published their treatise, *Permission to Destroy Life Unworthy of Life*, which insisted that the medical killing of the unfit, such as the feebleminded, was society's duty; but the extermination had to be overseen by doctors. Several subsequent publications endorsed the same view, making the topic *au courant* in German eugenic circles. Hitler, who had himself been hospitalized for battlefield gas injuries, wrote about gas in *Mein Kampf*. "If at the beginning of the War and during the War twelve or fifteen thousand of these Hebrew corrupters of the people had been held under poison gas, as happened to hundreds of thousands of our best German workers in the field, the sacrifices of millions at the front would not have been in vain. On the contrary: twelve thousand scoundrels eliminated in time might have saved the lives of a million real Germans, valuable for the future."[64]

On January 30, 1933, Adolf Hitler seized power following an inconclusive election. During the twelve-year Reich, he never varied from the eugenic doctrines of identification, segregation, sterilization, euthanasia, eugenic courts and eventually mass termination of germ plasm in lethal chambers. During the Reich's first ten years, eugenicists across America welcomed Hitler's plans as the logical fulfillment of their own decades of research and effort. Indeed, they were envious as Hitler rapidly began sterilizing hundreds of thousands and systematically eliminating non-Aryans from German society. This included the Jews. Ten years after Virginia passed its 1924 sterilization act, Joseph DeJarnette, superintendent of Virginia's Western State Hospital, complained in the *Richmond Times-Dispatch*, "The Germans are beating us at our own game."[65]

Most of all, American raceologists were intensely proud to have inspired the purely eugenic state the Nazis were constructing. In those early years of the Third Reich, Hitler and his race hygienists carefully crafted eugenic legislation modeled on laws already introduced across America, upheld by the Supreme Court and routinely enforced. Nazi doctors and even Hitler himself regularly communicated with American eugenicists from New York to California, ensuring that Germany would scrupulously follow the path blazed by the United States.[66] American eugenicists were eager to assist. As they followed the day-to-day progress of the Third Reich, American eugenicists clearly understood their continuing role. This was particularly true of California's eugenicists, who led the nation in sterilization and provided the most scientific support for Hitler's regime.[67]

In 1934, as Germany's sterilizations were accelerating beyond five thousand per month, the California eugenic leader and immigration activist C. M. Goethe was ebullient in congratulating E. S. Gosney of the San Diego–based Human Betterment Foundation for his impact on Hitler's work. Upon his return in 1934 from a eugenic fact-finding mission in Germany, Goethe wrote Gosney a letter of praise. The Human Betterment Foundation was so proud of Goethe's letter that they reprinted it in their 1935 Annual Report.[68]

"You will be interested to know," Goethe's letter proclaimed, "that your work has played a powerful part in shaping the opinions of the group of intellectuals who are behind Hitler in this epoch-making program. Everywhere I sensed that their opinions have been tremendously stimulated by American thought, and particularly by the work of the Human Betterment Foundation. I want you, my dear friend, to carry this thought with you for the rest of your life, that you have really jolted into action a great government of 60 million people."[69]

CHAPTER 15

Hitler's Eugenic Reich

On the evening of Friday, September 27, 1929, the upper echelon of eugenics met in majestic and Mussolini-ruled Rome, in the high-ceilinged library of the newly created Central Statistical Institute.[1]

They came from Sweden, Norway, Holland, Italy, England, Germany and the United States, gathering as the International Federation of Eugenic Organizations. Among this group, two men ruled supreme: Charles Davenport and Eugen Fischer. A large map dominated the room. This was no ordinary map, but an atlas of the defective populations on every inhabited continent.[2]

The men were flushed with excitement. Just two hours earlier, they had met personally with Mussolini at the Piazza Venezia, with a view of Trajan's Column of antiquity. Indeed, their mission was a return to hereditary antiquity. All were intensely aware that they were assembled for a sacred duty in a city they revered as "the oldest capital of the world." Davenport read the preliminary report of the Committee on Race Crossing. Entire populations of the unfit were designated. The eugenic atlas and other maps were scrutinized for the "regions in which the Committee had ascertained that tolerably pure races were intermarrying... [creating] first generation hybrids." These would be the first people subjected to eugenical measures.[3]

Jon Alfred Mjøen of Norway displayed a map of his country, pinpointing regions with high concentrations of tuberculosis; he proclaimed that the tubercular zones constituted "a map of race crosses in Norway." Mjøen wanted to target Lapp, Finn and Norwegian hybrids. Captain George Pitt-Rivers of England called for anthropologists to help catalog ethnographic statistics, asserting that the most dangerous effect of miscegenation was its disruption of "the ethnic equilibrium shown in the differential survival rate." The Dutch representative focused on the mixed breeds of the Java islands. In describing America's problem, Davenport spoke of U.S. Army intelligence testing that documented high levels of mental defectives. He also discussed tuberculosis rates in Virginia, comparing what he called "the Black Belt" against other areas in the state. Fischer insisted that the "whole weight of the Federation should be engaged in supporting this work." He suggested that "Jew-Gentile crosses providing excellent material were obtainable in most European countries, and that bastard twins would give splendid data."[4]

During the course of their deliberations, the eugenic leaders agreed that paupers, mental defectives, criminals, alcoholics and other inferior strains should be incarcerated *en masse*. They resolved that "all ... members [should] bring to the notice of their governments the racial dangers involved in allowing defective persons, after training and rehabilitation in institutions, to return to free life in the community." In other words, they were advocating permanent incarceration. Only later did someone think to amend the resolution to read, "whilst retaining their ability to procreate."[5]

The worldwide cataloging of the unfit was to begin at once. It would start on "the American continent and certain small and large islands in the oceans." At this point, America was still the only country with years of experience in state-sanctioned sterilization and other eugenic legislation. Fischer chimed in, however, that changes in the German criminal code were coming, and these would soon enable widespread sterilization and other eugenic measures there.[6]

Hitler's arrival on the eugenic scene changed the entire partnership between German and American eugenicists.

America had shown Germany the way during the first two decades of the twentieth century, treating the struggling German movement with both parental fascination and Nordic admiration. But when Hitler emerged in 1924, the relationship quickly shifted to an equal partnership. National Socialism promised a sweeping hereditary revolution, establishing dictatorial racial procedures American activists could only dream of. During the period between wars, the American movement viewed National Socialism as a rising force that could, if empowered, impose a new biological world order. Nazi eugenicists promised to dispense with the niceties of democratic rule. So even if America's tower of legislation, well-funded research and entrenched bureaucratic programs still monopolized the world of applied eugenics in the 1920s, National Socialism promised to own the next decade. American eugenicists welcomed the idea.

As early as 1923, Davenport and Laughlin decided that *Eugenical News* should add a subtitle to its name. It became *Eugenical News: Current Record of Race Hygiene*.[7] In doing so, the publication discarded any pretense that it might be anything other than a race science journal. Adding Germany's unique term for eugenics, *race hygiene*, was also a bow by the American movement to the Germans.

By 1923, articles from *Archiv für Rassen- und Gesellschaftsbiologie* (*Archives of Race Science and Social Biology*) were highlighted and summarized almost quarterly in *Eugenical News*. In fact, no longer did such reviews bear specific headlines about interesting articles; rather, the summaries appeared as though they were regular columns, often just headlined "Archiv für Rassen- und Gesellschaftsbiologie," and proceeded to explore the contents of the journal's latest issue. Articles by Lenz, Fischer and Baur were among those most frequently featured.[8]

In the 1920s, German raceologists became even more sought after as authors and topics for the *Journal of Heredity* and *Eugenical News*, thus increasing their influence in American eugenic circles. For instance, in May of 1924 Fritz Lenz authored a long article for the *Journal of Heredity* simply titled "Eugenics in Germany," with the latest news and historical reminiscences. California eugenicist Paul Popenoe, head of the Human Betterment Foundation, functioned as Lenz's principal translator in the United States. Similar articles were published from time to time as updates, thus keeping the American movement's attention riveted on the vicissitudes of the German school. A typically enthralled review of the latest German booklet on race hygiene ran in the October 1924 *Eugenical News* with the lead sentence: "It was a happy thought that led Dr. Lewellys F. Barker, a leading eugenicist as well as a physician, to translate the little book of Dr. H.W. Siemens, of Munich, into English."[9] Such fawning editorial treatment appeared in virtually every edition of American eugenic journals.

Nor was coverage of German race hygienists and their work limited to the eugenic press. They were reported as legitimate medical news in almost every issue of the *Journal of the American Medical Association*, chiefly by the journal's German correspondent. For example, in May of 1924, Erwin Baur's latest lecture to Berlin's local eugenics society was covered in great detail in a two-column story. *JAMA* repeated, without comment or qualification, Baur's race politics. "A person of moderate gifts may be educated to be very efficient," the article read, "but he will never transmit other than moderate gifts to his own offspring. The attempts to elevate the negroes of the United States by giving them the same educational advantages the white population receives have necessarily failed." The article also regurgitated Baur's contention that the Jukes family was proof positive of eugenically damaged ancestry. "Race suicide," *JAMA* continued from Baur's speech, "brought about the downfall of Greece and Rome, and Germany is confronted by the same peril." *JAMA* used no quotation marks and presented the statements as unredacted medical knowledge.[10]

Nor did the rise of Hitler in Weimar race politics, after 1924, diminish the frequency or prominence of German raceologists' exposure in the American eugenic press. The January 1926 issue of *Eugenical News* featured a long article, written by Lenz, entitled "Are the Gifted Families in America Maintaining Themselves?" Dense with statistics and formulas, Lenz's article analyzed recent California eugenic research with a German mindset, warning "the dying out of the gifted families ... of the North American Union [United States] proceeds not less rapidly; and also among us in Europe.... I think one ought not to look at the collapse of the best elements of the race without action."[11]

When Lehmann's fascist publishing house released a series of race cards, that is, popular trading cards depicting racial profiles—from the Tamils of India to the primitive Baskirs of the Ural Mountains—their availability was fondly reported in *Eugenical News*. Fascinated with the novelty, *Eugenical News* suggested, however, that the cards could be improved if the pictures would reveal more body features. German race cards, just like many baseball cards, came ten to a package.[12]

In May of 1927, *Eugenical News* reported the introduction of a German "race biological index," to eugenically rate different ethnic groups. The article repeated German warnings "of the danger of an eruption of colored races over Europe, through the French colonies and colonial troops." In the article, German researchers urged "further studies in America, both of Indians and American negroes, as compared with those still living in Africa."[13] German race analyses of American society were always well received.

Unqualified German racial references to Jews gradually became commonplace in American publications as well. For example, in the April 1924 issue of *Eugenical News*, an article reviewing a new German "racial pride" book published by Lehmanns mentioned, "In an appendix the Jews are considered, their history and their role in Germany." A German article on consanguineous marriages summarized in the November 1925 issue of *Eugenical News* stated, "Their evil consequences ... are pointed out [and] ... are commoner among Jews and royalty than elsewhere in the population." A December 1927 summary of a German article reported, "The social biology and social hygiene of the Jew is treated by the distinguished anthropologist, Wissenberg of Ukrania. This has largely to do with the vital statistics of the Jews in Odessa and Elizabethgrad, with special relation of the Jews to acute infection." In April of 1929, a *Eugenical News* book review entitled "Noses and Ears" informed readers, "The straight nose of Gentiles seems to dominate over the convex nose of Jews."[14] No explanation was necessary or offered for these out-of-context references to Jews. That Jews were eugenically undesirable was a given in German eugenics, and many American eugenicists adopted that view as well.

By the mid-twenties, Germany had achieved preeminence in both legitimate genetic research and racial biology. Germany's new status arose, in large measure, from its distinguished Kaiser Wilhelm Institutes. An outgrowth of the esteemed Kaiser Wilhelm Society, the Kaiser Wilhelm Institutes would over time develop a network of research institutions devoted to the highest pursuits of science. These included the Kaiser Wilhelm Institute for Physics, boasting a shelf of Nobel Prizes, a sister institute for chemistry, another for biology, another for pathology, and many more. The twenty-plus Kaiser Wilhelm organizations were easily confused and bore related names. But while they were related, they were independent and often located in different cities. In fact, at one point Davenport confessed to a London colleague, "There are so many Kaiser Wilhelm Institutes, that it is necessary to specify."[15]

Also among the Kaiser Wilhelm Institutes were several that would soon make their mark in the history of medical murder. The first was the Kaiser Wilhelm Institute for Psychiatry. The second was the Institute for Anthropology, Human Heredity and Eugenics. The third was the Institute for Brain Research. All received funding and administrative support from Americans, especially the Rockefeller Foundation.

James Loeb, an American banker and art lover of German-Jewish descent who lived in Europe, was among the first to subsidize the organizations that evolved into the Kaiser Wilhelm group. In early 1916, Loeb granted 500,000 marks to the German Psychiatric Institute in Munich.[16] Loeb's money, however, was quickly overshadowed by the Rockefeller Foundation's.

Rockefeller's connection to German biomedicine traced back to the early years of the twentieth century, when Germany's scientific preeminence was first challenged by America and its new system of corporate philanthropic funding begun by Carnegie, Rockefeller and Harriman. Medical educator Abraham Flexner was among the first to establish significant corporate philanthropic financial links with Germany. Flexner completed his monumental Carnegie Institution survey, *Medical Education in the United States and Canada*, in 1910. The prodigious report compared North America's medical inadequacy to Germany's excellence. Flexner next turned to Europe, creating the 1912 report, *Medical Education in Europe*. Soon Flexner was renowned for his pioneering reports and was invited to help lead medical efforts at Rockefeller's powerful new foundation.[17]

One of Flexner's first Rockefeller efforts yielded the 1914 study, *Prostitution in Europe*, which featured an introduction by John D. Rockefeller Jr. himself. Prostitution was a topic of recurring interest to both Rockefeller and his foundation. At about this time, 1914, German academicians began to realize that generous American-style philanthropy was a springboard to higher scientific achievement. Several esteemed German academicians and industrialists organized the Kaiser Wilhelm Society in this vein, with Kaiser Wilhelm II as its chief patron. The society sponsored the Kaiser Wilhelm Institutes, dedicated to a spectrum of new scientific disciplines. But the First World War, the Treaty of Versailles, and the crippling inflation of the early twenties paralyzed the KWI and German scientific progress.[18]

To literally save German science, Rockefeller money—guided by Flexner's recommendations—came to the rescue in November of 1922. Because anti-German feeling engendered by the war still roiled in America, and because Rockefeller, like many, distrusted German universities, viewing them as hotbeds of political agitation and warmongering academics, the Rockefeller Foundation circumvented the universities, the traditional channels of scientific funding. Instead, the foundation inaugurated its own special funding committee. Flexner selected his longtime Berlin friend Heinrich Poll to lead the committee. Poll had assisted Flexner during his earlier survey of German medical schools. Poll, also a leading eugenicist, advised the Prussian Ministry of Health and lectured extensively on hereditary traits and feeblemindedness. Since relations between Germany and the United States were still uneasy late into 1922, the foundation in large part administered the massive donations through its Paris office.[19]

Rockefeller Foundation money began to flow immediately. During the final weeks of 1922, 194 fellowships were awarded, totaling $65,000. The next year, 262 fellowships were awarded for a total of $135,000. By 1926, Rockefeller had donated some $410,000—almost $4 million in twenty-first-century money—to hundreds of German researchers, either directly or indirectly through international programs that passed funds through to German recipients.[20]

Quickly, Rockefeller's freely flowing money, distributed by Poll, became a forceful and intrusive factor in German research. Scientists across

Germany eagerly sent in reports of their worthiness, each hoping to be the next recipient. By March of 1923, leading German researchers, such as Fritz Haber, were grumbling to each other about "King Poll," whom they said exercised an intolerable control over Rockefeller grants and therefore German science itself.[21]

Ignoring any criticism, the Rockefeller Foundation only increased its extravagant spending. Loeb was instrumental in convincing Flexner to marshal Rockefeller millions for Loeb's favorite, the German Psychiatry Institute. Rockefeller officials were fascinated with the promise of psychiatry, and they began aligning themselves with German psychiatrists of all stripes. The German Psychiatry Institute was the first to receive big money. In May of 1926, Rockefeller awarded the institute $250,000 shortly after it amalgamated with the Kaiser Wilhelm Institute to become the Kaiser Wilhelm Institute for Psychiatry. The following November, Rockefeller trustees allocated the new institute an additional $75,000.[22]

Among the leading psychiatrists at the institute was Ernst Rüdin, who headed the genealogical and demographic department. Rüdin would soon become director of the institute. Later, he would become an architect of Hitler's systematic medical repression.[23]

Who was Rüdin? A founding father of German eugenics in the Weimar days, Rüdin was considered by American circles as among the most promising raceologists in Germany. In the 1890s, Rüdin joined Alfred Ploetz in a quest for utopian socialism. The two men became fast friends after Ploetz married Rüdin's sister. From the beginning, Rüdin's impulse was to stop dangerous human breeding. At the 1903 International Congress Against Alcoholism, Rüdin declared that the condition was an inherited trait. Alcoholics, he argued, should be segregated and allowed to marry only if they were first sterilized. In 1905, Rüdin cofounded the Society for Racial Hygiene (*Gesellschaft für Rassenhygiene*) with Ploetz. During the next several years, Rüdin pontificated against the unfit in articles and in his travels.[24]

After World War I, as the chief of the German Psychiatry Institute's genealogical and demographic department, Rüdin began assembling a massive catalog of family profiles from the records of prisons, churches, insane asylums, hospitals, and from family interviews. By 1926, Rüdin was granted special permission by the Reich Ministry of the Interior to consult criminal and institutional records and report back with his own findings. In other words, Rüdin's operation began forming the same types of discreet governmental relationships that the Eugenics Record Office had structured in the United States during the previous fifteen years.[25]

Rüdin, of course, was quite visible in America. Articles by and about him had run in the national eugenic press for years. In May of 1922, the *Journal of Heredity* published a brief about a discussion by Rüdin on the inheritance of mental defects. In June of 1924, *Eugenical News* informed its readership that Rüdin was building an extensive collection of family histories, and assured "a vast quantity of data has been obtained." Later that year, in the September issue, *Eugenical News* published a follow-up report, asserting that Rüdin's studies of the "inheritance of mental disorders are the most thorough that are being undertaken anywhere. It is hoped that they will be long continued and expanded." A 1925 *Eugenical News* article praising the family tree archives of the German Psychiatric Institute celebrated Rüdin, "whose dynamic personality infuses itself throughout the entire establishment." By this time Rüdin was the star of German eugenics. Later, the *Journal of the American Medical Association* also published a long report about Rüdin's work on heredity and mental disease.[26]

Davenport's efforts to bring the Germans back into the international movement were more than successful. In 1928, the International Federation of Eugenic Organizations met in Munich. Rüdin functioned as the gracious host when IFEO members, including the impressed American delegation, were treated to a guided tour of Rüdin's department at the Kaiser Wilhelm Institute for Psychiatry. The next year, the Kaiser Wilhelm Institute for Psychiatry was selected for IFEO membership. In 1932, Davenport consented to relinquish the presidency of the IFEO, and Rüdin was elected to succeed him. Laughlin was proud to offer the nomination. The vote was unanimous.[27] German race hygiene was now primed to seize the reins of the international movement and become senior in its partnership with the American branch.

In 1927, the Kaiser Wilhelm Institutes added another eugenic establishment, the Institute for Anthropology, Human Heredity and Eugenics (*Kaiser Wilhelm Institut für Anthropologie, menschliche Erblehre und Eugenik*), located in Berlin-Dahlem. The name itself symbolized the affinity between the American and German movements. Earlier, *Eugenical News* had adopted a subtitle in homage to the German term *race hygiene*; now the Kaiser Wilhelm Institutes reciprocated by including the term *eugenics* in tribute to the American movement.[28]

The first director of the Institute for Anthropology, Human Heredity and Eugenics was Eugen Fischer, a longtime Carnegie Institution associate and Davenport collaborator. This new institute was not funded by American capital, but rather by an assortment of German government agencies—local, Prussian and federal—to whom eugenics and race science were becoming increasingly important. The Ministry of the Interior provided the largest single donation: 500,000 marks. The Prussian Ministry of Science donated some 400,000 marks, including the land itself. Small amounts were also contributed by the provinces of Upper Silesia, the Rhine, Westphalia and the municipality of Essen. Funds from industrialists, such as the Thyssen brothers, comprised just token monies.[29] While the institute's initial funding was German, it enjoyed both the envy and unqualified support of the American eugenics establishment.

The grand opening of the Institute for Anthropology, Human Heredity and Eugenics took place in September of 1927 as an official function of the Fifth International Congress on Genetics in Berlin. Davenport was chairman of the human eugenics program and an honorary president of the congress. Baur was chairman of the local German eugenics committee. The congress was the first major international scientific event to be held in Germany since the Great War.[30]

The congress began on September 11, 1927, with approximately one thousand delegates from all over the world gathered in a gala Berlin setting. Registrants were first greeted with a Sunday dinner at the zoo, then a barrage of sumptuous banquets staged by the Berlin Municipality and formal dinner events enlivened by *divertimenti*, followed by the finest liquors and cigars. Museum tours were scheduled for the ladies, and everyone was invited to a special performance at the Opera House.[31] Germany was unfurling the red carpet to celebrate its regained scientific leadership.

Welcoming grandiloquence by both government officials and local academics eventually gave way to the real business of the conference: genetics. A procession of several dozen research papers and exhibits reported the latest developments in a spectrum of related disciplines, from genuine scientific revelations about the genetics of plants and animals, to the most recent advances in cytology, to the newest slogans and Mendelian math of traditional racial eugenics. A large Carnegie contingent was on hand to contribute its own research, proffering papers and delivering lectures.[32]

On the afternoon of September 27, Davenport and his colleagues traveled to Berlin-Dahlem for the much-anticipated grand opening of the new Institute for Anthropology, Human Heredity and Eugenics. Davenport had been eager to congratulate his friend Fischer in person from the moment he had learned about his appointment almost a year earlier. Situated on about an acre of land, with a museum in the basement and a complex of lecture rooms, measurement labs and libraries on most other floors, the institute was the new centerpiece of eugenic research in Germany. As the leader of American eugenics, Davenport proudly delivered one of the commemorating addresses at the grand opening. The next year, the IFEO added the new institute to its roster. Davenport was so impressed with Fischer's institute that he felt obliged to provide a brief history of eugenic progress in America to the institute's administration.[33]

The third Kaiser Wilhelm Institute in Germany's eugenic complex was the Institute for Brain Research. Like other Kaiser Wilhelm Institutes, this one grew out of a research operation created years earlier by the family of psychiatrist Oskar Vogt, which merged into the KWI in 1915. In those days the Institute for Brain Research was housed in a modest neurological laboratory also run by Vogt. Everything changed when the Rockefeller money arrived in 1929. A grant of $317,000 allowed the institute to construct a major building and take center stage in German race biology. Rockefeller funders were especially interested in the Institute's Department of Experimental Genetics, headed by Russian geneticist Nikolai Timoféeff-Ressovsky. The Institute for Brain Research received additional grants from the Rockefeller Foundation during the next several years.[34]

By the late twenties, Davenport and other Americans had created a whirlwind of joint projects and entanglements with German eugenics. No longer content to direct purely domestic efforts, the two schools now eyed the rest of the world. They graduated from discussion and philosophy to concrete plans and actions. Among the most ambitious of these was a project to identify and subject to eugenic measures every individual of mixed race, everywhere. The approach would be along the lines created in the United States. Identification was the first step. In 1927, Davenport proposed a systematic survey of mixed-race populations in every region of the world. It would cover all Africans, Europeans, Asians, Mexicans, indigenous peoples and others who had mixed during centuries of modern civilization.

The global search for hybrids originated around February of 1926. Davenport had made the acquaintance of wealthy raceologist Wickliffe Draper, who shared Davenport's anxiety about human hybridization. The plan was to conduct field surveys using questionnaires, just as eugenicists had done in various counties and remote areas around the United States. But this time they would cover not just a state, not just a nation, but eventually every populated region on earth.[35]

They needed a demonstration project. Davenport's first impulse was to survey New York City, but he thought mixed-race individuals would be easier to identify in foreign countries or colonies. "I am suggesting Jamaica,"

Davenport wrote Draper on February 23, 1926, "...because I take it that there is a larger proportion of mulattoes." Within three weeks, Draper wrote a check to the Eugenics Research Association for $10,000 to defray the costs of a two-year study of "pure-blooded negroes, as found in the western hemisphere ... and of white, as found in the same places with especial reference to inheritance of the differential traits in mulatto offspring."[36]

Over the next two years, Davenport's investigators deftly researched the family backgrounds of 370 individuals, taken from the local penitentiary and from the city center of Kingston. The American Consul in Jamaica interceded with the British Colonial Office to provide special access to the island's jails, schools and doctors. Some eight thousand sheets of information were generated by field workers and archived in the Eugenics Record Office.[37]

But the Jamaica project featured something totally new. For the first time, personal information and eugenic traits were punched into IBM's Hollerith data processing machines. International Business Machines would be a perfect match for eugenics. People tracking was the company's business. IBM's technology involved hundreds of thousands of custom-designed punch cards processed through punching, tabulating and sorting machines. Hollerith punch cards could store an almost unlimited amount of information on people, places and processes by virtue of the holes strategically punched into their columns and rows. Hollerith processors then read these holes and tabulated the results. Hollerith cards were originally developed for the U.S. Census, and IBM enjoyed a global monopoly on data processing. More than just counting machines, Hollerith systems could cross-tabulate all information on individuals and then match or cross-reference the data to their plain paper or already-punched street addresses or other geographic identifiers. Hence, people identified with certain traits could be easily located for additional eugenic action.[38]

For example, these high-speed tabulators could quickly identify a specific class of eugenic subjects, say, all first-generation morons of Mexican extraction with vision problems. All relatives across extended family trees could be connected to the selected individuals. Or the machines could identify all eugenically inferior residents in a single village, plus their descendants living elsewhere. At the rate of 25,000 cards per hour, IBM machines could rapidly search out the holes, stack the cards and provide seemingly miraculous results. Continuous refinements in high-speed Hollerith technology would soon permit alphabetizing and printouts. As massive numbers of individuals passed from identification to segregation to sterilization and beyond, even the workflow could be managed by IBM technology, using card designs, punching patterns and equipment arrays, each custom configured to a specific use. Mass eugenics required efficient systems.[39] IBM was willing.

IBM managers desired the lucrative ERO account, but the process of punching in the hundreds of thousands of existing index cards at Cold Spring Harbor was simply too massive and expensive an undertaking. But if brought into a project at the outset, IBM could cost-effectively tabulate all names, racial information, medical characteristics and other eugenic data. This required IBM engineers to confer with Davenport's eugenic investigators to jointly plan the program, ensuring that data was collected in a fashion that could be systematically coded and punched into Hollerith machines for later retrieval and management. To design the system correctly, the IBM engineers needed to know both the eugenic information that Carnegie researchers wanted to input as well as how they wanted the results retrieved. IBM always needed to know the end result in order to design the system. In a report on the Jamaica project, Davenport confirmed, "The test records were scored as received chiefly by Miss Bertha Jacobson. Codes for each of the traits to be tabulated were worked out, adapted to the Hollerith punch cards. Ratios were computed."[40]

IBM custom-designed the layout for at least forty-five variables to be punched in on the Jamaica project for later retrieval by eugenicists. Sex and race were to be punched into column 1. Age in column 2. Height in columns 3 and 4. Cranial capacity in column 18. Foot length in column 24. Army Alpha intelligence testing in column 33, and Beta testing in column 32. Information on fingerprints was punched into columns 44 and 45. At one point, Davenport considered securing data from banks about how much money was in each individual's account and cross-referencing this information against eugenic standards.[41]

The 1927–1928 Jamaica race-crossing investigation was the first time IBM devised a system to track and report racial characteristics. Five years later, IBM, under the leadership of its president, Thomas J. Watson, would adapt the same technology to automate the race warfare and Jewish persecution in Hitler's Reich. IBM custom-designed the indispensable systems that located European Jews and other undesirables, and then provided a multiplicity of custom-tailored punch card programs to help the Nazis trace family trees, index bank accounts and other property, organize eugenic campaigns and even manage extermination in death camps. Indeed, a decade later, the SS Race Office employed a punch card with physical attributes specified column-by-column in a fashion almost identical to those first worked out for the Jamaica study.[42]

The pilot investigation in Jamaica went well, so well that the Carnegie Institution proudly published a major research volume on the project. Even as the program was underway, in February of 1927, Davenport was confident enough to contact Fischer in Germany and discuss ideas with him. "No one has greater experience in the field than you," wrote Davenport, "and we shall of course want to get the benefit of that experience." A few days later, he notified the IFEO secretary in London that a race-crossing committee would be needed "in view of... the international nature of the problem." In short order, Fischer was invited to join the committee. Davenport would chair the panel.[43]

The campaign to identify mixed-race people of all varieties across America began on November 14, 1928, with one of the ERO's well-honed, massive letter-writing efforts. Beginning that day, scores of letters were mailed by Davenport to eugenic contacts at universities, prisons, agricultural colleges, as well as to members of the American Breeders Association and other interested parties in every state from California to Florida and even the Alaska territory. It was the first step in searching out the racially unacceptable. Davenport's letters were all variations on a few forms:

> The I.F.E.O. is making a survey of the points of contact of dissimilar human races in different parts of the world. In carrying out this program may I call upon you for some assistance? We should be glad if you would inform us if there are areas where widely different races of mankind have recently begun to come into contact in your state. By races we have in mind not only primary races, like white, negro, Indian and Orientals but also very dissimilar European races. Especially important would be localities where the first and second hybrid generations can be secured in considerable numbers.[44]

A letter went to sociologist Raymond Bellamy at the Florida State College for Women; Bellamy replied, "I am glad to do anything I can to help," and specified Negroes and Seminole Indians in South Florida, and Cubans in Tampa. A copy went to W. E. Bryan, a plant breeder at the University of Arizona in Tucson; Bryan reported race-mixing between American Indians and Mexicans, and suggested using a field worker who could speak Spanish. A letter went to J. S. Blitch, superintendent of the Florida State Reformatory; Blitch responded that of his 1,640 prisoners, fewer than a third were white, the rest being "plain negro stock." UCLA official Bennet Allen replied that Los Angeles was home to many ethnic groups, including Japanese, Mexican, Italian, and Portuguese. He also reported that the Mexicans and the Japanese rarely married outside their respective groups. Henry Bolley of the North Dakota Agricultural College's Botany Department reported "half-breeds among our North Dakota Indians, but I think largely of French origin," as well as farmers of Russian and possibly Polish heritage.[45]

On February 29, 1929, Davenport went global. He mass mailed letters to eugenic contacts and official sources in countries on every continent, signing them as president of the IFEO's Committee on Race Crossing. The letters all declared:

> The committee on race crossing of the Federation is seeking to plot the lines, or areas, where race crossing between dissimilar, more or less pure races is now occurring or has been occurring during the last two generations. The committee would appreciate very much your assistance. We should be glad to have a statement from you as to the location in your country or the principal regions of such race crossing, the races involved (e.g. European and negro, European and Amerindian, Chinese, Malay, North European and Mediterranean) together with the number of generations during which hybridization has been going on on a significant scale.[46]

In Norway, Dr. Halfdan Bryn focused on "the northern parts of the country," where, over the centuries, Laplanders and Alpines had mixed with pure Nordics; Bryn added that his forthcoming book, to be published by Lehmann in Munich, would include plenty of pictures of "Norwegian hybrids." In Moscow, Professor Bunak, director of the Institute of Anthropology, explained that the Eastern European plains, the Caucasus, Siberia and Turkistan all featured "numerous tribes, [such] as North European, Baltic, Mediterranean, Armenoid, Uralian (Ougrofinnic), Mongolic, Turck and others" who had intermingled during the past twenty to thirty centuries; more recently, Yakoutian-Russians and other "race-hybrids" had proliferated through the regions. In colonial Rhodesia, a museum zoologist acknowledged some Bantu and Asiatic mixtures, but he assured Davenport that "miscegenation is regarded by decent persons as severely as it probably is in the Southern States of the USA." Reports came from Brazil, China, Holland, France, Fiji, Chile and many more countries.[47]

In locations with no known eugenic contacts, Davenport resorted to Laughlin's network of American consuls. In the Azores, Vice-Consul

Prescott Childs demonstrated an excellent knowledge of eugenic principles and reported that due to the islands' remoteness, very few of Breton or Flemish blood had mixed with pure Portuguese; Childs added that his real "eugenic concern" was too much intermarriage, which he believed led to increased insanity. In Harbin, American Consul C. C. Hansen pointed out that a number of Russians had migrated into North Manchuria resulting in "intermingling of Chinese men with Russian women"; Hansen reported the villages along various rivers where "half-caste children... of the first generation" could be located. In Nairobi, American Consul Charles Albrecht outlined the geographic districts of Kenya and attached a list of photographers "who might be able to furnish you with photographs of race hybrids." In Estonia, Tahiti and other remote locations, American consuls pledged their assistance.[48]

At 6:15 P.M. on Friday, September 27, 1929, the International Federation of Eugenic Organizations met in Rome to consider the preliminary report of the Committee on Race Crossing. From their perspective, identification and eugenic countermeasures of all sorts were more than pressing—the world was in crisis, and they were in a race against time. Mussolini, a dictator, was not hampered by the checks and balances of democracy. The IFEO wanted to enlist him to help impose stern eugenic measures in Italy. Since the summer, Fischer and Davenport had been working on a special appeal to *Il Duce*. Now, in the Piazza Venezia, they and their colleagues would have an audience with Mussolini.[49]

Fischer stepped forward to read the long appeal. It was not lost on the delegation that they were in Rome, seat of the Catholic Church, which strenuously opposed all forms of eugenics. "It seems natural and desirable," Fischer read, "when considering eugenic problems, that some expression of our hopes and wishes should be addressed to the great statesman who... shows more than any other leader today... how much he has the eugenic problems of his people at heart." Fischer went on to label the effects of race mixing "catastrophes," and urged immediate measures to "[set] a model to the world by showing that energetic administration can make good the damage." In an emotional crescendo to his appeal, Fischer declared, "The urgency brooks no delay; the danger is imminent."[50]

Two hours later, the men retreated to the elegant library of the Central Statistical Institute where they huddled over maps, reports, tables and surveys as they plotted the course of their global eugenic action. Virginia, the Java Islands, Norway, Germany, all of Europe, all of the United States, all of the British Empire. The world. With trained field workers and Hollerith data processing equipment, the unfit could be quickly and methodically identified, quantified, qualified and prioritized for countermeasures—whether they resided in big cities, the hinterlands or island villages. Every delegate was instructed to lobby his government for cooperation.[51]

Davenport was encouraged. Fascism was on the rise in Europe, and he realized it was time to relinquish the reins. On December 2, 1929, Davenport wrote to Fischer asking him to assume chairmanship of the Committee on Race Crossing. Rüdin would soon replace Davenport as IFEO president as well. The Germans were the future. Davenport wrote Ploetz in Munich, "Personally, I am very glad that the Federation is now under the *Leitung* [leadership] of a German."[52]

Fischer was willing to assume leadership of the Committee on Race Crossing, but who would pay the postage and printing costs? Davenport replied that the IFEO treasury would, since "it is more important to spend our money that way than almost any other." Davenport and Fischer coauthored a questionnaire to be sent worldwide "to the persons living and working in foreign regions, physicians, missionaries, merchants, farmers and travelers," asking them to "send as detailed and significant data as possible." The questionnaires would be produced in English, Spanish and German. Davenport and Fischer reported in a joint memo that the data would eventually identify not only race-crossed individuals, "but entirely foreign people, that is the so-called colored ones."[53]

As the thirties opened, many key players in the American eugenics movement continued to support German raceology. In December of 1929, the Rockefeller Foundation began a five-year subsidy of Fischer's German national "anthropological survey" with a donation totaling $125,000. Although the study was labeled "anthropological," it was in fact racial, eugenic and, in part, directed at German Jewry. German officials who supported the proposal for the study made this clear in a letter to the foundation. They would not survey a single large sample of people "of an ancient type"; instead, they would select multiple smaller cross-sections of the general population, which would "be examined in its genealogical and historical relationships with the help of church records, place and family histories." The Germans specified, "In this way it is hoped to find new solutions about the appearance of certain signs of degeneration, especially the distribution of hereditary pathological attributes."[54]

The letter continued, "From the eugenic standpoint, questions will be submitted on the biological conditions of families, the number of births and abortions, succession and rate of births, and finally questions on the decline of births and birth registration in the region being investigated. . . . A determination of blood groups will also be undertaken. . . . There is also planned an investigation of the Westphalian aristocracy, of the old-established Jewish population of Frankfurt, and the so-called old lineage of some other towns. . . . For certain eugenic discussions it seemed of the greatest importance to obtain useful support for the question of . . . pathological lines of heredity among the population."[55]

Rockefeller executives quickly approved the idea, channeling the money through the Emergency Fund for German Science. Rockefeller trustees authorized the grant in the midst of the devastating worldwide depression ignited by the stock market crash of 1929. As breadlines stretched across American cities, the economic crisis also crippled the German economy.[56] German eugenicists needed all the financial assistance they could get.

In August of 1930, Germany's *Archiv für Rassen- und Gesellschaftsbiologie* ran a tribute to Ploetz on his seventieth birthday. Among those extending kudos were Davenport and Popenoe on behalf of the United States. In October of 1930, *Eugenical News* called the edition "a worthy tribute of esteem and affection for the genial and high-minded scholar whom it honors." In the same issue of *Eugenical News*, an article entitled "Jews in West Africa" reviewed a book claiming "evidence of Jewish infiltration" among the Masai tribes of Africa as a result of a "trek of Jews from Jerusalem to the Niger." The book was deemed "a good example of the deductive method . . . so great as to make the book a very valuable contribution." The next news item congratulated J. F. Lehmann, now openly Nazi, for being Germany's leading eugenic publisher. At about that time, the IFEO created a Committee on Racial Psychiatry under Rüdin's chairmanship.[57]

In December of 1930, *Eugenical News* reprinted Rüdin's long paper, "Hereditary Transmission of Mental Diseases." In it Rüdin declared, "Humanity demands that we take care of all that are diseased—of the hereditarily diseased too—according to our best knowledge and power; it demands that we try to cure them from their personal illnesses. But there is no cure for the hereditary dispositions themselves. In its own interest, consequently, and with due respect to the laws of nature, humanity must not go so far as to permit a human being to transmit his diseased hereditary dispositions to his offspring. In other words: Humanity itself calls out an energetic halt to the propagation of the bearer of diseased hereditary dispositions."[58]

Rüdin advocated sterilization of all members of an unfit individual's extended family. "It becomes clear," he argued, "that, in these cases, propagation ought to be renounced... for other degrees of relationship, e.g., for the nephews and nieces, grandchildren.... We must make the eugenic ideal a sacred tradition. It must be rooted so deeply in man, and at the right time, that the respect he owes it becomes a matter of course with him, and that he will find love without trespassing on the laws of eugenics."[59]

In 1931, Rockefeller approved an additional ten-year grant totaling $89,000 to Rüdin's Institute for Psychiatry. This grant funded research by two doctors into the links between blood, neurology and mental illness. It reflected a growing trend among some philanthropic foundations to avoid funding scientific organizations focused on eugenics, which in recent years had come under fire for being too political and too scientifically shoddy.[60] Genetics, psychiatry, brain research, anthropology and sociology were all preferable destinations for American biologic research dollars. One Rockefeller memo observed, "Race biology today suffers immensely from its mixture with political dogmas and drives"; in that instance, the foundation had granted $90,000 to a eugenic geneticist who had studied at Cold Spring Harbor, because they felt the recipient was worthy. Moreover, eugenicists were constantly seeking the "carriers"—the normal people who transmitted defective genes that might crop up once in several generations. Because of the bad publicity surrounding this idea, and the growing belief that eugenics was more racism than science, the new breed of eugenicists began looking for blood identifiers that seemed ethnically neutral. Even still, the searches remained race-specific.[61]

Whether under the banner of psychiatry, anthropology, genetics or race hygiene, American funding was still consciously promoting eugenic research. For example, in 1931, the Carnegie Institution contributed $5,000 for an international genetics congress and the separate Carnegie Endowment added $3,500. Davenport also contacted the Rockefeller Foundation to enlist their support for this event.[62]

Also in 1931, the famous Baur-Fischer-Lenz volume, *Foundation of Human Heredity and Race Hygiene* (*Grundriss der menschlichen Erblich-keitslehre und Rassenhygiene*), was translated into English. One chapter was entitled "Racial Psychology" and cited a study demonstrating that "the racial endowment of the Jews finds expression in the nature of the offences they commit." Another passage asserted that "fraud and the use of insulting language really are commoner among Jews," adding, "It is said that Jews are especially responsible for the circulation of obscene books and pictures, and for carrying on the White Slave Trade. Most of the White Slave traders are said to be Ashkenazic Jews." Another passage insisted, "The Jews could not get along without the Teutons." The term *Jewish Question* (*Judenfrage*), which was used throughout the book, required no explanation.[63]

A 1931 review of the newly translated book in *Eugenical News* lauded the work and declared, "the section on methodology is especially valuable," adding that it was now the "standard treatise" on the topic. The review concluded, "We welcome the English translation, which seems to have been well done.... We bespeak for it a wide circulation."[64]

During 1931 and 1932, Hitler became an increasingly loud and pernicious voice for persecution, fascist repression and warlike territorial occupation. In America he was heard on radios, seen in newsreels and read in newspapers. Virulent and very public anti-Semitism was sweeping across Germany.[65] None of this caused American eugenic circles to pause in their support of German eugenics.

In the March–April edition of *Eugenical News*, the long essay "Hitler and Racial Pride" heaped praise on the up-and-coming leader. One passage proclaimed, "The Aryans are the great founders of civilizations.... The mixing of blood, the pollution of race ... has been the sole reason why old civilizations have died out." The Hitlerite term *Aryan* was now becoming synonymous with the traditional *Nordic*. In another passage, the article cited an earlier *New York Times* report declaring, "The Hitlerites hold the Nordic race to be 'the finest flower on the tree of humanity'...It must be bred ... according to the 'criteria of race hygiene and eugenics.'"[66]

On May 13, 1932, the Rockefeller Foundation in New York dispatched a radiogram to its Paris office:

JUNE MEETING EXECUTIVE COMMITTEE NINE THOU-SAND DOLLARS OVER THREE YEAR PERIOD TO KWG INSTITUTE ANTHROPOLOGY FOR RESEARCH ON TWINS AND EFFECTS ON LATER GENERATIONS OF SUBSTANCES TOXIC FOR GERM PLASM. NATURE OF STUDIES REQUIRES ASSURANCE OF AT [Rockefeller's director of science in Europe, Augustus Trowbridge].[67]

At about that time, Fischer and other eugenicists were busy presenting drafts of compulsory sterilization laws to the Weimar authorities. During a committee meeting on the subject in the summer of 1932, Fischer shouted at the Nazi representative, "Your party has not been in existence nearly as long as our eugenic movement!" One leading eugenicist at the Kaiser Wilhelm Institute for Biology later bristled, "The Nazis took over the whole draft and they used the most inhumane and execrable methods to put the humane measures, which we had conscientiously and responsibly drafted, into everyday practice."[68]

The Third International Congress of Eugenics was held in New York City in August of 1932, once again at the American Museum of Natural History. Although organizations such as the Rockefeller Foundation were donating vast sums to German eugenics for research and travel, the grants were frequently limited to specific activities within Germany or neighboring countries. Hence there was no money for the German delegation to travel to Manhattan. Nor did Carnegie make up the shortfall. Davenport apologized in a letter to Fischer. "Of course, the depression at this time has interfered with our efforts to secure funds to help defray the expense of our foreign colleagues.... We are very much disappointed that you and other friends from Europe may not be able to...come to the United States and see the work going on there. We had hoped you would come and find your expenses paid by giving some lectures." But the German delegation did not come, and instead sent a few poster exhibits from the Kaiser Wilhelm Institute for Anthropology, Human Heredity and Eugenics. At the opening ceremonies Davenport lamented the absence of the German delegation and lauded their leadership.[69]

The September–October *Eugenical News* carried another long article praising Hitler and his eugenic ideas. It also explained how his ideology had been guided by such American authors as Lothrop Stoddard and Madison Grant. German elections were looming, and the article prophesied the results. "The Hitler movement sooner or later promises to give him full power, [and] will bring to the Nordic movement general recognition and promotion by the state." The article added, "When they [the Nazis] take over the government in Germany, in a short time there may be expected new race hygiene laws and a conscious Nordic culture and 'foreign policy.'"[70]

The next month, November of 1932, Germany held a fractious election. Hitler received twelve million votes, approximately a third, but no majority. A coalition government was out of the question because other parties refused to share power with Hitler and vice versa.[71]

January 30, 1933, as America awoke, swastikas flew above Berlin, Munich, Leipzig and the other strongholds of Nazi agitation. Brown-shirted mobs marched through the streets in celebration, swaggered in beer halls, rode their bicycles in tandem and joyously sang the "Horst Wessel Song." For years the Nazis had promised that upon assuming power they would rebuild Germany's economy, dismantle its democracy, destroy the German Jewish community and establish Aryans as the master race. On January 30, 1933, President Paul von Hindenburg, exasperated with fruitless all-night attempts to create a governing coalition, finally exercised his emergency powers. Hindenburg appointed Adolf Hitler interim chancellor. The Third Reich was born.[72]

*   *   *

Years later, many would deny knowledge of what Germany was doing, would claim they only discovered Hitler's merciless anti-Semitic and political repression, as well as the Reich's fascist medical programs, after the Allies triumphed in 1945. But in truth, Hitler's atrocities against Jews and others were chronicled daily on the pages of America's newspapers, by wire services, radio broadcasts, weekly newsreels, and national magazines.[73] Germany bragged about its anti-Jewish measures and eugenic accomplishments. An entire propaganda operation was established under Joseph Goebbels to publicize the information.[74] Simultaneously, American eugenicists kept day-to-day tabs on the Nazi eugenic program. As of January 30, 1933, however, the American-German eugenic partnership was obsolete. Germany was now completely leading the way, despite a hurricane of anti-Nazi denunciations and retaliatory economic boycotts.[75]

Once in power, Hitler's government immediately began issuing legal decrees to exclude Jews from professional and governmental life, and used other brutal methods—including condoned street violence—to eliminate political opponents. Dachau concentration camp opened on March 20, 1933, amid international news coverage of the event. Refugees, including many Jewish scientists, poured out of Germany. Their plight was visible in the cities of the world.[76]

It did not take Germany long to implement its eugenic vision. The first law was decreed July 14, 1933: Reich Statute Part I, No. 86, the Law for the Prevention of Defective Progeny. It was a mass compulsory sterilization law. Rüdin was coeditor of the official rules and commentary on the law.[77]

Nine categories of defectives were identified for sterilization. At the top of the list were the feebleminded, followed by those afflicted by schizophrenia, manic depression, Huntington's chorea, epilepsy, hereditary body deformities, deafness and, of course, hereditary blindness. Alcoholism, the ninth category, was listed as optional to avoid confusion with ordinary drunkenness. The Reich announced that 400,000 Germans would immediately be subjected to the procedure, beginning January 1, 1934.[78]

A massive sterilization apparatus was created: more than 205 local eugenic or hereditary courts would be ruled by a physician, a eugenicist and a panel chairman. For contested cases, there were at least twenty-six special eugenic appellate courts. Anyone could be reported for investigation. Doctors who failed to report their suspect patients would be fined. In hearings, physicians were obligated to provide confidential patient information. Fischer's institute was asked to quickly train the legion of race experts required for the task.[79]

Germany's program was immediately seized upon by the world's media as the latest example of Hitler's inhumane regime. Many eugenic leaders felt pressured into publicly disassociating themselves from Nazi barbarism, but their denunciations were only lip service. An anxious C. P. Blacker, director of Britain's Eugenics Society, watched as his own sterilization campaign lost public support as the obvious comparisons were made. "This Society deprecates the use of the term Eugenics to justify racial animosities," Blacker announced, adding that he condemned, "its misuse as an instrument of tyranny by racial or social majorities."[80]

While much of the world recoiled in revulsion, American eugenicists covered eugenic developments in Germany with pride and excitement. By the summer of 1933, *Eugenical News* had become bimonthly due to Depression-era finances, and had changed its subtitle again, this time to *Current Record of Genetic News and Race Hygiene*. Cold Spring Harbor quickly obtained a full copy of the eighteen-paragraph Nazi sterilization law from German Consul Otto Kiep, and rushed a verbatim translation into the next issue as its lead item. In accompanying commentary, *Eugenical News* declared: "Germany is the first of the world's major nations to enact a modern eugenical sterilization law for the nation as a unit.... The law recently promulgated by the Nazi Government marks several substantial advances. Doubtless the legislative and court history of the experimental sterilization laws in 27 states of the American union provided the experience, which Germany used in writing her new national sterilization statute. To one versed in the history of eugenical sterilization in America, the text of the German statute reads almost like the 'American model sterilization law.'"[81]

Proudly pointing out the American origins of the Nazi statute, the article continued, "In the meantime it is announced that the Reich will secure data on prospective sterilization cases, that it will, in fact, in accordance with 'the American model sterilization law,' work out a census of its socially inadequate human stocks."[82]

Countering criticism that Hitler's program constituted a massive human rights abuse, *Eugenical News* asserted, "To one acquainted with English and American law, it is difficult to see how the new German sterilization law could, as some have suggested, be deflected from its purely eugenical purpose, and be made 'an instrument of tyranny,' for the sterilization of non-Nordic races." The publication argued that in the 16,000 sterilizations performed in America over recent years, not a single "eugenical mistake" had been made. The publication concluded, "One may condemn the Nazi policy generally, but specifically it remained for Germany in 1933 to lead the great nations of the world in the recognition of the biological foundations of national character."[83]

Throughout 1933, American eugenic groups continued their enthusiastic coverage of and identification with German mass sterilization. *Birth Control Review* ran an extensive article entitled "Eugenic Sterilization, An Urgent Need," authored by Rüdin himself, and also reprinted a pamphlet he had prepared for British eugenicists. "Act without delay," urged Rüdin. By this time Margaret Sanger had left the publication, and *Birth Control Review* had relaxed its previous position that birth control was for everyone, not just the unfit, and that it was wrong to encourage greater birth rates for the eugenically preferred. Indeed, Rüdin's article did just that. "Not only is it our task to prevent the multiplication of bad stocks," he demanded, "it is also to preserve the well-endowed stocks and to increase the birth rate of the sound average population."[84]

Eugenic influence continued in mainstream medical publications. In 1933, the *Journal of the American Medical Association* reported on the new sterilization statute as if it were an almost routine health measure. *JAMA's* coverage included unchallenged data from Nazi eugenicists such as: "The fact that among the Jews the incidence of blindness is greater than among the remainder of the population of Germany (the ratio is 63 to 53) is doubtless due to the increased danger of hereditary transmission resulting from marriage between blood relatives."[85]

JAMA, in another 1933 issue, continued its tradition of repeating Nazi Judeophobia and National Socialist doctrine as ordinary medical news. For example, in its coverage of the German Congress of Internal Medicine in Wiesbaden, *JAMA* reported that the congress chairman "brought out the following significant ideas:...A foreign invasion, more particularly from the East, constitutes a menace to the German race. It is an imperative necessity that this menace be now suppressed and eliminated.... Racial problems and questions dealing with hereditary biology must receive special consideration." The article continued, "Eugenics and the influences of heredity must be the preferred topics [at future medical meetings]," and then warned of "the severity of the measures to be adopted for the preservation of the German race and German culture."[86]

Eugenical News spoke in similar terms. In a September–October 1933 review of yet another Lehmann-published anti-Semitic epistle, *Race Culture in the Nationalistic State (Rassenflege in Volkischenstaat)*, *Eugenical News* insisted in italics, "*There is no equal right for all....* Nature is not democratic, but aristocratic.... [German racial] demands appear harsh, but... the very existence of the race is at stake."[87]

Rockefeller money continued to stream across the Atlantic. The 1933 financial books of the Institute for Anthropology, Human Heredity and Eugenics reflected the foundation's continuing impact. Page four of the balance sheet: Rockefeller paid for a research assistant, a statistician, two secretaries and a gardener. Page six of the balance sheet: Rockefeller paid clerical costs associated with research on twins. Ironically, while Fischer remained in charge at the Institute for Anthropology, Human Heredity and Eugenics, he was being replaced at the Society for Racial Hygiene. He had taken over the society for Ploetz, but in 1933 Nazis overran the society and Fischer was considered too moderate. He was replaced by Rüdin, then president of the IFEO.[88]

Unlike eugenic leaders associated with *Eugenical News*, Rockefeller officials did not propagandize for Nazism, nor did they approve of the Reich's virulent repression. The Rockefeller Foundation's agenda was strictly biological to the exclusion of politics. The foundation wanted to discover the carriers of defective blood—even if it meant funding Nazi-controlled institutions. Moreover, Rockefeller executives knew their money carried power, and they used it to ensure that the most talented scientists continued at the various Kaiser Wilhelm Institutes, frequently shielding them from periodic Nazi purges.

For example, in early June of 1933, one of the foundation's favorite researchers, Oskar Vogt, head of the Institute for Brain Research, was threatened with removal because of his perceived socialist leanings. Rockefeller mobilized.[89] On June 7, 1933, H. J. Muller, a University of Texas geneticist working at the Institute for Brain Research, alerted Robert A. Lambert in Rockefeller's Paris office. Just days before, Lambert had toured various Berlin research facilities. In his letter, Muller warned Lambert, "If this director loses his position it is a foregone conclusion, and common knowledge, that the head of the genetics department and all other non-Germans, as well as Germans closely associated with the director, will also lose their positions. . . . I realize that the Rockefeller Foundation must preserve its neutrality so far as matters of politics are concerned. On the other hand, it wishes to have its funds used so that they can best serve the furtherance of truly scientific work."[90]

Muller asked Lambert and other Rockefeller executives to consider "the making of a statement, not necessarily a public one, but, it may be, one expressed in a letter to some responsible person, such as for example [physicist] Dr. [Max] Planck, which could then be shown to the authorities concerned, so that they could be informed of your policy, in advance. Some statement similar to that which you made orally to the director of the institute here, would suffice, namely, that the Rockefeller Foundation would not feel justified, from the point of view of the furtherance of scientific work, in sending additional funds to the support of institutions in Germany, (1) if, on grounds other than their scientific work, worthy scientists, not engaged in political activity, are dismissed from institutions which have been founded or supported in part by funds of the Foundation, or (2) if persons who have been assigned stipends from the Foundation are dismissed from such institutions."[91]

Oskar Vogt was not removed. He remained at his post until well after his Rockefeller funding had run its course.[92]

With each passing day, the world was flooded with more Jewish refugees, more noisy anti-Nazi boycotts and protest marches against any scientific or commercial exchanges with Germany, more public demands to isolate the Reich, and more shocking headlines documenting Nazi atrocities and anti-Jewish legislation. Still, none of this gave pause to America's eugenicists. Correspondence on joint research flowed freely across the Atlantic. American eugenicists, and their many organizations and committees, from New York to California and all points in between, maintained and multiplied their contacts with every echelon of official and semiofficial German eugenics. As the Reich descended into greater depths of depraved mistreatment and impoverishment of Jews, as well as territorial threats against its neighbors, these contacts seemed all the more insulated from the human tragedy unfolding within Europe. Eager and cooperative letters, reports, telegrams and memoranda did not number in the hundreds, but in the thousands of pages per month.

While concentration camps, pauperization and repression flourished in Nazi Germany, and while refugees filled ships and trains telling horrifying stories of torture and inhumanity, it was business as usual for eugenics.

Nor were the contacts and scientific support a secret. For example, in March of 1934, eugenicist W. W. Peter published a long article in the *American Journal of Public Health* defending Germany's sterilization program. Peter had traveled some 10,000 kilometers over the course of six months, visiting every region of Germany to study the Reich's plan. He gave it an unqualified endorsement, declaring, "This particular program which Germany has launched merits the attention of all public health workers in other countries."[93]

Sterilizations had begun January 1 of that year. Within forty-eight hours, the Reich Interior Ministry's eugenics expert announced that the list would include a vast cross-section of the population—from children as young as ten to men over the age of fifty. The ministry added that the first to be sterilized would not be residents of "institutions," but those who were "at large." Quickly, the procedure became known as the *Hitlerschnitte*, or "Hitler's cut." During 1934, the Third Reich sterilized at least 56,000 individuals—approximately one out of every 1,200 Germans.[94]

In mid-July of 1934 the IFEO met in Zurich, and congratulated Germany on a campaign being conducted "with characteristic thoroughness and efficiency... mainly on sound and truly eugenic lines." That conclusion was publicized in *Eugenical News*. The idea was to rebut mounting criticism that the Reich's mass sterilization program was not only a medical sham, but undisguised racial persecution. In Germany, "racial persecution" invariably meant "Jewish persecution." Newspapers around the world were filled with condemnation of Germany and its treatment of the Jews.[95]

Jews were indeed on the minds of the eugenicists at Cold Spring Harbor. For example, the *New York Times* of January 7, 1934, had run an article on Hitler's race policy headlined "NAZIS INSIST REICH BE RACE MINDED," and subheadlined "No One Knows Exactly What That Means There, Except That Jews Are Target." The article went into Laughlin's clipping folder. So did other *New York Times* articles from January and early February about German-Jewish refuges in Europe, as did articles about financial assistance to Jews in the United States.[96] The folder grew thick.

With so much anti-Nazi publicity in the air, putting a positive face on the Reich's conduct was a continuing priority at *Eugenical News*. Even as the *New York Times* was denigrating the Reich's eugenics as pure racial and religious oppression, and using quotes from Interior Minister Wilhelm Frick to illustrate the point, Laughlin was assuring colleagues that the Cold Spring Harbor publication would help counteract that impression among eugenicists. Laughlin's January 13, 1934, letter to Madison Grant explained, "We propose devoting an early number of the *Eugenical News* entirely to Germany, and to make Dr. Frick's paper the leading article. Dr. Frick's address sounds exactly as though spoken by a perfectly good American eugenicist in reference to what 'ought to be done,' with this difference, that Dr. Frick, instead of being a mere scientist is a powerful Reichsminister in a dictatorial government which is getting things done in a nation of sixty million people. Dr. Frick's speech marks a milepost in statesmanship. The new German attitude and resolution mean that in the future, regardless of nationality, every statesman, who takes the long view of his country's problems, will be compelled to look primarily to eugenics for their solution."[97]

In the very next issue, March–April 1934, the speech in question, delivered by Frick nine months earlier, led off an edition devoted to German eugenics. It included a detailed directory of the Third Reich's leading eugenicists, exuberant praise of the Nazi sterilization campaign, and one article describing the flood of Jewish refugees with the phrase, "it is 'raining' German Jews." Another article examined the destinations of some 60,000 German-Jewish refugees: 25,000 had fled to France, 6,500 to Palestine, 6,000 to Poland and so on.[98]

There was room in the issue to discuss other minorities as well. One article discussed the question of sterilizing some six hundred "negroid children in the Rhine and Ruhr districts—Germany's legacy from the presence of French colonial troops there during the war."[99] In a salute to the Führer, another article clearly suggested that Hitler's eugenics would soon be applied across all of Europe. "This State Cause does not only concern Germany but all European peoples. But may we be the first to thank this *one* man, Adolf Hitler, and to follow him on the way towards a biological salvation of humanity."[100]

Eugenical News was the official voice of the American eugenics movement. Its masthead declared it "the official organ of the Eugenics Research Association, the Galton Society, The International Federation of Eugenic Organizations, [and] the Third International Congress of Eugenics." It was published at the Carnegie offices in Cold Spring Harbor. A three-man editorial committee, listed on every masthead, tightly controlled all text: Harry Laughlin, Charles Davenport and Morris Steggerda (Davenport's assistant on the Jamaica project).[101]

Eugenical News was read by virtually the entire eugenics community in America and enjoyed an equally attentive overseas readership. In Nazi

Germany, race hygienists followed the publication closely. After the March–April 1934 issue, for example, Ploetz wrote a letter to the editor correcting several typos and adding a clarification. "The 60,000 Jews...were not expelled....Nobody chased them away....They went, frightened by the Jewish reports of horror." Ironically, in the same issue, *Eugenical News* ran a report headlined "Jewish Physicians in Berlin" that declared, "The city of Berlin quite logically is trying to reduce the number of its Jewish physicians, which is not in keeping with the racial composition of the general population." The article added that anti-Jewish laws were still not working and the numbers of Jewish doctors "were but slightly reduced."[102]

Rockefeller funding continued even as anti-Nazi protest groups complained directly to foundation executives. For example, shortly after Hitler attained power, Rüdin and the Kaiser Wilhelm Institutes became known as mere scientific fronts for Nazi ideology. The foundation's own best contact within Rüdin's institute, Dr. Walther Spielmeyer, confirmed in a November 3, 1933, letter, "Prof. Rüdin...also holds the post of Reichskomissar for Race Investigation." Once word surfaced in late 1933 of the foundation's ties to Rüdin and his Munich-based Kaiser Wilhelm Institute for Psychiatry, the anti-Nazi boycotters and protest movement mobilized. One typical complaint letter from *New Republic* editor Bruce Bliven to the Rockefeller Foundation, sent December 20, 1933, asked whether the reported link could be true. Concerned officials at the foundation jotted notes on Bliven's letter: "June 3, 1932 $9,000 3 for 3 yrs." Under that, someone wrote, "Inst for Anthro." Under that: "Sexuality & Genetics."[103]

On January 10, 1934, Rockefeller executive Thomas Appleget replied to Bliven that the foundation had indeed helped erect the building some years before, and had then approved another eight-year grant for two of its doctors. But, Appleget added, "Strictly speaking this [Rüdin's institute] is not an institute of the Kaiser Wilhelm Gesellschaft." A Rockefeller colleague who saw the falsity scribbled in the margin, "TBA—What basis for this?" On January 31, Appleget wrote to Bliven "in correction of my earlier communication" and admitted that the Institute for Psychiatry was indeed "one of the regular institutes."[104]

Protests did not subside. Two days later, Jewish newspapers across the country published notices similar to the one that appeared in the *American Hebrew*: "Recently the American Committee Against Fascist Oppression in Germany declared that the Kaiser Wilhelm Institute, a German institute for psychiatric research with headquarters in Munich, and subsidized by the Rockefeller Foundation is carrying on a bitter pro-Nazi agitation....

The Committee accuses the Institute of spreading Nazi propaganda under the cloak of science and paid for by the money of the Rockefeller Foundation.... One of the Institute's departments, devoted to the study of racial theories, has 'proved' through 'scientific claims' that Hitler's theory regarding the superiority of the 'Nordic race' and the inferiority of the Semitic and other races is altogether correct.... Dr. Theodore Lang, founder of the National Socialist Doctors' Association, is also a research worker at the Kaiser Wilhelm Institute; his Doctors' Association is carrying on a bitter campaign against Jewish physicians in Germany."[105]

The Federal Council of the Churches of Christ in America continued the pressure, sending the foundation the *American Hebrew* article and asking for an explanation. Worried Rockefeller officials sent a note to a foundation attorney explaining, "As a matter of fact, it is not research that would lend itself to propaganda purpose. Rüdin was, and continues to be, a member of the staff of the [Kaiser Wilhelm] Institute [for Psychiatry]. No grants have ever been made for his work or for the general budget of the Institute. Rüdin's present political affiliations are not under the control of the Institute or the Kaiser William Gesellschaft [Society]. Undoubtedly some of the [anti-Semitic] publications, which your correspondent describes, have been written in the building that we donated.... In the circumstances, I think it is quite untrue to say that Foundation funds are being used to subsidize race prejudice."[106]

Yet the protest letters still flowed in. "We are getting a number of inquiries from various liberal groups as to our connection with the Forschungsanstalt für Psychiatrie [Research Institute for Psychiatry] in Munich.... The principal complaint is that Professor Rüdin ... [is] apparently very active in the preparation of the anti-Jewish propaganda." Rockefeller officials tried to provide assurances to protestors that they were not funding Rüdin himself but rather two doctors working under his direction. But this hardly believable story was itself internally contradicted. A March 16, 1934, letter to Appleget by the foundation's Paris representative reminded, "There is however another grant of funds made through ... the Notgemeinschaft der [Deutschen] Wissenschaft [Emergency Fund for German Science] which at least in part is utilized by Professor Rüdin.... $125,000 over a period of five years." The sum of $125,000 equals more than a million dollars in twenty-first-century money.[107]

Despite anti-Nazi protests, Rockefeller continued its subventions to Germany. Indeed, the foundation made periodic increases to account for the fluctuating exchange rate. Moreover, it quickly learned that while its grants specified that money go to one project, Nazi science administrators were quite willing to divert it to another department with a greater ideological priority. For example, in October of 1934, Alan Gregg, director of the foundation's Division of Medical Sciences received a blunt letter from the foundation's most reliable contact in Rüdin's institute, Dr. Spielmeyer. "In the field of medicine," Spielmeyer unhappily conceded, "both practice and scientific research is concerned primarily with genetics and race hygiene, as you know. You convinced yourself of that this summer, during your visit." He went on to explain that the space and resources that the foundation financed for his blood chemistry research had been appropriated by Rüdin's race investigations. Rüdin, reported Spielmeyer, simply required more space, more stenographers and more race investigators. "For this reason, it was unfortunately not possible to maintain the chemical division properly.... The Rockefeller Foundation has, for the past four years, provided funds for the maintenance of the chemical division," said Spielmeyer, but those funds were now being used for "racial research."[108]

At about the same time, an internal note was circulated to Rockefeller Foundation officials informing them that a Jewish doctor at the Institute for Anthropology, Human Heredity and Eugenics had made clear to the foundation that, "In his lifetime, the Jews will not be permitted to return to Germany." Nonetheless, the foundation found additional recipients for its German research funding.[109]

The foundation began financing biologist Alfred Kühn's hormone studies on meal moths. German race hygienists had been actively researching moths for years, claiming they exhibited what Lenz in the *Archiv für Rassen- und Gesellschaftsbiologie* called "Mendelian segregation in later generations." As such, moths were an ideal species to study for "carriers" of defective genes. Rockefeller official Wilbur Tisdale commented on Kühn's 1934 grant, "However uncertain the political situation might make a large or longtime project, [we are] safe in dealing with sound men as Kühn on a year-to-year basis." Tisdale added, "Nowhere in the continent or England [does one] find chemists, embryologists, and geneticists willing to cooperate among themselves as are these German scientists."[110]

For Rockefeller, it was just eugenics. But for Hitler, science and technology were magical weapons to wield against the Jews and all other non-Aryan undesirables. Just after Hitler rose to power, IBM initiated an aggressive commercial compact with Nazi Germany, generating windfall profits as it organized and systemized the Reich's anti-Jewish and eugenic programs. As the Hitler regime took each step in its war against the Jews and all of Europe, IBM custom-designed the punch cards and other data processing solutions to streamline those campaigns into what the company described as "blitzkrieg efficiency."[111]

It began in 1933, when the company designed and executed Hitler's first census. From there, IBM's involvement with the Reich mushroomed. On January 8, 1934, IBM opened a million-dollar factory in Berlin to manufacture Hollerith machines and coordinate data processing functions. At the factory opening, the manager of IBM's German subsidiary, Willi Heidinger, spoke vividly about what IBM technology would do for Germany's biological destiny. Standing next to the personal representative of IBM president Thomas J. Watson, and with numerous Nazi Party officials in attendance at a ceremony bedecked by swastika flags and Storm Trooper honor guards, Heidinger emotionally declared that population statistics were key to eradicating the unhealthy, inferior segments of German society.[112]

"The physician examines the human body and determines whether... all organs are working to the benefit of the entire organism," asserted Heidinger to the crowd of Nazi officials. "We [IBM] are very much like the physician, in that we dissect, cell by cell, the German cultural body. We report every individual characteristic... on a little card. These are not dead cards, quite to the contrary, they prove later on that they come to life when the cards are sorted at a rate of 25,000 per hour according to certain characteristics. These characteristics are grouped like the organs of our cultural body, and they will be calculated and determined with the help of our tabulating machine.[113]

"We are proud that we may assist in such task, a task that provides our nation's Physician [Adolf Hitler] with the material he needs for his examinations. Our Physician can then determine whether the calculated values are in harmony with the health of our people. It also means that if such is not the case, our Physician can take corrective procedures to correct the sick circumstances.... Our characteristics are deeply rooted in our race. Therefore, we must cherish them like a holy shrine, which we will—and must—keep pure. We have the deepest trust in our Physician and will follow his instructions in blind faith, because we know that he will lead our people to a great future. Hail to our German people and *der Führer!*"[114]

Most of Heidinger's speech, along with a list of the invited Nazi Party officials, was rushed to Manhattan and immediately translated for Watson. The IBM leader cabled Heidinger a prompt note of congratulations for a job well done and sentiments well expressed.[115]

Following up, an August 1934 article in IBM's German customer newsletter, *Hollerith Nachrichten*, extolled the benefits of advanced data processing for eugenics. The article, entitled "An Improved Analysis of Statistical Interdependencies via Hollerith Punch Card Process," illustrated how complex data calculations could be better interpreted and predict probabilities. As a prime example, the journal cited "the field of medicine, and the science of genetics and race." Complex tabulations could be rendered, the article suggested, regarding "the size of fathers and their children, number of children and parents. Diphtheria and age, and the different racial characteristics."[116]

Medical questionnaires to be filled out by hand were jointly designed by IBM engineers and Nazi disability or welfare experts for compatibility with Hollerith cards. For example, diseases were coded: influenza was 3, lupus was 7, syphilis was 9, diabetes was 15; these were entered into field 9. As a notice from IBM's German subsidiary advised, the questionnaires would have to be adapted to the technical demands of IBM's Hollerith system, not the other way around. A vertical notice printed along the bottom left of typical welfare forms often indicated the information was to be processed "by the punch card office," generally an in-house bureau.[117]

Raceology in Nazi Germany was enabled as never before. Statistical official Friedrich Zahn extolled the fact that "registered persons can be observed continually, [through] the cooperation of statistical central offices... [so] other statistical population matters can be settled and regulated." Zahn proposed "a single file for the entire population to make possible an ethnic biological diagnosis [to] turn today's theory into tomorrow's practice. Such a file would serve both practical considerations as well as science." He added, "Clarified pictures of the volume of genetic diseases within the population... now gives science a new impetus to conduct research... which should promote good instead of bad genetic stock."[118]

Mathematic formulas and high-speed data processing of population and medical records would indeed become the key to Jewish persecution in Nazi Germany. In November of 1935, Germany took the next step.

Defining just who was a Jew was problematic, since so many of Jewish ancestry were practicing Christians or unaffiliated. Throughout 1935, German race specialists, bolstered by population computations and endless tabular printouts, proffered their favorite definitions of Jewishness. Some proposals were so sweeping as to include even those with the faintest Jewish ancestry—similar to the familiar "one drop" race purity laws in Virginia. But many tried to create complex pseudoscientific castes, comprised of "full Jews," who professed the religion or possessed four Jewish grandparents, as well as the so-called "three-quarter," "half," and "one-quarter" Jews with fewer Jewish ancestors.[119] Adolf Hitler was personally aware of preliminary findings showing that while only about a half million Germans had registered as Jews in the census, the veins of many more coursed with traces of Jewish blood. About a million more, he thought.[120] The Jews Hitler feared most were the ones not apparent—what eugenicists called the "carriers."

Suddenly, on September 13, 1935, *der Führer* demanded that a decree defining Jewishness be hammered out in time for his appearance two days later before the *Reichstag* (Parliament) at the culmination of Party Day festivities. Top eugenic experts of the Interior Ministry flew in for the assignment. Working with drafts shuttled between Hitler's abode and police headquarters, they finally patched together twin decrees of disenfranchisement and marriage restriction. The Law for the Protection of German Blood and a companion statute, the Reich Citizenship Law, deprived Jews of their German citizenship. These laws—the Nuremberg Laws—would apply not only to full Jews, but also to half and quarter Jews, all defined according to complex eugenic mathematics. Jewish hybrids were called *Mischling*, or mixed-breeds.[121] High-speed Hollerith systems offered the Reich the speed and scope that only an automated system could deliver to identify not only half and quarter Jews, but even eighth and sixteenth Jews. It was a new, automated system, yet applied to the well-developed, decades-old Cold Spring Harbor procedure of developing family pedigrees.[122]

The new formulaic approach to Jewish persecution exploded into world headlines. Under a page one banner story, the *New York Times*'s lead was typical: "National Socialist Germany definitely flung down the gauntlet before the feet of Western liberal opinion tonight... [and] decreed a series of laws that put Jews beyond the legal and social pale of the German nation." The newspaper went on to detail the legal import of the new ancestral fractions.[123]

The news was everywhere and inescapable. Centuries of religious prejudice had now been quantified into science. Even if Germans of Jewish ancestry had been practicing Christianity for generations—as many had—henceforth, they would all be legally defined as a race, without regard to religion. That was in 1935.

Eleven years earlier, Harry H. Laughlin's memo to Representative Albert Johnson's House Committee on Immigration and Naturalization regarding Jewish racial quotas read: "For this purpose, it would be neces-

313 II WAR AGAINST THE WEAK

sary to define a Jew. Tentatively, such a definition might read, 'A Jew is a person fifty percent or more of whose ancestry are generally recognized as being Jewish in race. The definition applies entirely to race and in no manner to religion.'"[124]

Shortly after the Nuremberg Laws were promulgated in 1935, and in view of the negative publicity race laws were receiving, Nazi eugenicist Ernst Rodenwaldt thought it might be helpful to give Laughlin special recognition for his contribution to Reich policy. Rodenwaldt suggested an honorary degree for Laughlin. In a December 1935 letter to Carl Schneider, dean of the University of Heidelberg's medical school, Rodenwaldt wrote, "Every race hygienist knows Laughlin as a champion of the eugenic sterilization. Thanks to his indefatigable studies and his indefatigable propaganda activity in America, there exist, since the end of the twenties, in several states of America, sterilization laws and we can report about 15,000 sterilizations until 1930, mainly in California. Professor Laughlin is one of the most important pioneers on the field of racial hygiene. I got to know him in 1927 in Cold Spring Harbor.... Heidelberg University honoring professor Laughlin's pioneer work would, in my opinion, make a very good and compensating impression in America, where racial hygienic questions are propagated in the same way as here, but where many questions of the German racial hygienic laws are mistrusted."[125]

Schneider gladly approved the honor. Laughlin could not travel to Heidelberg to accept, but he expressed his gratitude in a letter to Schneider. "I was greatly honored," Laughlin wrote, "to accept this degree from the University of Heidelberg which stands for the highest ideals of scholarship and research achieved by those racial stocks which have contributed so much to the foundation blood of the American people.... I consider the conferring of this high degree upon me not only as a personal honor, but also as evidence of a common understanding of German and American scientists of the nature of eugenics as research in and the practical application of those fundamental biological and social principles."[126]

Some three years after Laughlin's award, shortly after World War II broke out in September of 1939, the same Carl Schneider helped organize the gassing of thousands of adults adjudged mentally handicapped. The project was codenamed T-4 after the address of the staff, located at Tiergartenstrasse 4 in Berlin. Mass gassings with carbon monoxide, which began in January 1940 at locations across Germany, proved most efficient. Victims were told to undress and to enter a room resembling a shower complete with tiled surfaces, benches and a drain. Crematoria were erected nearby to dispose of the bodies.[127]

From 1936 to early 1939, Nazi Germany was considered a threat to the other countries of Europe, and indeed to all humanity. Refugees flooded the world. The Third Reich continued arming for war and demanded territorial concessions from its neighbors. In 1938 the Nazis annexed Austria, and then in early 1939 the Reich overran Czechoslovakia in prewar aggression and consolidation. Concentration camps of gruesome notoriety, from Dachau to Buchenwald, were established across Germany; the horror stories they inspired became common talk of the day. Nazi subversion was a new fear in American society.[128]

Certainly, there were many vocal Nazi sympathizers in America. But those who supported any aspect of the Hitler regime, from economic contacts to scientific exchanges, did so at a substantial moral risk. Genuine revulsion with Nazified eugenics was beginning to sweep over the ranks of previously staunch hereditarians who could no longer identify with a movement so intertwined with the race policies of the Third Reich. A group of longtime eugenicists and geneticists spoke of a resolution to disassociate eugenics from issues of race. Letters to Davenport calling for his support were unsuccessful. Institutions such as the Eugenics Research Association, the American Eugenics Society, the Eugenics Record Office and a labyrinth of related entities all remained intact in their support of Germany.[129]

Monthly coverage in *JAMA* became more skeptical and detached starting about 1936, with headlines such as "Strangulation of Intellectualism" placing the Nazi takeover of medical science into clearer perspective. One *JAMA* article unambiguously explained, "The president of the new [medical] society is no distinguished clinician; he is the Nazi district governor of Vienna, that is to say a politician who is also an official of the Nazi bureau of national health." *JAMA* also began inserting quotation marks around Nazi medical expressions and statements to differentiate them from ordinary medical discourse.[130]

After Raymond Fosdick assumed the presidency of the Rockefeller Foundation in 1936, the charitable trust became increasingly unwilling to fund any projects associated with the term *eugenics*, even Fischer's genealogical studies. The idea of investigating family trees was just too emblematic of repressive Nazi persecution. Funding was also curtailed for some of the foundation's traditional programs at the Kaiser Wilhelm Institutes. Money continued to flow for eugenic projects, but only when they were packaged as genetics, brain research, serology or social biology.

For example, Rockefeller fellowships and scholarships from 1936 through 1939 allowed German genetic researchers to travel to Cold Spring Harbor and California for further study. But the fact that Rockefeller executives became exceedingly cautious about their continued sponsorship of Nazi medicine was a testament to the controversial nature of any contact with the Third Reich.[131]

Indeed, on June 6, 1939, Fosdick circulated a pointed memo to Rockefeller Foundation executives. "I have read with a good deal of interest your Letter no. 40 of May 25th about our general relation with totalitarian countries, and particularly about the fellowship situation. The rumor which Mr. Kittridge brought back from Geneva to the effect that the Foundation was boycotting all requests from Germany is of course hardly correct.... I am frank in saying that at the present moment it would be not only embarrassing, but probably impossible, to make any major grants in Germany. There is a matter of public policy involved here which has to be taken into consideration, and I do not believe that this is the moment to consider any sizable requests for assistance from German sources." Fosdick added that individual fellowships to German scientists would still be possible, but only if "sifted with rigid scrutiny to make sure that we are not being used for ulterior purposes." He added, "I earnestly hope that this evil hour will soon pass."[132]

Despite Nazi Germany's descent into pariah status, core eugenic leaders were steadfast in their defense of, fascination with, and general admiration for Hitler's program. In late 1935, ERA president Clarence Campbell traveled to Berlin for the World Population Congress, an event staged under the patronage of Nazi Interior Minister Frick. Fischer was president of the congress. Campbell created a scandal back home when he loudly and passionately proclaimed his admiration for Hitler's policy. "The leader of the German nation, Adolf Hitler," declared Campbell, "ably supported by Frick and guided by this nation's anthropologists, eugenists and social philosophers, has been able to construct a comprehensive racial policy of population development and improvement that promises to be epochal in racial history. It sets a pattern which other nations and other racial groups must follow if they do not wish to fall behind in their racial quality, in their racial accomplishments and in their prospects for survival."[133]

Campbell's speech made headlines in the next morning's *New York Times*: "US EUGENIST HAILS NAZI RACIAL POLICY." When Campbell returned to America, he hit back at his critics in the lead article of the March–April 1936 issue of *Eugenical News*. "It is unfortunate that the anti-Nazi propaganda with which all countries have been flooded has gone far to obscure the correct understanding and the great importance of the German racial policy."[134]

Throughout 1936, the American eugenic leadership continued its praise for Hitler's anti-Jewish and racial policies. "The last twenty years witnessed two stupendous forward movements, one in our United States, the other in Germany," declared California raceologist C. M. Goethe in his presidential address to the Eugenics Research Association. He added with a degree of satisfaction, "California had led all the world in sterilization operations. Today, even California's quarter century record has, in two years, been outdistanced by Germany."[135]

Eugenicist Marie Kopp toured 15,000 miles across Nazi Germany, and with the assistance of one of the Kaiser Wilhelm Institutes, was able to undertake extensive research on the Nazi program in cities and towns. Kopp was even permitted access to the secret Nazi Heredity Courts. Throughout 1936, Kopp wrote articles for eugenic publications, participated in promotional roundtables with such luminaries as Margaret Sanger, and presented position papers praising the Nazi program as one of "fairness." Kopp was able to assure all that "religious belief does not enter into the matter," because Jews were defined not by their religious practices, but by their bloodlines.[136]

At one American Eugenics Society luncheon, Kopp emphasized, "Justice Holmes, when handling down the decision in the Buck versus Bell case, expressed the guiding spirit.... 'It is better for all the world, if instead of waiting to execute degenerate offspring for crime or let them starve for their imbecility, society can prevent those who are manifestly unfit from continuing their kind. Three generations of imbeciles are enough.'"[137]

In 1937, Laughlin and his Cold Spring Harbor office became the U.S. distributor of a two-reel Nazi eugenic propaganda film entitled *Erbkrank* (*The Hereditarily Diseased*). *Erbkrank* began with scenes of squalid German slums where superior Nordic families were forced to live because so much public money was spent on bright, well-constructed institutions to house the feebleminded. Laughlin loaned the film to high schools in New York and New Jersey, to welfare workers in Connecticut, and to the Society for the Prevention of Blindness. Although he acquired the film from the Race Policy Office of the Nazi Party (*Rassenpolitisches Amt der NSDAP*), he assured, "There is no racial propaganda of any sort in the picture; it is [simply] recognized that every race has its own superior family-stocks and its own degenerate strains."[138]

Yet in fact the film declared, "Jewish liberal thinking forced millions of healthy *volk*-nationals into need and squalor—while the unfit were overly coddled." In another frame the movie explained, "The Jewish people has a particularly high percentage of mentally ill." Indeed, one archetypal defective citizen was a mental patient described as a "fifty-five year old Jew—deceitful-rabble-rouser."[139]

No matter how dismal the plight of the Jews in Germany, no matter how horrifying the headlines, no matter how close Europe came to all-out war, no matter how often German troops poured across another border, American eugenicists stood fast by their eugenic hero, Adolf Hitler.

In 1938, Germany accelerated the humiliation of the Jews, as well as the Aryanization and confiscation of their property. On November 10, 1938, the world was shocked by the German national anti-Jewish riots and pogroms known as *Kristallnacht*. Over one hundred synagogues were burned across the Reich, and thousands of Jews were marched off to concentration camps. The Gestapo and SS had by now subsumed the Kaiser Wilhelm Institutes, the Society for Racial Hygiene and indeed all of German medicine.[140]

Fischer, Lenz, Rüdin and the other stalwarts became the medical generals of Hitler's campaign against humanity. In 1936, Rüdin assumed leadership of the Institute for Racial Hygiene in Munich, one of the main centers tasked with deciding which German citizens possessed Jewish blood, and how much. In 1937, Lenz and Rüdin, in a joint operation with the Gestapo, orchestrated the identification and rounding-up of some five hundred to six hundred "Rhineland bastards," the offspring of Black French colonial soldiers; they were all secretly sterilized. Some 200,000 Germans of all backgrounds had been sterilized by 1937. After that the records were not published.[141]

Fischer was increasingly accompanied by SS officer Wolfgang Abel, who was usually dressed in a typical black Nazi uniform. The two could be seen in each other's company even when visited by American eugenicists at the Kaiser Wilhelm Institute for Anthropology, Human Heredity and Eugenics. Together, Fischer and Abel manufactured fictitious eugenic profiles of Jews, Gypsies and other non-Aryan undesirables, accusing them of numerous hereditary afflictions. In order to justify their eugenic persecution, the Reich falsely ascribed flat feet, mental illness and an assortment of other maladies to those the Reich wanted to eliminate.[142]

In one lecture, Fischer declared, "When a people wants, somehow or other, to preserve its own nature, it must reject alien racial elements, and when these have already insinuated themselves, it must suppress them and eliminate them. The Jew is such an alien and, therefore, when he wants to insinuate himself, he must be warded off. This is self-defense. In saying this, I do not characterize every Jew as inferior, as Negroes are, and I do not underestimate the greatest enemy with whom we have to fight. But I reject Jewry with every means in my power, and without reserve, in order to preserve the hereditary endowment of my people."[143]

The concept of describing people as leading a "life unworthy of life," sometimes known as "worthless eaters," rose to the fore.[144] Eugenic terminology and conceptualizations such as *subhuman* and *bacterium* were becoming more than jargon. They were becoming policy guidelines. Leon Whitney, executive secretary of the American Eugenics Society, declared, "While we were pussy-footing around . . . the Germans were calling a spade a spade." Goddard expressed his frustration another way: "If Hitler succeeds in his wholesale sterilization, it will be a demonstration that will carry eugenics farther than a hundred Eugenics Societies could. If he makes a fiasco of it, it will set the movement back where a hundred eugenics societies can never resurrect it."[145]

On September 1, 1939, Germany launched its *blitzkrieg* against Poland, beginning Word War II. The Reich needed hospital beds, and had to ration its wartime resources. Now the medical men of German eugenics would graduate from sterilization to organized euthanasia. Lenz helped draft euthanasia guidelines whereby a patient could be killed "by medical measures of which he remains unaware." The continued existence of those classed defective could no longer be justified in Hitler's war-strapped Reich. Beginning in 1940, thousands of Germans taken from old age homes, mental institutions and other custodial facilities were systematically gassed. Between 50,000 and 100,000 were eventually killed. Psychiatrists, steeped in eugenics, selected the victims after a momentary review of their records, jotted their destinies with a pen stroke, and then personally supervised the exterminations.[146]

With the war raging, Lothrop Stoddard, a leader of the Eugenics Research Association, traveled to Nazi Germany. His 1940 book, *Into the Darkness*, celebrated Hitler and Nazi eugenics. "Nothing is so distinctive in Nazi Germany as its ideas about race," wrote Stoddard. "Its concept of racial matters underlies the whole National Socialist philosophy of life and profoundly influences both its policies and practices. We cannot intelligently evaluate the Third Reich unless we understand this basic attitude of mind.[147]

"As is well known, the Nazi viewpoint on race and the resultant policies are set forth by Adolf Hitler himself in the pages of *Mein Kampf*, the Bible of National Socialism. The future Fuehrer therein wrote: 'It will be the duty of the People's State to consider the race as the basis of the community's existence. It must make sure that the purity of the racial strain will be preserved.... In order to achieve this end the State will have to avail itself of modern advances in medical science. It must proclaim that all those people are unfit for procreation who are afflicted with some visible hereditary disease, or are the carriers of it... having such people rendered sterile."[148]

Focusing on Hitler's Jewish policy, Stoddard observed, "The relative emphasis which Hitler gave racialism and eugenics many years ago foreshadows the respective interest toward the two subjects in Germany today. Outside Germany, the reverse is true, due chiefly to Nazi treatment of its Jewish minority. Inside Germany, the Jewish problem is regarded as a passing phenomenon, already settled in principle and soon to be settled in fact by the physical elimination of the Jews themselves from the Third Reich."[149]

Stoddard was so favored by Hitler that *der Führer* granted him a rare, exclusive audience. In a chapter entitled "I See Hitler," Stoddard wrote of the moment of his encounter in these words, "At that moment I was bidden to the Presence."[150]

Goebbels's ministry escorted Stoddard around Berlin and arranged access to other senior Reich officials, especially those concerned with race policy. The Eugenics Courts, normally conducted in secret, granted Stoddard extraordinary permission to sit on the bench next to the judges and observe their racial judgments of Jews and non-Jews alike. His courtroom experiences were recounted in a chapter entitled "In a Eugenics Court," in which he bemoaned the race tribunals for being "almost too conservative."[151]

As Hitler's divisions smashed through Europe, his eugenic ideal would be enforced not only against those in Germany, but also against those in conquered or dominated countries. In country after country, Hitler rounded up the defective Jews and other subhumans, systematically making one region after another *judenrein*—Jew free. As Hess insisted, "National Socialism is nothing but applied biology."[152]

For decades, Hitler's bloody regime, the Holocaust and the Second World War would be perceived as merely the outgrowth of the unfathomable madness and blind hatred of one man and his movement. But in fact Hitler's hatred was not blind; it was sharply focused on an obsessive eugenic vision. The war against the weak had graduated from America's slogans, index cards and surgical blades to Nazi decrees, ghettos and gas chambers.

CHAPTER 16

Buchenwald

Buchenwald concentration camp near Weimar. The "Little Camp"—the isolation and quarantine section of Buchenwald. Block 57. One morning in late May of 1944.[1]

Three-tiered geometric boxes lined the barrack. Each housed as many as sixteen emaciated humans per shelf. A thirsty and exhausted Frenchman named Oliv struggled to climb down from the top level for his day's work. But he was too weak to climb out and negotiate the eight feet down. As Oliv lay limp, a fat, well-fed inmate doctor walked in. The other French prisoners pleaded with the doctor that Oliv was too ill and suffered from severe rheumatism, making his every movement painful. The frail man needed medical attention. A small infirmary, stocked with medicines and called "the hospital," had been established in the Little Camp. The doctor controlled access to the facility and the drugs. Those admitted to the hospital could be excused from work until nursed back to working strength—and thereby live another day.[2]

But the doctor, himself a prisoner yet reviled as a barbaric stooge of the SS, was known for refusing admission to the hospital except to those he favored—or those who could bribe their way in by turning over their relief packets. Most of all, the doctor hated the French communists. They—and their diseases—were everywhere in the Little Camp. The doctor believed that each inferior national group was a carrier of its own specific set of diseases. Frenchmen, he thought, brought in diphtheria and related throat diseases as well as scarlet fever. Simply put, the Little Camp doctor was unwilling to use his limited hospital to lessen the prisoners' loads, extend their lives or relieve their suffering. The prisoners' job was to work. His job was to ensure they kept working—until they could work no more.[3]

Furious and impatient, the Little Camp doctor pushed the others out of the way, stepped onto the lowest of the three tiers, reached up and grabbed Oliv's emaciated foot as it dangled over the edge. He then yanked Oliv over the short sideboard and down the eight feet to the floor. Oliv tumbled to the floor like a doll and cracked his skull. Blood soaked down the back of his shirt. As the life seeped out of Oliv, his comrades hauled him onto the lowest bunk, and then hurried out to their backbreaking labors at the quarry. When they came back to Block 57 that night, Oliv was dead. Next to the bathroom was a makeshift morgue; they moved his body there. Later, Oliv's body waited its turn at the crematorium.[4]

The French inmates of the Little Camp never forgot the brutality the doctor showed them, while exhibiting seemingly incongruous medical compassion to others. They never forgot that while most of them were worked and starved into skeletons, the doctor ate well. Many prisoners lost 40 percent of their weight shortly after arriving in the Little Camp. But the doctor arrived at Buchenwald fat and stayed fat. No one could understand how a talented physician could render his skills so effectively to some, while allowing others to die horrible deaths. After Buchenwald was liberated in April of 1945, the stories about Dr. Edwin Katzen-Ellenbogen emerged in French reports and then in occupation German newspapers and the Allied armed forces media. Katzen-Ellenbogen was accused of murdering a thousand prisoners by injection.[5]

The United States military conducted war crimes trials at Dachau for a variety of lesser-known concentration camp Nazis and their inmate collaborators, especially the medical killers. Katzen-Ellenbogen was among them, and was found guilty of war crimes, right along with the other so-called "butchers of Buchenwald." He was sentenced to a long term in prison. The court finding, however, was not an easy one. It was complicated by conflicting stories of Katzen-Ellenbogen's outstanding academic background and prewar record.[6]

Many found Dr. Katzen-Ellenbogen and the many lives he led incomprehensible. How could he alternately function as a gifted psychiatrist and as a murderous man of medicine? At the time, none understood that Katzen-Ellenbogen viewed humanity with multiple standards. He was an American eugenicist. Nor was he just any eugenicist. Katzen-Ellenbogen was a founding member of the Eugenics Research Association and the chief eugenicist of New Jersey under then-Governor Woodrow Wilson.[7]

Viewing humanity through a eugenic prism, Katzen-Ellenbogen was capable of exhibiting great compassion toward those he saw as superior, and great cruelty toward those he considered genetically unfit. In Buchenwald, the French, with their Mediterranean and African hybridization, were eugenically among the lowest. They were not really worthy of life. At the same time, in Katzen-Ellenbogen's view, those of Nordic or Aryan descent were treasured—to be helped and even saved. It all followed classic eugenic thought. But in Buchenwald, it was the difference between life and death.

How did one of America's pioneer eugenicists wend his way from New Jersey to Buchenwald's notorious Little Camp? The story begins in late nineteenth-century Poland. Katzen-Ellenbogen was the name of a famous line of Polish and Czech rabbis going back centuries. However, as the doctor's life was built, he—or perhaps his immediate branch of the family—obscured any connection with a Jewish heritage. Like many European Jews who had drifted from tradition, he spelled his last name numerous ways, hyphenated and unhyphenated, and sometimes even signed his name "Edwin K. Ellenbogen." He was probably born as Edwin Wladyslaw Katzen-Ellenbogen in approximately 1882, in Stanislawow, in Austrian-occupied Poland.[8]

As a youth, Katzen-Ellenbogen developed severe vision problems. But he achieved academic success despite the affliction, attending fine schools and developing extraordinary powers of observation and ratiocination. First, he studied at a Jesuit high school in Poland. Then he attended the University of Leipzig, where he secured his medical degree in 1905. While in medical school, he became engaged to a girl from Massachusetts, Marie A. Pierce, daughter of a judge and scion of a prominent family of Americans dating back to the Minutemen. In 1905, Katzen-Ellenbogen sailed for America, settling briefly in Massachusetts, where he married Marie. He added "Marie" to his various middle names, and utilized her family's connections to further his academic pursuits. Various letters of introduction were provided, as was the money Katzen-Ellenbogen needed to continue his university work in Europe. There he studied psychiatry with some of the best names in the field, during the formative years of the profession, and he also learned the mystifying medical art of hypnosis.[9]

In 1907, Katzen-Ellenbogen returned to the United States, where he was naturalized as a citizen and started work in state institutions, such as the Danvers State Hospital of Massachusetts. One of the early exponents of Freud in America, Katzen-Ellenbogen became a Harvard lecturer in abnormal psychology. He developed expertise on fake symptoms. He authored an article in the *Journal of Abnormal Psychology* on "The Detection of a Case of Simulation of Insanity by Means of Association Tests."[10]

Katzen-Ellenbogen began to specialize in epilepsy, especially with regard to mental deficiency. His expert testimony was pivotal in convicting a murderer who claimed diminished mental capacity due to an epileptic attack; the convicted man was electrocuted in 1912. He authored numerous articles on the subject and became a coeditor of the international quarterly, *Epilepsia*. One of his articles asserted that different races should have their own standards for imbecility. A child, he posited, "may be inferior as to race, but be up to the mark for its own racial standards... especially... in America."[11]

In 1911, Woodrow Wilson became governor of New Jersey. Katzen-Ellenbogen was asked to become scientific director of the State Village for Epileptics at Skillman, New Jersey. It was there that he would develop his eugenic interests. "While there," recalled Katzen-Ellenbogen, "I particularly studied... the hereditary background of epilepsy." As the state's leading expert, Katzen-Ellenbogen was then asked by Wilson to draft New Jersey's law to sterilize epileptics and defectives. In the process, he became an expert on legal and legislative safeguards and jurisprudence.[12]

As a leading member of the National Association for the Study of Epilepsy, Katzen-Ellenbogen delivered an address on epilepsy and feeble-mindedness at Goddard's Vineland Training School. In 1913, Katzen-Ellenbogen became charter member #14 of the Eugenics Research Association at Cold Spring Harbor. The doctor continued his active membership even after he sailed for Russia in 1915, never to return to the United States.[13]

Katzen-Ellenbogen bounced around the capitals of Europe for the next few years. He was about to board a ship in Holland when he received a telegram informing him that his only son had died in America after falling from a roof. Katzen-Ellenbogen was never the same. He became morose and introspective, questioning the value of human life, at least his own. "I contemplated to offer myself as physician to the leprosy colony in the upper State of New York," he recounted. He also considered suicide. At the same time, Katzen-Ellenbogen deepened his fascination with things Catholic, purchasing a valued copy of a rare Madonna.[14]

As Katzen-Ellenbogen wandered through Europe, he impressed many people as a kind humanitarian. He met one woman briefly on a train in 1921 and discussed his favorite Madonna. More than two decades later, even after learning of his notorious war crimes, she wrote him, "I cannot believe that anyone who likes a picture of the Madonna can be entirely bad." Years later, another woman, recalling their fond encounter in Germany, insisted, "There still are people in this world who believe in you."[15]

In 1925, Katzen-Ellenbogen developed a relationship with a woman named Olga. She described him as "the companion of my life." He described her as "my old housekeeper." By any measure, Katzen-Ellenbogen developed deep parental feelings for Olga's two orphaned grandsons, and raised them as though they were his own. Together with his daughter, Katzen-Ellenbogen led an *ad hoc* family of five.[16]

They were living in Germany when Hitler rose to power. Despite his Catholic observances, after the 1935 Nuremberg Laws Katzen-Ellenbogen found himself defined as Jewish and subject to encircling anti-Jewish decrees. Like many practicing Christians of Jewish ancestry, he fled across the Czech border in 1936, establishing a clinic in Marienbad. When anti-Jewish agitation spread into Czechoslovakia, Katzen-Ellenbogen moved again, this time to the democratic stronghold of Prague, where in 1938 he began working with refugee groups.[17]

After Hitler invaded Czechoslovakia in March of 1939, Katzen-Ellenbogen followed a typical route of flight. First, he crossed into Italy. After war broke out in September of 1939, he escaped to France. But when the Nazis bifurcated France in 1940, Katzen-Ellenbogen was caught in the occupied zone in Paris. As a result of his many recent relocations, he was a suspicious refugee in a city teeming with Gestapo agents. In 1941 he was arrested by Gestapo counter-intelligence corps, but he was soon released. Like many foreigners living in Nazi-occupied Paris, Katzen-Ellenbogen was ultimately arrested several times for questioning or detention. He was denied permission to leave for neutral Portugal. Finally, just as he was planning to leave for Prague in the late summer of 1943, Nazi security agents came for him. The knock on the door came at six in the morning.[18]

Many eugenicists considered Nazi racial policies a biological ideal. Katzen-Ellenbogen discounted his Jewish ancestry, considering himself a eugenicist first and foremost. This made him different, and almost appealing to the Gestapo, especially under the circumstances.

Although a prisoner, he was given access to top Nazi generals in Paris to discuss his detention status. The war-stretched Nazis needed doctors, especially in occupied lands. As a distinguished physician and psychiatrist who spoke German and also enjoyed American citizenship, Katzen-Ellenbogen became very useful to both the Gestapo and the Wehrmacht. Twice he was brought to the Reich military prison in France to examine a German soldier suffering from mental problems. Katzen-Ellenbogen even testified as an expert at the soldier's court martial.[19]

Katzen-Ellenbogen found himself in a somewhat unique position. "I was the only doctor in France, a psychiatrist," he recalled, "who was [also] qualified in Germany as a doctor, and they didn't have anybody [with those skills] in the army." Eventually, the overworked regular German army doctor visiting the military prison asked Katzen-Ellenbogen, "As you speak French anyway and other languages, relieve me here. And when something very important happens, they can telephone for me." Thus, Katzen-Ellenbogen became a general practitioner for the German military in Paris even as he remained in custody. Eventually, Katzen-Ellenbogen's services were requested for German military men outside the prison. For all intents and purposes, he was at the disposal of the German medical staff. But in September of 1943, when orders came from Berlin to transfer prisoners in France to slave labor camps in the Reich, Katzen-Ellenbogen was put on a train and shipped to the dreaded Buchenwald.[20]

Buchenwald functioned for two purposes: to inflict cruelty on the Nazis' enemies and to systematically work its inmates to death in service of the Reich—in that order. In the hierarchy of hell, Buchenwald was considered among the worst of Nazi labor camps. Hundreds to thousands of people died within its confines each week from beatings, disease, starvation, exhaustion or execution.[21]

Cruel and painful medical experiments were conducted at Buchenwald, especially in Block 46, known for its frosted windows and restricted access. Nazi doctors deliberately infected prisoners with typhus, converting their bodies into so many living test tubes, kept alive only as convenient hosts for the virus. Doctors then carefully observed the progress of the disease in order to help evaluate potential vaccines. Some six hundred men died from such infections. In addition, Russian POWs were deliberately burned with phosphorus to observe their reactions to drugs. As part of the Reich's program to develop mass sterilization techniques, fifteen men were castrated to observe the effects. Two died from the operation. Experimental Section V employed gland implants and synthetic hormones on homosexuals to reverse their sex drive; the SS officers delighted in joking about the men. Those who survived these heinous tests, or otherwise outlived their usefulness, were often murdered with injections of phenol.[22]

Horrible punishments were everyday occurrences. Many were hung from their wrists with their hands tied behind their backs, thus painfully tearing arms from their sockets. Weakened inmates who did not die quickly enough were bludgeoned with a large blood-encrusted club. Russian POWs were systematically shot in the back of the neck through a small hole as they stood at the height-measuring wall.[23]

Large electric lifts continuously shuttled corpses to waiting crematoria, which operated ten hours a day and produced prodigious heaps of white ash. Death was an hourly event at Buchenwald—ultimately more than 50,000 perished. More French died than any other national group. But before the victims were burned, they performed additional service to the Reich. Pathologists in Block 2 dissected some 35,000 corpses so their body parts could be studied and then stored in various jars on shelves. Tattooed prisoners were especially prized. In Block 2, their skins were stripped off, tanned and stretched into lampshades and other memorabilia.[24]

Nuremberg Trial judges denounced "conditions so ghastly that they defy description. The proof is overwhelming that in the administration of the concentration camps the German war machine, and first and foremost the SS, resorted to practices which would shame the most primitive race of savage barbarians. All the instincts of human decency which distinguished men from beasts were forgotten, and the law of the jungle took command. If there is such a thing as a crime against humanity, here we have it repeated a million times over."[25]

In assessing Buchenwald just after liberation, a British Parliamentary delegation declared, "We have endeavored to write with restraint and objectivity, and to avoid obtruding personal reactions or emotional comments. We would conclude, however, by stating...that such camps as this mark the lowest point of degradation to which humanity has yet descended. The memory of what we saw and heard at Buchenwald will haunt us ineffaceably for many years."[26]

Most new arrivals at Buchenwald were instantly shocked by the camp's brutality and the physical cruelty heaped upon them by the guards. Upon initial entry, it was common for new prisoners to run a two-hundred-meter gauntlet of guards, who viciously beat them with clubs and truncheons as they passed. But Katzen-Ellenbogen seemed fascinated. Recalling his first moments in the camp, he said, "I was really amazed about the efficiency and quickness about everything that happened there." He added, "We were treated not badly there...." Katzen-Ellenbogen was in fact privileged from the moment he entered the camp. While other prisoners at that time were forced into tattered zebra-stripe uniforms, the doctor was permitted to wear civilian attire, including a three-piece suit and tie. But he complained that the shirt with its button-down collar was too small, and the trousers too long. His warm furry hat and medical armband gave him a distinctive look as he toured the barracks.[27]

Early on, Buchenwald administrators learned through the prisoner grapevine of Katzen-Ellenbogen's helpfulness to the Gestapo in France. He quickly became a trusted prisoner to the camp's medical staff as well as its SS officers, especially chief camp doctor Gerhard Schiedlausky. Katzen-Ellenbogen announced to everyone that he was an American doctor from New Jersey, and a skilled hypnotist to boot. None of this failed to impress the camp administrators, who often referred to him by the name Dr. K. Ellenbogen. One senior Nazi medic dared Katzen-Ellenbogen to demonstrate his skill as a hypnotist. A test subject was brought over, and within five minutes Katzen-Ellenbogen successfully placed him in a trance.[28]

Thereafter, Katzen-Ellenbogen was assigned to the hospital at the Little Camp, which functioned as the segregated new prisoner intake unit. Unlike the other inmates who slept sixteen-deep on stark wooden shelves and were fed starvation rations, Katzen-Ellenbogen enjoyed a private room with a real bed that he shared with only one other block trustee. He ate plenty of vegetables and even meat purchased through black market sources in Weimar. From time to time he cooked his own meals, an almost unimaginable prisoner luxury. The doctor was able to count SS and Gestapo officers among his friends even as fellow prisoners detested him and despised their Nazi taskmasters. He was widely believed to be a Gestapo spy.[29]

One day in mid-1944, the camp doctor, Schiedlausky, summoned Katzen-Ellenbogen to the SS hospital. "You're a hypnotizer," said Schiedlausky with distress, "You're a psychotherapist. Save me." In the midst of the human depravity he oversaw, Schiedlausky had become unable to sleep. Self-administered drugs were no help. Katzen-Ellenbogen replied, "I can help you only, Doctor, if you will forget that I am a prisoner and you are the SS doctor." Schiedlausky collegially replied, "Naturally."[30]

As Katzen-Ellenbogen analyzed Schiedlausky's dreams, he concluded that the SS doctor's mind was troubled by a great burden. "Unless you are willing to tell me what it is," Katzen-Ellenbogen told him, "no further treatment would be of value." Schiedlausky answered, "You're right, but I can't tell you." At one point Katzen-Ellenbogen came upon Schiedlausky weeping uncontrollably and consoled the man. Katzen-Ellenbogen continued to treat Schiedlausky, whose mental state deteriorated. Soon Katzen-Ellenbogen was exercising great influence over the camp doctor.[31]

Schiedlausky was so impressed with Katzen-Ellenbogen that he asked him to treat other SS men unable to sleep because of their murderous deeds. Even though Katzen-Ellenbogen was a prisoner, the Nazis opened up to him. For example, a bloodthirsty Austrian-born SS lieutenant named Dumböck admitted to Katzen-Ellenbogen that he was haunted—day and night—by the ghosts of at least forty men he had personally beaten to death. As though confessing to a priest, Dumböck admitted that sometimes when he caught someone stealing vegetables from the garden, he just "[couldn't] control himself." It would typically begin as an urge to only slap the prisoner, but then Dumböck would begin jumping on the man's body until his ribs caved in. Katzen-Ellenbogen helped Dumböck realize why he could not sleep: the killings. "That's it exactly," Dumböck agreed. Dumböck was so grateful that he granted Katzen-Ellenbogen special privileges—ironically, to the vegetables in the garden. [32]

Katzen-Ellenbogen proudly remembered that the SS men "trusted me as a doctor very much."[33]

Back at the Little Camp, Katzen-Ellenbogen administered cruel medicine. He forced Frenchmen to exercise in the frigid outdoors without their scarves and often without their shirts—this to "cure" infected throats. He smuggled in needed medicines through the SS medics but then sold them for money or favors. Such extortions allowed him to deposit some 50,000 francs into a camp bank account. He also cached large quantities of Danish food, medicines and cigarettes in his bedroom, mainly pilfered from the Danish Red Cross packets turned over by the sick and injured.[34]

Denying medical treatment was an entrenched eugenic practice at the state institutions Katzen-Ellenbogen was familiar with, from Danvers in Massachusetts to Skillman and Vineland in New Jersey. In those institutions, eugenic psychiatrists felt that medical care only kept alive those whom nature intended to die off. Katzen-Ellenbogen applied the same principles in Buchenwald.

Katzen-Ellenbogen capriciously decided who entered the hospital. Another camp doctor confirmed in court, "It depended on Katzen-Ellenbogen whether a certain person would be admitted into the little hospital... or in the main hospital." A Czech doctor added, "If he [Katzen-Ellenbogen] found a man with appendicitis or pneumonia and said, 'I will not send you to the hospital,' then the man would not get through because he, Dr. Katzen-Ellenbogen, was the only medical liaison [in the Little Camp]."[35]

Katzen-Ellenbogen himself casually admitted at his trial, "We selected.... Let's say there were 35 [needing hospitalization, and I was told] there are only 17 free [beds]. Which 17 should have preference for immediate hospitalization?" He held the power of life and death over those who desperately needed his help, and he sadistically exercised this power every day.[36]

In 1944, for instance, two French arrivals—a Protestant minister named Roux and a doctor named Rodochi—suffered greatly during the horrific railroad trip to Buchenwald. Upon entering the Little Camp, compatriots asked that Roux and Rodochi be admitted to the hospital. Katzen-Ellenbogen refused the first day. Even as they became weaker, he continued his refusals for two more days. On the fourth day, the two died during roll call, having never been seen by any doctor.[37]

After the war, a French physician internee identified as Denis told investigators that many men died who might have recovered had they been admitted to the hospital. But when French prisoners approached, Katzen-Ellenbogen often chased them away, slapped and punched them, or simply "beat them with any instrument handy." Other inmates who were physicians would sometimes complain that Katzen-Ellenbogen stocked the necessary medicines, but that the Little Camp doctor would snarl that they were in Buchenwald to "die like dogs—not to be cured."[38]

At his trial, prosecutors demanded answers.

PROSECUTOR: Isn't it also a fact, doctor, that many a prisoner died while he was waiting his turn to be examined there at the dispensary?

KATZEN-ELLENBOGEN: . . . When patients arrived he [a medical staffer] went always outside and looked who was the most ill and needs immediate attention or in a dangerous condition, to get them there first.

Q: Just answer the question please.

A: . . . If you want me to answer the question *yes* or *no*, then I will have to answer *no*.

Q: All right then your answer is: at no time did any prisoner die while waiting his turn to be examined in the dispensary.

A: You say those questions [as though] with a revolver with "hands up." It is impossible to answer whether yes or no.

Q: You were there were you not?

A: I was there.

Q: You know whether a man is living or dead, don't you?

A: Yes.

Q: All right. Did any man die while he was awaiting his turn in that line?

A: Sure he did.

Q: I though you said a moment ago that he didn't.

A: Yes, that is what I said—that is "a revolver," a little so—*yes*, but not while he was awaiting his turn [and] *because* of waiting, but because he was in a condition that a few minutes later while they brought him in he was dead.

Q: Just listen to my questions please, Doctor. I did not ask you *because* he was waiting in that line?

A: I know. That is what I said: *yes*.[39]

Failure to be hospitalized also bestowed a death sentence because it often facilitated assignment to the fatal work details at the nearby Dora works. At Dora, slave laborers were systematically worked to death tunneling into a mountain, constructing the secret German V-2 missile facilities. Dora's death rate was among the highest of any of the thousands of labor camps and subcamps in all of Nazi-occupied Europe. Many of Dora's victims were shuttled in from Buchenwald. Transports regularly delivered thousands of prisoners at a time, and some twenty thousand of them died in backbreaking labor. In fact, for the Nazi campaign known as Extermination by Labor, Dora was a convenient final destination to extract a prisoner's final ergs of energy.[40]

The weakened inmates whom Katzen-Ellenbogen callously refused to exempt from Dora work transports were essentially sentenced to death. In one typical transport of 1,000 to 1,200 French workers whom Katzen-Ellenbogen reviewed, only 97 came back alive. Indeed, the *Dora Kommando*, or work detail, was known everywhere as a "death kommando." One Frenchman, when condemned to duty at Dora, turned to Katzen-Ellenbogen and declared, "*Caesar, morituri te salutant.*" ("We who are about to die salute you.") Katzen-Ellenbogen recalled jocundly that the man "still had a sense of humor."[41]

At his trial Katzen-Ellenbogen was asked by prosecutors, "The personnel in the Medical Department... certainly knew that Dora was a death commando, isn't that so?" Katzen-Ellenbogen replied, "I should guess so."[42]

Prisoners reported that Katzen-Ellenbogen actually encouraged unsuspecting French inmates to volunteer for "death details." In one instance, a Frenchman discovered the ruse and warned comrades to remove their names from the volunteer roster. Katzen-Ellenbogen reported the Frenchman who spread the warning and the prisoner was brutally punished.[43]

Certainly, many concentration camp trustees, *capos* and block elders curried favor by demonstrating heightened brutality toward the inmates under their authority. But many used their trusted positions to subtly connive and cajole the SS, in small ways helping others survive. For example, Austrian journalist Eugen Kogon worked as a clerk in Buchenwald's hospital under the notorious Dr. Erwin Ding-Schuler. It was Ding-Schuler who in 1941 wrote in his diary, "Since tests on animals are not of sufficient value, tests on human beings must be carried out." When testifying against Katzen-Ellenbogen, Kogon explained to prosecutors that it was not necessary to be merciless even when working for the most depraved doctors. "I worked in exactly the opposite way," he said. "I made Major Dr. Ding-

Schuler a tool of the prisoners and all this only in a positive manner from the beginning to the end. . . . That's the difference." Kogon went on to write numerous articles and books on the inhumanity of concentration camps such as Buchenwald.[44]

Camp medical men did more than just withhold treatment. Many actively participated in the murder process itself. Katzen-Ellenbogen was publicly accused of finishing off a thousand men with injections. The fact that thousands were killed by an instantly-acting injection—20cc of phenol—was amply proved. But there were no witnesses to corroborate that Katzen-Ellenbogen was among the medics who wielded the hypodermics. He never directly denied being involved in injections, although he asserted he was unaware of Schiedlausky's mass injection campaign in Block 61. When the subject of injections was brought up in court, Katzen-Ellenbogen nonchalantly testified that the allegation against him was just that—an allegation in the newspapers that could not be proved.[45]

However, Katzen-Ellenbogen's guilt-ridden colleague, camp doctor Schiedlausky, did admit his involvement in the injections as well as the other medical atrocities that took place in Block 61. Katzen-Ellenbogen denied claims that he exercised a "sinister influence" over Schiedlausky that could have made a difference. Prosecutors charged, "You could have stopped it, is that correct?"[46]

With typical insouciance, Katzen-Ellenbogen replied, "Not that I could stop it, but that I would do my best, and I think that I would have succeeded to persuade Schiedlausky not to burn his fingers." Prosecutors shot back, "Well, isn't it a fact, doctor, that you [previously] testified that you would have had enough influence that his extermination of prisoners in Block 61 would never have happened?" Katzen-Ellenbogen admitted, "Yes, I said it before. It is the same thing I just said."[47]

Q: Well, then, you certainly were able to exercise a considerable power over Schiedlausky, is that not correct?
A: I wouldn't use the word "power." Influence, yes.
Q: Well, was there any other man in Buchenwald that could exercise that same influence over Schiedlausky?
A: Probably not, because Schiedlausky was a very secretive man, who, for instance, didn't say anything to anybody, even his colleagues. . . . Due to the fact that he was a patient of mine—I have a certain influence of psychoanalysis which is exercised over a patient."[48]

But ghastly science continued in Block 61. Heinous surgical procedures involving eye color and corneas were among the experiments performed by Nazi eugenicists operating in concentration camps. At Auschwitz, chemicals were injected into the eyes of children to observe color changes. At Buchenwald, trachoma was among the eye diseases investigated.[49]

Katzen-Ellenbogen claimed that he did not participate in the deliberate infections, painful experiments and euthanasia at Buchenwald, only pure research. One Nazi doctor, Werner Greunuss, received life imprisonment for his activities at Buchenwald. While admitting that he assisted Greunuss, Katzen-Ellenbogen explained, "I conducted with him scientific research about vision, and the experiments were made by [prisoner medical assistants] Novak and Sitte on rabbits." He added, "I worked on literature, particularly as my doctor thesis was in this region. Dr. Greunuss was able to read all my work which was then in German, and furnish me books from Jena University Library."[50] Nothing further was proved about Katzen-Ellenbogen's involvement with eye research.

Katzen-Ellenbogen did engage in other experimental medical activity, however. He regularly applied his skills as an accomplished hypnotist, including posthypnotic suggestions. There were the bedwetters, for example. In a hell where Katzen-Ellenbogen regularly ignored the severest diseases, injuries and afflictions, the doctor took an inexplicably keen interest in *enuresis*, or bedwetting. Many young boys, gripped by fright and mistreatment, urinated uncontrollably at night. These boys were brought to the doctor, who placed them under hypnotic suggestion to cure their problem. But prisoners openly accused Katzen-Ellenbogen of using his hypnotic skills to extract information and confessions for the SS and Gestapo. Katzen-Ellenbogen was proud of his work. In one case, a young man between eighteen and twenty years old was brought in at 4 P.M. on a Sunday afternoon; he was placed under a trance in the presence of other SS doctors. On this point, Katzen-Ellenbogen in open court denied that he "was hypnotizing people in order to extort confession of political prisoners and deliver them to the Gestapo." Yet he was never able to explain why he rendered service for bedwetters when he denied medical attention to so many others who were dying.[51]

Eugenics was always an undercurrent at Buchenwald. One block was known as the *Ahnenforschung* barrack, or ancestral research barrack. It was worked by a small detachment known as Kommando 22a, mainly Czech prisoners, researching and assembling family trees of SS officers. SS officers were required to document pure Aryan heredity. In addition, the SS

Race and Settlement Office was systematically sweeping through Poland looking for *Volksdeutsche*, that is, persons of any German ancestry. When this agency discovered Polish children eugenically certified to have Aryan blood, the youngsters were kidnapped and raised in designated Nazi environments. This program was called "Germanization." As a skilled and doctrinaire eugenicist, Katzen-Ellenbogen was assigned to perform eugenic examinations of Polish prisoners, seeking those fit for Germanization. Eugenic certification saved them from extermination.[52]

In describing Katzen-Ellenbogen's duties, one Buchenwald medical colleague, Dr. Horn, said, "The first one, he was consulting psychiatrist. That is, later on they were Germanizing Poles. For that reason you had to examine the Poles somatically and psychically and since later on the SS used us for this delicate mission, I used Katzen-Ellenbogen to write the psychiatric reports. It was a pretty difficult job to talk about the intelligence of a Polish farm worker who didn't even speak German and Katzen-Ellenbogen speaks some sort of Slavic Esperanto very well and in all the cases that he wrote for me, and there were at least 60 cases which he did, he recommended that for every one of them that they should be Germanized, so none of them were hanged."[53]

To protect those fit for Germanization, Katzen-Ellenbogen engaged in all manner of medical charades. "So I manufactured all kinds of new forms of insanity and made false reports about their condition," he recalled. "As the invalids were not sent out at that time, they were probably saved from being gassed at one of the extermination camps. In many cases, similar cases, particularly when Rogge, one of the SS Doctors, was making selections for the transport, I trained them to throw a fit, epileptic fit, and I don't think that so many epileptics were ever in one place at one time as in Buchenwald." Katzen-Ellenbogen did not save others in a similar fashion, just the fifty or so Polish prisoners he eugenically certified as possessing Aryan qualities, in spite of their mental or intellectual conditions.[54]

Katzen-Ellenbogen was an expert at faking symptoms. While on the witness stand at his trial, he was asked if someone could be trained to feign symptoms. He bragged, "To throw a fit? With training, he could do it. I myself, for instance, could give a wonderful performance in that respect." Asked if a specialist could be fooled, Katzen-Ellenbogen rejoined, "To fool [SS] Dr. Rogge [who was making selections], yes. But not a real specialist." Asked again, Katzen-Ellenbogen repeated, "Not a real specialist."[55]

Katzen-Ellenbogen was very sure of himself. When called to testify against other doctors in the so-called "Doctors Trial" at Nuremberg, his usual brashness was more than evident. When a prosecutor asked when he had joined the Nazi Party, Katzen-Ellenbogen snapped back, "When I was in America, I never asked a nigger *whether* he had syphilis, only *when* he got syphilis." Later he explained, "That's about the same [as the] question he put to me."[56]

By any measure, the forgotten story of Katzen-Ellenbogen, an expert American eugenicist in Buchenwald, is one that stands alone. Kogon recalled it this way for prosecutors: "Katzen-Ellenbogen's power in the Little Camp was an entirely extraordinary one. An extraordinarily large one, it should be. He was the man who was feared by the prisoners in the little camp as 'the man in the background.' He had under his command the block doctors . . . and his influence upon them was considerable."[57]

When it came time to bring Katzen-Ellenbogen to justice, prosecutors found his record filled with contradictions. He saved Polish men with German blood, he let Frenchmen die before his eyes, and he sent thousands to their deaths by not exempting them from death kommandos. He was a Nazi collaborator; he was an eminent New Jersey doctor with Harvard credentials. The haze around Katzen-Ellenbogen's record grew thicker in the postwar chaos. The witnesses were gone—either returned to their homes or incinerated—the evidence was burned, and Nazi medical cohorts were quick to support each other with glowing affidavits.

Moreover, Katzen-Ellenbogen was an expert on the fine points of American jurisprudence—the standard that applied to his trial for war crimes. His court record is riddled with procedural jousting as he corrected prosecutors on what questions they were allowed to ask, and how questions should be phrased. At one point the prosecutor asked, "So that everything else, other than what you have qualified, has been of your own personal knowledge?" The defendant replied, "Most of the things I testified to was of my own personal knowledge. Still, I did not say that everything I said is correct, because I know too well the psychology of testimony, and I think you know it too, from your point of view that every witness tells objectively spoken truth."[58]

In one tense exchange, a prosecutor failed to establish the proper legal foundations for a fact; in other words he did not introduce the particulars first and then ask the defendant's relation to it. "As a matter of fact," the prosecutor asked, "do you not know that the treatment that was given him was this: that you had him stretched and spread-eagled out on one of those bunks?" Katzen-Ellenbogen rebutted the prosecutor's form, "Are you testifying again yourself or are you—"[59]

Q: You answer my question, Doctor?...Is it not fact that you let him lay there for approximately three days without any food, any water or any treatment at all?

A: That new case that you are *testifying* about....

Q: Answer my questions, is it or is it not a fact?

A: No. If you want a case like that, I answer you no....

Q: Did he or did he not die?

A: I am not an author of fiction, Mr. Prosecutor.

Q: Is your answer yes or no?

A: Mr. Denson [the prosecutor], you are the author. You must have known whether you killed in the fiction that patient or not? I don't know.[60]

In another exchange, Prosecutor William Denson attempted to poke holes in Katzen-Ellenbogen's stories.

Q: Is it not a fact, doctor, that they were beaten two to three hours later at Schebert's order?

A: I couldn't say yes or no to that. I refer once more to the well known psychology of the testimony that if a man, month after month, tells the same story, then he is lying.

Q: That is the reason you are not telling the same story?

A: Maybe so, because if everybody—I heard here so many testimonies, I am influenced. I made in Harvard experiments of students [who] wanted to kill somebody and they made a statement immediately and four weeks later. You would see the discrepancy between the first and second statement. I am not above that myself.[61]

When it finally came time to sum up, Katzen-Ellenbogen virtually commanded the judges to take the contradictions and inconsistencies into account. From the witness box, he reminded the judges: "It is a legal principle of all courts of all nations, the Romans as well in that time, *in dubio pre vero*, which in the English says: 'give them the benefit of the doubt.' That means if you are in doubt about my guilt, you have to acquit me."[62]

Then he actually invited the judges to commit a reversible error. "[But] I reverse that case," he continued. "If you are in any doubt that I am not guilty, convict me because I would have a chance then in higher court or any other place to defend myself in a way that I perhaps didn't do here."[63]

On August 14, 1947, in a Dachau barrack set up for war crimes trials, Katzen-Ellenbogen stood, somewhat disheveled, before the military tribunal. Flanked by three shiny-helmeted MPs, his shoelaces removed to prevent suicide, bright lights above to aid the photographers, Edwin Marie Katzen-Ellenbogen awaited his judgment.[64]

Without evidence of specific murders, he could not be hanged, as were other medical war criminals at Buchenwald. Instead, the tribunal used the legal theory that applied to so many Nazi conspirators. This theory was called "common design," meaning that Katzen-Ellenbogen joined "a common design" to perpetrate the horrors of Buchenwald on the inmates. "It is clear," concluded the tribunal, "that the accused, although an inmate, cooperated with the SS personnel managing the camp and participated in the common design."[65]

Judgment: Guilty. Sentence: Life imprisonment.[66]

Katzen-Ellenbogen appealed, issuing a *pro se* cascade of letters, petitions and motions, stressing his American citizenship and desire to help mankind. Upon review, his sentence was commuted to fifteen years. Katzen-Ellenbogen then appealed for special clemency on the grounds of poor health. In July of 1950, a clemency board comprised of three civilian attorneys reduced his sentence to just twelve years, concluding, "Katzen-Ellenbogen's health is poor. He is suffering from a coronary insufficiency causing severe myocardic damage, and a chronic congestive heart failure."[67]

He had all the symptoms.

CHAPTER 17

Auschwitz

After two or three days of terror in a sealed train, the Jews of Europe arrived at their eugenic apocalypse: Auschwitz.

Suddenly the wooden boxcar doors would growl open. The stifling stench inside from the sick and dying and the overflowing bucket of defecation would be replaced by the throat-stinging pungency of burning flesh as the victims glimpsed Hitler's sprawling extermination center. SS troops, backed up by barking German shepherds, would begin shouting for the eighty or ninety people in each boxcar to jump down from the train and onto the ramp.

Quick! *Schnell!* Terrified, the helpless Jews massed into orderly groups, unaware they were being assembled for eugenic selection. Teams of doctors swarmed everywhere, organizing people into lines. Two groups would be selected: those strong enough to be worked to death, and those to be gassed immediately. Women and children under fourteen to one side. Men to the other.[1]

Then camp doctor Josef Mengele, the Angel of Death, would review the frantic lines: one by one, Jew by Jew. Then with the power of his thumb, he pointed to the left, to the left, to the left, to the right, to the right, to the left. As he condemned and spared, moment-to-moment, he whistled, as though conducting a Devil's orchestra.[2]

Jews sent to the left were hustled to the showers for gassing, a procedure completely administered and supervised by doctors from start to finish. Once doctors gave the all-clear signal, groups of prisoners called *Sonderkommandos* were compelled to scavenge piles of corpses for gold teeth and rings. Only then were bodies carted off for cremation to destroy the evidence.[3]

Those sent to the right could live another day and in the process endure their own brutalities and degradation. The living were registered and tattooed. The exterminated required no registration.[4] Subject to this selection, many survived and perhaps 1.5 million at this camp complex alone were murdered—some quickly, and some very slowly.[5]

Among those selected for death at Auschwitz, several hundred, mostly children, were briefly exempted. Some even lived to tell their stories. These lucky albeit misfortunate few were chosen for cruel medical experiments conducted by Mengele. First these children were coddled and fed well to keep them in pristine shape. Then they were subjected to painful procedures. Often they were murdered as soon as the tests were completed, so they could be fastidiously dissected.[6]

After the war Mengele's sadistic experiments were considered by many to be the inexplicable actions of a scientist gone utterly mad. But in fact Mengele was following a fascinating research topic that was continuously discussed among eugenicists going back to Galton. This topic was as important to the researchers at Cold Spring Harbor and the funders at the Rockefeller Foundation as it was to Nazi medical murderers in Berlin, Munich and Frankfurt.

No words will ever capture the inhumanity of Auschwitz. But one word does explain why Auschwitz was the last fanatic stand of the eugenic crusade to create a super race, a superior race—and finally a master race. As the cattle cars emptied their human cargo onto the ramp, as the helpless millions lined up for selection, they all heard one word, shouted twice. One word shouted twice could help them live as those next to them were sent to the gas chambers. One word shouted twice would link the crimes of Mengele to the war against the weak waged by the eugenics movement.

*　*　*

Dr. Otmar Freiherr von Verschuer was crucial to the work at Auschwitz.

Verschuer lived the Nazi ideal long before Hitler emerged. A virulent anti-Semite and a violent German nationalist, he was among the student *Freikorps* militia that staged the Kapp Putsch in March of 1920. Two years later, Verschuer articulated his eugenic nationalist stance in a student article entitled "Genetics and Race Science as the basis for *Völkische* [People's Nationalist] Politics." "The first and most important task of our internal politics is the population problem.... This is a biological problem which can only be solved by biological-political measures."[7]

In 1924, at about the time Hitler staged his Beer Hall Putsch in Munich, Verschuer lectured that fighting the Jews was integral to Germany's eugenic battle. He was speaking on race hygiene to a nationalist student training camp when the question of Jewish inferiority came up.

"The German, *Völkische* struggle," he told the students, "is primarily directed against the Jews, because alien Jewish penetration is a special threat to the German race." The next year, he helped found the Tübingen branch of Ploetz's Society for Racial Hygiene and became its secretary. In 1927, Verschuer distinguished himself among German race hygienists when he was appointed one of three department heads at the Kaiser Wilhelm Institute for Anthropology, Human Heredity and Eugenics. Verschuer chaired its Human Heredity department.[8]

In 1933, Verschuer published numerous tables setting forth the exact ratios of environmental influences to human heredity. Later that year, when the State Medical Academy in Berlin offered its initial course on genetics and racial hygiene, Verschuer was one of the featured lecturers. He joined other eminent Nazi eugenicists in the program, such as Eugen Fischer and Leonardo Conti, who was a chief Nazi Party health officer and would later become Hitler's main demographic consultant when the 1935 Nuremberg Laws were being formulated. Later, Conti was put in charge of the 1939 euthanasia program.[9]

In June of 1934, Verschuer launched *Der Erbarzt* (*The Genetic Doctor*) as a regular supplement to one of Germany's leading physicians' publications, *Deutsches Ärzteblatt*, published by the German Medical Association. In it, Verschuer asked all physicians to become genetic doctors, which is why his eugenic publication was a supplement to the German Medical Association's official organ. Sterilization of the unfit was of course a leading topic in *Der Erbarzt*. Eugenic questions from German physicians were answered in a regular "Genetic Advice and Expertise" feature. In the first issue, Verschuer editorialized that *Der Erbarzt* would "forge a link between the ministries of public health, the genetic health courts, and the German medical community." Henceforth, he insisted, doctors must react to their patients not as individuals, but as parts of a racial whole. A new era had arrived, in Verschuer's view: medical treatment was no longer a matter of doctor and patient, but of doctor and state.[10]

After the Nazi sterilization law took effect in 1934, German eugenicists were busy creating national card files, automated by IBM, to cross-index people declared unfit. A plethora of eugenic research institutes were established at various German universities to advance the effort. Their researchers scoured the records of the National Health Service, hospitals and hereditary courts, and then correlated health files on millions of Germans. In this process, Verschuer considered himself nothing less than a eugenic warrior. In 1935, he left the Institute for Anthropology, Human

Heredity and Eugenics to found Frankfurt University's impressive new Institute for Hereditary Biology and Racial Hygiene. Boasting more than sixty rooms, including labs, lecture halls, libraries, photography sections, ethnic archives and clinical rooms, the new institute was the largest of its kind in Germany. The institute's mission, according to Verschuer, was to be "responsible for ensuring that the care of genes and race, which Germany is leading worldwide, has such a strong basis that it will withstand any attacks from the outside." More than just a research institute, Verschuer's institution held courses and lectures for the SS, Nazi Party members, public health and welfare officials, as well as medical instructors and doctors in general to indoctrinate them with scientific anti-Semitism and eugenic theory.[11]

Soon the Institute for Hereditary Biology and Racial Hygiene had surpassed the Kaiser Wilhelm Institute in race biology and race politics, becoming the new model for German eugenic centers. Verschuer was doing his part to ensure that racial eugenics, the fulcrum of which was rabid Jew-hatred, became the standard for all medical training in Germany. He would soon boast that eugenics had become completely integrated into "the normal course of studies of medical students." In a report to the Nazi Party, he advocated registering all Jews and half-Jews. Hitler, said Verschuer, was "the first statesman to recognize hereditary biology and race hygiene."[12]

By 1937, Verschuer had gained the trust of the highest Nazi authorities and was beginning to eclipse his colleagues, and by 1939 he was describing his personal role as pivotal to Nazi supremacy. "Our responsibility has thereby become enormous," said Verschuer. "We continue quietly with our research, confident that here also, battles will be fought which will be of greatest consequence for the survival of our people." In an article for a series called *Research into the Jewish Question* (*Forschungen zur Judenfrage*), Verschuer wrote, "We therefore say no to another race mixing with Jews just as we say no to mixing with Negroes and Gypsies, but also Mongolians and people from the South Sea. Our *völkisch* attitude to the biological problem of the Jewish Question ... is therefore completely independent of all knowledge of advantages or disadvantages, positive or negative qualities of the Jews.... Our position in the race question has its foundation in genetics." In another article he insisted, "The complete racial separation between Germans and Jews is therefore an absolute necessity."[13]

Quickly, Verschuer became a star in American eugenic circles as well. His career and his writings fascinated the U.S. movement. When he became secretary of the Tübingen branch of the Society for Race Hygiene in 1925, *Eugenical News* announced it. His 1926 article on environmental influences for *Archiv für Rassen- und Gesellschaftbiologie* (*Archives of Race Science and Social Biology*) was promptly summarized in *Eugenical News*. The publication also noted Verschuer's 1927 appointment as one of three department heads at the Institute for Anthropology, Human Heredity and Eugenics. In 1928, Verschuer's presence as a guest at an International Federation of Eugenic Organizations meeting was mentioned in *Eugenical News*. In the years leading up to the ascent of Hitler, his articles continued to be cited in *Eugenical News*.[14]

Even after the Nazis assumed power in 1933, the American eugenic and medical media kept Verschuer in the spotlight. In January of 1934, the *Journal of the American Medical Association* cited a paper he presented at the German Congress of Gynecology. That same month, *Journal of Heredity* reviewed his book on the relationship between eugenics and tuberculosis. In the spring of that year, both *Eugenical News* and *American Journal of Obstetrics and Gynecology* highlighted him as a leader for his work in developing more than a thousand Nazi marriage screening centers. In September of 1934, *JAMA* questioned Verschuer's estimate that the frequency of hereditary blindness in vulnerable populations was a full third, but this only confirmed his status as a major voice in genetic science. That same month, *Eugenical News* published an article entitled "New German Etymology for Eugenics" and cited two definitions for *Rassenhygiene*; Verschuer's definition ran first, and Ploetz's second. In *Eugenical News*'s next issue, November–December, Verschuer was listed in a feature titled "Names of Eminent Eugenicists in Germany."[15]

By 1935, Verschuer was so admired by American eugenicists that *Eugenical News* heralded the opening of his Institute for Hereditary Biology and Racial Hygiene with the simple headline "Verschuer's Institute." The publication's ecstatic article asserted that Verschuer's new facility was the culmination of decades of preliminary research by Mendel, race theorist Count Gobineau, Ploetz and even Galton himself. Suggesting the far-reaching nature of his enterprise, *Eugenical News* made clear that Verschuer's mission was not merely the "individual man" but "mankind" itself. Among the new institute's several dozen rooms, the paper reported, were a number for "special investigators." *Eugenical News* was so enamored that it departed from its usual text-only format and included two photographs: a picture of the building's exterior plus one of an empty, nondescript corridor. The article closed, "Eugenical News extends best wishes to

Dr. O. Freiherr von Verschuer for the success of his work in his new and favorable environment."[16]

Goodwill among American eugenicists toward Verschuer was ceaseless. On April 15, 1936, Stanford University anatomist C. H. Danforth wrote to Verschuer offering to translate abstracts of one of Verschuer's journals. On July 7, 1936, Goddard, now located at Ohio State University, sent Verschuer several of his publications hoping that they might be useful to experiments at the new institute. On July 16, 1936, Popenoe wrote from the Human Betterment Foundation asking for statistics to rebut negative publicity about German sterilizations, saying, "We are always anxious to see that the conditions in Germany are not misunderstood or misrepresented." E. S. Gosney, Popenoe's partner at the Human Betterment Foundation, sent Verschuer three letters and two pamphlets in two months with the latest information on California's sterilization program.[17]

Laughlin himself sent two letters, one in German offering reprints of his own articles and a second in English conveying salutations from America on Germany's accomplishment. Writing on Carnegie Institution ERO letterhead, Laughlin stated, "The Eugenics Record Office and the Eugenics Research Association congratulate the German people on the establishment of their new Institute for the Biology of Heredity and Race Hygiene.... We shall be glad indeed to keep in touch with you in the development of eugenics in our respective countries."[18]

Verschuer sent back an effusive letter of appreciation. He congratulated Laughlin on his recent honorary degree from the University of Heidelberg, adding, "You have not only given me pleasure, but have also provided valuable support and stimulus for our work here. I place the greatest value on incorporating the results of all countries into the scientific research that takes place here at my Institute, since this is the only way of furthering the construction of the edifice of science. The friendly interest that you take in our work gives me particular pleasure. May I also be allowed to express my pleasure that you have been awarded an honorary doctorate from the University of Heidelberg and congratulate you on this honor? You have surely concluded from this that we German hereditarians and race hygienists value the pioneering work done by our American colleagues and hope that our joint project will continue to progress in friendly cooperation."[19]

Verschuer and his institute remained prominent in the American medical and eugenic press. When in mid-1935, Verschuer's new institute began deploying a force of young women as field workers to assemble family trees, *Eugenical News* reported it. *JAMA* covered the new institute in-depth in its September 1935 issue, specifying that cards on individuals arising from the investigations were being sent to other Reich health bureaus. *JAMA* reported on Verschuer's work again a few months later in 1936, focusing on his desire to engage in mass research on heredity and illness.[20]

Verschuer's well-received book, *Genetic Pathology (Erbpathologie)*, claimed that Jews disproportionately suffered from conditions such as diabetes, flat feet, deafness, nervous disorders and blood taint. In its January–February 1936 edition, *Eugenical News* enthusiastically reviewed *Genetic Pathology* and parroted Verschuer's view that a physician now owed his first duty to the "nation," adding, "The word 'nation' no longer means a number of citizens living within certain boundaries, but a biological entity." Verschuer's language on citizenship was a clear precursor to the Reich's soon-to-be-issued decree declaring that Jews could no longer be citizens of Germany, even if they resided there. Stripping German Jews of their citizenship was the next major step toward mass ghettoization, deportation and incarceration. *Eugenical News* closed its review of *Genetic Pathology* with this observation: "Dr. von Verschuer has successfully bridged the gap between medical science and theoretical scientific research."[21]

Verschuer's popularity with American eugenicists had soared by 1937. Senior U.S. eugenicists were clamoring for his attention. Anti-Semite and Nazi sympathizer Charles M. Goethe sent a letter introducing himself. "I am National President of the Eugenics Research Association of the United States," Goethe wrote. "I have heard much of your work at Frankfurt.... May I ask whether I could visit your Institution? I feel, because of the violent anti-German propaganda in the United States, our people know almost nothing of what is happening in Germany."[22]

Later that year, Goethe sent an equally fawning correspondence, apologizing for not visiting Germany but appealing to Verschuer's anti-Jewish sentiment. "It was with deep regret that I was unable to come to Frankfurt this year," he wrote. "Dr. Davenport and Dr. Laughlin of the Carnegie Institute have told me so much about your marvelous work.... I feel passionately that you are leading all mankind herein. One must exercise herein the greatest tact. America is flooded with anti-German propaganda. It is abundantly financed and originates from a quarter which you know only too well [Jews].... However, this ought to not blind us to the fact that Germany is advancing more rapidly in Erbbiologie than all the rest of mankind."[23]

By 1938, the plight of the Jews in Germany and thousands of refugees had become a world crisis, prompting the Evian Conference. Hitler's Reich had become identified in the media with brutal concentration camps.

Germany was again menacing its neighbors' territory. Yet Goethe continued his zealous propagandizing for Nazism. "Again and again," Goethe wrote Verschuer in early 1938, "I am telling our people here, who are only too often poisoned by anti-German propaganda, of the marvelous progress you and your German associates are making." In November of 1938, less than two weeks after the *Kristallnacht* riots, Goethe again wrote Verschuer, this time to lament, "I regret that my fellow countrymen are so blinded by propaganda just at present that they are not reasoning out regarding the very fine work which the splendid eugenists of Germany are doing.... I am a loyal American in every way. This does not, however, lessen my respect for the great scientists of Germany."[24]

Clyde Keeler, a Harvard Medical School researcher at Lucien Howe's laboratory, visited Verschuer's swastika-bedecked institute at the end of 1938. There he was able to see the center's anti-Jewish program and its devotion to Aryan purity. Upon his return to the United States, Keeler gave fellow eugenicists a glowing report. On February 28, 1939, Danforth of Stanford wrote Verschuer to applaud him, adding that Keeler "thinks that you have by all means the best equipped and most effective establishment of the sort that he has seen anywhere. May I extend my congratulations and express the hope that your group will long continue to put out the same excellent work that has already lent it distinction."[25]

Davenport was equally inspired by Verschuer. On December 15, 1937, he asked Verschuer to prepare a special summary of his institute's work for *Eugenical News*, "to keep our readers informed." Davenport also asked Verschuer to join three other prominent Nazi eugenicists on *Eugenical News*'s advisory committee. Falk Ruttke, Eugen Fischer and Ernst Rüdin were already members. With a letter of gratitude, Verschuer agreed to become the fourth.[26] Verschuer was now an essential link between American eugenics and Nazi Germany.

Otmar Freiherr von Verschuer had an assistant. His name was Josef Mengele.

* * *

Mengele began his career as a doctrinaire Nazi eugenicist. He attended Rüdin's early lectures and embraced eugenic principles as part of his fanatic Nazism. Mengele became a member of the SA, also known as the Storm Troopers, in 1934. His first academic mentor was the anti-Semitic eugenicist Theodor Mollison, a professor at Munich University. Just as Goddard claimed he could identify a feebleminded individual by a mere glance,

Mollison boasted that he could identify Jewish ancestry by simply examining a person's photograph. Under Mollison, Mengele earned his Ph.D. in 1935. His dissertation on the facial biometrics of four racial groups—ancient Egyptians, Melanesians and two European types—asserted that specific racial identification was possible through an anthropometric examination of an individual's jawline. Medical certification in hand, Mengele became a practicing doctor in the Leipzig University clinic. But this was only temporary. Mengele's dream was research, not practice. In 1937, on Mollison's recommendation, Mengele became Verschuer's research assistant at the Institute for Hereditary Biology and Racial Hygiene in Frankfurt. Here Mengele's eugenic knowledge could be applied. Some of Mengele's work involved tracing cranial features through family trees.[27]

Verschuer and his new assistant quickly bonded. Mengele had applied for Nazi Party membership as soon as the three-year ban was lifted in 1937. He and Verschuer made a good professional team. Together the two wrote opinions for the Eugenic Courts enforcing anti-Jewish Nuremberg Laws. In one case, a man suspected of having a Jewish father was prosecuted for engaging in sexual relations with an Aryan woman. Under the Nuremberg Laws, this was a serious criminal offense calling for prison time. As the prosecution's eugenic consultants, Mengele and Verschuer undertook a detailed examination of the suspect's family tree and carefully measured his facial features. Their eugenic report declared the man to be fully of Jewish descent.[28]

However, the accused man provided convincing evidence that he was in fact the illicit offspring of Christians. His father was indeed Jewish, but his mother was not. The man claimed to be the product of his non-Jewish mother's illicit affair with a Christian; hence he was no Jew. Illegitimacy was a common refrain of Jews seeking safe harbor from the Nuremberg statutes. The court believed the man's story and freed him. The decision outraged Mengele and Verschuer, who wrote a letter to the Minister of Justice complaining that their eugenic assessment had been overlooked. Approximately 448 racial opinions were ultimately offered by Verschuer's institute; these were so doctrinaire that Verschuer frequently appealed when the opinions were not accepted.[29]

Mengele's relationship with Verschuer was more than collegial. Staff doctors at the institute recalled that Mengele was Verschuer's "favorite." Verschuer's secretaries enjoyed Mengele's constant visits to the office, and nicknamed him "Papa Mengele." He would drop by the Verschuer home for tea, sometimes bringing his family. Mengele even made an impression on Verschuer's children, who years later remembered him in friendly terms.[30]

In 1938, Mengele joined the SS and received his medical degree, yet continued his close association with Verschuer. In fact his SS personnel file, number 317885, listed his employment in 1938 as an assistant doctor at the Institute for Hereditary Biology and Racial Hygiene. In the fall of that year, preparing for field assignment with an SS unit, Mengele underwent three months of rigorous basic training. Afterwards, he returned to Verschuer's institute in Frankfurt to resume eugenic research. For example, he examined the inheritance of ear fistulas and chin dimples, and then published the results. In a summary of 1938 projects for the German Research Society, Verschuer listed Mengele's work on inherited deformities and cited two of Mengele's papers, including one he completed for another doctor.[31]

In December of 1938, Mengele and Verschuer, as well as two other Nazi doctors associated with the institute, requested a grant from the Ministry of Science and Education to attend the International Congress of Genetics in Edinburgh, scheduled for the last week of August 1939. All four men secured initial authorization to attend as part of a large Nazi delegation, approved by the Party. Train and ferry schedules were researched. But after further review, the ministry lacked the funds to send them all. Ministry officials decided Mengele could not go. Germany began World War II on September 1, 1939. England and Germany were now enemies, so Nazi conferees returned in the nick of time.[32]

Mengele wanted to get into the war, but a kidney condition prevented him from joining a combat unit. He continued working with Verschuer and in early 1940 was still listed on Institute for Hereditary Biology and Racial Hygiene rosters as being on Verschuer's staff. An internal list of publications and papers, dated January 1939, listed two papers written by Verschuer with the help of assistants including Mengele. One was entitled "Determination of Paternity," recalling their days providing genealogical testimony for the Eugenic Courts. Mengele authored a third paper on the list with two of Verschuer's other assistants.[33]

Mengele also contributed several book reviews to Verschuer's publication, *Der Erbarzt*, in 1940. One review covered a book called *Fundamentals in Genetics and Race Care*, in which Mengele criticized the author for failing to adequately describe "the relationship between the principal races that are to be found in Germany and the cultural achievements of the German people." In another review critiquing a book about congenital heart defects, Mengele complained, "Unfortunately the author did not use subjects where the diagnosis could be verified by an autopsy."[34]

By June of 1940, when Germany was advancing on Western Europe, Mengele could no longer wait to enter the battle. He joined the Waffen SS and was assigned to the Genealogical Section of the SS Race and Settlement Office in occupied Poland. He undoubtedly benefited from Verschuer's March 1940 letter of recommendation averring that Mengele was accomplished, reliable and trustworthy. At the SS Race and Settlement Office, his mission was to seek out Polish candidates for Germanization. He would perform the racial and eugenic examinations. Eventually, in 1941, he was transferred to the Medical Corps of the Waffen SS, and then to the elite Viking unit operating in the Ukraine, where he rendered medical assistance under intense battlefield conditions. He was awarded two Iron Crosses and two combat medic awards. The next year, 1942, as the Final Solution was taking shape, Verschuer arranged for Mengele to transfer back to the SS Race and Settlement Office, this time to its Main Office in Berlin.[35]

By 1942, an aging Fischer was preparing to retire from the Kaiser Wilhelm Institute for Anthropology, Human Heredity and Eugenics in Berlin. His replacement was a major source of debate within eugenic and Nazi Party circles. By this time, Hitler's war against the Jews had escalated from oppressive disenfranchisement to systematic slaughter.[36]

Fischer had emerged as a major advocate of "a total solution to the Jewish question." His view was that "Bolshevist Jews" constituted a dangerous and inferior subspecies. At a key March 1941 conference on the solution to the Jewish problem held in Frankfurt, Fischer had been the honored guest. It was at this meeting that Nazi science extremists set forth ideas on eliminating Jews *en masse*. A leading idea that emerged was the gradual extinction (*Volkstod*) of the Jewish people by systematically concentrating them in large labor camps to be located in Poland. Later, Fischer specified that such labor must be unpaid slave labor lest any "improvement in living standards...lead to an increase in the birth rate."[37]

Given Fischer's high profile in Nazi Party extermination policies, his successor would have to be selected carefully. Lenz was considered for the job, but Fischer worked behind the scenes with the Nazi Party to have Lenz passed over. Fischer thought Lenz was too tutorial, and not bold enough for the challenges ahead. Instead, Fischer's hand-picked successor would be Verschuer—something Fischer had actually planned on for years.[38]

In 1942, Verschuer wrote in *Der Erbarzt* that Germany's war would yield a "total solution to the Jewish problem." He wrote a friend, "Many important events have occurred in my life. I received an invitation, which I accepted, to succeed Eugen Fischer as director of the Dahlem Institute

[Kaiser Wilhelm Institute for Anthropology, Human Heredity and Eugenics at Berlin-Dahlem]. Great trust was shown toward me, and all my requests were granted with respect to the importance and authority of the institute....I will take almost all my coworkers with me, first Schade and Grebe, and later Mengele and Fromme." Even though Mengele was still technically attached to the Race and Settlement Office, he was still Verschuer's assistant. Mengele's name was even added to the special birthday list for the institute's leading staff scientists.[39]

In January 25, 1943, with Hitler's extermination campaign in full swing, Verschuer wrote to Fischer, "My assistant Mengele...has been transferred to work in an office in Berlin [at the SS Race and Settlement Office] so that he can do some work at the Institute on the side."[40]

On May 30, 1943, Mengele arrived at Auschwitz.

* * *

Eugenics craved one type of human being above all others to answer its biological questions and to achieve its ultimate biological goal. The quest to locate this type of human being arose at the dawn of eugenics, and continued ceaselessly for four decades, throughout the voluminous discourse, research and publishing of the worldwide eugenic mainstream. To the eugenic scientist, no subject was of greater value. Young or old, healthy or diseased, living or dead, they all wanted one form of human—twins.

Twins were the perfect control group for experimentation. How people developed, how they resisted or succumbed to disease, how they reacted to physical or environmental change—all these questions could be best answered by twins precisely because they were simultaneous siblings. While fraternal twins sprang from two separate eggs fertilized at the same time, identical twins were, in fact, one egg split in two. Identical twins were essentially Nature's clones.[41]

Twins were valued for a second eugenic reason: Nature itself could be outmaneuvered if desirable individuals could be biologically enabled to spawn twins—or even better, triplets, quadruplets and quintuplets. In other words, a world of never-ending multiple births was the best assurance that the planned super race would remain super.

About a decade before Galton coined the term *eugenics*, he was convinced he could divine the secret of human breeding by studying twins. In 1874 and 1875, he published various versions of a scientific essay entitled "The History of Twins as a Criterion of the Relative Powers of Nature and Nurture." In analyzing whether environment or heredity was responsible for an individual's success, Galton complained that his investigations were always hampered by the unending variables—that is, until he located biological comparables. "The life history of twins supplies what I wanted," he wrote. Galton had closely studied some eighty sets of twin children by the time he wrote that essay. These included twins of the same and different gender as well as identical and non-identical twins.[42]

Cold Spring Harbor's handwritten outlines for key Mendelian traits listed twinning as one of the ten salient physical characteristics to explore. Davenport's 1911 textbook, *Heredity in Relation to Eugenics*, included a section on twins with the introduction, "It is well known that twin production may be an hereditary quality." Three years later, Heinrich Poll, Rockefeller's first fund administrator in Germany, published a major volume on twin research; Poll's interest in the topic dovetailed with the Rockefeller Foundation's years-long support of the subject.[43]

American eugenic publications constantly dotted their pages with the latest twin theory and research. Identifying the mechanism governing the creation and development of twins quickly became a major pursuit for eugenics. In 1916, *Eugenical News* published three articles on the subject, including one that examined a recent article in *Biological Bulletin* on armadillo quadruplets, hoping to apply the principle to multiple births in humans. One of the 1917 articles on twins in *Eugenical News* indicated that in about a quarter of same sex twins, "there is some factor that definitely forces the two children to be of the same sex." A second article in 1917 announced that a doctor in a Michigan institution for the feebleminded was searching the nation for mongolism in twins, especially cases in which only one of the siblings manifested the condition.[44]

The problem with studying twins was that in adulthood most twins lived separate lives, often in separate cities and even in different countries. It was hard to locate them, let alone bring them together for examination. In 1918, the American Genetic Association, the renamed American Breeders Association, announced that it desired to "communicate with twins living in any part of the world." The AGA explained, "It has been discovered that twins are in a peculiar position to help in the elucidation of certain problems of heredity. . . . 'Duplicate' twins have a nearly (though never an absolutely) identical germ plasm. . . . It is fortunate for our knowledge . . . on account of the chance it gives [us] to study the relative importance of heredity and of environment." Within a year of its announcement, the AGA had identified some six hundred twins, and by soliciting photos it had assembled a photo archive of several hundred.[45]

The ERO initiated its own twin study with a detailed four-page questionnaire. Among its numerous questions: "What is your favorite fruit?" and "Do you prefer eggs boiled soft or hard?" It also provided a place for each twin's fingerprints and the names and addresses of family members. ERO investigators located one especially fertile family in Cleveland that had repeatedly produced multiple births. When Davenport wrote up the case for *Journal of Heredity* in 1919, he explained that it had taken more than six visits by field workers to determine the full scope of the original couple's fecundity. Later, *Eugenical News* announced that Columbia, Missouri, was home to more twins than any other city in the nation—one pair for every 477 people.[46]

Hereditarians sought twins of all ages—not just children—for proper study. The family tree of a New England family of twins, including one pair ninety-one years of age, fascinated eugenicists. Geneticists excavated old journals to discover even earlier examples, such as a seventeenth-century Russian woman who gave birth twenty-seven times, each time producing twins, triplets or quadruplets, yielding a total of sixty-nine children.[47]

Race and twins quickly became an issue for American eugenicists. In a 1920 lecture series, Davenport raised the issue of "racial difference in twin frequency" in the same geographic area. He pointed out that from 1896 to 1917, in Washington, D.C., the "negro rate [of twins] is 20 percent higher than the white rate." For whites in the nation's capital, it was 1.82 pairs of twins per hundred births, while blacks had 2.27 per hundred. At about the same time, *Eugenical News*, analyzing recent census data, claimed that twin births overall still occurred at a frequency of approximately 1 percent nationwide; but the percentage of multiple births among Blacks was almost one-fifth greater than among whites. Davenport followed up such observations in his Jamaica race-crossing study, which featured in-depth studies of three sets of twins.[48]

Diagnostic and physiological developments in twin studies from any sector of the medical sciences were of constant interest to eugenic readers. So *Eugenical News* regularly summarized articles from the general medical literature to feed eugenicists' unending fascination with the topic. In 1922, when a state medical journal reported using stethoscopes to monitor a twin pregnancy, it was reported in *Eugenical News*. When a German clinical journal published a study of tumors in twins, this too was reported in *Eugenical News*.[49]

With each passing issue, *Eugenical News* dedicated more and more space to the topic. The list of such reports became long. By the early 1920s, articles on twins became increasingly instructive. One typical article explained how to more precisely verify the presence of identical twins using a capillary microscope. *Journal of Heredity* also made twins a frequent subject in its pages. For example, it published Popenoe's article entitled "Twins Reared Apart," and Hermann Muller's article "The Determination of Twin Heredity," and regularly reviewed books about twins.[50]

Every leading eugenic textbook included a section on twins. Popenoe's *Applied Eugenics* explained that identical twins "start lives as halves of the same whole" but "become more unlike if they were brought up apart." Baur-Fischer-Lenz's *Foundation of Human Heredity and Race Hygiene* cited several studies including those written by Popenoe in *Journal of Heredity*. The German eugenicists wrote, "Of late years, the study of twins has been a favorite branch of genetic research" and thanked Galton for his "flash of genius" in "[recognizing] this a long while ago."[51]

In a similar vein, most international eugenic and genetic conferences included presentations or exhibits on twins—their disparity or similarity, their susceptibility to tuberculosis, their likes and dislikes. R. A. Fisher opened one of his lectures to the Second International Congress of Eugenics with the phrase: "The subject of the genesis of human twins... has a special importance for eugenicists." The third congress offered an exhibit on mental disorders in twins, an exhibit illustrating fingerprint comparisons, a third juxtaposing identical and fraternal twins, and a fourth offering an array of fifty-nine anthropometric photos.[52]

The quest for a superior race continued to intersect with the availability of twins. In the July–August 1935 edition of *Eugenical News*, Dr. Alfred Gordon published a lengthy article entitled "The Problems of Heredity and Eugenics." His first sentence read: "Regulation of reproduction of a superior race (eugenics) is fundamentally based on the principles of heredity." Gordon went on to explain, "The role of heredity finds its strongest corroboration in cases of psychoses in twins." He then gave an example of just two case studies of twins. Such enthusiastic coverage in the biological and eugenic media was prompted a few months before by the extensive examination of just a single pair of twins undertaken at New York University's College of Dentistry, this to identify pathological dentition.[53]

There were so few twins to study that surgeons in the eugenics community passed along their latest discoveries, one by one, to advance the field's common knowledge. In one case, Dr. John Draper of Manhattan wrote to Davenport, "Last Thursday, I opened the abdomen of twin girls, fourteen years old. They presented very similar physical characteristics and the psychoses so far as could be determined were identical." Davenport replied, "Your observations upon the internal anatomy of the twin girls is exceedingly important, as very few observations of this type have been made upon twins." He offered to dispatch a field worker to make facial measurements. Such random reports were precious to eugenicists because physical experimentation on large groups was essentially impossible.[54]

All that changed when Hitler came to power in 1933. Germany surged ahead in its study of twins. The German word for twins is *Zwillinge*. There were tens of thousands of twins in the Reich. In 1921 alone, 19,573 pairs were born, plus 231 sets of triplets. In 1925, 15,741 pairs of twins were born, as well as 161 sets of triplets. Twins were now increasingly sought to help combat hereditary diseases and conditions, real and imagined. Verschuer's book, *Twins and Tuberculosis*, was published in 1933 and received a favorable review in *Journal of Heredity*. In 1934, a Norwegian physician working with Verschuer and Fischer published in a German anthropology journal his analysis of 116 pairs of identical twins and 127 pairs of fraternal twins for their inheritance of an ear characteristic known as Darwin's tubercle.[55]

But many more twins would be needed to accomplish the sweeping research envisioned by the architects of Hitler's master race. In early December of 1935, Verschuer told a correspondent for the *Journal of the American Medical Association* that eugenics had moved into a new phase. Once Mendelian principles of human heredity were established, the correspondent wrote, "Further progress was achieved with the beginning of research on twins, by means of which it is possible to measure hereditary influence even though the hereditary processes are complicated....Many of these researches, however, as Freiherr von Verschuer recently pointed out, are of questionable value....What is absolutely needed is research on series of families and twins selected at random...examined under the same conditions, a fixed minimum of examinations being made in all cases." The article went on to cite Verschuer's view that meaningful research would require entire families—from children to grandparents.[56] In plain words, this meant gathering larger numbers of twins in one place for simultaneous investigation.

To attract more twins, the Nazi Party and the National Socialist Welfare League promoted "twin camps" for the holidays. Verschuer circulated handy text references for all German physicians who might encounter twins. When Verschuer opened his Institute for Hereditary Biology and Racial Hygiene in 1936, the event created such fanfare in *Eugenical News* partially because, "Dr. Verschuer states that the object of his investigation is mankind, not the individual man, but families and twins; and in this work there will not [only] be investigated...interesting twins, but all twins and families of definite geographical origin."[57]

At about that time, German neuropsychiatrist Heinrich Kranz of the University of Breslau published extensive genealogical details about seventy-five pairs of twin brothers and fifty pairs of opposite gender twins, seeking correlations on criminal behavior. In a *Journal of Heredity* essay, Popenoe lauded Kranz's investigation and predicted that such efforts would help identify "born criminals." Popenoe welcomed more such German research because "it has become one of the most dependable methods of studying human heredity."[58]

Indeed, a plethora of Nazi scientific journals were brimming with regular coverage of eugenic investigations of twins. Several publications were devoted solely to the subject, such as *Zwillingsforschungen* (*Twin Research*) and *Zwillings- und Familienforschungen* (*Twin and Family Research*). Verschuer frequently wrote for these journals. In some cases Mengele coauthored the articles, including an article on systemic problems and cleft palate deformation published in *Zwillings- und Familienforschungen*. Some published twin research credited Mengele as the principal investigator, such as an article on congenital heart disease, also for *Zwillings- und Familienforschungen*.[59]

Verschuer's preoccupation with twin studies expanded feverishly. He required more and more twins. In a September 1938 application for funds from the German Research Society, Verschuer explained his plans. "Large-scale research on twins is necessary to explore the question of the hereditary aspects of human characteristics, especially illnesses. This research can take two paths: 1. Testing of all twins in a specific geographic area, done at our institute by Miss Liebmann. All twins in the Frankfurt district back to 1898 have been listed and almost all have been examined; she discussed some interesting cases in several articles and a comprehensive summary is being done. 2. Listing of series of twins. Based on cases in over 100 hospitals in west and southwest Germany, the number of twins among them were determined and the cases were examined according to illnesses." He listed rheumatism, stomach ulcers, cancer, heart defects, anemia and leukemia as the conditions he was focusing on. Verschuer assured, "A good deal of material has been collected."[60]

In 1939, Interior Minister Frick issued a public decree compelling all twins to register with their local Public Health Office and make themselves available for genetic testing. The Reich Statistics Bureau would cooperate in the identification campaign. The announcement in the Nazi medical publication *Ziel und Weg* (*Goal and Path*) was published with a lengthy quotation from *Mein Kampf* on the cover: "We must differentiate most stringently between the state as a mere *container* and race as its *contents*. This container is meaningful only when it has the ability to preserve and protect the contents; otherwise it is worthless."[61]

American eugenicist T. U. H. Ellinger was in Germany shortly after the decree to visit with Fischer at the Kaiser Wilhelm Institute for Anthropology, Human Heredity and Eugenics. In a *Journal of Heredity* essay on his visit, Ellinger flippantly reported to his colleagues, "Twins have, of course, for a long time been a favorite material for the study of the relative importance of heredity and environment, of nature and nurture. It does, however, take a dictatorship to oblige some ten thousand pairs of twins, as well as triplets and even quadruplets, to report to a scientific institute at regular intervals for all kinds of recordings and tests."[62]

When twins did report to the Institute for Anthropology, Human Heredity and Eugenics, they were often placed in small, specially-constructed examination rooms, each lined with two-way mirrors and motion picture camera lenses camouflaged into the wallpaper. The staff proudly showed Ellinger all of these facilities.[63] However, eugenicists at the institute could only go so far with mere observations.

Reich scientists needed more if they were to take the next step in creating a super race resistant to disease and capable of transmitting the best traits. Autopsies were required to discover how specific organs and bodily processes reacted to various experiments. Verschuer needed more twins and the freedom to kill them. The highest ranks of the Hitler regime agreed, including Interior Minister Frick, who ran the concentration camps, and SS Chief Heinrich Himmler.[64] Millions of dispensable human beings from across Europe—Jews, Gypsies and other undesirables—were passing through Hitler's camps to be efficiently murdered. Among these millions, there were bound to be thousands of twins.

Shortly after Verschuer took over for Fischer at the Institute for Anthropology, Human Heredity and Eugenics, he proposed a *Zwillingslager*, or "twins camp," within Auschwitz. He applied to the German Research Society, which between July and September of 1943 passed his application through the various steps needed for approval and funding. The grant covered a six-month period beginning in October 1943 under contract number 0296/1595. The camp was approved and was bureaucratically filed under the keyword "Twins Camp."[65]

At the end of May 1943, Mengele arrived in Auschwitz, where he took control of the ramps where Jews were brought in. Verschuer notified the German Research Society, "My assistant, Dr. Josef Mengele (M.D., Ph.D.) joined me in this branch of research. He is presently employed as Hauptsturmführer [captain] and camp physician in the Auschwitz concentration camp. Anthropological testing of the most diverse racial groups in this concentration camp are being carried out with permission of the SS Reichsführer [Himmler]."[66]

Nazi Germany had now carried eugenics further than any dared expect. The future of the master race that would thrive in Hitler's Thousand-Year Reich lay in twins. For this reason, there would now be a special class of victims at Auschwitz. There would be a special camp, special medical facilities and special laboratories—all for the twins.

After the locomotives lurched to a final stop at Auschwitz, after the whistle shrieked and the doors rolled open, after the bewildered masses tumbled out of the boxcars and onto the ramp, above the tumult of their own fear and the incessant barking dogs, all of them heard one word, and they heard it shouted twice.

As the SS passed through the trembling crowds lining up for the gas chambers, they cried out for all to hear:

Zwillinge! Zwillinge!
Twins! Twins!

LEA LORINCZI: *"When we got off the trains, we could hear the Germans yelling, 'Twins, twins!'"* Lea and her brother were spared.[67]

MAGDA SPIEGEL: *"SS guards were yelling, 'Twins, twins, we want twins.' I saw a very good-looking man coming toward me. It was Mengele."* They were also spared.[68]

JUDITH YAGUDAH: *"When it was our turn, Mengele immediately asked us if we were twins. Ruthie and I looked identical. We had similar hairdos. We were wearing the same outfits. Mengele ordered us to go in a certain direction—and our mother, too."* Judith and Ruthie were spared.[69]

EVA MOZES: *"As I clutched my mother's hand, an SS man hurried by shouting, 'Twins! Twins!' He stopped to look at us. Miriam and I looked very much alike. We were wearing similar clothes. 'Are they twins?' he asked my mother. 'Is that good?' she replied. He nodded yes. 'They are twins,' she said."* Eva and Miriam were also pulled out of the gas chamber line.[70]

ZVI KLEIN: *"My twin brother and I were marching toward the gas chambers when we heard people yelling, 'Twins! Twins!' We were yanked out of the lines and brought over to Dr. Mengele."* Zvi and his brother were spared.[71]

MOSHE OFFER: *"I heard my father cry out to them he had twins. He went over personally to Dr. Mengele and told him, 'I have a pair of twin boys.'... But we did-n't want to be separated from our mother, and so the Nazis separated us by force. My father begged Mengele ... As we were led away, I saw my father fall to the ground."* The Offer boys lived. Their parents disappeared into the selection.[72]

HEDVAH AND LEAH STERN: *"Some prisoners told [my mother] in Yiddish, 'Tell them you have twins. There is a Dr. Mengele here who wants twins. Only twins are being kept alive.'"* The Stern sisters lived to tell their story.[73]

All of them lived through the *Selektion*. But now they lived in Mengele's world of torture and testing, electroshock and syringes, eye injections and other hideous experiments—where live children and fresh cadavers were equally prized—all to achieve the eugenic ideal of a superior race in a place where mankind had sunk to the nadir of humanity.

*   *   *

Sadistic science at Auschwitz was part of Nazi Germany's eugenic desire to create its master race.

Like Verschuer, Mengele considered himself a warrior in the battle for eugenic supremacy. In an autobiographical account, Mengele spoke of his desire to create a super race as his initial motive for becoming a doctor. He traced his own family pedigree—pure Aryan stock—back four generations. An inmate anthropologist, Martina Puzyna, saved from death in order to work with Mengele, recalled, "He believed you could create a new super-race as though you were breeding horses.... He was mad about genetic engineering." A prisoner pathologist forced to work closely with Mengele wrote that the Angel of Death was obsessed with "the secret of the repro-duction of the race. To advance one step in the search to unlock the secret of multiplying the race of superior beings destined to rule was a 'noble goal.' If only it were possible, in the future, to have each German mother bear as many twins as possible."[74]

Shortly after arriving at Auschwitz, Mengele established Verschuer's twin camp at Barrack 14 in Camp F. Mengele had his pick of assistants from the finest doctors and pathologists in Europe, who came to Auschwitz con-demned in sealed boxcars. One whom he selected from the ramp was a Hungarian Jewish pathologist named Miklos Nyiszli, a graduate of Friedrich Wilhelm University medical school in Breslau. He became one of

Mengele's favorite assistants. Nyiszli's task was to dissect the endless torrent of special corpses and create meticulous postmortem reports. For this process, Mengele would not settle for a typical ramshackle, makeshift concentration camp facility. Instead, amid the filth and squalor of Auschwitz, Mengele requisitioned and created a modern well-equipped pathology lab.[75]

The lab had everything needed for perfect autopsies. It was eerily professional, with light green painted walls surrounding a red concrete floor. A polished marble dissection table with fluid drains abutted a utility basin with shiny nickel faucets. Three white porcelain sinks lined the wall. Mosquito screens covered the windows. In the adjacent room, Nyiszli found a well-stocked library with the latest publications, three microscopes, and a closet full of mortuary supplies—everything from aprons to gloves. Nyiszli recalled it as "the exact replica of any large city's institute of pathology."[76]

Dina, a Czech inmate known for her skillful paintings, was selected at the ramp to become Mengele's anthropological artist. She would create anatomical drawings of the twins' features: noses, ears, mouths, hands, feet and skulls. Her artwork would accompany the experimentation data in each patient's folder.[77]

Mengele was happy in his work, frequently whistling as he selected human guinea pigs, discarded others to the gas chambers, inflicted his experiments and then reviewed the autopsies. A broad smile lit up his face as he surveyed his precious subjects, especially the children. "Almost like he had fun," one surviving twin recalled, adding, "He was very playful." Diligent and detailed, he once noticed a smudge on a bright blue file cover and sternly turned to Nyiszli, asking, "How can you be so careless with these files, which I have compiled with so much love!"[78]

Love was a corrupted word for Mengele. He certainly loved his work. At times, he seemed to love the youngest twins. All of Mengele's twins were better fed than other prisoners and even allowed small personal freedoms, such as roaming around the camp. Sometimes he served the children chocolates, patted them on the head affectionately, chaperoned them to camp concerts and made them feel as though he were a father figure looking after them. Eva Kupas remembered that once, when she wanted to see her twin brother, Mengele personally escorted her and "held my hand the whole way." He seemed to identify with one very young boy who somewhat resembled him, and actually trained the child to say "My name is 'Mengele.'"[79]

But without warning Mengele could fly into uncontrollable murderous frenzies. One teenage girl wept and begged when she was separated from her mother and sisters. She recounted that Mengele "grabbed me by the hair, dragged me on the ground and beat me." When the girl's mother pleaded, Mengele brutally beat her with his riding crop. In one case, a frantic mother fought to remain with her younger daughter. Mengele simply drew his pistol and shot the woman and her daughter, then waved the entire transport to the gas chambers, remarking, "Away with this shit!" Another time he caught a woman named Ibi, who had cleverly evaded the gas chambers six times by jumping off the truck just in time. A suddenly enraged Mengele shrieked, "You want to escape, don't you. You can't escape now!...Dirty Jew!" As he screamed, Mengele viciously beat the woman to death and kept beating her until her head resembled a bloody, formless mass. After these savage incidents, Mengele could immediately Jekyll-Hyde back to the charming, whistling clinician enchanted with his subjects and his science.[80]

In fact, Mengele loved his twins not because he thought they should be preserved, but only because they briefly served his mad scientific quest. Nyiszli recounted that siblings were subjected "to every medical examination that can be performed on human beings," from blood tests to lumbar punctures. Each was rigorously photographed naked, and calipered from head to toe to complete the record. But these were only the baselines and vital signs. Then came the actual experiments. The Reichenberg boys, mistakenly thought to be twins because they so closely resembled each other, piqued Mengele's interest because one possessed a singer's voice while the other couldn't carry a tune. After crude surgery on both boys' vocal chords, one brother lost his speech altogether. Twin girls were forced to have sex with twin boys to see if twin children would result. Efforts were made to surgically change the gender of other twins.[81]

One day, Mengele brought chocolates and extra clothing for twin brothers, Guido and Nino, both popular with the medical personnel. A few days later the twins were brought back, their wrists and backs sewn together in a crude parody of Siamese twins, their veins interconnected and their surgical wounds clearly festering. The boys screamed all night until their mother managed to end their agony with a fatal injection of morphine.[82]

Mengele suspected that two Gypsy boys, about seven years of age and well-liked in the lab, carried latent tuberculosis. When prisoner doctors offered a different opinion, Mengele became agitated. He told the assembled staff to wait a while. An hour later he returned and sedately declared, "You are right. There was nothing." After a brief silence, Mengele acknowledged, "Yes, I dissected them." He had shot both in the neck and autopsied them "while they were still warm."[83]

It was imperative that twins be murdered simultaneously to analyze them comparatively. "They had to die together," Nyiszli recounted. For example, the bodies of four sets of Gypsy twins under the age of ten were delivered to Nyiszli for autopsy in one shipment. Twelve sets of gassed twins were diverted from the furnace so they could be dissected as a group; to facilitate identification among the hundreds of twisted corpses, the twelve had been coded with chalk on their chests before they entered the chamber. One girl recovered from an implanted infection too soon; he killed her quickly so both siblings would be freshly deceased.[84]

If one of Mengele's precious human guinea pigs was harmed before he could complete his work, he became incensed. Guards were under strict instructions to keep Mengele's twins alive, or face his wrath if they died during the night prior to his handling. Some 1,500 twins were subjected to Mengele's atrocities. Fewer than two hundred survived.[85] Those who lived had simply not yet been killed.

Mengele also sought dwarfs and the physically deformed—really any specimen of interest. He ghoulishly and capriciously explored the effects of genetics, disease and mass breeding. In one case, Mengele removed part of a man's stomach without administering anesthesia. To investigate the pathology of dysentery, Mengele told Nyiszli to prepare for 150 emaciated corpses, and to autopsy them at the rate of seven per day; Nyiszli protested that he could only complete three per day if he was to be thorough. Eye color was a favorite subject for experimentation. Eager to discover if brown eyes could be converted to Nordic blue, Mengele would introduce blue dyes, sometimes by drops, sometimes by injection. It often blinded the subjects, but it never changed their eye color.[86]

While evidence of mass murder in the trenches of Russia and the gas chambers of Poland was systematically destroyed, Mengele's murders were enshrined in the protocols of science. Mengele's ghastly files did not remain his private mania, confined to Auschwitz. Every case was meticulously annotated, employing the best scientific method prisoner doctors could muster. Then the files were sent to Verschuer's offices at the Institute for Anthropology, Human Heredity and Eugenics in Berlin-Dahlem for study.

An adult prisoner, chosen to help care for the youngest twins, recounted, "The moment a pair of twins arrived in the barrack, they were asked to complete a detailed questionnaire from the Kaiser-Wilhelm Institute in Berlin. One of my duties as [the] 'Twins' Father' was to help them fill it out, especially the little ones, who couldn't read or write. These forms contained dozens of detailed questions related to a child's background, health, and physical characteristics. They asked for the age, weight, and height of the children, their eye color and the color of their hair. They were promptly mailed to Berlin."[87]

Nyiszli, who had to fill out voluminous postmortem reports, recalled Mengele's warning: "'I want clean copy, because these reports will be forwarded to the Institute of Biological, Racial and Evolutionary Research at Berlin-Dahlem.' Thus I learned that the experiments performed here were checked by the highest medical authorities at one of the most famous scientific institutes in the world."[88]

The reports, countersigned by Mengele and sent to Berlin, were not just received and warehoused, they were carefully reviewed and discussed. A dialogue developed between Verschuer's institute and Mengele. Another prisoner assistant recounted that Mengele "would receive questions about the twins from the Kaiser Wilhelm Institute in Berlin, and he would send them the answers."[89]

The volume of exchange was massive. In a March 1944 memo from Verschuer to the German Research Society, which financed his work, he asked for more clerical assistance and supplies for the Auschwitz project. The memo, entitled "On the continuation of hereditary-psychological research" and filed under the keyword "Twins camp," was coded G for *geheime*, or "secret." Verschuer explained, "Analysis of material obtained from the twins camp continued during the half-year reporting period October 1943 to March 15, 1944. Some 25 psychological analyses, each of which consisted of about 200 pages, were dictated during this period, continuing to round out the overall description of the experiences gained through the twins camp. These analyses were continued, following the same methods as those analyses which began in the summer of 1943. The evaluation system employed has proven useful and was developed further. Several secretaries will be necessary in order to continue the evaluation, as well as sufficient amounts of typing paper, steno blocks and other writing equipment. Some 10,000 sheets of paper will be needed for the coming quarter-year."[90]

More than just reports, Nyiszli sent body parts. "I had to keep any organs of possible scientific interest," he remembered, "so that Dr. Mengele could examine them. Those which might interest the Anthropological Institute at Berlin-Dahlem were preserved in alcohol. These parts were specially packed to be sent through the mails. Stamped 'War Material—Urgent,' they were given top priority in transit. In the course of my work at the crematorium I dispatched an impressive number of such packages. I received, in reply, either precise scientific observations or instructions. In order to classify this correspondence I had to set up special files. The directors of the Berlin-Dahlem Institute always warmly thanked Dr. Mengele for this rare and precious material."[91]

Among his many grisly memories, one case especially haunted Nyiszli. Mengele spotted a hunchbacked Jew, a respected cloth merchant from Lodz, Poland, and his teenage son, handsome but with a deformed foot supported by an orthopedic shoe. Mengele ordered his slave pathologist, Nyiszli, to interview the father and son for the file. Nyiszli did so, not in the dissecting room, which reeked of formaldehyde, but in an adjacent study hall, trying his best not to alarm them. After the interview, the father and son were shot. Nyiszli performed detailed autopsies, complete with copious notes. Mengele was fascinated with the eugenic potential of the information, since each individual carried his own deformity. "These bodies must not be cremated," Mengele ordered. "They must be prepared and their skeletons sent to the Anthropological Museum in Berlin." After some discussion, Nyiszli began the gruesome chore of creating two lab-quality skeletons. This involved cooking the corpses to detach all flesh. During the long cooking process in the courtyard, four starving Polish slave workers mistook the contents of the vats and began eating. Nyiszli ran out to stop them. The cooled and treated skeletons were then wrapped in large sacks, labeled "Urgent: National Defense," and mailed to the Institute for Anthropology, Human Heredity and Eugenics.[92]

In the depths of his misery, Nyiszli wondered if he had witnessed too much. "Was it conceivable," he wrote, "that Dr. Mengele, or the Berlin-Dahlem Institute, would ever allow me to leave this place alive?"[93]

Like many eugenic research organizations, the Institute for Anthropology, Human Heredity and Eugenics valued twins' eyes. For decades, American eugenicists had stressed the research importance of twins' eyes, and the German movement naturally adopted the precept. Indeed, typical enthusiasm for the topic was evident in the March–April 1933 edition of *Eugenical News* in an article headlined "Hereditary Eye Defects," which reviewed a newly released book that included a chapter on "eyes of twins." *Eugenical News* closed its review with the comment, "We have nothing but praise for the assiduity in the gathering of the data.... We are happy to have this long needed work done and so well done." Similarly enthusiastic reviews and articles on the subject of twins' eyes and vision were published in *Eugenical News* during the latter 1930s.[94]

In 1936, a colleague had sent Laughlin a request to expand the eye color question of the ERO's Twin Schedule. The new instructions would read:

"Look at the colored part of the eye carefully in a good light with the help of a mirror. Is there any difference that you can see in the color or pattern of marks in the right and left eyes? Blue and gray eyes have brownish streaks, sometimes a few, which can be easily counted and usually more in one eye than in the other. Please describe any such difference between your eyes."[95]

Like his American colleagues, Verschuer was long interested in twin eye color. He wanted eye color studies included in his Auschwitz experiments, and the German Research Society funded one such project in September of 1943. Mengele was careful to gather all the eyes Verschuer needed.[96]

Inmate doctor Jancu Vekler never forgot what he saw when he entered one room at the Gypsy camp. "There I saw a wooden table with eyeballs laying on it. All of them were tagged with numbers and little notes. They were pale yellow, pale blue, green and violet." Vera Kriegel, another slave doctor, recalled that she walked into one laboratory and was horrified to see a collection of eyeballs decorating an entire wall, "pinned up like butterflies. . . . I thought I was dead," she said, "and was already living in hell."[97]

One day a prisoner transcriptionist was frantic because while a family of eight had been murdered, only seven pairs of eyes were found in the pathology lab. "You've given me only seven pairs of eyes," the assistant exclaimed. "We are missing two eyes!" He then scavenged similar eyes from other nearby corpses to complete the package for Verschuer's institute—without Mengele being the wiser.[98]

Chief recipient of the eyes was Karin Magnussen, another Verschuer researcher at the institute who was investigating eye anomalies, such as individuals with irises of different colors. In a March 1944 update subheaded "Work on the Human Eye" and submitted to the German Research Society, Magnussen reported, "The first histological work, which was concluded in the fall, 'On the Relationship Between Iris Color, Histological Distribution of Pigment and Pigmentation of the Bulb of the Human Eye,' to be published in the *Zeitschrift für Morphologie und Anthropologie* [*Journal for Morphology and Anthropology*], is currently in press. Material for a second series of experiments is currently being prepared for histological examination. The article on the determination of iris color, which was intended for publication in *Erbarzt* in December 1943, was printed but destroyed by enemy attacks and is now being reprinted. Observations continue on links among certain anomalies in humans. Other observations of humans had to be temporarily suspended for war-related reasons, but are to resume in summer if possible. Material is constantly being collected and evaluated for the expert opinions."[99]

Among the several scholarly articles on eyes from Auschwitz that Magnussen was authoring was one intended for the journal *Zeitschrift für Induktive Abstammungslehre und Vererbungsforschung* (*Journal for Inductive Genealogical Science and Hereditary Research*). Editorial board member Professor George Melchers, who reviewed the submission draft, remembered, "I was struck by the fact that the whole family—grandparents, parents and children—had died at the same time. I could only assume they had [all] been killed in a concentration camp." The war was coming to an end, so Melchers never submitted Magnussen's article to the full board.[100]

Magnussen later told her denazification tribunal, "I became acquainted with Dr. Mengele, who had been inducted as a medical officer, in [Berlin-] Dahlem during the war, when he visited the institute while on leave. I spoke with him a few times during such visits to the institute about scientific projects and scientific problems.... I completed my research, although after [a Gypsy] clan with heterochromatic eyes was imprisoned in Auschwitz, I was refused all access to these family members. Completion of my research was only possible through the help given me by Dr. Mengele, who coincidentally had been transferred to the camp. At that time, he helped me trace the hereditary path by determining eye color and family relationships. Through him I also learned that one of the most important families in the clan was contaminated with tuberculosis. I then asked him if he could send me the autopsy and pathological tissue from the eyes if someone from this family should die." She added, "The impression I received from the cases of illness and from the very responsible and very humane and very decent behavior exhibited by Dr. Mengele toward his imprisoned patients and subordinates... was such that I would never have thought that anything could have happened in Auschwitz that violated laws of the state, medicine or of humanity."[101]

In addition to eyes, Verschuer wanted blood. Liters of it. For decades, eugenicists had sought the genetic markers for "carriers," or people who appeared normal but were likely to transmit a Mendelian predisposition for a range of defective traits from pauperism to epilepsy. This effort was at first bogged down in early attempts to assemble race-based family trees and to create pseudoscientific ethnic and class countermeasures. But by the twenties, the most talented eugenicists and geneticists were working hard to analyze blood serum to solve the question of defective germ plasm. They weren't sure whether they were seeking a specific hormone, an enzyme, a protein, genetic material or other blood molecule. They only knew that mankind's eugenic destiny was lurking in the blood and waiting to be discovered.[102]

In 1924, Davenport had told the Second International Congress of Eugenics, "The hormones that determine our personality, constitute the bridge that connects this *personality* on the one hand, with the *specific enzymes* packed away in the chromosomes of the germ cells, on the other." Davenport went on to explain, "You and I differ by virtue of the... atomic activity of the enzymes and hormones which make up that part of the stream of life-yeast which has got into and is activating our protoplasm and will activate that of the fertilized egg that results from us and our consorts." He stressed that a human being was dictated "by virtue of the peculiar properties of those extraordinary activating substances, which are specific for him and other members of his family and race or biotype. The future of human genetics lies largely in a study of these activities.... Of these [studies], one of the most significant is that of twin-production."[103]

The *Eugenical News* report on the 1927 grand opening of the Kaiser Wilhelm Institute for Anthropology, Human Heredity and Eugenics pointed out, "In the section on human genetics, twins and the blood groups were specially considered." On May 13, 1932, the Rockefeller Foundation's Paris office dispatched a radiogram to its New York headquarters asking for funds to support Verschuer's research while he was at the Institute for Anthropology, Human Heredity and Eugenics. The foundation approved a three-year grant totaling $9,000 to "KWG Institute [for] Anthropology for research [on] twins and effects on later generations of substances toxic for germ plasm."[104]

At the same time, the foundation was already funding an array of vocal German anti-Semites in a five-year $125,000 study. Internal foundation reports described the study as "the racial or biological composition of the German people and of the interaction of biological and social factors in determining the character of the present population." Twin research was repeatedly cited as a key facet of the research. Among the scientists listed on the foundation's roster was Rüdin in project items 9 and 10; project item 16 was Verschuer. This $125,000 grant was not made directly, but channeled through the Emergency Fund for German Science (*Notgemeinschaft der Deutschen Wissenschaften*), which evolved into the German Research Society (*Deutsche Forschungsgemeinschaft*).[105]

When Hitler came to power the next year, Rockefeller did not cease its funding of race biology in Germany. However, unlike many American eugenic leaders, Rockefeller officials were more circumspect. Rockefeller executives did not propagandize for Nazism, nor did they approve of the Reich's virulent repression. The foundation's agenda was strictly biological to the exclusion of politics. It wanted to discover the specific genetic components of the blood of the unfit—even if that meant funding Nazi-controlled institutions.

Rockefeller's seed money was not wasted. In 1935, *Eugenical News* published a notice entitled "Blood Groups of Twins," which summarized a Nazi medical journal article based on Verschuer's research. "*The Kaiser-Wilhelm Institute für Anthropologie Menschliche Erblehre und Eugenik*, at Dahlem-Berlin," reported *Eugenical News*, "is conducting, through Dr. O. v. Verschuer, studies on twins. Of 202 one-egg twins on whom the blood group was determined, in every case the serologic findings were the same; that is, both fell into the same blood group, just as both are of the same sex. On the other hand, in the case of two-egg twins the blood groups of the twins, whether of same or opposite sex, were frequently unlike."[106]

After attorney Raymond Fosdick assumed the presidency of the Rockefeller Foundation in 1936, the charitable trust became increasingly reluctant to fund any projects associated with the term *eugenics*. Rockefeller money continued to flow into prewar Nazi Germany to fund eugenic projects, but only when the proposals were packaged as genetics, brain research, or serology investigations attempting to locate the specific substances in the blood. However, Rockefeller financing was often too slow for Verschuer, who now sought faster and closer funding through the Reich Research Fund in Berlin, which in the thirties continued to enjoy annual Rockefeller monies. In June of 1939, when the Rockefeller Foundation tried to convince protestors that it was not financing Nazi science, Fosdick was forced to remind his colleagues that such denials were "of course hardly correct." Rockefeller money was still flowing through the Emergency Fund for German Science, now the German Research Society.[107]

A cascade of German Research Society grants financed Verschuer's continuing heredity research, including a 1935 grant for twin studies. In 1936 and 1937, Verschuer again received funding for twin research and his search for the specific components in blood. The grants continued through the war years, supporting a broad array of concentration camp experimentation.[108]

In the late summer of 1943, Verschuer received German Research Society funding for serology experiments filed under the keyword *Spezifische Eiweisskörper*, alternately translated as "Specific Proteins" or "Specific Albuminous Matter." His project would require voluminous blood samples, as he was seeking the specific blood proteins or albuminous matter that carried genetic traits, from epilepsy to eye color. Verschuer explained in a memo that the blood would come from the Twins Camp at Auschwitz. Mengele, wrote Verschuer, would supervise the operation with the explicit permission of Himmler. "The blood samples are being sent to my laboratory for analysis."[109]

Victim after victim, Mengele extracted large amounts of blood from twins and gypsies. He siphoned it from their arms, sometimes both arms, from the neck, sometimes from fingers. Hedvah and Leah Stern recalled, "We were very frightened of the experiments. They took a lot of blood from us. We fainted several times." One twin survivor remembered years later, "Each woman was given a blood transfusion from another set of twins so Mengele could observe the reaction. We two each received 350 cc of blood from a pair of male twins, which brought on a reaction of severe headache and high fever."[110]

Mengele returned to Berlin from time to time. On one of these trips, he visited his mentor Verschuer for a cozy family dinner. Mengele was asked whether his work at Auschwitz was hard. Years later, Verschuer's son recalled Mengele's reply to his mother: "It's dreadful," Mengele said. "I can't talk about it."[111]

Nevertheless, Mengele was tireless in his bloodletting, his eyeball extractions, his infecting, his autopsying and his selecting, most to the left and some to the right. In mid-August of 1944, his superior filed a letter of commendation. "During his employment as camp physician at the concentration camp Auschwitz," Verschuer asserted, "he has put his knowledge to practical and theoretical use while fighting serious epidemics. With prudence, perseverance and energy, he has carried out all tasks given him, often under very difficult conditions, to the complete satisfaction of his superiors and has shown himself able to cope with every situation."[112]

Years later, Verschuer's medical technician, Irmgard Haase, was interviewed about the work at Auschwitz. She admitted, "There was the research work, which included enzymes in the blood of Gypsy twins and of Russian prisoners of war.... From the middle of 1943 onwards, there were several consignments of 30 ml samples of citrated blood." Asked where the blood had come from, she replied, "I don't know. The specimens were in boxes, which had been opened. I never saw the sender's name." She added, "I thought that they were from a camp for prisoners." Auschwitz? "I never heard the word at that time."[113]

Mengele? "Never heard of him." She emphasized, "Specific enzymes in the blood were being investigated by means of... protective enzyme reactions." Were there any misgivings? Haase responded no: "It was science, after all."[114]

*   *   *

Mengele was not alone. Hitler's doctors operated a vast network of experi-mentation in Nazi concentration camps, euthanasia mills and other places in the territories it occupied. Much of that experimentation was eugenic and genetic, such as the work of Mengele. Much of it was strictly medical, such as the testing at Buchenwald designed to find cures or medicines for well-known diseases. Much of it was simply strategic, such as the cruel ice water and high altitude tests at Dachau intended to benefit *Luftwaffe* pilots bailing out over the North Atlantic.[115]

But even when strictly medical or military testing was inflicted on help-less subjects, it was most often imposed along eugenic lines. More specifi-cally, many Aryans—such as habitual criminals, Jehovah's Witnesses and socialists—were imprisoned in camps under beastly conditions. Mostly, it was the worthless and expendable—Jews, Gypsies, Russians and other "sub-human" prisoners—who were victimized as medical fodder. The exceptions were those Germans considered hereditary misfits, such as homosexuals and the feebleminded. All of it was in furtherance of Hitler's biological rev-olution and his quest for a master race in a Thousand-Year Reich.

Hitler's master race would be more than just chiseled blond and blue-eyed Nordics. Special breeding facilities were established to mass-produce perfect Aryan babies.[116] They would all be closer to super men and women: taller, stronger and in many ways disease-resistant. Therefore Verschuer was the vanguard of a corps of Nazi medical men who saw the struggle against infirmity and sickness as consonant if not intrinsic to their struggle for eugenic perfection. Nazi Germany was indeed engaged in advanced medical genetics, now amply funded by the Reich's plunder, and militarized and regimented by the fascist state.

Therefore, even as Verschuer and the Kaiser Wilhelm Institute for Anthropology, Human Heredity and Eugenics were supervising the eugenic murders at Auschwitz, they enjoyed military contracts and German Research Society funding to attack a gamut of dreaded inherited diseases. This research could be conducted in concentration camps such as Buchenwald and Birkenau, or in Kaiser Wilhelm's grandiose complex of centers for higher learning.

For example, Hans Nachtsheim, who also worked under Verschuer, investigated epilepsy and other illnesses under German Research Society aegis and military contract SS 4891-5376, filed under "Research into Heredity Pathology." One typical status memo in October of 1943 reported that, "Experiments on the significance of a lack of oxygen for the triggering of epileptic seizures in epileptic rabbits, which were carried out jointly with Dr. Ruhenstroth-Bauer from the Kaiser Wilhelm Institute for Biochemistry...have essentially been concluded. A preliminary report of the research is currently being printed in the journal *Klinische Wochenschrift* [*Clinical Weekly*]; a comprehensive report is in the process of being drawn up to be published in the journal *Zeitschrift für menschliche Vererbungs- und Konstitutionslehre* [*Journal for Science of Human Genetics and Constitution*]."[117]

The depth of Nachtsheim's learning was evident. "Further experiments," he continued, "are concerned with the effect of the epilepsy gene in association with other genes [*Gengesellschaft*]. It has been determined that a single dosage of the epilepsy gene may suffice to induce epilepsy in combination with certain other genes, although the epilepsy gene is usually recessive, meaning that it must be present in a double dosage in order to become effective. Thus, a carrier of two albino genes and a single epilepsy gene can become an epileptic. The albino gene is the most extreme and most recessive allele [chromosomal pair] of a series of 6 alleles. In order to understand the essence of genes and their interaction, it is important to know how the other alleles act in combination with the epilepsy gene. Up to now, it could be proven that the allele most closely related to the albino gene...reacts just as the albino factor, while the normal allele, which is dominant over all other alleles in the series, suppresses the outbreak of epilepsy even in a single dosage in the presence of even one epilepsy gene. Experiments with the other alleles remain to be done."[118]

Verschuer studied tuberculosis in rabbits under German Research Society aegis and contract SS 4891-5377. One typical report explained that, "In addition to crossbreeding, pure breeding continued; in particular, the attempt was made to determine why the members of one family were always killed by lung tuberculosis while this form did not develop in the other family. The attempt was made to change the way in which tuberculosis presented in the various breeds. This was done by means of sac blockage, reinfections and organ implants. These experiments have not yet been concluded, but it appears that the development of tuberculosis in the breeds is extremely resistant. It will be necessary to expand these experiments, since their results could be of fundamental significance for the treatment of tuberculosis in humans."[119]

Similar genuine science could be seen in the other reports of the various Kaiser Wilhelm Institutes. One of them was the Institute for Brain

Research, an organization financed by Rockefeller money from the ground up starting in the late 1920s. Senior researchers Drs. Julius Hallervorden and Hugo Spatz published their pioneering work on a form of inherited brain degeneration, which was eventually named Hallervorden-Spatz Syndrome. After Institute for Brain Research founder Oskar Vogt was removed for his lack of Nazi activism, Spatz took his place and the organization was fully integrated into the Nazi killing process. While Hallervorden held the neuropathology chair at the Institute for Brain Research, he was also appointed senior physician at Brandenburg State Hospital, one of six institutions operating gas chambers under the T-4 euthanasia program. Ultimately, more than 70,200 Germans classed feebleminded were gassed under T-4. In 1938, four autopsies were performed at the Brandenburg facility. During the next five years, 1,260 would be completed. The brains—nearly 700—went to Hallervorden.[120]

Hallervorden to his interrogators after the war: "I heard that they were going to do that, and so I went up to them and told them, 'Look here now, boys, if you are going to kill all those people, at least take the brains out so that the material could be utilized.'...There was wonderful material among those brains, beautiful mental defectives, malformations and early infantile disease.... They asked me: 'How many can you examine?' and so I told them an unlimited number—the more the better.... They came bringing them in like the delivery van from the furniture company. The Public Ambulance Society brought the brains in batches of 150-250 at a time....I accepted the brains, of course."[121]

Direct Rockefeller funding for Hallervorden and Spatz's Institute for Brain Research during the Hitler regime stopped in 1934, and funding for Rüdin's Kaiser Wilhelm Institute for Psychiatry ended in 1935. However, there were undoubtedly additional Rockefeller funds made available to institute researchers through the German Research Society. Rockefeller also provided the seed money for research at the Kaiser Wilhelm Institute for Biology until the war broke out. Moreover, the foundation continued to fund individual physicians, such as Tübingen forensic psychiatrist Robert Gaupp, Breslau patho-psychologist Kurt Beringer, Munich psychiatrist Oswald Bumke and Freiburg neurologist Werner Wagner, each affiliated with his own institution. During these years, Rockefeller also subsidized social scientists in Nazi-annexed Vienna. Much of this money continued until 1939. During the thirties, millions in Rockefeller Foundation grants also flowed to other Kaiser Wilhelm Institutes devoted to the physical sciences. One such was the

Kaiser Wilhelm Institute for Physical Chemistry and Electrochemistry, which was engaged in weapons research.[122]

The mentality behind the foundation's biological funding could best be seen in the words of Rockefeller Natural Science Director Warren Weaver. Just a few months after Hitler came to power in 1933, Weaver circulated a report to the trustees entitled "Natural Sciences—Program and Policy: Past Program and Proposed Future Program." That report asserted, "Work in human genetics should receive special consideration as rapidly as sound possibilities present themselves. The attack planned, however, is a basic and long-range one." A year later, Weaver asked "whether we can develop so sound and extensive a genetics that we can hope to breed, in the future, superior men?"[123]

In pursuing its breeding goals, the Rockefeller Foundation could reassure itself and others that it was not actually furthering the pseudoscience of eugenics. In fact, that 1933 report to the trustees specifically stated, "The attack [for heredity research] planned, however, is a basic and long-range one, and such a subject as eugenics, for example, would not be given support." After rejecting eugenics by name, the report went on to advocate that "support should be continued and extended to include the biochemical, physiological, neurological and psychological aspects of internal secretions in general."[124]

But while openly eschewing eugenics with statements and memos, Rockefeller in fact turned to eugenicists and race scientists throughout the biological sciences to achieve the goal of creating a superior race.

Rockefeller never knew of Mengele. With few exceptions, the foundation had ceased all eugenic studies in Nazi-occupied Europe when the war erupted in 1939. But by that time the die had been cast. The talented men Rockefeller financed, the great institutions it helped found, and the science it helped create took on a scientific momentum of their own.

What could have stopped the race biologists of Berlin, Munich, Buchenwald and Auschwitz? Certainly, the Nazis felt they were unstoppable. They imagined a Thousand-Year Reich of super-bred men. Hence when the twins, the prisoner doctors and those selected for the gas chamber looked at Mengele, time after time they reported the piercing look in his eyes. That look—Mengele's glare—was the Nazi vision wedded to a fanatical science whose soul had been emptied, its moral compass cracked; a science backed not merely by iron dogma but by men wielding machine guns and pellets of Zyklon B. All of them were versed in the polysyllabics of cold clinical murder. Surely, to the victims of Auschwitz, it must have seemed like nothing could stop Nazi science from its global biological triumph.

But something did defeat Mengele and his colleagues. Not reason. Not remorse. Not sudden realization. Nazi eugenicists were impervious to those powers. But two things did stop the movement. On June 6, 1944, the Allies invaded at Normandy and began defeating the Nazis, town by town and often street by street. They closed in on Germany from the west. The Russian army overran the Auschwitz death camp from the east on January 27, 1945. Mengele fled.[125]

Hence, Auschwitz was indeed the last stand of eugenics. The science of the strong almost completely prevailed in its war against the weak. Almost.

PART THREE

Newgenics

CHAPTER 18

From Ashes to Aftermath

On January 17, 1945, as the Russian army approached Auschwitz, Mengele went from office to office methodically gathering his research materials. "He came into my office without a word," recounted pathologist Martina Puzyna. "He took all my papers, put them into two boxes, and had them taken outside to a waiting car." Mengele and the documents fled first to Gross-Rosen concentration camp, and then into Czechoslovakia. There he joined up with Hans Kahler, a close friend, coauthor and one of Verschuer's twins researchers. The Russians liberated Auschwitz on January 27, at about 3 P.M., and Mengele's horrors were quickly discovered. International commissions listed him as a war criminal. But Mengele slipped through the Allied manhunt and eventually escaped to South America.[1]

Even as the Allies closed in, Verschuer still hoped he and Hitler's Reich would prevail in its war against the Jews. Just months before Mengele abandoned Auschwitz, Verschuer published part of a lecture proclaiming, "The present war is also called a war of races when one considers the fight with World Jewry.... The political demand of our time is the new total solution [*Gesamtlösung*] of the Jewish problem." By the beginning of 1945, the Reich was collapsing. On February 15, 1945, amid the chaos of Berlin's last stand, Verschuer found two trucks with which to ship his lab equipment, library, and several boxes of records to his family home in Solz.[2]

Nazi eugenicists continued their cover-up, in progress since the Normandy invasion. On March 12, 1945, Hans Nachtsheim, assistant director at the Institute for Anthropology, Human Heredity and Eugenics, wrote Verschuer in Solz. "A mass of documents have been left here which should be or have to be destroyed should the enemy ever come close to here.... We should not choose a moment... too late to destroy them."[3]

In the first days of May, the Reich was reduced to rubble and *der Führer* had killed himself.[4] Nazism and its eugenics were defeated. But now its architects and adherents would reinvent its past.

In April of 1946, the military occupation newspaper in Berlin, *Die Neue Zeitung*, published an article on various doctors who had fled Germany, and followed it up on May 3 with specific accusations against Verschuer. In the article, Robert Havemann, a communist and chemist who had resisted the Nazis, expressed out loud what many knew. He openly accused Verschuer of using Mengele in Auschwitz to obtain blood samples and eyeballs from whole murdered families.[5]

A nervous Verschuer reacted at once. He sent a sworn statement to Otto Hahn, the occupation-appointed administrator of the Kaiser Wilhelm Institutes, insisting that he had always opposed racial concepts. "Even before 1933," averred Verschuer, "but also after, I took personal risks and attacked, as a scientist, in speeches and in writing, the race concept of the Nazis.... I argued against attributing values to races, I warned against the high estimation of the Nordic race, and I condemned the misuse of the results of anthropology and genetics to support a materialistic and racial point of view of life and history."[6]

He went on to concede his relationship with Mengele, referring to him only as "Dr. M.," and insisting it was totally innocent. Verschuer stated, "A post-doc of my former Frankfurt Institute, Dr. M., was sent against his will to the hospital of the concentration camp in Auschwitz. All who knew him learned from him how unhappy he was about this, and how he tried over and over again to be sent to the front, unfortunately without success. Of his work we learned that he tried to be a physician and help the sick.... [7]

"After I went to Berlin [from Frankfurt]," Verschuer continued, "I began research on the individual specificity of the serum proteins ·and the question of their heredity.... For these experiments I needed blood samples of people of different geographic background.... At that time my former post-doc Dr. M. visited me and offered to obtain such blood samples for me within the context of his medical activity in the camp Auschwitz. In this manner I received—during this time, certainly not regularly—a few parcels of 20–30 blood samples of 5–10 mls."[8]

Verschuer then asked Hahn to give him a character reference, and even drafted a statement for Hahn to sign: "Professor von Verschuer is an internationally known scientist who has kept away from all political activity.... Professor von Verschuer had nothing to do with the errors and misuses of the Nazis, by which his scientific field was particularly hit. He kept his distance from them and, whenever he was confronted by them, he criticized them courageously." Hahn would not sign such a document.[9]

So Verschuer sought support from his allies in American eugenics. Shortly after Havemann's exposé, Verschuer wrote to Paul Popenoe in Los Angeles, hoping to reestablish cooperative ties. On July 25, Popenoe wrote back, "It was indeed a pleasure to hear from you again. I have been very anxious about my colleagues in Germany.... I suppose sterilization has been discontinued in Germany?" Popenoe offered tidbits about various American eugenic luminaries and then sent various eugenic publications. In a separate package, Popenoe sent some cocoa, coffee and other goodies.[10]

Verschuer wrote back, "Your very friendly letter of 7/25 gave me a great deal of pleasure and you have my heartfelt thanks for it. The letter builds another bridge between your and my scientific work; I hope that this bridge will never again collapse but rather make possible valuable mutual enrichment and stimulation." Seeking American bona fides, Verschuer tried to make sure his membership in the American Eugenics Society was still active. "In 1940, I was invited to become a member of the American Eugenics Society," Verschuer wrote. "Now that this calamitous war has ended, I hope that this membership can be continued. I would be grateful if you might make a gesture in this matter. In this context, I would like to mention that in recent months a former employee, a person devoid of character, has made extremely defamatory statements about me, which have also found their way into the American press. Therefore, it is possible that persons who do not know me better might have formed a wrong opinion of me. You will surely understand that it is important to me that any damage to my reputation be repaired and I would be very grateful for your kind help in doing so."[11]

Verschuer wrote again at the end of September 1946, requesting Popenoe's help. Because Verschuer was considered part of the Nazi medical murder apparatus, the Americans had halted his further work. "Since I wrote you," said Verschuer, "I have learned that the American military government does not intend to permit the continuation of my scientific work. This attitude can only be due to the spread of false information about me and my work. I have regularly sent you all of my scientific publications and you have known me for many years through correspondence. Therefore, may I ask for two things? 1. For a letter of recommendation from yourself and other American scientists who know me, stating that you know me as a serious scientific researcher and that you value my continued scientific work; 2. I ask you and other American geneticists and eugenicists who know me to undertake steps with the American military government in Germany to bring about the granting of permission for me to continue my life's work as a scientific researcher. It is my urgent wish that I be able to rebuild genetic and eugenic science from the ruins we stand upon in every area in Germany, a science that—free of the misuse of past years—may again attain international renown."[12]

Popenoe, who had also been corresponding with Lenz, was eager to be helpful, but uncomfortable standing up for an accused Nazi doctor. "I am distressed to hear that you may not be allowed to go ahead with your scientific work," Popenoe replied to Verschuer on November 7, 1946, "but it is hard for me to see how any of us over here could give any evidence that would be of value to you, even if we knew where to send it. Of course we could all testify that your scientific work before the war was objective and maintained very high standards. But if you have been 'denazified,' as I take to be the case from what you say, it was certainly not for that work, which is the only work I know about. None of us over here knows anything about what was going on in Germany from about 1939 onwards, but I suppose the action taken against you is due to your prominence in public life, as the successor of Eugen Fischer (who has been attacked bitterly in this country), etc. I could say nothing that would be pertinent, because I don't know anything about it. I am being perfectly frank with you, as you see. . . . But as it stands now, all I could say is: 'All his work that I saw before the war was of high quality,' and the authorities would presumably reply, 'That has nothing to do with it.'"[13]

Correspondence bounced back and forth between the two until Popenoe finally sent a brief letter of endorsement, limited to the prewar years. Verschuer then asked if he could be invited to join the faculty of an American university. "I have inquired from some leaders in American genetics," Popenoe replied, "and they all feel that it will be a long time before any university here is ready to offer a position to any German scientist who occupied an important position in Germany during the war years. As you perhaps know, our army brought over a number of physicists and other specialists, and their presence in this country has led to many protests and recriminations. I think it is out of the question, therefore, for you to look forward to any scientific activity here in the next few years—much as I myself should like to have a visit from you."[14]

Throughout late 1947 and 1948, Verschuer continued corresponding with leading eugenicists and geneticists at American institutions, seeking to reestablish academic exchanges and professional standing. He submitted one of his older books for a new review by the American Eugenics Society. Popenoe promptly assured he would review it in a new eugenic publication called *Family Life*, and then bemoaned the loss of German eugenic publications. "It is sad to think," Popenoe wrote, "that the scientific journals, and even the publishing houses that produced them no longer exist!" Verschuer also began exchanges with scientists at the University of Michigan and the University of Minnesota. These were received with goodwill and even enthusiasm. When Nazi agitator C. M. Goethe of California received Verschuer's letter, he replied that he was "thrilled."[15]

While Verschuer was busy reestablishing his support in America, he was rehabilitating himself in occupied Germany as well. After making his accusations public, Havemann organized a committee of Kaiser Wilhelm Institute scientists to examine the evidence against Verschuer. They ruled that Verschuer indeed had engaged in despicable acts in concert with Mengele at Auschwitz, but their report was kept secret for fifteen years. In 1949, while the first report remained under lock and key, a second board of inquiry was urged to reexamine the issue. This second board unanimously ruled that he had committed no transgressions involving Auschwitz, and indeed that "Verschuer has all the qualities which qualify him to be a researcher and teacher of academic youth." Virtually comparing Verschuer to Christ being crucified, the esteemed panel of German scientists declared they could not sit in judgment of him as "Pharisees" (*Pharisäerhaft*).[16]

Soon, Verschuer once again became a respected scientist in Germany and around the world. In 1949, he became a corresponding member of the newly formed American Society of Human Genetics, organized by American eugenicists and geneticists. Hermann Joseph Muller of Texas, a Rockefeller fellow who had worked at the Kaiser Wilhelm Institute for Brain Research during 1932, served as the first president of the American Society of Human Genetics.[17]

In the fall of 1950, the University of Münster offered Verschuer a position at its new Institute of Human Genetics, where he later became a dean. At about that time he helped found the Mainz Academy of Sciences and Literature, which later published his books, including one on cancer. In the early and mid-1950s, Verschuer became an honorary member of numerous prestigious societies, including the Italian Society of Genetics, the Anthropological Society of Vienna, and the Japanese Society for Human Genetics.[18]

A later president of the American Society of Human Genetics, Kurt Hirschhorn, remembered his own encounter with Verschuer in about 1958. An Austrian Jew, Hirschhorn had come to the United States as a refugee during the Hitler era. Hirschhorn became a genetic researcher and, while on a fellowship to Europe, he had visited Verschuer at the University of Münster. "Verschuer was partly responsible for the whole extermination," Hirschhorn related emphatically during a February 2003 interview. "He was the one that gave the Nazis the pseudo-genetic rationale to destroy the Jews and Gypsies. He was part of the organization [American Society of Human Genetics] in 1949 because in those days...it was all covered up. No one really knew. But I'll never forget. I was sitting in his university office in Münster as a young man, and he asked a lot of personal questions about my background, and so forth, until he found out I was Jewish. I knew who he was by that time. I took a great deal of pleasure in telling him that I came to the United States from Austria, and when I turned eighteen, I enlisted in the army and went over there and fought the Nazis—and went right through Münster. He was taken aback."[19]

In the 1960s, Frankfurt prosecutors were obliged by international pressure to continue their hunt for Nazis. The same prosecutors who investigated Mengele examined his relationship to Verschuer but concluded there was no connection between the two. Benno Müller-Hill, a German geneticist, later investigated Verschuer's activities. Müller-Hill reviewed Verschuer's many written defenses, including the one in which Verschuer claimed that while in Auschwitz, Mengele "tried to be a physician and help the sick." Writing in the journal *History and Philosophy of Science*, Müller-Hill described Verschuer's account as "Lies, lies, lies."[20]

Verschuer was never prosecuted. In 1969, he was killed in an automobile accident. But the legacy of his torturous medicine, twisted eugenics and conscious war crimes lives on.

*   *   *

As the ashes of Jews and Gypsies wafted into the air of Europe and were dumped into the Vistula River coursing through the heart of Europe, so their victimization flowed into the mainstream of modern medical literature. Medical literature evolves from decade to decade. As American eugenic pseudoscience thoroughly infused the scientific journals of the first three decades of the twentieth century, Nazi-era eugenics placed its unmistakable stamp on the medical literature of the twenties, thirties and forties.

The writings of Nazi doctors not only permeated the spectrum of German medical journals, they also appeared prominently in American medical literature. These writings included the results of war crime experimentation at concentration camps. Verschuer's own bibliographies, circa 1939, enumerated a long list of Nazi scientific discoveries, authored by

him, his colleagues and assistants, including Mengele. Such scientific publication continued right through the last days of the Third Reich. The topics included everything from rheumatism, heart disease, eye pathology,
blood studies, brain function, tuberculosis, and the gastric system to endless permutations of hereditary pathology.[21] Much of it was sham science.
Some of it was astute. Both types found their way into the medical literature of the fifties and sixties. Hence, Nazi victimization contributed significantly to many of the modern medical advances of the postwar period.

For example, the Nazis at Dachau, using ice water tests, were the first
to experimentally lower human body temperature to 79.7 degrees
Fahrenheit—this to discover the best means of reviving *Luftwaffe* pilots
downed over the North Sea. Nazi scientists learned that the most effective
method was rapid rewarming in hot water. Nuremberg testimony revealed
that Dr. Sigmund Rascher, who oversaw these heinous hypothermia tests,
prominently reported his breakthroughs at a 1942 medical symposium with
a paper entitled "Medical Problems Arising from Sea and Winter."[22]

After the war, Rascher's conclusions were gleaned from Nazi reports
and reluctantly adopted by British and American air-sea rescue services. A
Nuremberg war crimes report on Nazi medicine summed up the extreme
discomfort of Allied military doctors: "Dr. Rascher, although he wallowed
in blood...and in obscenity...nevertheless appears to have settled the
question of what to do for people in shock from exposure to cold....The
method of rapid and intensive rewarming in hot water...should be immediately adopted as the treatment of choice by the Air-Sea Rescue Services
of the United States Armed Forces."[23]

Rascher reported to Hubertus Strughold, director of the Luftwaffe
Institute for Aviation Medicine. Strughold attended the Berlin medical
conference that reviewed Rascher's revelations. A Nazi scientist wrote at
the time that there were no "objections whatsoever to the experiments
requested by the Chief of the Medical Service of the Luftwaffe to be conducted at the Rascher experimental station in the Dachau concentration
camp. If possible, Jews or prisoners held in quarantine are to be used."[24]

After the war, Strughold was smuggled into the United States under the
infamous Operation Paperclip project, which offered Nazi scientists refuge
and immunity in exchange for their scientific expertise. Once in the U.S.,
Strughold became the leader in American aviation medicine. His work was
directly and indirectly responsible for numerous aeromedical advances,
including the ability to walk effortlessly in a pressurized air cabin—now
taken for granted—but which was also developed as a result of Dachau

experiments. He was called "the father of U.S. Space Medicine," and
Brooks Air Force Base in Texas named its Aeromedical Library in his
honor. A celebratory mural picturing Strughold was commissioned by
Ohio State University. When Jewish and Holocaust-survivor groups, led
by the Anti-Defamation League, discovered the honors extended to
Strughold, they objected. Ohio State University removed its mural in
1993. The U.S. Air Force changed its library's name in 1995.[25]

In 2003, the state of New Mexico still listed Strughold as a member of
its International Space Hall of Fame. But on February 13, 2003, when this
reporter asked about their honoree's Nazi connection, a startled museum
official declared, "If he was doing experiments at Dachau, it would give one
pause why anyone would ever nominate him in the first place." Museum
officials added they would immediately look into removing his name.[26]

Another case involved Nazi doctors Hallervorden and Spatz. In 1922,
the two had successfully identified a rare and devastating brain disease
caused by a genetic mutation. The disease came to be known as
Hallervorden-Spatz Syndrome in their honor. During the Hitler era, while
working at the Kaiser Wilhelm Institute for Brain Research, Hallervorden
and Spatz furthered their research by utilizing hundreds of brains har-
vested from T-4 victims. Right through the 1960s, Hallervorden authored
numerous influential scientific papers on the subject. For decades, the
name Hallervorden-Spatz has been used by the leading medical institutions
in the world, honoring the two Nazis who discovered the disorder.
Thousands of articles and presentations have been made on the topic, using
the name Hallervorden-Spatz. Medical investigators created an "International Registry of Patients with Hallervorden-Spatz Syndrome and
Related Disorders."[27]

Leading family support groups involved with the disorder have also
taken their organizational names from the two Nazi doctors. But the news
about Hallervorden and Spatz's Nazi past recently became known to many
in the field. In 1993, two doctors expressed the view of many in a letter to
the editor of the journal *Neurology*. "It is also time to stop using the term,
'Hallervorden-Spatz disease' whose only purpose is to honor Hallervorden
by using his name." Another journal, *Lancet*, expressed a similar view in
1996, describing the continued honorary use of the name "Hallervorden-
Spatz" as "indefensible" because "both Hallervorden and Spatz were
closely associated with the Nazi extermination policies."[28]

In January of 2003, the Hallervorden-Spatz Syndrome Association
renamed itself the NBIA Disorders Association; the acronym was derived

from "neurodegeneration with brain iron accumulation." Just after the announcement, the newly-renamed association's president, Patricia Wood, told this reporter that the name change was certainly due to the legacy of Nazi experiments attached to Hallervorden and Spatz. The association's website confirmed that the name change was driven by "concerns about the unethical activities of Dr. Hallervorden (and perhaps also Dr. Spatz) involving euthanasia of mentally ill patients during World War II."[29]

The National Institutes of Health also adopted the Hallervorden-Spatz appellation for its research into the disease. NIH convened a two-day workshop on the disorder in May of 2000. As of March 2003, the National Institutes of Health continues to maintain a Hallervorden-Spatz Disease Information web page. On February 13, 2003, an NIH spokesman said that the institute was becoming aware of the Hallervorden-Spatz Nazi legacy and monitoring name changes in the field. "It is unfortunate that the two people who have discovered and researched this disease have undergone political scrutiny," the spokesman said, "but I don't see any name change at this time." The spokesman stressed that the problem was mere "political scrutiny." The spokesman did confirm that the institute would adjust its website's search engine to permit the term "NBIA" to reach its Hallervorden-Spatz information sites.[30]

Nazi medical victims suffered torture to substantially advance Reich scientific knowledge and modern medicine. Then the murdered specimens were delivered to the likes of Verschuer and Hallervorden and their eugenic institutions. But then what? After the war, victims' remains were transferred to or maintained by some of Germany's leading medical research facilities. Hence the exterminated continued to provide organic service to German medicine. In 1989, the Max Planck Institute for Brain Research, the successor to Hallervorden's center, admitted that it still possessed thirty tissue samples in its files. That same year, tissue samples and skeletons were also found in universities in Tübingen and Heidelberg. In 1997, investigators confirmed that the University of Vienna's Institute of Neurobiology still housed four hundred Holocaust victims' brains. The University of Vienna had functioned as part of the Reich after Austria's union with Germany in 1938. Similar discoveries have been made elsewhere in former Nazi-occupied Europe.[31]

In many cases, local officials, acting nearly a half-century after the fact, have elected to cremate the remains respectfully and bury them in memorial cemeteries. At one such burial service, conducted by Eberhard-Karls University in Tübingen, Professor Emeritus of Neuropathology Jürgen

Peiffer spoke solemnly. "We must remember," he eulogized, "that there is a dangerous possibility that we may bury our bad consciences together with these tissue remains, thereby avoiding the necessity of remembering the past.... I know that there are those who think we are acting out of faint-heartedness and anxiety; some ask whether 'dust to dust' really applies to glass slides and whether this act is the appropriate answer?" He answered his own question when he read aloud the inscription on the tablet.[32]

Displaced, oppressed, maltreated,
Victims of despotism or blind justice,
They first found their rest here.
Science, which did not respect
Their rights and dignity during life,
Sought even to use their bodies after death.
Be this stone a reminder to the living.[33]

CHAPTER 19

American Legacy

A merica's retreat from eugenics was precipitated by the convergence of two forces: Hitler's ascent in Germany and the climactic exit of the pseudoscience's founding fathers from Cold Spring Harbor. But it was not a moment of truth that finally convinced the Carnegie Institution and the eugenic establishment to turn away from their quest for a superior Nordic race. Rather, the end was an inexorably slow process devoid of *mea culpas*, one that saw the major players withdraw only with great reluctance.

The real father of eugenics was of course Charles Benedict Davenport. Galton was merely the grandfather. It was Davenport who twisted Galton's stillborn Victorian vision into self-righteous social-biological action. Eugenics always risked veering completely out of control. It did in Nazi Germany.

During the twelve-year Hitler regime, Davenport never wavered in his scientific solidarity with Nazi race hygiene. Nor did he modify his view that the racially robust were entitled to rule the earth. But Germany's triumph in the thirties wielding his principles did not bring Davenport the personal fulfillment he craved. During all his years at the pinnacle of international eugenic science, Davenport remained the same sad, embittered, intellectually defensive man who had first embarked upon a biological crusade at the turn of the century. As one lifelong friend remembered, Davenport remained "a lone man, living a life of his own in the midst of others, feeling out of place in almost any crowd." Davenport could acquire international celebrity, but never personal happiness.[1]

Correction. Davenport did find personal joy in one thing: his children, especially his son Charlie, born January 8, 1911. Little Charlie unlocked the affectionate quality guarded deep within men like Davenport. Proudly, Davenport would call out through the neighborhood for Charlie to come back for dinner after a day's play. A family friend remembers the intense "pride and devotion" Davenport felt when it came to little Charlie.[2]

The same year Charlie was born, Davenport published his cornerstone volume, *Heredity in Relation to Eugenics*, which explained the biological basis of the superior family. Even as millions were devastated by crippling diseases, such as tuberculosis and polio, Davenport's answer was to blame their ancestry, or more precisely, unsound protoplasm. "It is an incomplete statement," asserted Davenport's book, "that the *tubercle bacillus* is the cause of tuberculosis, or alcohol the cause of *delerium tremens* or syphilis the cause of *paresis*. Experience proves it. . . . In general, the causes of disease as given in the pathologies are not the real causes. They are due to inciting conditions acting on susceptible protoplasm. The real cause of death of any person is his inability to cope with the disease germ or other untoward conditions." Fatal epidemics did not kill, preached Davenport, only defective germ plasm.[3]

On the evening of September 5, 1916, Davenport came face to face with his own dogma. That night, young Charlie was stricken by polio. Death entered the Davenport household quickly; within hours of showing symptoms, Charlie was dead. The next day the boy was interred in the family plot of a Brooklyn cemetery. Davenport never recovered from the loss. A close associate recounted a broken man, a man absolutely "prostrated." After the funeral, both he and his wife retired to a sanitarium for several weeks. When he emerged, Davenport became even more cloistered and relentless in his work.[4]

For years, Davenport uncompromisingly continued to seek out the imperfect, the inferior, the weak and the susceptible, demanding their elimination. In 1934, at age sixty-eight, after a three-decade crusade, Davenport retired from the Carnegie Institution. Officials at the Washington office allocated a small room at the Eugenics Record Office to him, along with clerical help. On June 28, he delivered his final official address, "Reminiscences of Thirty Years." The next day, Davenport began the remainder of his joyless life. The letter he dictated to his secretary almost stoically informed the Carnegie Institution: "I am now getting settled in a corner of the south room, second floor, of the Eugenics Record Office, and am looking forward to a chance of uninterrupted research."[5]

Davenport of course continued to be active as the elder statesman of eugenics into the 1940s, even as the Nazis assumed international leadership and swept Davenport's principles into a brutal war. As late as 1943, Davenport was protesting, in *Eugenical News*, the widespread opposition to stern racial policies. But during his retirement years, Davenport mostly busied himself with continuous private investigations of mice, children, and other organisms.[6]

In January of 1944, Davenport became fixated on a killer whale that had beached itself off Long Island. He was determined to have its skull to exhibit at his new whaling museum at Cold Spring Harbor. Night after night, in a steam-filled but uninsulated shed, Davenport boiled the whale's head in a great cauldron. It was a slow process. The enormous orca was tough and resistant. Even as the weather became more and more brutal, Davenport would not give up. He fought the elements and the whale skull for two weeks, determined to beat them both. He became weaker and weaker.[7]

Colleagues remembered that one night Davenport appeared at an ERO staff meeting reeking of blubber. He sat off by himself, seventy-eight years old and still unshakable. Shortly thereafter, Davenport came down with a severe case of pneumonia. On February 18, 1944, Davenport died, not of old age, but of germs.[8]

*   *   *

The Carnegie Institution continued to back eugenics long after its executives became convinced it was a worthless nonscience based on shabby data, and years after they concluded that Harry Hamilton Laughlin himself was a sham.

Laughlin and eugenics in general had become the butt of jokes and the object of reprehension as far back as 1912, when the world learned that its proponents planned to sterilize millions in America and millions more in other nations. Scientists from other disciplines ridiculed the movement as well. Despite the widespread derision, eugenics persevered as a science under siege, battling back for years, fortified by its influential patrons, the power of prejudice and the big money of Carnegie. But the Carnegie Institution's patience began to erode as early as 1922, when Laughlin became a public font of racist ideology during the Congressional immigration restriction hearings.[9]

Carnegie president John C. Merriam continued to be embarrassed by Laughlin's immigration rantings throughout the 1920s. But he tolerated them for the greater agenda of the eugenics movement. However, Laughlin struck a particular nerve in the spring of 1928, while Merriam and a U.S. government official were touring Mexican archaeology sites. During the tour, Mexican newspapers splashed a story that Merriam's Carnegie Institution was proposing that Congress severely limit immigration of Mexicans into the United States. It was Laughlin who prompted the story.[10]

Merriam immediately instructed Davenport to muzzle Laughlin. "He [Merriam] feels especially that you ought not go further," Davenport wrote Laughlin, "...helping the [House] committee on a definition of who may be acceptable as immigrants to the United States from Spanish America. The Spanish Americans are very sensitive on this matter.... It will not do for the Carnegie Institution of Washington, or its officers, to take sides in this political question." Anticipating Laughlin's predictable argument, Davenport continued, "I know you regard it properly as more than a political question and as a eugenical question—but it is in politics now, and that means that the institution has to preserve a neutrality."[11]

Yet Laughlin did nothing to restrict his vocal activities. By the end of 1928, Merriam convened an internal committee to review the value of the Eugenics Record Office. In early February of 1929, the committee inspected the Cold Spring Harbor facility and concluded that the accumulation of index cards, trait records and family trees amounted to little more than clutter. They "are of value only to the individual compiling them," the committee wrote, and even then "in most cases they decrease in importance in direct proportion to their age." Some of the files were almost two decades old, and all of them reflected nineteenth-century record-keeping habits now obsolete. The mass of records yielded much private information about individuals and their families, but little hard knowledge on heredity.[12]

Nonetheless, with Davenport and Laughlin lobbying to continue their work, the panel rejected any "radical move, such as relegating them [the files] to dead storage." Instead, Carnegie officials decided a closer affiliation with the Eugenics Research Association would help the ERO achieve some approximation of genuine science. Hence the Carnegie Institution would continue to operate the ERO under Carnegie's Department of Genetics.[13]

Genetics, however, was not the emphasis at Cold Spring Harbor. Laughlin and his ERO continued their race-based political agitation unabated. Moreover, once Hitler rose to power in 1933, Laughlin forged the ERO, the ERA and *Eugenical News* into a triumvirate of pro-Nazi agitation. But things changed when Davenport retired in June of 1934. Laughlin lost his greatest internal sponsor, and with Davenport out of power, Carnegie officials in Washington quickly began to move against Laughlin. They pointedly questioned his race science and indeed the whole concept of eugenics in a world where the genuine science of genetics was now emerging.

Carnegie officials first focused on *Eugenical News*, which had become a compendium of American raceology and Nazi propaganda. Although *Eugenical News* was published out of the Carnegie facilities at the ERO, by a Carnegie scientist, and functioned as the official voice of Carnegie's eugenic operations, the Carnegie Institution did not legally own or control *Eugenical News*. It was Laughlin's enterprise. Carnegie wanted an immediate change and made this clear to Laughlin.[14]

Laughlin became very protective. He had always chosen what would and would not run in *Eugenical News*, and he even authored much of the text. In a September 11, 1934, letter to Davenport's replacement, Albert F. Blakeslee, Laughlin rebuffed attempts to corral *Eugenical News*, defensively insisting, "In this formative period of making eugenics into a science, the ideals of the Eugenics Record Office, of the Eugenics Research Association, of the International Congresses and Exhibits of Eugenics, and of the *Eugenical News* are identical. I feel that the position of the *Eugenical News* as a scientific journal is quite unique, in that eugenics is a new science, and that the trend and rate of its development, and its ultimate character, will be influenced substantially by the *Eugenical News*."[15]

Laughlin made clear to Carnegie officials that they simply could not control *Eugenical News*, because it was legally the property of the Eugenics Research Association—and Laughlin was the secretary of the ERA. To drive home his point, a Laughlin memo defiantly included typed-in excerpts from committee reports and letters to the printer, plus sample issues going back to 1916—all demonstrating the ERA's legal authority over *Eugenical News*. "I feel that the Institution should go into the matter thoroughly," insisted Laughlin, "and make a clean-cut and definite ruling concerning the relationship of the Carnegie Institution (represented by the Eugenics Record Office) to the *Eugenical News*."[16]

By now, Carnegie felt it was again time to formally revisit the worth of Laughlin and eugenics. A new advisory committee was assembled, spearheaded by archaeologist A.V. Kidder. He began assembling information on Laughlin's activities, and Laughlin was only too happy to cooperate, almost boastfully inundating Kidder with folder after folder of material. With Davenport in retirement, Laughlin undoubtedly felt he was heir to Cold Spring Harbor's throne. He sent Washington a passel of demands about revamping Cold Spring Harbor's administrative structure, renovations of its property and new budget requests for 1935.[17]

Kidder was not encouraging. He wrote back, "I think I ought to tell you that I feel quite certain that the administrative and financial changes which you advocate are extremely unlikely, in my opinion, to be carried into effect in 1935." Kidder was virtually besieged with Laughlin's written and printed submissions to support his requests for a sweeping expansion of the ERO. On November 1, 1934, Kidder acknowledged, "I am at present reviewing all the correspondence and notes in my possession relative to the whole Cold Spring Harbor situation and in the course of a few days I shall prepare a memorandum for Dr. Merriam." But within two days, Kidder conceded that he was overwhelmed. "I have read all the material you sent me with close attention," he wrote Laughlin. "I have also read all the Year Book reports of the Eugenics Record Office. ... I am now trying to correlate all this information in what passes for my brain."[18]

On Sunday, June 16 and Monday, June 17, 1935, the advisory committee led by Kidder visited Cold Spring Harbor, touring both the ERO and the adjacent Carnegie Station for Experimental Evolution. Laughlin's residence, provided by the Carnegie Institution, was one of the buildings in the compound, and Mrs. Laughlin graciously prepared Sunday lunch and Monday dinner for the delegation. The men found her hospitality delightful, and Laughlin's presentations exhaustive. But after a thorough examination, the advisory committee concluded that the Eugenics Record Office was a worthless endeavor from top to bottom, yielding no real data, and that eugenics itself was not science but rather a social propaganda campaign with no discernible value to the science of either genetics or human heredity.[19]

Almost a million ERO records assembled on individuals and families were "unsatisfactory for the scientific study of human genetics," the advisory committee explained, "because so large a percentage of the questions concern ... traits, such as 'self-respect,' 'holding a grudge,' 'loyalty,' [and] 'sense of humor,' which can seldom truly be known to anyone outside an individual's close associates; and which will hardly ever be honestly recorded, even were they measurable, by an associate or by the individual concerned."[20]

While much ERO attention was devoted to meaningless personality traits, key physical traits were being recorded so sloppily by "untrained persons" and "casually interested individuals" that the advisory committee concluded this data was also "relatively worthless for genetic study." The bottom line: a million index cards, some 35,000 files, and innumerable other records merely occupied "a great amount of the small space available ... and, worst of all, they do not appear to us really to permit satisfactory use of the data."[21]

The advisory committee recommended that all genealogical and eugenic tracking activities cease, and that the cards be placed in storage until whatever bits of legitimate heredity data they contained could be properly extracted and analyzed using an IBM punch card system. A million index cards had accumulated during some two decades, but because of the project's starting date in 1910 and Laughlin's unscientific methodology, the data had never been analyzed by IBM's data processing system. This fact only solidified the advisory committee's conclusion that the Eugenics Record Office was engaged in mere biological gossip backed up by reams of worthless documents. The advisory committee doubted that the demographic muddle would "ever be of value," and added its hope that "never again . . . should records be allowed to bank up to such an extent that they cannot be kept currently analyzed."[22]

The advisory committee vigorously urged that "The Eugenics Record Office should engage in no new undertaking; and that all current activities should be discontinued save for Dr. Laughlin's work in preparation of his final report upon the Race Horse investigation." Moreover, the advisory committee emphasized, "The Eugenics Record Office should devote its entire energies to pure research divorced from all forms of propaganda and the urging or sponsoring of programs for social reform or race betterment such as sterilization, birth-control, inculcation of race or national consciousness, restriction of immigration, etc. Hence it might be well for the personnel of the Office to discontinue connection with the *Eugenical News*." Committee members concluded, "Eugenics is by generally accepted definition and understanding not a science." They insisted that any further involvement with Cold Spring Harbor be devoid of the word *eugenics* and instead gravitate to the word *genetics*.[23]

Geneticist L. C. Dunn, a member of the advisory committee traveling in Europe at the time, added his opinion in a July 3, 1935, letter, openly copied to Laughlin. Dunn was part of a growing school of geneticists demanding a clean break between eugenics and genetics. "With genetics," advised Dunn, "its relations have always been close, although there have been distinct signs of cleavage in recent years, chiefly due to the feeling on the part of many geneticists that eugenical research was not always activated by purely disinterested scientific motives, but was influenced by social and political considerations tending to bring about too rapid application of incompletely proved theses. In the United States its [the eugenics movement's] relations with medicine have never been close, the applications having more often been made through sociology than through medicine, although the basic problems involved are biological and medical ones."[24]

Dunn wondered if it wasn't time to shut down Cold Spring Harbor altogether and move the operation to a university where such an operation could collaborate with other disciplines. "There would seem to me to be no peculiar advantages in the Cold Spring Harbor location." As it stood, "'Eugenics' has come to mean an effort to foster a program of social improvement rather than an effort to discover facts." In that regard, Dunn made a clear comparison to Nazi excesses. "I have just observed in Germany," he wrote, "some of the consequences of reversing the order as between program and discovery. The incomplete knowledge of today, much of it based on a theory of the state, which has been influenced by the racial, class and religious prejudices of the group in power, has been embalmed in law, and the avenues to improvement in the techniques of improving the population have been completely closed."[25]

Dunn's July 3 letter continued with even more pointed comparisons to Nazi Germany. "The genealogical record offices have become powerful agencies of the [German] state," he wrote, "and medical judgments even when possible, appear to be subservient to political purposes. Apart from the injustices in individual cases, and the loss of personal liberty, the solution of the whole eugenic problem by fiat eliminates any rational solution by free competition of ideas and evidence. Scientific progress in general seems to have a very dark future. Although much of this is due to the dictatorship, it seems to illustrate the dangers which all programs run which are not continually responsive to new knowledge, and should certainly strengthen the resolve which we generally have in the U.S. to keep all agencies which contribute to such questions as free as possible from commitment to fixed programs."[26]

Carnegie's advisory committee could not have been more clear: eugenics was a dangerous sham, the ERO was a worthless and expensive undertaking devoid of scientific value, and Laughlin was purely political. But as Hitler rose and the situation of the Jews in Europe worsened, and the plight of refugees seeking entry into the United States became ever more desperate, the Carnegie Institution elected to ignore its own findings about Cold Spring Harbor and continue its economic and political support for Laughlin and his enterprises. Shortly after Merriam reviewed the advisory committee's conclusions, the Reich passed the Nuremberg Laws in September of 1935. Those of Jewish ancestry were stripped of their civil rights. Laughlin, *Eugenical News* and the Cold Spring Harbor eugenics establishment propagandized that the laws were merely sound science. *Eugenical News* even gave senior Nazi leaders a platform to justify their decrees. The Carnegie Institution still took no action against its Cold Spring Harbor enterprise.

In 1936, the brutal Nazi concentration camps multiplied. Systematic Jewish pauperization accelerated. Jews continued fleeing Germany in terror, seeking entry anywhere. But American consulates refused them visas. In the face of the humanitarian crisis, Laughlin continued to advise the State Department and Congress to enforce stiff eugenic immigration barriers against Jews and other desperate refugees. The Carnegie Institution still took no action against its Cold Spring Harbor enterprise.[27]

In 1937, Nazi street violence escalated and Germany increasingly vowed to extend its master race to all of Europe—and to completely cleanse the continent of Jews. Laughlin, *Eugenical News* and the eugenics establishment continued to agitate in support of the Reich's goals and methods, and even distributed the anti-Semitic Nazi film, *Erbkrank*. The Carnegie Institution still took no action against its Cold Spring Harbor enterprise.[28]

In 1938, as hundreds of thousands of new refugees appeared, an emergency intergovernmental conference was convened at Evian, France. It was fruitless. Germany then decreed that all Jewish property was to be registered, a prelude to comprehensive liquidation and seizure. In November, *Kristallnacht* shocked the world. Nazi agitation was now spreading into every country in Europe. Austria had been absorbed into the Reich. Hitler threatened to devour other neighboring countries as well. Laughlin, *Eugenical News* and the eugenics establishment still applauded the Hitler campaign. By the end of 1938, however, the Carnegie Institution realized it could not delay action much longer.[29]

On January 4, 1939, newly installed Carnegie president Vannevar Bush put Laughlin on notice that while his salary for the year was assured, Bush was not sure how much funding the ERO would receive—if any. At the same time, Jews from across Europe continued to flee the Continent, many begging to enter America because no other nation would take them. In March of 1939, the Senate Immigration Committee asked Bush if Laughlin could appear for another round of testimony to support restrictive "remedial legislation." Bush permitted Laughlin to appear, and only asked him to limit his unsupportable scientific assertions. But Laughlin was not prohibited from again promoting eugenic and racial barriers as the best basis for immigration policy. Indeed, the Carnegie president reminded him, "One has to express opinions when he appears in this sort of inquiry, and I believe that yours will be found to be a conservative and well-founded estimate of the situation facing the Committee." Bush added that he had personally reviewed Laughlin's prior testimony and felt it was "certainly well handled and valuable."[30]

After testifying, Laughlin received a postcard at the Carnegie Institution in Washington from an irate citizen in Los Angeles. "As an American descendant of Americans for over 300 years, I'd like to learn what prompted you to supply [the Senate Immigration Committee] ... with so much material straight from Hitler's original edition of *Mein Kampf*."[31]

At about this time, Laughlin was also permitted to testify before the Special Committee on Immigration and Naturalization of the New York State Chamber of Commerce. In May of 1939, Laughlin's report, *Immigration and Conquest*, was published under the imprimatur of the New York State Chamber of Commerce and "Harry H. Laughlin, Carnegie Institution of Washington." The 267-page document, filled with raceological tenets, claimed that America would soon suffer "conquest by settlement and reproduction" through an infestation of defective immigrants. As a prime illustration, Laughlin offered "The Parallel Case of the House Rat," in which he traced rodent infestation from Europe to the rats' ability "to travel in sailing ships."[32]

Laughlin then explained, in a section entitled "The Jew as an Immigrant Into the United States," that Jews were being afforded too large a quota altogether because they were being improperly considered by their nationality instead of as a distinct racial type. By Laughlin's calculations, no more than six thousand Jews per year ought to be able to enter the United States under the existing national quota system—the system he helped organize a half-decade earlier—but many more were coming in because they were classified as German or Russian or Polish instead of Jewish. He asked that Jews in the United States "assimilate" properly and prove their "loyalty to the American institutions" was "greater than their loyalty to Jews scattered through other nations." *Immigration and Conquest*'s precepts were in many ways identical to Nazi principles. Laughlin and the ERO proudly sent a copy to Reich Interior Minister Wilhelm Frick, as well as to other leading Nazis, including Verschuer, Lenz, Ploetz and even Rüdin at a special address care of a university in occupied Czechoslovakia.[33]

In late 1938, the Carnegie Institution finally disengaged from *Eugenical News*. The publication became a quarterly completely under the aegis of the American Eugenics Society, published out of AES offices in Manhattan, with a new editorial committee that did not include Laughlin or any other Carnegie scientist. The first issue of the reorganized publication was circulated in March of 1939. Shortly thereafter, the Carnegie Institution formalized Laughlin's retirement, effective at the end of the year. On September 1, 1939, the Nazis invaded Poland, igniting World War II. Highly publicized atrocities against Polish Jews began at once, shocking the world. Efforts by Laughlin in the final months of 1939 to find a new sponsor for the ERO were unsuccessful. On December 31, 1939, Laughlin officially retired. The Eugenics Record Office was permanently closed the same day.[34]

Laughlin and his wife immediately moved back to Kirksville, Missouri. The last years of his life were uneventful, and he died in Kirksville on January 26, 1943. Davenport eulogized him in *Eugenical News* as a great man whose views were opposed by those of "a different social philosophy which is founded more on sentiment and less on a thorough analysis of the facts." Davenport saluted his protégé, predicting that within a generation Laughlin's work would be "widely appreciated" for what it really was: "preservation...from the clash of opposing ideals and instincts found in the more diverse racial or geographical groups."[35]

Strangely enough, Laughlin, the staunch defender of strong germ plasm and warrior against the feebleminded and the hereditarily defective, left no children. The family kept it a secret, but the rumor was that Laughlin himself suffered from an inherited disease that made him subject to uncontrollable seizures. These seizures had occasionally occurred in front of his colleagues at the ERO. Laughlin's condition had been discovered in the 1920s upon his return from Europe. During one episode, Laughlin reportedly drove off the road near Cold Spring Harbor and almost ran into the water. An obstruction stopped his vehicle. Laughlin nearly died that night, and his wife reportedly never allowed him to drive a car again.[36]

Among his many crusades, Laughlin may best be remembered for his antagonism toward epileptics. He claimed that epilepsy was synonymous with feeblemindedness, and that people with epilepsy did not belong in society. He fought to keep such people out of America and demanded their sterilization and even their imprisonment in segregated camps. No wonder the family kept his condition a secret. Childless and frustrated, Harry Hamilton Laughlin reportedly suffered his genetic disease in silence and died under its grip. The disease: epilepsy.[37]

* * *

Once Laughlin retired on December 31, 1939, Carnegie began the immediate and systematic dismantling of the ERO, abandoning three decades of support for racial eugenics. Mail addressed to the ERO, and even letters specifically addressed to Laughlin or Davenport, were not forwarded to either man. Instead, a series of standard responses were typed up for clerical staff to utilize in replying to all correspondents. The message: work at the office had been suddenly discontinued and no questions could be answered.[38]

Personal correspondents were told to contact Laughlin or Davenport directly at their home addresses. But if a letter involved even the slightest reference to eugenics or the Cold Spring Harbor installation, it was answered with a vague customized form letter. For example, on February 19, 1940, the widow of Lucien Howe sent a handwritten personal note to Davenport lamenting the news that the ERO had been discontinued. An officer of the Carnegie Institution replied for him, writing back to the aging Mrs. Howe, "Your letter of the 19th to Dr. Davenport has been turned over to me for reply" and so on.[39]

When eugenic enthusiasts earnestly mailed in their family trees or genealogical trait records, or requested copies of their files or pertinent information from them, they were deftly answered with noncommittal form letters. When a Texas man offered family information, he received a curt note, "Doctor Laughlin has resigned, and for the time being at least, the Genetics Record Office is not in a position to file and index family records." The same type of reply was mailed out time and time again. The ERO had operated under the name "Eugenics Record Office" until 1939, when Carnegie officials insisted on a cosmetic name change to "Genetics Record Office." From 1939 on, Carnegie Officials consistently referred to the ERO as the "Genetics Record Office" or sometimes simply the "Record Office," avoiding any use of the word *eugenics*.[40]

Letters came in for years. Carnegie officials generally acted as though they had no access to Laughlin's files and therefore could not answer specific questions. But in fact Carnegie administrators kept the files close at hand and quietly checked them in some cases. For example, when Jane Betts in Wichita asked about record #51323 on February 29, 1944, a Carnegie official quickly plucked her record out of a million files and replied about its status. With few exceptions, however, questions addressed to the Eugenics Record Office were generally answered with no real information except that the office was closed and no data was available.[41]

After World War II, when the magnitude of Hitler's eugenicide became apparent, the Carnegie Institution decided to get rid of its records. It sold the ERO building at Cold Spring Harbor but retained the rest of the facilities. Officials destroyed many of Laughlin's years-old unpublished worksheets on horse racing and breeding (an adjunct to his investigations in human heredity), but finding recipients for the rest of the ERO's enormous and controversial collection was not easy. In May of 1947, a leading heredity clinic at the University of Michigan was offered the files but wondered whether Carnegie would provide a stipend to house the materials. Carnegie would not. So Carnegie kept searching for someone to take the files.[42]

In September of 1947, a Carnegie administrator overseeing the dismantled Cold Spring Harbor operation wrote to the Dight Institute, an independent eugenic research organization at the University of Minnesota. "If any institution is interested in the records of the Genetics Record Office, I am confident that arrangements could be made...to transfer them." But, the note added, "there is very little chance that those funds [formerly used to run the ERO] would be transferred with the records."[43]

Dight director Sheldon Reed, an ardent eugenicist, replied, "It seems a great pity to me that the work must be abandoned." As for transferring the voluminous files to Dight, Reed posed a number of questions about the size and breadth of the collection and the cost of the transfer. Dight did not want to pay any of the moving expenses. As Dight officials pondered the usefulness of a collection they termed "colossal," Reed was frank with the Carnegie Institution. "I am sorry to take up your time with this business [the many logistical questions]," he wrote, "but it may be that you are even more interested in getting rid of records than I am in obtaining them."[44]

Eventually, Carnegie officials decided the best idea was to disperse the ERO records. In January of 1948, the Dight Institute agreed to house the ERO's extensive individual trait and family documents if Carnegie would defray the expected $1,000 shipping costs. Some six months later the Minnesota Historical Society agreed to take a half-ton of biographical jubilee books, family genealogical volumes and related materials. At the same time, the New York Public Library received a thousand ERO volumes of family genealogical books and local histories. Horse racing and stud breeding publications were handed over to the family that had originally sponsored the research. Carnegie donated Davenport's voluminous papers and Laughlin's ERO operational papers to the American Philosophical Society in Philadelphia, while maintaining some documents at a Cold Spring Harbor archive and retaining some others in Washington. When the Dight Institute closed its doors in the 1990s, its ERO papers were also sent to the American Philosophical Society, which now holds the largest consolidated eugenic collection anywhere.[45]

The dispersal of the records of the Cold Spring Harbor enterprise did not end the flow of letters to the ERO. For decades, people continued to send requests for eugenic information, updates of their pedigrees, and proof of their family's biological worth. In 1952, a dozen years after the ERO's closure, Clifford Frazier, an attorney in Greensboro, North Carolina, wrote offering to "bring my family data heretofore furnished up to date." In 1953, James Brunn, a realtor in Kansas City, Missouri, wrote requesting information to help trace his lineage back to the Revolutionary War. In 1959, Minnie Williams of Harrison, Ohio, wrote to say that she had finally assembled as much information as she could about her family pedigree; she had been working at it for years. In 1966, Elsie Van Guilder addressed a letter to "American Breeders Association, Eugenics Section, Cold Spring Harbor" seeking to trace her family. In 1976, E. Taylor Campbell of St. Joseph, Missouri, explained that he had been working on his family tree for fifty-one years, and he still needed nine more forms.[46]

Indeed, eugenic enthusiasts continued remitting family traits and proffering inquiries for decades. Letters continued into the 1980s, forty years after the ERO was dismantled. They probably never stopped. In February of 2003, a North Carolina attorney told this reporter than he had just discovered old ERO forms from his father's day; the attorney said his daughter was working with them to advance the family genealogy. Laughlin's work was that engrained in America. It persevered—not only in the mindsets of generations of Americans, but also in America's laws.[47]

Although the ERO stopped functioning in 1939, America's eugenic laws did not. Tens of thousands of Americans continued to be forcibly sterilized, institutionalized and legally prevented from marriage on the basis of racial and eugenic laws. During the 1940s, some 15,000 Americans were coercively sterilized, almost a third of them in California. In the fifties, about ten thousand were sterilized. In the sixties, thousands more were sterilized. All told, an estimated 70,000 were eugenically sterilized in the first seven decades of the twentieth century; the majority were women. California consistently outdistanced every other state.[48]

Victims, especially those who only discovered their sterilizations years after the fact, eventually began to initiate litigation. One such victim was Joseph Juhan, a Tennessee war veteran with little formal education but with a pointed message for the Carnegie Institution. In late 1976, he penned a letter filled with poorly formed characters and numerous misspellings, randomly employing parentheses for emphasis, that nevertheless poignantly asserted his legal rights. The letter was addressed to "Dr. Charles Davenport, Dept of Experimental Evolution" at Cold Spring Harbor.

Dear Sir: I write to "request" your help. In the year of "1954" while a patient at the (State Hospital), at Milledgeville, Ga, a visectomy or sterilization operation was performed upon me, by orders of a state (eugenics board). A mental (deficiency dygnoses was made of my case. At the time I was only 18 years old.

I was wondering as the (Carnegie Instutions Dep. of experimental evolution or (eugenics studies) have have been ingaged in the study of (state mental inistutions records of (certain mental deficiency cases, if to your "knowledge" there has been in (eugenic's studys connected with the (Carnegie Inistutions at the (Milledgeville State Hosp in the State of Ga, in 1954.

The purpose of this "inquirey" is to obtain records for the American Civil Liberty's Union, in order to present befor a (U.S. Court of Law the (circumstances of my case, in 1954, whereby a (State Hospital acting under orders of a (Eugenics) Board did cause a (vocectomy) or sterlization operation, upon me at the age of 18.

I feel this was (uncessary, in violation of the (Fundimental, or basic freedoms guaranteed under the (U.S, Contitution) as no (mental deficiency of a genetic nature has ever exzisted in my case.

Your help in this matter will be greately appriecated.

I am Sincerely
Joseph Juhan
c/o U.S., VA Hospital
Murfreesboro, Tenn 37130[49]

A response came from Agnes Fisher, the Record Office's secretary.

Dear Mr. Juhan,
I am writing in reply to your letter addressed to Dr. Charles Davenport. (Dr. Davenport retired from the Carnegie Institution in 1934, and died in 1944.)

You inquired about the possibility that eugenic studies were made by the Carnegie Institution at the Milledgeville State Hospital in 1954.

The Eugenics Record Office, formerly connected with the Department of Genetics in Cold Spring Harbor, was closed in 1939 upon the retirement of its director, Dr. H. H. Laughlin. At that time all studies and activities carried on by the Record Office or its staff were discontinued. Therefore no such studies could have been made in 1954.[50]

The American Civil Liberties Union never filed a sterilization suit in Georgia. But a few years later, in 1980, the ACLU in Richmond did file a historic suit against the state of Virginia on behalf of the victims of the Lynchburg Training School where Carrie Buck was sterilized. The ACLU ultimately forced Virginia to confront its history. In May of 2002, the governor of Virginia formally apologized to victims living and dead for decades of eugenic sterilizations. The governors of California, Oregon, North Carolina and South Carolina have followed suit.[51]

Nonetheless many of the laws are still on the books. For example, North Carolina's eugenic sterilization law, although not used for years, remains in force and was even updated in 1973 and 1981. Chapter 35, Article 7 still allows for court ordered sterilization for moral as well as medical improvement. While most states stopped enforcing sterilization statutes in the sixties and seventies, the practice did not stop everywhere. Across the country, additional thousands of poor urban dwellers, Puerto Rican women and Native Americans on reservations continued to be sterilized—not under state laws, but under special federal provisions.[52]

In the seventies, for example, a group of Indian Health Service physicians implemented an aggressive program of Native American sterilization. According to a U.S. General Accounting Office study, hospitals in just four cities sterilized 3,406 women and 142 men between 1972 and 1976. The women widely reported being threatened with the loss of welfare benefits or custody of their children unless they submitted to sterilization. A federal court ordered that all future Indian Health Service sterilizations employ the proper safeguards of legitimate therapeutic procedures, and that "individuals seeking sterilization be orally informed at the outset that no Federal benefits can be withdrawn because of failure to accept sterilization." During the same four-year period, one Oklahoma hospital alone sterilized nearly 8 percent of its fertile female patients. No one will ever know the full scope of Indian sterilization in the postwar period because medical records were either not kept or were incomplete.[53]

Eugenics left behind more than sterilization laws. Marriage prohibitions remained in force. For example, Walter Plecker's Racial Integrity Act and numerous similar state statutes endured long after the ERO and Plecker disappeared. These laws potentially affected millions in ways that society can never measure. In 1958, two Virginians, a black woman named Mildred Jeter and a white man named Richard Loving, were married in Washington, D.C., to avoid violating Plecker's law. Upon their return to Virginia, they were arrested and indicted by the Caroline

County grand jury. The trial judge suspended their one-year jail sentence on the condition that they leave Virginia and not return together for twenty-five years.[54]

From their new residence across the river in Washington, D.C., the Lovings appealed the infringement of their civil rights. Appellate courts, one after another, affirmed Virginia's law and the couple's conviction. Finally, almost nine years later in 1967, the United States Supreme Court considered the case.[55]

Writing for the majority, Chief Justice Earl Warren declared: "There can be no doubt that restricting the freedom to marry solely because of racial classifications violates the central meaning of the Equal Protection Clause.... The freedom to marry has long been recognized as one of the vital personal rights essential to the orderly pursuit of happiness by free men. Marriage is one of the 'basic civil rights of man,' fundamental to our very existence and survival.... To deny this fundamental freedom on so unsupportable a basis as the racial classifications embodied in these [Virginia] statutes, classifications so directly subversive of the principle of equality at the heart of the Fourteenth Amendment, is surely to deprive all the State's citizens of liberty without due process of law.... These convictions must be reversed. It is so ordered."[56]

After the Lovings' victory in 1967, other states' racial integrity laws became unenforceable. In 2000, Alabama became the last state in the union to repeal its antimiscegenation statute.[57]

With the science stripped away, all that remained to justify eugenic legislation was bigotry. Late in the twentieth century, in an enlightened postwar era, the eugenic notions that gripped a nation and then a world were finally understood. It had all just been colossal academic hubris masquerading as erudition.

* * *

By the late 1920s, the Carnegie Institution had confirmed by its own investigations what many in the scientific world and society at large had long been saying: that the eugenic science it helped create was a fraud.[58] Nevertheless, Carnegie allowed its Cold Spring Harbor enterprise to supply the specious information needed to validate Virginia's legal crusade to sterilize Carrie Buck. Relying on Laughlin's pseudoscience and his own prejudices, U.S. Supreme Court Justice Oliver Wendell Holmes had established the law of the land. In 1927, Holmes' famous opinion decreed:

It is better for all the world, if instead of waiting to execute degenerate off-spring for crime, or to let them starve for their imbecility, society can prevent those who are manifestly unfit from continuing their kind.... Three generations of imbeciles are enough.[59]

With Holmes' decision in hand, Carnegie's Cold Spring Harbor enterprise had unleashed a national campaign to reinforce long dormant state laws, enact new ones and dramatically increase the number of sterilizations across America. Sterilizations multiplied, marriage restrictions were broadened. Hundreds of thousands were never born. Untold numbers never married. The intent had been to stop the reproduction of targeted non-Nordic groups and others considered unfit. It continued into the 1970s, probably even later. It was all said to be legal, based on science, sanctioned by the highest courts. But what was it really?

As early as December of 1942, the Nazi plan was obvious. In a highly-publicized warning simultaneously broadcast in more than twenty-three languages the world over, the Allies announced that the Nazis were exterminating five million Jews and murdering millions of other national peoples in a plan to perpetrate a master race. The Allies vowed to hold war crimes trials to punish the Nazis and all those who abetted them.[60] Ultimately, the trials would bring to justice more than just the executioners, but those who ordered them, financed them, inspired them, facilitated their crimes and gave them scientific and medical support. These war crimes trials would ultimately include bankers, industrialists, philosophers, a newspaper editor, a radio propagandist, and many doctors and scientists.

By 1943, humanity needed a new word for the Third Reich's collective atrocities. The enormity of Nazi butchery of whole peoples by physical extermination, cultural obliteration, biological deracination and negative eugenics defied all previous human language. Nothing like it on so sweeping a scale had ever occurred in history.

Raphaël Lemkin, a Jewish refugee at Duke University, formerly a prosecutor from Warsaw and an expert on international law, was commissioned by human rights organizations to study the crime. After a few months fighting as a *partisan*, Lemkin had fled Poland for Sweden and ultimately settled in the United States. His new word describing the overall Nazi campaign in Europe sprang from the same Greek root Galton had used. *Eugenics* was the study of "well-born life." Lemkin's new word, contemplated by him since 1940, encompassed the systematic destruction of an entire group's life. His new word was *genocide*.[61]

On October 30, 1943, as Lemkin was finalizing his study, the Allies met in Moscow and issued a joint declaration reconfirming that there would be war crimes trials for Nazi perpetrators, to be conducted in both the victimized countries and in Germany. The Allies demanded that all such crimes cease during the final turbulent days of Europe's liberation. "Let those who have hitherto not imbrued their hands with innocent blood beware lest they join the ranks of the guilty, for most assuredly the three Allied powers will pursue them to the uttermost ends of the earth and will deliver them to their accusors in order that justice may be done." The declaration was signed by Franklin Delano Roosevelt, Winston Churchill, and Josef Stalin.[62]

Days later, on November 15, 1943, Lemkin completed his study, *Axis Rule in Occupied Europe*, which was published a year later. In a chapter entitled "Genocide," Lemkin listed the several physical and administrative "techniques of genocide." Among the techniques was a section labeled "Biological." Lemkin later explained the principle: "The genocidal policy [of the Nazis] was far-sighted as well as immediate in its objectives. On the one hand an increase in the birth rate, legitimate or illegitimate, was encouraged within Germany and among *Volksdeutsche* in the occupied countries.... On the other hand, every means to decrease the birth rate among 'racial inferiors' was used. Millions of war prisoners and forced laborers from all the conquered countries of Europe were kept from contact with their wives. Poles in incorporated Poland met obstacles in trying to marry among themselves. Chronic undernourishment, deliberately created by the occupant, tended not only to discourage the birth rate but also to an increase in infant mortality. Coming generations in Europe were thus planned to be predominantly of German blood, capable of overwhelming all other races by sheer numbers."[63]

Axis Rule in Occupied Europe even quoted a relevant Hitler speech: "We are obliged to depopulate as part of our mission of preserving the German population. We shall have to develop a technique of depopulation. If you ask me what I mean by depopulation, I mean the removal of entire racial units. And that is what I intend to carry out.... Nature is cruel, therefore we, too, may be cruel.... I have the right to remove millions of an inferior race that breeds like vermin! And by 'remove,' I don't necessarily mean destroy; I shall simply take the systematic measures to dam their great natural fertility.... There are many ways, systematical and comparatively painless, or at any rate bloodless, of causing undesirable races to die out."[64]

Some five months later, Lemkin's chapter on genocide was popularized in an article entitled "Genocide—A Modern Crime," appearing in *Free*

World, a new United Nations multilingual magazine. In *Free World*, Lemkin again cited "Biological" techniques as a means of genocide. By this time Lemkin had become an advisor to the Judge Advocate General of the U.S. Army, and military tribunal planners were working with him and his concepts as they prepared to bring Nazi war criminals to justice.[65]

Within a month of the publication of "Genocide—A Modern Crime," the Third Reich fell. Lemkin's codified principles of genocide, war crimes and crimes against humanity became pivotal. In August of 1945, the victorious Allies met in London and chartered an international military tribunal to bring the highest-ranking Nazi war criminals to justice. The so-called Nuremberg Trials began just three months later. The dock was hardly limited to those Nazis who pulled triggers and ordered murders—such as Interior Minister Wilhelm Frick and Governor-General of Poland Hans Frank—but also included key propagandists and facilitators, such as newspaper editor Julius Streicher and radio director Hans Fritzsche. At the same time, international justice groups continued to further define the prior acts of genocide in anticipation of more war crimes tribunals, these for individuals of lesser stature who were nonetheless instrumental in Nazi genocide. These additional trials would prosecute doctors, scientists and industrialists. Many of these tribunals would be conducted exclusively by the United States.[66]

On December 11, 1946, as the United States was readying its own prosecutions, the United Nations approved Resolution 96 (I), which embedded the concept of "genocide" into international law. It proclaimed: "Genocide is a denial of the right of existence of entire human groups, as homicide is the denial of the right to live of individual human beings; such denial of the right of existence shocks the conscience of mankind, results in great losses to humanity in the form of cultural and other contributions represented by these human groups, and is contrary to moral law and the spirit and aims of the United Nations."[67]

Shortly thereafter, the articles of a forthcoming Treaty Against Genocide were formulated and later adopted through a succession of resolutions, conventions and treaties to become settled international law. The international convention enumerated crimes against humanity and crimes of genocide in five categories; the last two categories—in subsections (d) and (e)—squarely confronted eugenic policies: sterilization and the kidnapping of eugenically qualified children to be raised as Aryans. Article II stated: "In the present Convention, genocide means any of the following acts committed with intent to destroy, in whole or in part, a national, ethnical, racial or religious group, as such:

(a) Killing members of the group;
(b) Causing serious bodily or mental harm to members of the group;
(c) Deliberately inflicting on the group conditions of life calculated to bring about its physical destruction in whole or in part;
(d) Imposing measures intended to prevent births within the group;
(e) Forcibly transferring children of the group to another group."[68]

Article III assigned equal guilt to those who were responsible for "direct and public incitement" to commit the crimes described as genocide, and those who in other ways become complicit. Article IV declared that the law could punish anyone in any country, "whether they are constitutionally responsible rulers, public officials or private individuals." American prosecutors at the subsequent Nuremberg Trials took their cue from the treaty.[69]

In early July of 1947, the Allies indicted the leaders of the Reich's militarized eugenics umbrella organization, the SS Race and Settlement Office, which forcibly sterilized thousands, kidnapped Polish children with Nordic racial features, organized the Nordic breeding program known as *Lebensborn*, developed extensive genealogy files on millions and conducted eugenic examinations of prisoners before deciding if they should be saved or exterminated. For these activities, SS Race and Settlement Office leader General Otto Hofmann stood among those in the dock.[70]

The indictment clearly enumerated the various aspects of Nazi eugenics as genocide: "Kidnapping the children of foreign nationals in order to select for Germanization those who were considered of 'racial value.'... Encouraging and compelling abortions on Eastern workers.... Preventing marriages and hampering reproduction of enemy nationals."[71]

A week after the indictment was served on the accused, the military occupation's semiofficial newspaper, *Die Neue Zeitung*, drove home the point to the German people, publishing extracts of the U.N. Treaty on Genocide. The newspaper announced: "On 10 June the Secretary's Office of the United Nations completed the first draft of an international convention for the punishment of government officials who attempted to exterminate racial, religious, national, or political groups.... Three distinct types of 'genocide' are listed." The paper then itemized actions that qualified as genocide, including "open mass murder" and housing people in conditions calculated to kill. *Die Neue Zeitung* explained that the other of the three most significant forms of genocide was "sterilization of large groups and forcible separation of families as 'biological genocide.'" The article itself was entered into the Nuremberg Trial record.[72]

During the long trial, which lasted almost a year, prosecutors outlined a lengthy bill of eugenic particulars, including the murder of those who did not pass eugenic tests. "The SS Race and Settlement Main Office (RuSHA) was responsible," prosecutors declared, "among other things, for racial examinations. These racial examinations were carried out by RuSHA leaders or their staff members, called racial examiners." Prosecutors charged that as part of the Reich's genocidal campaign, RuSHA was continually engaged in "classification of people of German descent." It added, "RuSHA, in carrying out racial investigations and examinations, took a leading part in the accomplishment of the [extermination] program. Since negative results of racial investigations and examinations led to the extermination or imprisonment in concentration camps of the individuals concerned, the Staff Main Office . . . acted in close cooperation with the SS Reich Security Main Office [the chief SS agency overseeing physical extermination]. The Reich Security Main Office imposed capital punishment and imprisonment in concentration camps upon individuals designated by RuSHA."[73]

An entire portion of the prosecutors' case, "Section 4: Sterilization," presented documents and evidence concerning the mass sterilization of unfit individuals by Nazis throughout Europe during the Reich's twelve-year reign of terror. Leaving no doubt, prosecutors declared, "The fundamental purpose . . . was to proclaim and safeguard the supposed superiority of 'Nordic' blood, and to exterminate and suppress all sources which might 'dilute' or 'taint' it. The underlying objective was to assure Nazi dominance over Germany and German domination over Europe in perpetuity."[74]

Eugenics was also pivotal to a gamut of other war crimes. Often before burning a town or murdering an entire community, Nazis identified and kidnapped the eugenically fit Nordic children so they could be raised in Aryan institutions. This was done, prosecutors stated, "in accordance with standards . . . [of] Nazi racial and biological theories." What had occurred in Lidice, Czechoslovakia, was read into the record as an example. After Lidice was selected for obliteration, every adult man in the village was executed and most of the village's women were deported to Ravensbrück concentration camp. But the village's children were dispatched to Poland for a thorough "medical, eugenic, and racial examination carried out by the physicians of the health offices." Those deemed sufficiently Nordic were sent to live with Aryan families where they would undergo Germanization. Those deemed unfit were "deported." The prosecutor stated, "Here ends all traces of these 82 children of Lidice."[75]

"And so," prosecutors solemnly explained, "the final balance gives us these terrible facts: 192 men and 7 women shot; 196 women taken into concentration camps, of whom 43 died from torture and maltreatment; 105 children kidnapped.... The village was burned, buildings leveled, streets taken up and all other signs of habitation completely erased." To protest the utter eugenic extermination of Lidice, many small towns later adopted the name of the village. Hence the people are gone, but the memory of Lidice lives on.[76]

Count after count recited the fact that "racial value" following a eugenic analysis made all the difference between life and death, genocide and survival.[77] Prosecutors sorted Germany's many eugenic atrocities into specific categories of war crimes. Point 15, entitled "Hampering Reproduction of Enemy Nationals," specified sterilization and marriage restriction: "To further weaken enemy nations, both restrictive and prohibitive measures were taken to discourage marriages and reproduction of enemy nationals. The ultimate aim and natural result of these measures was to impede procreation among nationals of Eastern countries." Point 18, entitled "Slave Labor," explained that through the racial examinations of RuSHA, "foreign nationals without any German ancestry were sent to Germany as slave labor," where they were worked to death.[78]

Point 21, "Persecution and Extermination of Jews," explained how genealogy offices were critical to Hitler's war against the Jews across Europe. "RuSHA also participated extensively in the persecution and extermination of Jews. The Genealogy Office (*Ahnentafelamt*) of RuSHA prepared and retained in its files the names of all Jewish families in the Reich and persons having any Jewish ancestry. This office also participated in preparing similar files in the Netherlands, Belgium, Norway, Denmark, Danzig, and France where it worked together with the SS Reich Security Main Office. These files were used for enforcing discriminatory measures against Jews and preparing transport lists of Jews to be taken from Germany and the occupied countries to the extermination camps in the East."[79]

On January 20, 1942, SS Race and Settlement Office leader Hofmann had attended the infamous Wannsee Conference, the planning session associated with the Final Solution. The Wannsee Protocol produced after the conference made the eugenic guidelines clear. Mixed Jews of the "first degree," that is, Jews with substantial German blood in their ancestry, could be exempted from "evacuation," the code word for extermination, but only if they were sterilized. The Wannsee Protocol recorded: "Hofmann is of the opinion that extensive use must be made of sterilization." The protocol also recorded that "[Persons of mixed blood] exempted from evacuation will be sterilized in order to obviate progeny and to settle the [mixed blood] problem for good. Sterilization is voluntary, but it is the condition for remaining in the Reich."[80]

Confronted by prosecutors at his trial with charges of eugenic extermination, Hofmann said little in his own defense, and openly admitted he was a Nazi eugenicist.

PROSECUTOR: When did you become chief of the Eugenics Office in RuSHA?
HOFMANN: At the beginning of 1939 I was appointed to this task....
Q: What were your duties there?
A: The Eugenics Office was responsible for carrying out the betrothal and marriage order which Himmler had issued on 31 December 1931 to the SS.... The RuSHA leader had to look after the eugenics research offices of the SS, regiments, and, according to his qualifications and talents, he influenced cultural life within the areas of the main district.[81]

Hofmann could not understand why the United States thought his actions were crimes against humanity. He placed into evidence a special report on America produced by the Nazi Party's Race-Political Office years before on July 30, 1937. "The United States," asserted the report, "however, also provides an example for the racial legislation of the world in another respect. Although it is clearly established in the Declaration of Independence that everyone born in the United States is a citizen of the United States and so acquires all the rights which an American citizen can acquire, impassable lines are drawn between the individual races, especially in the Southern States. Thus in certain States Japanese are excluded from the ownership of land or real estate and they are prevented from cultivating arable land. Marriages between colored persons and whites are forbidden in no less than thirty of the Federal States. Marriages contracted in spite of this ban are declared invalid." Typical laws were recited from Alabama, Arizona, Arkansas, California and Florida.[82]

The special report added, "Since 1907, sterilization laws have been passed in twenty-nine States of the United States of America."[83]

Hofmann's document made one other point. It offered the following justification, originally translated from English into German and then back into English for the trial:

In a judgment of the [U.S.] Supreme Court...it says, among other things: "It is better for everybody if society, instead of waiting until it has to execute degenerate offspring or leave them to starve because of feeble-mindedness, can prevent obviously inferior individuals from propagating their kind."[84]

Hofmann was sentenced to twenty-five years imprisonment.[85]

For three—perhaps four—decades after the Treaty Against Genocide was adopted, the United States continued to sterilize targeted groups because of their eugenic or racial character, real or supposed; continued to prevent marriages because of their eugenic or racial character, real or supposed; and continued to hamper reproduction, interfere with procreation, and prevent births in targeted groups. After the Hitler regime, after the Nuremberg Trials, some twenty thousand Americans were eugenically sterilized by states and untold others by federal programs on Indian reservations and in U.S. territories such as Puerto Rico.

They said it was legal. They said it was science. What was it really?

CHAPTER 20

Eugenics Becomes Genetics

After Hitler, eugenics did not disappear. It renamed itself. What had thrived loudly as eugenics for decades quietly took postwar refuge under the labels *human genetics* and *genetic counseling*.

The transition was slow and subtle and spanned decades. Some defected from American eugenics as early as the twenties, prompted by a genuine revulsion over a movement that had deteriorated from biological utopianism into a campaign to destroy entire groups. For others who defected in the thirties and early forties, it was the shock of how Adolf Hitler applied eugenics. For America's eugenic holdouts, it was only the fear of guilt by scientific association with genocide that reshaped their memories and guided their new direction. It took a Holocaust, a continent in cinders and a once great nation bombed and battled into submission to force the issue.

Originally, human genetics and eugenics were one and the same. At the turn of the twentieth century, American breeders of plants and animals had turned their hybridizing skills and social prejudices on their fellow man, trying to manage humanity the same way they managed crops and herds. The American Breeders Association created its Eugenics Committee in 1903. In 1904, the Carnegie Institution founded its eugenic installation at Cold Spring Harbor.[1] The word *genetics* did not exist at the time.

In England, meanwhile, research into Mendel's decades-old discovery of cellular "elements" had advanced and was sorely in need of a new dedicated field of study. By 1905, William Bateson, the man who several years earlier had promulgated the rediscovery of Mendel's theories, was now privately referring to the new science of heredity as "genetics," from the same Greek root Galton employed. Bateson publicly announced the new science during his inaugural address during the Royal Horticultural Society's Third International Conference on Hybridization in 1906. "The science itself is still nameless," declared Bateson. "...I suggest for the consideration of this Congress the term *Genetics*, which sufficiently indicates that our labors are devoted to the elucidation of the phenomena of heredity and variation . . . and [their] application to the practical problems of breeders, whether of animals or plants." When the conference proceedings were published, the society renamed the event the Third International Conference on Genetics.[2] Genetics was born.

Shortly thereafter, students of genetics began referring to the transmittable cellular elements as "genes." By 1912, Cambridge University received a sizeable endowment for genetic studies and in 1914 established the world's first chair in genetics. Mainstream European and American geneticists were primarily devoted to the study of hereditary mechanisms, probing the structure and interactions of enzymes, proteins and other cellular components. Plant and animal geneticists zealously explored the protoplasm of fruit flies, maize, sheep and other species, hoping to understand and manage the lower life-forms. They understood that man was a more complex animal that had both conquered, and was conquered by, his environment. In Europe, human studies of cellular mechanisms were undertaken, but slowly. Not so in America, where breeders distorted Mendelian principles into eugenics and then subsumed nascent human genetics. The two words were synonymous in the United States.[3]

In 1914, the American Breeders Association changed its name to the American Genetic Association, and its publication from *American Breeders Magazine* to *Journal of Heredity*. The organization and its publication functioned as a scientific jumble, combining the best efforts of good agronomy and zoology with tainted, ill-advised and racist social engineering. The Carnegie Institution ran the Eugenics Record Office under its Department of Genetics, with Davenport as its director. Many of the nation's leading geneticists, such as W. E. Castle and Raymond Pearl, were among the earliest dues-paying members of the Eugenics Research Association. Genetics and biology departments across America taught eugenics as part of their curriculums. In 1929, *Eugenical News* changed its subtitle once again, this time to "Current Record of Human Genetics and Race Hygiene."[4]

However, by the late twenties and early thirties many human geneticists who had joined the eugenic charge were defecting. L. C. Dunn exemplified this growing trend. In 1925, he had coauthored *Principles of Genetics*, asserting in typical eugenic rhetoric that "even under the most favorable surroundings there would still be a great many individuals who are always on the borderline of self-supporting existence and whose contribution to society is so small that the elimination of their stock would be beneficial."[5] But in 1935, two years after the rise of Hitler, Dunn formally suggested that the

Carnegie Institution shut down its Cold Spring Harbor eugenic enterprise. "With genetics," Dunn told Carnegie officials, "its relations [with eugenics] have always been close, although there have been distinct signs of cleavage in recent years, chiefly due to the feeling on the part of many geneticists that eugenical research was not always activated by purely disinterested scientific motives, but was influenced by social and political considerations." Dunn later became an outspoken critic of both Nazi eugenics and the American movement.[6]

In 1937, Laurence Snyder, the incoming president of the Eugenics Research Association and chairman of its Committee on Human Heredity, became convinced it was time for a break with the past. In a lengthy report to Laughlin and the Carnegie Institution, Snyder's committee concluded that the end for organized eugenics was near. "The recent attacks upon orthodox eugenics," the committee declared, "and indeed upon the whole present social set-up . . . emphasize more than ever the need for accurate facts and information on basic human genetics. These attacks, it may be stated in passing, come not from irresponsible nor untrained minds, but from some who have the authority of long and honorable scientific achievements behind them."[7]

Referring to the worries over a Europe in political turmoil and preparing for war, the committee report continued, "In these days when the social outlook of whole nations is undergoing far-reaching changes, any fact contributing to our knowledge of basic human welfare becomes of especial importance. The science of human genetics, judged by its past achievements and by what we may reasonably expect in its future developments, is more certainly basic to any well-formulated plan of human welfare."[8]

Unfortunately, noted Snyder, in America the concept of "human genetics" had itself become as tarnished as eugenics. "The interest of American geneticists in human genetics," the committee reported, "appears to have been waning of late, as evidenced by the almost complete absence of papers on human heredity at the various scientific meetings. This state of affairs in America, in contrast to the condition in some of the European countries, is to be deplored. It has come about, in the opinion of your committee, because of two main reasons. First, there has appeared from time-to-time a good deal of unscientific writing on the subject of eugenics. Since the terms 'eugenics' and 'human genetics' are in the minds of many persons synonymous, human genetics has suffered a loss of prestige as a result."[9]

In his June 1938 presidential address to the Eugenics Research Association, Snyder boldly laid the framework for a transition to genuine human genetics programs. In doing so, he first admitted that much of the vocabulary and theory of eugenics was little more than polysyllabic non-sense. "When the Mendelian laws were rediscovered," began Snyder, "and especially when the more modern complicated extensions of genetic theory became understood by research workers in the field of heredity, geneticists spoke a language largely unintelligible to the psychologist, the sociologist and the layman. At that time it was possible, by invoking a phraseology mysterious and somewhat awe-inspiring, to make generalizations regard-ing racial degeneration, the inheritance of personality, character, insanity and criminality, which could not be analyzed immediately by the sociolo-gists and the psychologists because of their unfamiliarity with the 'rules of the game.'"[10]

Snyder knew he was speaking to a constituency of longtime ardent eugenicists, and proceeded cautiously. "This does not mean that the eugenicist must completely renounce a eugenic program," he stated. "It does mean, however, that the immediate and imperative need is for more facts about human inheritance, specifically, facts about socially significant traits and their possible genetic backgrounds."[11]

Nonetheless, the voices of reform were generally drowned out by race-ology and eugenics from the entrenched ranks and longtime leaders, such as Davenport, Laughlin and Popenoe. Organized eugenics remained com-mitted to the Nazi program through much of the Reich years. After the war, geneticists would claim they had no affinity with their Nazi counter-parts. But that was not the case.

For example, in April 1942, amid worldwide charges of mass extermina-tion, the American Genetic Association's *Journal of Heredity* published a long, flippant, almost cheery assessment of Nazi eugenics and genetics. American geneticist Tage U. H. Ellinger's article entitled "On the Breeding of Aryans and Other Genetic Problems of War-time Germany" recounted his exciting visit to the Kaiser Wilhelm Institute for Anthropology, Human Heredity and Eugenics. Institute officials granted him an insider's tour of the Reich's twins lab and other advanced genetic projects.[12] Ellinger's stun-ning article was an American geneticist speaking about Nazi genetics to fel-low geneticists.

"I had an opportunity to meet some of my fellow geneticists," began Ellinger, "who seemed to be working undisturbed by the campaign and the 'mopping up' in Poland, and by the hectic preparations for the assaults on a great many peaceful countries such as Denmark, Norway, Holland, and Belgium. The following unpretentious notes, written for laymen, may per-haps interest some of their many American friends.[13]

"Quite a few of them were busy treating or rather mistreating the sex cells of animals and plants in order to produce new varieties. I was introduced to all kinds of extraordinary creatures produced in that way, mice without toes or with corkscrew tails, flies that violated the very definition of a fly by having four wings instead of two, funny-looking moths, and strange plants. Radiation, especially with X-rays, is the principal means of producing such new kinds, or rather monsters, of animals and plants."[14]

Kaiser Wilhelm Institute officials made Ellinger privy to their surreptitious surveillance methods and government procedures. In his article, Ellinger jocundly reported, "Twins have, of course, for a long time been a favorite material for the study of the relative importance of heredity and environment, of nature and nurture. It does, however, take a dictatorship to oblige some ten thousand pairs of twins, as well as triplets and even quadruplets, to report to a scientific institute at regular intervals for all kinds of recordings and tests."[15]

As for Jews, Ellinger told his fellow geneticists, "In itself, the problem is a fairly simple one when it is first understood that the deliberate eradication of the Jewish element in Germany has nothing whatever to do with religious persecution. It is entirely a large-scale breeding project, with the purpose of eliminating from that nation the hereditary attributes of the Semitic race. Whether this be desirable or not is a question that has nothing to do with science. It is a matter of policy and prejudice only. It is a problem similar to that [which] Americans have solved to their own satisfaction with regard to their colored population. The story of the cruel ways in which life has been made unbearable for millions of unfortunate German Jews belongs exclusively in the shameful realm of human brutality. But when the problem arises as to how the breeding project may be carried out most effectively, after the politicians have decided upon its desirability, biological science can assist even the Nazis."[16]

Ellinger elaborated on Nazi eugenic examinations. "It is a problem," he wrote, "of exactly the same nature as if you were asked to record the exact hereditary differences between a bird dog and a hound. It has nothing whatever to do with your personal preference for one or the other. It is a matter of common knowledge that anybody can immediately recognize many Jews by simply looking at them. In other words, the Jew has a number of characteristic bodily features not often combined in a non-Jew or 'Aryan.' In addition, he may display certain mental characteristics you would soon notice by personal association...."[17]

"An amazing amount of unbiased information has accumulated dealing, for instance, with such features as the position of the ears, the shape of the nostrils, etc. As a result, it is quite possible, by studying the bodily features of a person and his relatives, to state, with considerable likelihood of being right, whether this person has Jewish ancestors. . . . If it be decided by the Nazi politicians that persons with Jewish ancestors shall be prevented from mating with those who have not such ancestors, science can undoubtedly assist them in carrying out a reasonably correct labeling of every doubtful individual. The rest remains in the cruel hands of the S.S., the S.A., and the Gestapo."[18]

As for the fate of the Jews, Ellinger wrote, "What I saw in Germany often made me wonder whether the subtle idea behind the treatment of the Jews might be to discourage them from giving birth to children doomed to a life of horrors. If that were accomplished, the Jewish problem would solve itself in a generation, but it would have been a great deal more merciful to kill the unfortunates outright." Ellinger's article candidly admitted, "As things are run in Nazi Germany, it is obviously a matter almost of life and death whether you carry the label Aryan or Jew."[19]

Summing up, Ellinger attested that, "Genetics really seems to have an unlimited field of practical applications, but I am sure that the old priest Mendel would have had the shock of his life had he been told that seventy-five years after he planted his unpretentious peas in the monastery garden of Brünn, his new science would be called upon to 'grade up' the 'scrub' population of Greater Germany to new 'standards of Aryan perfection.'"[20]

A year later, in 1943, *Eugenical News* projected the future of eugenics. An article entitled "Eugenics After the War" cited Davenport's work at Carnegie's Department of Genetics. Davenport envisioned a new mankind of biological castes with master races in control and slave races serving them. He compared the coming world order to "colonies of bees and termites. . . . All the bees in a hive, including the queen, are full sisters and have been for uncounted generations. Each one is hatched with a set of instincts, which enables it, in machine-like fashion, to do the proper thing at the proper time for the existence of the colony. In human communities, also, the more uniform the instincts and ideals the less friction and the less need for government control with its vast system of law, law enforcement and punishment.[21]

"Contrariwise the more mixed the population from the standpoint of instincts and physical and mental capacity, the more badly does the machine work and the more need of constant repair and adjustment." Davenport added that additional worker strains might be imported to help serve America's coming biological order. "It is quite possible," wrote Davenport, "that some tens of thousands of 'Black fellows' [aborigines] from central Australia might be induced to come to this country." But he added that he hoped America would forgo any further opportunities for race-mixing.[22]

But by 1943, reformers were shouting down diehard Nazi supporters such as Davenport. In the same issue in which Davenport forecast a new biological order, other *Eugenical News* correspondents were condemning Hitler's eugenics, and negative eugenics in general. Following Davenport's remarks, another article entitled "Eugenics in 1952" prophesied various views of eugenics some nine years ahead. One writer urged new thinking on the subject, insisting, "The history of the Nazi movement in Germany proves... [that] unless the new brain functions in an emotional climate of decent social mindedness, it is going to breed a race of madmen rather than of supermen."[23]

Another commentator insisted that any fascism in the United States a decade hence would fail because it "will be shown to belong to the discredited Nazi ideology." A third writer, obviously repulsed by the death and desolation in Nazi-occupied Europe, simply hoped for better times: "A new era is dawning.... Hatred, hostility, and homicide, so recently ended, gives way to love, understanding and growth."[24]

The next 1943 issue of *Eugenical News* published a scathing denunciation of Adolf Hitler for decimating Europe's families. "Hitler, who has torn children from the heart of the family and sent them to the four corners of the earth, without any identification; Hitler, who has torn brothers from sisters, husbands from wives, sons from mothers... and planted them among strangers; Hitler, who by his plans attacked the sacred tie of marriage; Hitler, who believed he could do this and so establish his new order, now sees that it is just this eternal tradition and sanctity of marriage and the family that cannot break, and that will ultimately bring his end."[25]

Eugenical News had changed. Its readers had changed. For some the change was reluctant. For many others it was genuine. Within the smoke of Nazi eugenics, many saw a frightful image. Perhaps they saw themselves.

The transformation of eugenics into human genetics accelerated after the war. By 1944, the American Eugenics Society informed its membership that it now defined eugenics as "genetics plus control of physical and social environment." Meanwhile, *Eugenical News* was publicly debating whether eugenics would even exist after the war. The June 1945 edition, released just after the fall of Germany, admitted, "The question as to what the AES should do after the war is a difficult one. The times will not be very favorable."[26]

The September 1945 issue of *Eugenical News* decried the "Perversions of Eugenics," declaring, "Galton regarded eugenics as a means by which persons with valuable inborn qualities could make a larger contribution to posterity than persons less well endowed.... Galton's view has been perverted by German race superiority, by irresponsible and unimportant racial agitators in America, and by cranks with various plans for breeding a better race." The publication called for a revamped "eugenic policy which is socially acceptable."[27]

Months later, American Eugenics Society President Frederick Osborn prepared a crestfallen lead story for the September 1946 edition of *Eugenical News*. His confession-like epistle, "Eugenics and Modern Life: Retrospect and Prospect," admitted everything. "The ten years, 1930 to 1940 marked a major change in eugenic thinking," Osborn began. "Before 1930, eugenics had a racial and social class bias. This attitude on the part of eugenists was not based on any scientific foundation. It had developed naturally enough out of the class-conscious society of Galton's England, and out of the racial problems presented so vividly to the United States by the great immigration of the early part of the century. The ruling race and the ruling class seemed, to the members of the ruling race and class, to be evidently superior to the non-ruling races and classes...."[28]

Without naming names, Osborn conceded, "A few of the older pioneers never accepted the change and eugenics lost some followers." He counseled, "Population, genetics, psychology, are the three sciences to which the eugenist must look for the factual material on which to build an acceptable philosophy of eugenics and to develop and defend practical eugenics proposals." But he cautioned, "We do not want to repeat in some new form the mistake of the earlier eugenists who declared for race and social class, and thereby set back the cause of eugenics for a generation."[29]

* * *

Beyond mere commentary and condemnations, the incremental effort to transform eugenics into human genetics forged an entire worldwide infrastructure. In 1938, for example, the Institute for Human Genetics opened in Copenhagen. It became a leader in genetic research under the leadership of the Danish biologist and geneticist Tage Kemp. Kemp, however, was actually a Rockefeller Foundation eugenicist. The Institute for Genetics was established by Rockefeller's social biology dollars. Moreover, the Rockefeller effort in Denmark would serve as a model for what it would do elsewhere in Europe.

Kemp's relationship with Rockefeller's eugenics program began in 1932, when Rockefeller officials granted Kemp a fellowship to travel to Cold Spring Harbor and study alongside Davenport and Laughlin. In his report to Rockefeller's Paris office, Kemp related, "To begin with, I endeavored to gain a thorough knowledge of the working methods of the Eugenics Record Office.... In connection with my studies at the Eugenics Record Office, I pursued study of the heredity of sporadic goiter, carrying out examinations amongst the population of Long Island and, in certain cases, also amongst the patients of the U.S. Veteran Hospital, Northport, L.I., and Kings Park State Hospital, L.I." During his U.S. tour, Kemp also attended the Third International Congress of Eugenics in New York City, and presented a paper on "A Study of the Causes of Prostitution, Especially Concerning Hereditary Factors."[30]

Kemp became a rising star at Rockefeller and was utilized as an advance man and confidential source for the foundation as it sought to create a eugenic infrastructure throughout Europe. On June 29, 1934, Daniel O'Brien, who ran Rockefeller's Paris office, notified Kemp, "It is a pleasure to inform you that, at the last meeting of our Committee, a special fellowship was granted to you in order to permit you to spend three months on visits to various European institutes of genetics." O'Brien's letter continued, "I should like to have your comments on individuals who might be helped by means of a fellowship of approximately one year.... It would be particularly helpful to receive your personal impressions of the able men you come into contact with.... It would of course be understood that any information you may give would be considered strictly confidential."[31]

Kemp's itinerary included Holland, England, France, Austria, Switzerland, Russia, Germany and several other nations. His extensive report to Rockefeller included a significant section on Germany, which included summaries on the leading race hygienists and their institutions. For example, in Munich he met with Rüdin and reported: "On the whole, I am finding the work going on there rather important and serious, and it is supported by enormous means." Kemp then rated the leading scientists under Rüdin, indicating which ones spoke English, and the nature of their projects. Bruno Schultz, for example, was "doing a great deal of statistical work concerning mental diseases of practical value for the sterilization law and the eugenical legislation in Germany."[32]

In Berlin, Kemp toured the Institute for Brain Research, which Rockefeller had built. Kemp was impressed, writing back to Rockefeller officials, "I learned all concerning the anatomical, physiological and clinical work going on at this immense, remarkable and rather complicated institution." He also spent time at the Institute for Anthropology, Human Heredity and Eugenics, "which I am finding one of the best centers in the world for the study of normal and morbid inheritance by human beings." Kemp was also impressed with Verschuer, whom he described as "a keen National Socialist, completely honest, however, I feel, so one can rely upon his scientific results as being objective and real. He works especially with twin investigations and is doing this research very thoroughly and systematically."[33]

In Munich, Kemp also met with Theodor Mollison, Mengele's first advisor. He described Mollison as "a very fine and charming personality." Kemp reported, "He is especially working on the specificity of the proteins of various human races."[34]

Rockefeller continued granting Kemp funds for eugenic work, albeit always calling it "genetics." Indeed, just after his report about European genetics, discussions were launched to build the institute in Copenhagen, which Kemp would lead. Previously, Kemp's fledgling studies were confined to one or two small rooms at the University of Copenhagen. That would all change once the spacious new Institute of Human Genetics was erected.[35]

Although Kemp's new institute was packaged as genetics, its eugenic nature was never in doubt. For example, within Denmark, directors of two existing centers for the feebleminded, as well as other local eugenicists, hoped Rockefeller's new institute would bolster the "scientific foundation for eugenic sterilization." Indeed, at times the project was described in Rockefeller memorandums as the institute for "Human Genetics and Eugenics." Once plans became final, Rockefeller officials confirmed their plans had been developed "on the basis of his [Kemp's] experiences gathered in studies in 1932 and 1934 partly at Eugenics Record Office and Department of Genetics in Cold Spring Harbor, USA," as well as at leading eugenic centers in Uppsala, Austria and Munich.[36]

The University of Copenhagen and the local government planned to contribute land and financial support. But executives at the Rockefeller Foundation clearly understood, as their memos on the proposal reflected, "It will be impossible to have this plan realized at present without a grant from the Rockefeller Foundation." The foundation committed $90,000, and the new Institute for Human Genetics opened to much fanfare in 1938. After the war, the Bureau of Human Heredity, another Danish eugenic agency, transferred its operations to the institute and the personal direction of Kemp.[37]

Thus Rockefeller inaugurated another eugenic outpost in Europe. It was not Germany; it was Denmark. It was not eugenics; it was genetics.

* * *

While human genetics was becoming established in America, eugenics did not die out. It became quiet and careful. The American Eugenics Society inherited the residuum of the movement.

The AES assumed primacy in organized eugenics in the late thirties. It established a relationship with the Carnegie Institution just as the ERO was being dismantled. In 1939, Carnegie awarded the AES its first grant of $5,000 for genetic research. Additional grants in 1941 allowed the AES to help establish the Department of Medical Genetics at what became Wake Forest Medical School, the first such medical genetic chair in the United States. The Eugenics Research Association's vice president, William Allan, was chosen to lead the new department. Allan had previously studied eugenic defects of people in the Appalachians, and now he would head the new $50,000 project funded by Carnegie. Writing in *Eugenical News*, Allan urged county-based "Family Record Offices" in North Carolina to assist in identifying the unfit and screening marriages. Such record offices would integrate marriage records and birth and death registries with family information going back more than a century. The undertaking could be implemented easily, he stated, because, "We already have a small army of men, our County Health Officers." Allan himself was experienced in assembling family pedigrees.[38]

When Allan suddenly died two years later, fellow eugenicist C. Nash Herndon took over. Herndon advocated forced sterilization. Emulating the technique of the Kaiser Wilhelm Institute, Herndon's Department of Medical Genetics provided what he called the "genetic work-ups and medical affidavits" for the county to sterilize dozens of it citizens. Blacks were mainly targeted. He described the campaign in a 1943 university report: "This project consists of a gradual, but systematic effort to eliminate certain genetically unfit strains from the local population. About thirty operations for sterilization have been performed."[39]

Writing in *Eugenical News* years after he joined the Wake Forest staff, Herndon also urged genetic counseling to encourage the fit to marry the fit. In addition, he called for educational efforts for the feebleminded to be reduced, declaring "It is of course an obvious waste of time to attempt to teach calculus to a moron." Under Herndon, Wake Forest Medical School became one of America's premier genetic research establishments. In late

2002, the *Winston-Salem Journal* published a five-part investigation of North Carolina's eugenics program and the university's involvement. The newspaper quoted the record of one woman who in 1945 pleaded with the eugenics board: "I don't want it. I don't approve of it, sir. I don't want a sterilize operation.... Let me go home, see if I get along all right. Have mercy on me and let me do that." A shocked Wake Forest Medical School announced an internal investigation to discover the extent of the school's connection to North Carolina's eugenics program. In February of 2003, some two months after the articles ran, a spokesman told this reporter that the university still did not understand the historical facts or context of eugenics, but was determined to be thorough in its investigation.[40]

The AES was making some progress launching human genetic programs like the one at Wake Forest, but when America went to war, the nation's priorities dramatically changed. By 1942, the AES had virtually disbanded. Its office closed, and its papers were shipped to the home of *Eugenical News* editor Maurice Bigelow. The publication continued during the war years, but circulation dwindled to just three hundred.[41]

After the war, it took Frederick Osborn to salvage the organization. He became president of the AES in 1946. Osborn was a former president of the Eugenics Research Association and the nephew of eugenic raceologist Henry Fairfield Osborn, who was cofounder of the AES and president of the Second International Congress of Eugenics. The younger Osborn was determined to continue the eugenics movement, but under the name of "genetics." Constantly introspective about eugenics' calamitous past, Osborn wondered why "the other organizations set up in this country under eminent sponsorship have long since disappeared.... Was it because...some of the early eugenicists placed a false and distasteful emphasis on race and social class? ... Was it because of the emotional reaction to Hitler's excesses and his misuse of the word 'eugenics'? Or did it go deeper."[42] He concluded that the public was not ready to cope with eugenic ideals, especially in the absence of irrefutable science.

In 1947 the remnant board of directors unanimously agreed, "The time was not right for aggressive eugenic propaganda." Instead, the AES continued quietly soliciting financial grants from such organizations as the Dodge Foundation, the Rockefeller-funded Population Council, and the Draper Fund. The purpose: proliferate genetics as a legitimate study of human heredity.[43]

During the fifties, Osborn took extraordinary pains to never utter a provocative eugenic word. In a typical 1959 speech on genetics at Hunter

College, Osborn was explicit, "We are not speaking here of any manipulation of the genes to produce a superior race. This would require a knowledge of human genetics we do not at present possess, and changes in our social mores which would be presently unacceptable." He merely insisted, "Medical genetics has recently become an accepted field of study; the larger medical schools are developing departments of human genetics and setting up heredity counseling clinics."[44]

At the same time, Osborn and his colleagues were searching for a new socially palatable definition of eugenics that would promote the same ideals under a new mantle. One Osborn cohort, Frank Lorimer, wrote Osborn, "Personally, I would redefine 'eugenics' to include concern with all conditions affecting the life prospects of new human beings at birth." He added the caveat, "This is a matter of strategy rather than ideology."[45]

The AES knew that reestablishing eugenics was an uphill battle. Osborn's draft address for the 1959 board of directors meeting outlined an ambitious campaign of behind-the-scenes genetic counseling, birth control, and university-based medical genetic programs. At the same time, Osborn conceded that the movement's history was too scurrilous to gain public support. "Lacking a scientific base," wrote Osborn, "the eugenics movement was taken over successfully by various special interests. The upper social classes assumed that they were genetically superior and that eugenics justified their continuing position. People who thought they belonged to a superior race assumed that the purpose of eugenics was to further their interests. ... The worst in all these movements found their climax under Hitler who combined them for political motives. It is no wonder that for a long time afterwards eugenics had few followers among thoughtful people." But, he concluded, "With the close of World War II, genetics had made great advances and a real science of human genetics was coming into being. ... Eugenics is at last taking a practical and effective form."[46] For Osborn, eugenics and genetics were still synonymous.

Osborn's warnings notwithstanding, some AES members were eager to resume their former propaganda campaigns against the unfit. "The Society is torn," one member wrote Osborn. "Is it to be a 'scientific' society or is it to be a 'missionary' or 'educational' society?"[47]

In 1961, geneticist Sheldon Reed wrote to an AES official, "It seems to me that there is considerable schizophrenic confusion as to whether eugenics exists or not." He wondered if perhaps "the society should disband." Reed added defiantly that the AES should cast off any guilt about the Holocaust. "My final point," Reed declared, "is concerned with the allocation of guilt for the murder of the Jews. Was this crime really abetted by the eugenics ideal? One should remember that the Jews and other minorities have been murdered for thousands of years and I suspect that motives have been similar on all occasions, namely robbery with murder as the method of choice in disposing of the dispossessed individuals.... I do not wish to make Charles Davenport my scapegoat for this, as seems to be the fashion these days. As far as I can see, the motives behind the liquidation of the Jews were not eugenic, not genocide... but just plain homicidal robbery."[48]

But Osborn felt, "We have to take into account that Europeans under Hitler suffered almost a traumatic experience." He had already cautioned, "We must not put out anything that would upset the best of the scientists." On another occasion, he warned, "This question of how to make selection an effective force is the crux of any eugenics program. It is completely irrelevant to get involved in red herrings regarding 'breeding of supermen.'" To dampen his colleagues' ardor, Osborn constantly reminded AES members, "The purpose of eugenics is not to breed some... superior being, but to provide conditions... for each succeeding generation to be genetically better qualified do deal with its environment."[49] Such remarks were made even as the AES continued to promote the gradual development of a superior race, albeit under the guise of genetic counseling and human genetics and with the full participation of hard science.

Eschewing high-profile agitation, Osborn insisted that only quiet work with scientists could accomplish the goal. In a candid 1965 letter, he wrote, "I started hopefully on this course thirty-five years ago and some day would be glad to tell you all of the steps we took—the work we did, the conferences we held, and the money we put into the then *Eugenical News*—about $30,000 a year, to propagandize eugenics. It got us no where, probably because we did not have the backing of the scientific world."[50]

That same year, after numerous genetic counseling and human heredity programs had been established, Osborn was able to confidently write to Paul Popenoe, "The term medical genetics has taken the place of the term negative eugenics." Keeping a low profile had paid off. On April 12, 1965, Osborn wrote a colleague at Duke University somewhat triumphantly, "We have struggled for years to rid the word eugenics of all racial and social connotations and have finally been successful with most scientists, if not with the public."[51]

Indeed, by 1967, Osborn's society had become a behind-the-scenes advisor for other major foundations seeking to grant monies to genetic research. Even the National Institutes of Health sought their advice in parceling out major multiyear grants for what was called "demographic-genetics." By 1968, a pathologist at Dartmouth Medical School was asking the Carnegie Institution if he could access the ERO's trait records on New Englanders for his "medical genetics project."[52]

During the sixties, seventies and eighties, the racist old guard of eugenics and human genetics died out, bequeathing its science to a new and enlightened generation of men and women. Many entities changed their names. For example, the Human Betterment League of North Carolina changed its name to the Human Genetics League of North Carolina in 1984. In Britain there were name changes as well. The *Annals of Eugenics* became the *Annals of Human Genetics* and is now a distinguished and purely scientific publication. The University College of London's Galton Chair of Eugenics became the Chair of Genetics. The university's Galton Eugenics Laboratory became the Galton Laboratory of the Department of Genetics. The Eugenics Society changed its name to the Galton Institute.[53]

In 1954, *Eugenical News* changed its name to *Eugenics Quarterly* and was renamed again in 1969 to *Social Biology*. Later the AES renamed itself the Society for the Study of Social Biology. As of March 2003, both the organization and its publication are operating out of university professors' offices. *Social Biology* editors and the leaders of the society are aware of their society's history, but are as far from eugenic thought as anyone could be. The group is now researching genuine demographic and biological trends. Professor S. Jay Olshansky of the University of Illinois at Chicago and *Social Biology*'s associate editor as of March 2003, denounced eugenics and his journal's legacy during an interview with this reporter. "You couldn't find anyone better to run this society," he insisted. "I carry a potentially lethal genetic disorder. Plus, I'm a Jew. I would be the exact target of any eugenics campaign. I hate what eugenics and the Nazis stood for."[54]

The American Genetic Association, formerly the American Breeders Association, also continues today. As of March 2003, it was headquartered out of a scientist's home office in Buckeystown, Maryland. In the 1950s, the American Genetic Association still listed its three main endeavors at the top of its letterhead: "Eugenics-Heredity-Breeding." As of 2003, most of the organization's early twentieth-century papers were in storage. As of early 2003, AGA leaders knew little of the association's past. But the group still publishes *Journal of Heredity*. Once a font of eugenic diatribe, it is now a completely different journal with a different and enlightened mission. Its editor as of March 2003, Stephen O'Brien, is a distinguished government geneticist who has been featured in documentaries for his efforts to help develop countermeasures to fight plague-like diseases.[55]

Planned Parenthood went on to promote intelligent birth control and family planning for people everywhere, regardless of race or ethnic background. It condemns its eugenic legacy and copes with the dark side of its founder, Margaret Sanger. Planned Parenthood exists in a community of other population-control groups, such as the Population Council and the Population Reference Bureau, many of which sprang from eugenics.[56]

Cold Spring Harbor stands today as the spiritual epicenter of human genetic progress. Following the war, it devoted itself to enlightened human genetics and became a destination for the best genetic scientists in the world. In the summer of 1948, a visionary young geneticist named James Watson studied there. He returned in 1953 to give the first public presentation on the DNA double helix, which he had codiscovered with Francis Crick. Watson became director of Cold Spring Harbor Laboratory in 1968, and president in 1994. In February of 2003, the lab hosted an international celebration of the fiftieth anniversary of the discovery of the double helix.[57]

The world is now filled with dedicated genetic scientists devoted to helping improve all mankind. They fight against genetic diseases, help couples bear better children, investigate desperately-needed drugs, and work to unlock the secrets of heredity for the benefit of all people without regard to race or ethnicity. Every day, more eager scientists join their ranks, determined to make a contribution to mankind. Genetics has become a glitter word in the daily media. Most of the twenty-first century's genetic warriors are unschooled in the history of eugenics. Most are completely divorced from any wisp of eugenic thought.

Few if any are aware that in their noble battle against the mysteries and challenges of human heredity, they have inherited the spoils of the war against the weak.

CHAPTER 21

Newgenics

What now? The short answer is nobody knows. The world will not discover the latest human genetic trends in books like this one, but rather in the morning paper and on the evening news. Almost as soon as any author's page is typed, genetic advances redefine the realities, the language and the timelines. By creep and by leap, the world will be alternately shocked and lulled—and then shocked again—to learn how rapidly humanity and nature are changing.

Today's headline is tomorrow's footnote. In 1978, Louise Brown became the world's first test-tube baby and a braver new world shuddered. Since then, *in vitro* fertilization has become common reproductive therapy. In 1997, Dolly the cloned Scottish sheep captured cover stories and stirred acrimonious debate across the world. Shortly after that, several cows were cloned in Japan, but the news merely flashed across CNN as a fleeting text report behind the comical headline "Udderly Amazing." In 1998, the Chinese government launched a program to clone its pandas. Shortly thereafter, Spanish authorities approved cloning of a bucardo, a recently extinct mountain goat. In 2000, Virginia scientists cloned five pigs. Entire menageries are in various stages of being cloned, from monkeys to mastodons to family pets.[1]

Human clones are next. In late 2001, when editors were discussing this book, the experts insisted we were decades away from the first human clone. As chapters were being submitted, the prediction of "decades" shortened to "years." By the end of 2002, those same experts were debating whether any of several competing scientists had already successfully created the first cloned babies. There is no shortage of willing donors or parents, nor rumors to supply the field. Legislation enacted in several countries cannot address the international dimensions of the where, who and how of impregnation, gestation and conception itself.[2]

Predictions and timelines are little more than well-intentioned self-delusion. However, this much is certain: a precocious new genetic age has arrived. This genetic age, morphing at high velocity, can barely be comprehended by a world that doesn't even speak the language of genetic engineering. Certainly, the latest developments continuously flood a spectrum of scientific journals and symposia, prominent and obscure. Yet few can keep up with the moral, legal and technological implications, especially since much of the information is so technical.

At the same time, the consequences of genetic advance are obscured by hype and conspiratorial clamors. Adding more fog, human genetics is now in many ways dominated by capital investment, and many revelations are subject to the eighteen-month initial secrecy of patent applications, the protracted strictures of Wall Street financing and the permanence of corporate nondisclosure agreements. Many areas of human science are now trade secrets. Twentieth-century corporate philanthropy has given way to twenty-first-century corporate profits. Information is often controlled by public relations officers and patent attorneys. It takes a profoundly trained professional eye and a clear mind to separate fact from fantasy and blessings from menaces.

No one should fear the benefits of human reengineering that can obliterate terrible diseases, such as cystic fibrosis and Tay-Sachs. The list is long and genetic researchers are constantly laboring toward the next breakthrough. Every such medical advance is a long overdue miracle. Society should welcome corrective genetic therapies and improvements that will enhance life and better mankind.

Yet humanity should also be wary of a world where people are once again defined and divided by their genetic identities. If that happens, science-based discrimination and the desire for a master race may resurrect. This time it would be different. In the twenty-first century it will not be race, religion or nationality, but economics that determines which among us will dominate and thrive. Globalization and market forces will replace racist ideology and group prejudice to fashion mankind's coming genetic class destiny. If there is a new war against the weak it will not be about color, but about money. National emblems would bow to corporate logos.

Newgenics may rise like a phoenix from the ashes of eugenics and continue along the same route blazed in the last century. If it does, few will be able to clearly track the implications because the social and scientific revolutions will develop globally and corporately at the speed of a digital signal.

The process will manifest as gradual genetics-based economic disenfran-chisement. First, newgenics will create an uninsurable, unemployable and unfinanceable genetic underclass.

The process has already started.

* * *

Like eugenics, newgenics would begin by establishing genetic identity, which is already becoming a factor in society, much like ethnic identity and credit identity. DNA identity databanks are rapidly proliferating. The largest group of databanks warehouse the genetic identities of criminals, suspects, arrestees and unidentified individuals whose DNA is found at crime scenes. The Federal Bureau of Investigation's Combined National DNA Index System (CODIS) was inaugurated in 1990 and has been steadily databasing DNA from criminal encounters. All fifty states have now passed laws creating state databanks that feed CODIS using the FBI's software. By March of 2003, these state databanks were just becoming operational, but legal reviewers have already pointed out the state-to-state inconsistencies in collection and dissemination standards, as well as stor-age protocols. The FBI's databank, which in March of 2003 maintained more than 1.5 million profiles, is growing by some 100,000 profiles a month, and the Department of Justice has asked the FBI to prepare for up to fifty million.[3]

England's rapidly expanding National DNA Database is expected to hold DNA "prints" on three million individuals by 2004. Canada's newly-created databank stored some 23,000 samples as of March 2003, and adds more than a thousand profiles a month. Canada is also pioneering total robotic management and retrieval. China is building extensive databanks, employing more than a hundred DNA laboratories to process the samples. By March of 2003, national DNA databases had become active in Austria, Holland, Germany, Australia and many other countries. Local DNA drag-nets in Germany, England, Australia and the United States have been launched by police to snare offenders who would otherwise never be identi-fied. Such dragnets, which typically ask every citizen of a certain profile or geographical area to provide a DNA sample, are becoming more common.[4]

Police DNA databanks are a powerful and needed tool to help thwart crime and terrorism. They have not only trapped many criminals, they have also prompted the release of many wrongfully arrested or convicted. A number of death row inmates and long-term convicts have been freed only because of DNA analysis of previously untested evidence. Moreover, helpful medical information on individuals is already being discerned from police DNA "fingerprints." For example, British police DNA specialists have concluded that one of the ten DNA markers they analyze for criminal identification also carries information about diabetes. Information about various types of cancer has been derived from DNA fingerprints as well.[5]

The network of DNA databases will soon be global. Interpol conducts a regular International DNA Users Conference to proliferate and link police DNA databank systems worldwide. Soon every nation from Argentina to Zambia, and every local jurisdiction in between, will be able to tap into the international genetic network.[6]

While police DNA databanks are a necessity, they carry twenty-first-century problems. Each country will develop its own rules and regulations about storage, handling and access. There is as yet no body with the authority to set global standards for collection, maintenance or dissemination of DNA data. Quickly, society has learned that crime fighting is no longer the only reason to collect and organize DNA fingerprints. Identification itself is a compelling issue. Military organizations now record DNA fingerprints of their soldiers. America's Armed Forces Repository of Specimen Samples, located in a facility outside Washington, maintains hundreds of thousands of profiles. The tomb of the unknown soldier will soon be a thing of the past.[7]

States are discussing local genetic identification banks for ordinary citizens as well. Connecticut's Department of Social Services already operates a special Biometric ID Project that stores digital fingerprints of its welfare recipients to combat widespread interstate welfare fraud. The Connecticut program currently only records digital scans of traditional fingerprints, but the agency has publicly indicated that stored biometric data could also include retinal scans and facial imaging.[8] Eventually, each state will probably develop its own biometric methodology, which would almost certainly include genetic identification. Such systems would ultimately proliferate down to the county and municipal levels, creating a diverse interoperable national network.

The events of September 11 only accelerated fascination with genetic identification. The technique is now widely studied as a weapon in the war against global terrorism. Think tanks have discussed a wide range of biometric recognition systems and smart cards to secure our society. Biometric databanks—to include DNA fingerprints—have been proposed for airports, immigration bureaus, customs stations, passport offices and even university programs for foreign students. Such systems would be deployed worldwide and could be used at airline counters and visa offices in countries across the world.

Genetic identification has also become a consumer commodity. Paternity suits, cultural and family ancestry claims, inheritance disputes and the simple fear of losing loved ones in terrorist attacks or massive calamities have caused many to obtain their own DNA information and store it personally or in private repositories. Genetic counseling is commonly advised for many couples who may be troubled by hereditary diseases or conditions. Such genetic screens are imperative for those carrying dreaded inherited diseases, such as Huntington's chorea, sickle-cell anemia, Tay-Sachs or a history of breast cancer. Registries are being built. Private labs now market their genetic testing.[9] The field is proliferating in a global community, employing the Internet to enable all citizens from any country to contribute to and access various labs in Australia, the United States and England.

Soon DNA fingerprints will become as common as the traditional fingerprints first discovered by Galton.[10] He suspected they might reveal much about an individual. But he probably never expected that within a century his term for the unique sworls on one's fingertips would expand into the name for genetic identification that would reveal the secrets of a person's biological past and future.

Eventually, genetic databases will go far beyond the identification of mere individuals. The science will create family genetic profiles for use in litigation, health and employment that may function as credit bureaus do today. The day is coming when such family information will be routinely sought in conjunction with employment, insurance and credit granting.

The Medical Information Bureau (MIB) is the American insurance industry's massive databank that dispenses coded medical information and certain lifestyle traits on the millions of individuals who have applied for health and medical insurance. More than sixteen million individual records are stored at any given time. Records are retired after seven years. In their constant battle against fraud, the MIB enables insurance companies to double-check the veracity of applications. Like a credit bureau, the MIB collects information its insurance company members report, and dispenses it to them when they inquire. Since the 1970s, the MIB has included two codes to signify hereditary conditions, this reporter has learned. One code is for hereditary cardiovascular conditions, and the other is a general code to designate "other family hereditary medical conditions," according to MIB officials. As of March 2003, neither hereditary code is subcoded for any specified condition such as epilepsy, congestive heart failure or clinical anxiety, officials said. Instead, the codes are designed to alert insurers to seek additional information from their applicants.[11]

In a group interview with the databank's counsel, marketing director and manager, MIB officials repeatedly insisted the two codes did not signify a genetic predisposition to a health problem, but instead merely "a family hereditary" trait. Family hereditary codes, once gathered, are reported whether or not an individual applicant has shown any symptoms. The family's medical history itself, not the individual's condition, is the determining factor. MIB officials also insisted they would never search out and link other family members based on hereditary conditions.[12]

No DNA repository—police, medical or governmental—is currently linking family members. To do so would create modern-day, genetically-stigmatized Jukes or Kallikaks. It would be the first giant step down the road of newgenics. The financial ramifications are extraordinary and the potential for targeted exclusion is manifest. If the world sees such exclusions, it will probably see them first and most dramatically in the insurance industry.

Insurance companies vigorously claim they do not seek ancestral or genetic information. This is not true. In fact, the international insurance field considers ancestral and genetic information its newest high priority. The industry is now grappling with the notion of underwriting not only the individual applicant, but his family history as well. Insurers increasingly consider genetic traits "pre-existing conditions" that should either be excluded or factored into premiums. A healthy individual may be without symptoms, or asymptomatic, but descend from a family with a history of a disease. In the industry's view, that individual presumably knows his family history; the insurance company doesn't. Insurers call this disparity "asymmetrical information," and it is hotly discussed at numerous industry symposiums and in professional papers. Governments and privacy groups worldwide want to prohibit the acquisition and use of genetic testing. Many in the insurance world, however, argue that their industry cannot survive without such information, and the resulting coverage restrictions, exclusions and denials that would protect company liquidity.[13]

A June 2000 American Academy of Actuaries industry-only monograph entitled "Genetic Information and Medical Expense," obtained by this reporter, cautiously addressed the question. In a section headed "Asymmetrical Information," the monograph asked: "Would a ban on the use of genetic information merely prohibit insurers from asking for genetic tests, or would they also be barred from obtaining test results already known to the applicant? While a more encompassing ban may remove applicants' fears of genetically based denial of coverage, the imbalance of information would leave insurers at a disadvantage." The section concluded, "...biased selection would have a direct impact on premium rates, ultimately raising the cost of insurance to everyone."[14]

In the next section, entitled "Pre-existing Conditions," the monograph argued, "Such a ban [on genetic testing] could have more severe consequences over time, as genetic technology advances." In a series of attached potential "market scenarios," the monograph speculated about individuals with healthy heredities subsidizing those destined to become ill. In one scenario, the monograph stated, "The ultimate character of the market depends on the relative number of these 'genetically blessed' individuals."[15]

A Spring 2002 American Academy of Actuaries briefing paper entitled "The Use Of Genetic Information In Disability Income And Long-Term Care Insurance," obtained by this reporter, suggests that the insurance industry could become insolvent without the benefits of predictive testing. In a section labeled "Adverse Selection," the briefing paper declared, "Insurers maintain that the view of the consumer advocates conflicts with the economic realities of the voluntary insurance market. Insurers are concerned that if they were prohibited from obtaining genetic information from the medical records of applicants, then those applicants would know more about their genetic predisposition than the insurance company (asymmetric information), and more substandard and uninsurable individuals would qualify for insurance. Premiums could not be adjusted adequately to cover the deterioration of the insured population because the higher prices would drive out the healthy. As the insured population disproportionately became weighted toward those who were predisposed to certain genetic defects, experience would worsen and premiums would increase. The increase in premiums would further reduce the number of healthy policy-holders and could eventually cause the insurers to become insolvent."[16]

Insurance discrimination based on genetics has already become the subject of an active debate in Great Britain. British insurers were widely employing predictive genetic testing by the late 1990s to underwrite life and medical insurance, and utilizing the results to increase premiums and deny coverage. The science of such testing is by no means authoritative or even reliable, but it allows insurers to justify higher prices and exclusions. Complaints of genetic discrimination have already become widespread. A third of those polled from genetic disorder support groups in Britain reported difficulties obtaining insurance, compared to just 5 percent from a general population survey. Similarly, a U.S. study cited by the American insurance publication *Risk Management* found that 22 percent of nearly one thousand individuals reported genetic discrimination. A *British Medical Journal* study paper asserted, "Our findings suggest that in less clear cut instances, where genes confer an increased susceptibility rather than 100% or zero probability, some people might be charged high premiums that cannot be justified on the actuarial risk they present."[17]

Nearly three-quarters of a group surveyed by Britain's Human Genetics Commission (HGC) objected to insurer access to genetic testing. One man who tested positive for Huntington's told of being denied insurance when his genetic profile became known; later, when he did obtain a policy, it was five times more expensive. One forty-one-year-old London woman recalled that after her genetic report showed a gene associated with breast cancer, she was unable to buy life insurance. In consequence, when she attempted to purchase a home in 1995, it was more costly. Chairman of the HGC Helena Kennedy said: "Most of us are nervous and confused about where technology might be leading, and the potential challenges to privacy and confidentiality. We know from our survey that people are worried that these developments might lead to discrimination or exploitation, and are skeptical of the law's ability to keep up with human genetics."[18]

A Code of Practice for genetic testing by British insurers was established in 1997, but in 2001, Norwich Union Insurance, among other firms, admitted it had been using unapproved genetic tests for breast and ovarian cancers, as well as Alzheimer's. British insurers began widely utilizing genetic tests after a leading geneticist consulting for the industry's trade association recommended the action, a Norwich Union executive explained. The widespread concern in England is generation-to-generation discrimination pivoting not on race, color or religion, but on genetic caste. "We are concerned, of course," warned Dr. Michael Wilks, of the British Medical Association's Medical Ethics Committee, "that the more we go down the road of precision testing for specific patients for specific insurance policies the more likely we are to create a group who simply will not be insurable." Wilks called such a group a genetic "underclass." A member of Parliament characterized Norwich Union's actions as an attempt to construct a "genetic ghetto."[19]

The British government ultimately imposed an industry-wide moratorium permitting the use of just one type of test. In the subsequent three-year period, out of 800,000 Norwich policies, only 150 involved genetic tests. But British insurance industry sources argue that unless widespread genetic testing and access is restored, the industry and the health service will be overrun with claims.[20]

Moreover, some insurers may also want genetic data so they can use the information to rescind insurance, claiming that an individual fraudulently or even inadvertently omitted ancestral information from an application—even if the insurance claim is unrelated to the medical condition. Precedents abound for such retroactive invalidations, albeit based on family health history rather than genetic testing. In a 1990 Quebec case, a man was killed in a car crash. He carried the gene for a degenerative disease, a form of myotonic dystrophy, and knew his father had suffered from the malady but omitted the information from his 1987 application for a $30,000 policy. His widow was denied a policy payment when Industrial Alliance, one of Quebec's largest insurers, prevailed in court, claiming fraud by omission. An Industrial Alliance attorney told this reporter that the company was aware the man came from a region known for a great deal of consanguinity and where myotonic dystrophy is common. Hence, the company's postcrash investigation bore fruit.[21]

The Industrial Alliance attorney added that such policy invalidations, based on applicants' statements, are common in Canada. A company attorney explained that his firm had even invalidated one car crash death when they learned the applicant indicated he was not a smoker, and a postcrash investigation revealed the man had actually smoked within the previous year. "Even my mother was angry at me for that one," the company attorney admitted. "She said, 'What does cigarettes have to do with the car crash?'" But, explained the attorney, under Quebec law, within the first two years of a life insurance policy, any material omission, deliberate or accidental, can be investigated to invalidate a life insurance claim. After two years, Quebec insurance companies are allowed to invalidate a policy if they can prove a deliberate omission.[22]

The Quebec precedent, which is now spreading to other countries, means that if a person does not possess his genetic information—even innocently—he is being omissive. On the other hand, possessing it makes the data automatically disclosable to the company at the point of application. Insurers worldwide argue that if they cannot require testing, they should be permitted access to the genetic information individuals will increasingly feel obligated to gather. Either way, genetics will soon be an underwriting factor in everyone's personal insurance.

Information from America's MIB, and repositories like it, is often used by insurers to detect omissive statements, this as a basis for denying claims and invalidating policies. The MIB cites combating application fraud as its chief mission. Ironically, many applicants simply do not know their ancestors' health conditions. For example, many American Jews descended from Europe do not know the exact health conditions of ancestors killed in the Holocaust or Eastern Europe's pogroms. Many African-Americans know little of their ancestors reared in slavery or abject twentieth-century poverty. Our mobile society includes many single-parent families where little is known about ancestral health problems. The paucity of genetic information is all the more reason for insurers to press for genetic bureaus to emulate the medical and credit bureaus they currently employ.

A cross-referenced genetic information bureau would permit insurers and financial institutions to create the commercial "genetic underclass" envisioned by critics. Insurers deny that such databanks are in the offing or even desired. Many continue to argue that the insurance community is simply not interested in genetics.

Yet the worldwide insurance industry is indeed rushing to integrate advanced genetics into their everyday business. In England, an insurance industry program called the UK Forum for Genetics and Insurance regularly brings genetic scientists and insurance executives together. The debate is an international one because all insurance is global. All risk—no matter how local—is studied, shared and reinsured by worldwide layers of the insurance industry. The International Actuarial Association's 2002 colloquium in Cancun highlighted genetics as one of its four main agenda items. "Are we expecting trouble for the insurance industry from genetic information?" an IAA program memo pointedly asked. MIB's industry intelligence website, as of March 2003, featured a "Special Section: Genetics" offering an in-depth survey of genetics and insurance, including writings on genetic discrimination, "Balancing Interests in the Use of Personal Genetic Data," and one major reinsurer's article entitled "The Future Will Not Wait for Us."[23]

For decades, insurers, realtors and financial institutions engaged in lucrative racial, sexual and geographic discrimination and preferential treatment known as *redlining* and *greenlining*. The terms derive from the colored lines drawn on maps by insurers and realtors to select neighborhoods for discrimination or preference. Such practices are now outlawed in many countries. But for *genelining*, the laws in various countries are vague, insufficient or nonexistent. Entire extended families of undesirable insureds could be identified with the same subtlety and secrecy with which geographic and ethnic undesirable insureds were identified a few decades ago. Corporate newgenics, blind to the color of one's flag, skin or religious creed, would be driven only by profit.

While insurers and banks may create a genetic underclass in finance, employers may create a genetic underclass among workers. As early as the 1960s, Dow Chemical undertook long-term genetic screening in search of mutagenic effects arising from its workplace. A 1982 federal government survey of several hundred U.S. companies found that 1.6 percent admitted they were utilizing genetic testing, mainly for hazardous workplace monitoring and screening new hires. In 1997, an American Management Association survey reportedly indicated that 6–10 percent of employers polled had asked their employees to submit to voluntary genetic testing. By and large, such screening was conducted openly and was necessary to protect workers from hazardous employment environments.[24] The increase in employer testing since the Human Genome Project was completed in June 2000 can only be imagined. How each company will use its information is neither standardized nor regulated.

In 1994, investigators discovered that the University of California's Lawrence Berkeley Lab went further than simply monitoring the workplace. At the suggestion of the U.S. Department of Energy, which largely funds the lab, medical officers tested employees' blood and urine samples for syphilis, sickle-cell and pregnancy. African-Americans and Latinos were often repeatedly tested for syphilis. The one white employee repeatedly tested for syphilis was married to an African-American. Employees sued. When asked by U.S. News & World Report why only minorities were singled out for repeated syphilis testing, a Berkeley Lab medical officer reportedly replied: "Because that's where the prevalence of the disease is. How come only people over a certain age would get an EKG? See the logic?" The man reportedly later denied he had made the inflammatory statement to U.S. News & World Report.[25]

A landmark federal court ruling in 1998 in favor of the Berkeley Lab employees established the Constitutional right of citizens to their genetic privacy. The court's opinion declared, "One can think of few subject areas more personal and more likely to implicate privacy interests than that of one's health or genetic makeup." The lab settled for $2.2 million in 2000 and deleted the employee information from its computers.[26]

Burlington Santa Fe, one of North America's largest railroads, went a step further in an attempt to stem soaring carpal tunnel claims by employees. Its medical director had read two medical journal articles on carpal tunnel, including one that indicated a genetic predisposition for the syndrome. In March of 2000, Burlington launched a program of surreptitious genetic testing of thirty-five employees making claims for carpal tunnel to determine whether they possessed genetic predisposition. Tests on some twenty employees were actually completed. The intent was to help the company deny carpal tunnel claims.[27]

Burlington's medical director selected Athena Diagnostics, the nation's premier genetic testing laboratory, to analyze the tests. Athena annually performs some 70,000 doctor-referred genetic tests for conditions such as hearing loss, movement disorders, epilepsy, mental retardation and carpal tunnel, a lab source told this reporter. The lab did not understand the purpose of Burlington's testing, a lab source said. Once they learned it was not for therapeutic but insurance purposes, "we were dismayed," a senior Athena executive told this reporter. Burlington was sued on a Friday afternoon in February 2001. Senior executives spent a frantic weekend reviewing the charges and settled by Monday with a $2.2 million payout to employees. Athena soon implemented safeguards such as requiring a signed patient authorization. But according to a company source, Athena still accepts genetic test requests from any licensed physician—whether on behalf of an individual, insurance company or attorney—and from any licensed lab in the U.S. or overseas.[28]

In the late nineties, government officials in Hong Kong refused to hire two men and fired a third after learning that each had a schizophrenic parent. The men had variously worked as a fireman, an ambulance worker and a customs officer. At first, the men were not told why the actions were taken. Government officials claimed the men were not fit for work because their parentage suggested a 10 percent chance they would also become schizophrenic. In fact, the officials had misread a genetic textbook; in reality there was only a 4 percent chance the employees would develop schizophrenia at their ages, compared to 1 percent for the general population. The three men sued. The judge stated that the "genetic liability to develop the disease their parent suffers from does not present a real risk to safety at either place of employment" and awarded the three $2.8 million in damages.[29] There was no genetic test involved in these three cases, just a review of the employees' written personnel files. But the incident again illustrates the danger of genetic information being misinterpreted and abused by local officials and corporate executives who have the power to discriminate.

*   *   *

Are national genetic databanks of all citizens coming? Sir Alec Jeffreys, the founder of DNA fingerprinting, originally believed that DNA fingerprints should be limited to criminals. But late in 2002, he changed his mind and declared that every person's profile should be added to the databank. Former New York City mayor Rudolph Giuliani has urged that a DNA fingerprint be recorded for every American at birth.[30] That day is coming.

In 1998, Iceland created the world's first national DNA database of its citizens. Almost all of its 275,000 citizens trace their lineages to the original Nordic Vikings of a prior millennium. In a unique arrangement, Iceland's national genetic code was sold to the genetic research and pharmacogenetic industries through an entity called deCODE Genetics. Less than 8 percent of Iceland's population opted out of the voluntary program, hence deCODE possesses virtually a complete national genetic and hereditary portrait of Iceland. Scientists at deCODE are currently utilizing the information in their study of a range of debilitating conditions, including respiratory and muscular diseases. Safeguards have been built into the program to conceal individuals' names. But at least one Icelander has sued the government to have her father's genetic history removed. As of March 2003, the case is still winding its way through Icelandic courts. Iceland's national genomic information will be made available to a wide variety of scientific, commercial and governmental entities in an Internet-based system that employs massive data storage drives codenamed "Shark."[31]

One main company manages and controls Iceland's genetic data. That company is already positioned to become the worldwide manager and disseminator of all genomic information globally. In anticipation of that day, the company currently operates genomic offices in California, New York, Zurich, Haifa, New Delhi and Tokyo. The name of the company is IBM. Its Iceland project operates under a division known as "Life Services—Nordic."[32]

Estonia became the second nation to databank its entire population. In 2001, Estonia created the Estonia Genome Project to capture the genetic profiles of its 1.4 million citizens. A biotechnology industry article cited by the government's website explains, "Unlike remote Iceland, Estonia has long been a European stomping ground, ruled by a succession of Russian, Swedish, German, and Danish invaders who left their genetic heritage. Estonia's ethnic mix thus could be a major draw for pharmaceutical companies that want to find disease genes common to most Europeans."[33]

The tiny Polynesian nation of Tonga sold the information on its unique gene pool to Autogen, an Australian genetic research firm, in 2000. Tonga's 170 islands host a group of some 108,000 natives isolated for more than three millennia. Autogen was quoted as explaining its interest in Tonga's population: "The less mixture of inter-racial marriage, the more likely you are to be able to determine a particular gene that may be responsible for a particular disease, whether it's breast cancer or whether it's kidney disease."[34]

After reevaluation the arrangement between Tonga and Autogen was cancelled. Autogen instead focused on a Tasmanian genetic repository. "Tasmania is one of only a few populations in the world where up to seven-generation family pedigrees are available," the company announced. "This makes it an ideal location to study the genetics of complex diseases such as obesity and diabetes."[35]

In England, the UK Biobank recently opened as a repository for the medical information and genetic data of a half million volunteers. More commercial initiatives are underway to secure national genetic information around the world using ethnic, national, racial and even religious parameters. The pharmaceutical companies, governmental agencies and research foundations that operate these databanks will interconnect them globally. The devoted men and women laboring on these national projects are joining research hands to create new disease-fighting drugs, unlock the mysteries of hereditary disease and improve the quality of human life. In the process, prodigious masses of individual genetic information are being gathered. This data can be exchanged and retrieved at the speed of light from a computer and even downloaded to a cell phone.[36]

Lawmakers worldwide recognize both the great potential to mankind and the profound dangers. In America, the Genetic Anti-Discrimination Bill, which would prohibit genetic testing in group insurance and employment, has been percolating in Congress in various forms for years. In previous anti-discrimination laws, Congress has sought to remedy entrenched injustice. But in this case one of the bill's sponsors, U.S. Representative Louise Slaughter of New York, described the proposed legislation as "prophylactic," since Congress can hardly imagine what genetic misuses are in store. As of March 2003 the bill was stalled.[37]

Other countries are also grappling with protective legislation. As of March 2003, Finland and Sweden have been debating legislation for years. Denmark, however, has already banned insurance companies from utilizing genetic information. Employers in Austria are prohibited from utilizing employee genetic data obtained from any source. French bioethics legislation prohibits access by employers and insurance companies.[38]

But in reality, there are so many uses for genetic information—proper and improper, obtainable from so many globalized sources, in so many formats, employing such diverse and fast-moving technical and scientific jargon—that drafting genuine protective legislation is frustrating to lawmakers and genetic privacy groups alike. Is a paper notation of a history of heart disease in a family the same as a genetic predisposition? Is a cholesterol test genetic? Is bloodwork genetic testing? Is information imported from one country governed by another country's laws? Japanese employers utilize genetic labs in America; whose safeguards on access, dissemination and use govern? What if the origin and destination is cyberspace? If an individual knows certain genetic information, why shouldn't he disclose it to insurance companies and employers like any other required medical information?

The problem is growing exponentially. "We need to stop genetic discrimination before it becomes widespread," Representative Louise Slaughter told this reporter. "The U.S. Congress has been debating my legislation for over seven years. Genetic discrimination is already occurring. If we can't pass a ban on these practices today, what are we going to do as the science becomes more complex? It is crucial that we, as a nation, state unequivocally that genetic discrimination is wrong and will not be tolerated."[39] Like-minded legislators and advocates in many countries echo those words.

* * *

Prominent voices in the genetic technology field believe that mankind is destined for a genetic divide that will yield a superior race or species to exercise dominion over an inferior subset of humanity. They speak of "self-directed evolution" in which genetic technology is harnessed to immeasurably correct humanity—and then immeasurably enhance it. Correction is already underway. So much is possible: genetic therapies, embryo screening in cases of inherited disease and even modification of the genes responsible for adverse behaviors, such as aggression and gambling addiction. Even more exotic technologies will permit healthier babies and stronger, more capable individuals in ways society never dreamed of before the Human Genome Project was completed. These improvements are coming this decade. Some are available now.

But correction will not be cheap. Only the affluent who can today afford personalized elective health care will be able to afford expensive genetic correction. Hence, economic class is destined to be associated with genetic improvement. If the genetically "corrected" and endowed are favored for employment, insurance, credit and the other benefits of society, then that will only increase their advantages. But over whom will these advantages be gained? Those who worry about "genelining," "genetic ghettos" and a "genetic underclass" see a sharp societal gulf looming ahead to rival the current inequities of the health care and judicial systems. The vogue term *designer babies* itself connotes wealth.

The term *designer babies* is by and large just emblematic of the idea that genetic technology can do more than merely correct the frail aspects of human existence. It can redress nature's essential randomness. Purely elective changes are in the offing. The industry argues over the details, but many assure that within our decade, depending upon the family and the circumstances, height, weight and even eye color will become elective. Gender selection has been a fact of birth for years with a success rate of up to 91 percent for those who use it.[40]

It goes further—much further. A deaf lesbian couple in the Washington, D.C., area sought sperm from a deaf man determined to produce a deaf baby because they felt better equipped to parent such a child. A child was indeed born and the couple rejoiced when an audiology test showed that the baby was deaf. A dwarf couple reportedly wants to design a dwarf child. A Texas couple reportedly wants to engineer a baby who will grow up to be a large football player. One West Coast sperm bank caters exclusively to Americans who desire Scandinavian sperm from select and screened Nordics.[41]

All of us want to improve the quality of our children's futures. But now the options for purely cosmetic improvements are endless. A commercialized, globalized genetic industry will find a way and a jurisdiction. It will be an international challenge to successfully regulate such genetic tampering and the permutations possible because few can keep up with the moment-to-moment technology.

It goes much further than designer babies. Mass social engineering is still being advocated by eminent voices in the genetics community. Celebrated geneticist James Watson, codiscoverer of the double helix and president of Cold Spring Harbor Laboratories, told a British film crew in 2003, "If you are really stupid, I would call that a disease. The lower 10 per cent who really have difficulty, even in elementary school, what's the cause of it? A lot of people would like to say, 'Well, poverty, things like that.' It probably isn't. So I'd like to get rid of that, to help the lower 10 per cent."[42] For the first half of the twentieth century, Cold Spring Harbor focused on the "submerged tenth"; apparently, the passion has not completely dissipated.

Following in the footsteps of Galton, who once amused himself by plotting the geographic distribution of pretty women in England, Watson also told the film crew, "People say it would be terrible if we made all girls pretty. I think it would be great." Watson gave no indication of what the standard for beauty would be.[43]

Some who speak of human cloning speak of mass replication of a perfected species. That is nothing less than a return to the campaign to create a master race—but now aided by computers, digital communications and a globalized commercial infrastructure to accelerate the process. Some of America's leading thinkers on genetic evolution believe that within a few hundred years, the world will indeed be divided into the "genetically endowed"—or "GenRich" as some call them—and those who will serve them, almost like the worker bees Davenport envisioned.[44] Advocates of the genetic divide encourage it as a matter of personal choice, and argue that the same man who purchases eyeglasses, tutors his child or seeks medical attention to conquer his biological limitations is destined to take the next step and achieve genetic superiority. This is not the philosophy as much as the *raison d'être* of newgenics.

It will transform the human species as we know it. Transgenic creatures—created from two or more species—are now commonplace. Genomic engineers have implanted a human embryo in a cow. In British Columbia, fish hatcheries have engineered an oversized salmon dubbed "Frankenfish" that is more profitable to raise. Geneticists have inserted the jellyfish's gene for luminescence into rhesus monkey DNA, creating a monkey that glows in the dark; the creature was named ANDi for "inserted DNA" in reverse. No one can successfully legislate or regulate experimentation on monkeys. In the suburbs of Washington, D.C., J. Craig Venter, one of the scientists who led the efforts to map the human genome, has announced plans to create a new form of bacterial life to aid in hydrogen energy production.[45]

Bioethicists are of little help in this hurtling new world. The still emerging field of bioethics includes self-ordained experts who grant interviews to television talk shows and newspapers even as they consult as scientific advisors to the very corporations under question. The do's and don'ts of genetic tinkering are being revised almost daily as the technology breeds an ever-evolving crop of moral, legal and social challenges that virtually redefine life itself.

It will take a global consensus to legislate against genetic abuse because no single country's law can by itself anticipate the evolving inter-collaborative nature of global genomics. Only one precept can prevent the dream of twentieth-century eugenics from finding fulfillment in twenty-first-century genetic engineering: no matter how far or how fast the science develops, nothing should be done anywhere by anyone to exclude, infringe, repress or harm an individual based on his or her genetic makeup. Only then can humankind be assured that there will be no new war against the weak.

Notes

CHAPTER ONE
1. See Jill Durance and William Shamblin, ed., *Appalachian Ways* (Washington D.C.: The Appalachian Regional Commission, 1976), pp. 8-9, 18-19, 24, 32, 79-80. Also see Carolyn and Jack Reeder, *Shenandoah Heritage: The Story of the People Before the Park* (Washington D.C.: The Potomac Appalachian Trail Club, 1978).
2. "Welfare Cause For Sterilization," *Richmond Times-Dispatch*, 6 April 1980.
3. "Welfare Cause For Sterilization," *Richmond Times-Dispatch*, 6 April 1980.
4. Charles B. Davenport, *Heredity In Relation To Eugenics*, p. 257-258; see Bleecker Van Wagenen, chairman, *Preliminary Report of the Committee of the Eugenic Section of the American Breeder's Association to Study and to Report on the Best Practical Means for Cutting Off the Defective Germ-Plasm in the Human Population* (ABA), p. 4; also see Paul Popenoe and Roswell Hill Johnson, *Applied Eugenics*, rev. ed. (New York: Macmillan Company, 1935), p. 396-397 as compared to Frederick Osborn, *Preface to Eugenics* (New York: Harper & Brothers, 1940) p. 14; also see J. David Smith, *Minds Made Feeble: The Myth and Legacy of the Kallikaks* (Rockville, MD: Aspen Systems Corporation, 1985) p. 21-36, 83-114.
5. *The Lynchburg Story*, dir. Stephen Trombley, prod. Bruce Eadie, Worldview Pictures, 1993, videocassette. *Poe v. Lynchburg Training School and Hospital*, 518 F. Supp. 789 (W.D. Va. 1981).
6. "Welfare Cause For Sterilization," *Richmond Times-Dispatch*, 6 April 1980.
7. "Welfare Cause For Sterilization," *Richmond Times-Dispatch*, 6 April 1980.
8. "Patient 'Assembly Line' Recalled By Sterilized Man," *Richmond Times-Dispatch*, 24 February 1980.
9. "Patient 'Assembly Line' Recalled By Sterilized Man," *Richmond Times-Dispatch*, 24 February 1980.
10. "Patient 'Assembly Line' Recalled By Sterilized Man," *Richmond Times-Dispatch*, 24 February 1980.
11. "Patient 'Assembly Line' Recalled By Sterilized Man," *Richmond Times-Dispatch*, 24 February 1980.
12. "Patient 'Assembly Line' Recalled By Sterilized Man," *Richmond Times-Dispatch*, 24 February 1980.
13. *The Lynchburg Story*.
14. *The Lynchburg Story*.
15. *The Lynchburg Story*.
16. *The Lynchburg Story*.
17. See Lothrop Stoddard, *The Rising Tide of Color* (New York: Charles Scribner's Sons, 1926) p. xxix-xxxi, p. 306-308.
18. "Delegates Urge Wider Practice of Sterilization," *Richmond Times-Dispatch*, 16 January 1934.
19. International Military Tribunal, *Nuremberg Military Tribunal, Green Book, Volume V*, p. 159. See International Military Tribunal, *Nuremberg Military Tribunal, Green Book, Volume IV*, p. 609-617, 1121-1127, 1158-1159. See United Nations Resolution 95 (I), "Affirmation of the Principles of International Law Recognized by the Charter of the Nürnberg Tribunal." United Nations Archives. See United Nations Resolution 96 (I), "The Crime of Genocide." United Nations Archives. See Office of the High Commissioner for Human Rights, "Convention on the Prevention and Punishment of the Crime of Genocide," at www.unhchr.ch.

CHAPTER TWO
1. *Code of Hammurabi*, trans. L. W. King, item #48 at www.wsu.edu.
2. See Henry Hazlitt, *The Conquest of Poverty* (New Rochelle, NY: Arlington House, 1973), Chapter 6.
3. Deuteronomy 15:11 NIV Study Bible.
4. Luke 7:22; Matthew 10:6-8, 11:4. Matthew 5:5.
5. Catholic Encyclopedia, s.v., "Hospital."
6. Catholic Encyclopedia, 1967, s.v., "Orphan (In the Early Church)." English Heritage, "Hospitals," at www.eng-h.gov.uk.
7. E. M. Leonard, *The Early History of English Poor*

Relief (Cambridge: Cambridge University Press, 1900; London: Frank Cass & Co., 1965) pp. 3-5. Encyclopedia Judaica, s.v., "Black Death."

8. Leonard, pp. 16-17.

9. Catholic Encyclopedia, 1967, s.v., "Henry VIII." Paul Slack, *The English Poor Law 1531-1782*, (London: Macmillan Education Ltd., 1990), pp. 16-17.

10. See John Bohstedt, *Riots and Community Politics in England and Wales 1790-1810* (Cambridge: Harvard University Press, 1983).

11. Slack, p. 17. Hazlitt, Chapter 7. Leonard, pp. 10-11.

12. Slack, pp. 18, 25. Hazlitt, Chapter 7.

13. Charles L. Brace, "Pauperism," *North American Review* 120 (1875) as cited by Elof Axel Carlson, *The Unfit: A History of a Bad Idea* (Cold Spring Harbor, NY: Cold Spring Harbor Laboratory Press, 2001), p. 76. Carlson, p. 77. Hazlitt, Chapter 7.

14. James Greenwood, *The Seven Curses of London* (London: S. Rivers and Co., 1869) Chapter XXIII.

15. Thomas R. Malthus, *An Essay on the Principle of Population*, as selected by Donald Winch (Cambridge: Cambridge University Press, 1992) pp. 19, 100-101, 221.

16. Charles Darwin, *The Origin of the Species* (New York: D. Appleton & Co., 1881), chapter 3. Herbert Spencer, *Social Statics*, (New York: Robert Schalkenback Foundation, reprint, 1970), pp. 58-60, 289-290, 339-340.

17. Darwin, *The Origin of the Species*, Chapter 3.

18. See Robert C. Bannister, *Social Darwinism: Science and Myth in Anglo-American Social Thought* (Philadelphia: Temple University Press, 1979), p. xii. See Carlson, pp. 124. See Daniel Kevles, *In The Name of Eugenics* (Cambridge: Harvard University Press, 1985), pp. 20-21.

19. Genesis 30: 38-42. Matthew 7: 18-19.

20. Herbert Spencer, *The Principles of Biology* (New York: D. Appleton and Company, 1884) Vol. I, p. 183.

21. V. Kruta and V. Orel, "Johann Gregor Mendel," *Dictionary of Scientific Biography*, (New York: Scribner's, 1970-1980), Vol. IX, p. 277-283, as cited by Kevles, p. 41. Vitezslav Orel, *Gregor Mendel: The First Geneticist* (Oxford: Oxford University Press, 1996) p. 169.

22. Charles Darwin, *The Variation of Animals and Plants under Domestication* (London: John Murray, 1868; reprint, New York: D. Appleton & Co., 1883), vol. 2, p. 370.

23. Francis Galton, *Memories of my Life*, (London: Methuen & Co., 1908), pp. 46-47, 58. Kevles, p. 5.

24. Letter, Francis Galton to Samuel Galton, 5 December 1838 and Letter, Francis Galton to Samuel Galton 10 November 1838, as cited by

Kevles, p. 303, footnote 10. Copperplate prepared for Biometrika, circa 1888, at www.mugu.com.

25. Karl Pearson, *The Life, Letters, and Labours of Francis Galton* (Cambridge: Cambridge at the University Press, 1930), Vol. I, p. 232. Galton, *Memories of my Life*, p. 315.

26. Pearson, Vol. II, p. 340.

27. Galton, *Memories of my Life*, pp. 232, 325.

28. Francis Galton, *Finger Prints* (New York: Da Capo Press, 1965), p. iv.

29. Francis Galton, *Hereditary Genius: An Inquiry Into Its Laws And Consequences Second Edition* (London: Macmillan & Co., 1892; reprint, London: Watts & Co., 1950), p. 1. "Sir Francis Galton F.R.S. 1822-1911," at www.mugu.com.

30. Galton, *Hereditary Genius*, p. 1. Francis Galton, *Restrictions in Marriage* (American Journal of Sociology, 1906), p. 50.

31. Pearson, Vol. I, p. 32.

32. Pearson, Vol. IIIA, p. 348.

33. Personal scrap of paper: Galton Papers 138/1, UCL. Francis Galton, *Inquiries Into Human Faculty And Its Development* (London: JM Dent & Co., 1883), p. 17.

34. Personal scrap of paper.

35. Francis Galton, *Natural Inheritance* (London: Macmillan & Co., 1889), pp. 72-79. Francis Galton, "On The Anthropometric Laboratory at the Late International Health Exhibition," *Journal of the Anthropological Institute*, 1884: pp. 205-206.

36. August Weismann, *Essays Upon Heredity and Kindred Biological Problems* (Oxford: Clarendon Press, 1889), pp. 190-191.

37. Galton, *Natural Inheritance* (London: Macmillan, 1889), pp. 2, 192-197. Francis Galton, "Regression Towards Mediocrity in Hereditary Stature," *Journal of the Anthropological Institute* (1885), p. 261. See Francis Galton, "A Diagram of Heredity," *Nature* (1898).

38. Galton, *Hereditary Genius*, p. xviii.

39. Galton, *Hereditary Genius*, p. xx.

40. Francis Galton, "Index To Achievements of Near Kinsfolk of Some Of The Fellows Of The Royal Society" (Unrevised proof, 1904 papers), p. 1: UCL.

41. Pearson, vol. IIIA, p. 349.

42. Francis Galton, "Eugenics: Its Definition, Scope, and Aims," *The American Journal of Sociology* Vol. X, No. 1 (July 1904).

43. "Notes On The Early Days Of The 'Eugenics Education Society,'" p. 1: Wellcome Library SA/EUG/B11.

CHAPTER THREE

1. Gary B. Nash, *Red, White, and Black: The Peoples of Early America* (Englewood Cliffs, NJ: Prentice-Hall, Inc., 1974), pp. 168-169, 186. See Library of Congress, *Images of African-American Slavery and Freedom* at www.loc.gov.

2. Daniel J. Kevles, *In The Name of Eugenics,* (Cambridge, MA: Harvard University Press, 1985), p. 21. Mark H. Haller, *Eugenics: Hereditarian Attitudes in American Thought* (New Brunswick, NJ: Rutgers University Press, 1963), pp. 37-38.

3. Michael W. Perry, ed., *The Pivot of Civilization: In Historical Perspective* (Seattle, WA: Inkling Books, 2001), p. 31.

4. Israel Zangwill, "The Melting Pot: Drama in Four Acts" (New York: The Macmillan Company, 1909; reprint, 1919), pp. 215-216.

5. U.S. Department of Commerce, *Historical Statistics of the United States, Colonial Times to 1970,* (Washington D.C.: U.S. Department of Commerce, 1976).

6. See Paula Mitchell Marks, *In a Barren Land: American Indian Dispossession and Survival* (New York: William Morrow and Company, Inc., 1998). See Carey McWilliams, *North From Mexico: The Spanish-Speaking People of the United States,* (New York: Greenwood Press, 1968), pp. 51, 112-113. See Dr. David Pilgrim, "Jim Crow: Museum of Racist Memorabilia" at www.ferris.edu. See Immigration and Naturalization Service, *Chinese Exclusion Act of May 6, 1882 (22 Statutes-at-Large 58)* at www.ins.usdoj.gov. See Immigration and Naturalization Service, *Act of April 29, 1902 (32 Statutes-at-Large 176)* at www.ins.usdoj.gov.

7. Edward Alsworth Ross, "The Value Rank of the American People," *The Independent,* pp. 57, 1063.

8. National Association for the Advancement of Colored People, *Thirty Years of Lynching in the United States, 1889-1918* (New York: NAACP, 1919; reprint, New York: Negro Universities Press, 1969) pp. 7, 30-31, 45, 51, 58, 70.

9. Dr. Cecil E. Greek, Lecture Notes, *The Positive School: Biological and Psychological Factors* at www.criminology.fsu.edu.

10. Author's interview with Robin Walsh, Local History Librarian with SUNY Ulster, 13 November 2002. See Alf Evers, *Woodstock: History of an American Town,* (Woodstock, New York: The Overlook Press, 1987).

11. Richard L. Dugdale, *The Jukes* (New York: Putnam, 1910), pp. 1-15. "Bad Seed or Bad Science?" *New York Times,* 8 February 2003. See Oscar C. McCulloch, "The Tribe of Ishmael: A Study In Social Degradation," *Proceedings of the National Conference of Charities and Correction* (Boston: George H. Ellis, 1888), p. 154. See Norbert Vogel, "Die Gippe Delta," *Ziel und Weg,* vol. 7 (1937), no. 4. pp. 85-88. See Dr. Daniel R. Brower, "Medical Aspects of Crime," *Journal of the American Medical Association* vol. 32 (1899), p. 1283.

12. Dugdale, pp. 62, 65-66, 72. Richard L. Dugdale, "Origin of Crime in Society," *The Atlantic Monthly,* Vol. 48, Issue 288 (October 1881), p. 462.

13. Edward S. Morse, "Natural Selection and Crime," *Popular Science Monthly,* Vol. 41 (1892), pp. 433-446, as cited by Elof Alex Carlson, *The Unfit* (Cold Spring Harbor, NY: Cold Spring Harbor Press, 2001), p. 171.

14. Diane B. Paul, *Controlling Human Heredity* (Atlantic Highlands, NJ: Humanities Press International, Inc., 1995), p. 44. Carlson, p. 172. McCulloch, pp. 154-155. The Columbia Encyclopedia, 6th Ed., s.v. "Jackson Whites."

15. McCulloch, pp. 154-155.

16. Vitezslav Orel, *Gregor Mendel: The First Geneticist* (Oxford: Oxford University Press, 1996) pp. 2, 256-257.

17. Orel, pp. 99, 102, 104-105, 120-121.

18. Orel, pp. 270-271. Carlson, p. 137.

19. Orel, pp. 283-288, 291. Caleb Saleeby, "The Discussion of Alcoholism at the Eugenics Congress," *British Journal of Inebriety,* October 1912, p. 6.

20. Letter, Francis Galton to William Bateson, 8 September 1904: Galton Papers, University College London 245/3. Letter, Francis Galton to William Bateson, 12 June, 1904: Galton Papers, University College London 245/3.

21. Karl Pearson and Ethel M. Edlerton, *A Second Study of the Influence of Alcoholism on the Physique and Ability of the Offspring* (London: Dulau and Co., 1910) pp. 39-40.

22. Galton to Bateson, 8 September 1904. Francis Galton, *Index To Achievements of Near Kinsfolk* (Unrevised proof, 1904), p. iii: Galton Papers, University College London 245/3.

23. Francis Galton, *Restrictions in Marriage* (American Journal of Sociology, 1906), p. 3. Francis Galton, *Memories of My Life* (London: Methuen & Co., 1908), p. 310.

24. Galton, *Restrictions in Marriage,* pp. 7, 12-13.

25. Galton, *Memories, p.* 322. "Eugenics: Its Definition, Scope, and Aims," *The American Journal of Sociology* Vol. X, No. 1.

26. John Franklin Bobbitt, "Practical Eugenics," *The Pedagogical Seminary* vol. XVI (1909), p. 388.

27. Bobbitt, pp. 385, 387, 391.

28. Bobbitt, p. 388. Madison Grant, *The Passing of the Great Race* (New York: Charles Scribner's Sons, 1936), p. 167.

29. Lothrop Stoddard, *The Rising Tide of Color* (New York: Charles Scribner's Sons, 1926), p. 165. Grant, p. 65.

30. Stoddard, pp. 165-166, 167.

31. Grant, pp. 19-20, 188-212.
32. Grant, pp. 29, 60-64.
33. Harry H. Laughlin, secretary, *Bulletin No. 10A: The Report of the Committee to Study and to Report on the Best Practical Means of Cutting Off the Defective Germ-Plasm in the American Population* (Cold Spring Harbor, NY: Cold Spring Harbor, 1914), p. 16.
34. Grant, p. 18.
35. Biography of Andrew Carnegie at www.carnegie.org. Eugenics Record Office, "Official Record of the Gift of the Eugenic Record Office" (Cold Spring Harbor, NY: Cold Spring Harbor), p. 3.
36. Eugenics Record Office, "Official Record of the Gift of the Eugenic Record Office," pp. 5-6, 12.
37. See Bleecker Van Wagenen, Chairman, *Preliminary Report of the Committee of the Eugenic Section of the American Breeder's Association to Study and to Report on the Best Practical Means for Cutting Off the Defective Germ-Plasm in the Human Population*, ABA. See Laughlin, *Bulletin No. 10A.*
38. E. Carlton MacDowell, "Charles Benedict Davenport, 1866-1944: A Study of Conflicting Influences," *BIOS* vol. XVII no. 1, pp. 4, 8.
39. MacDowell, p. 5.
40. MacDowell, pp. 4-7.
41. MacDowell, pp. 4-5.
42. MacDowell, p. 5.
43. MacDowell, pp. 8, 10.
44. MacDowell, p. 12. Carnegie Institution of Washington Administrative Files, *Biography of Charles Davenport*, pp. 1-2.
45. MacDowell, pp. 19, 27. See also autographed photograph, c. 1928 in March 1944 Eugenical News.
46. MacDowell, pp. 8, 14, 33. Kevles, p. 52.
47. Letter, George Macon to Charles B. Davenport, 24 June 1899: APS. Letter, C.H. Walters to Charles B. Davenport, 24 May, 1898: APS B-D27. Letter, American Net & Twine Co. to Charles B. Davenport, 27 July 1899: APS B-D27. Letter, American Net & Twine Co. to Charles B. Davenport, 1 August, 1899: APS B-D27. Letter, University of Minnesota to Charles B. Davenport, 1 September, 1898: APS B-D27.
48. Letter, Walter Rankiss to Charles B. Davenport, 6 June 1898: APS B-D27. Letter, Rudolph Hering to Charles B. Davenport, 28 March, 1898: APS B-D27. Letter, Katherine Hobach to Franklin Hooper, 16 April, 1898: APS B-D27. Letter, C.O. Townsend to Charles B. Davenport, 2 April, 1898: APS B-D27. Letter, Dudley Greene to Charles B. Davenport, 11 May, 1898: APS B-D27. Letter, C.O. Townsend to Charles B. Davenport, 14 June, 1898: APS.
49. Letter, Francis Galton to Charles B. Davenport,

6 April, 1897: APS: B-D27 Galton, Sir Francis. Letter, Francis Galton to Charles B. Davenport, 5 May, 1897: APS: B-D27 Galton, Sir Francis.
50. Francis Janet Hassencahl, "Harry H. Laughlin, "Expert Eugenics Agent" for the House Committee on Immigration and Naturalization" (Ph. D. diss., Case Western Reserve University, 1970), p. 53. Letter, Francis Galton to Charles B. Davenport, 20 October, 1899: APS: B-D27 Galton, Sir Francis. Letter, Francis Galton to Charles B. Davenport, 19 November, 1903: APS: B-D27 Galton, Sir Francis.
51. See *State Laws Limiting Marriage Selection*, Eugenics Record Office (Cold Spring Harbor: Cold Spring Harbor, 1913), pp. 31-36. Also see Charles B. Davenport, *Race Crossing In Jamaica*, (Washington: Carnegie Institute of Washington, 1929). Charles B. Davenport, "Heredity and Race Eugenics," p. 10: APS: B-D27.
52. Charles B. Davenport, *Heredity in Relation to Eugenics* (New York: Arno Press & The New York Times, 1972), pp. 213, 214, 218.
53. Stoddard, p. 165.
54. Letter, Charles B. Davenport to Professor V. L. Kellogg, 30 October 1912: APS: B-D27- Kellogg, Professor V.L.
55. Margaret Sanger, *Margaret Sanger: An Autobiography*, (New York: W. W. Norton & Company, 1938; reprinted by Dover Publications, Inc., 1971) p. 374.
56. Letter, Charles B. Davenport to Franklin Hooper, 21 April 1902: APS B-D27 Cold Spring Harbor Beginnings Correspondence #3.
57. Letter, Charles B. Davenport to the Trustees of the Carnegie Institution, 5 May 1902: APS B-D27 Cold Spring Harbor Beginnings Correspondence #3.
58. Charles B. Davenport, "A Summary of Progress in Experimental Evolution," p. 5: APS B-D27 Cold Spring Harbor Beginnings Correspondence #2. Letter, Franklin Hooper to Charles B. Davenport, 23 May 1902: APS B-D27 Cold Spring Harbor Beginnings Correspondence #3.
59. Letter, Charles B. Davenport to Henry Osborn, 30 May 1902: APS B-D27 Cold Spring Harbor Beginnings Correspondence #3. Letter, Henry Osborn to Charles B. Davenport, 25 July 1902: APS B-D27 Cold Spring Harbor Beginnings Correspondence #3. Letter, Franklin Hooper to Charles Wolcott, 24 July 1902: APS B-D27 Cold Spring Harbor Beginnings Correspondence #3.
60. Davenport to Osborn, 30 May 1902.
61. Osborn to Davenport, 25 July, 1902. Davenport, "A Summary of Progress," pp. 4-5.
62. MacDowell, pp. 19-21. Letter, Francis Galton

to Charles B. Davenport, 28 September 1902: APS: B-D27 Galton, Sir Francis. Letter, Charles B. Davenport to Trustees of the Carnegie Institution, 5 March 1903: APS B-D27 Cold Spring Harbor Beginnings Correspondence #3.

63. Letter, Charles B. Davenport to John S. Billings, 3 May 1903: APS B-D27 Cold Spring Harbor Beginnings Correspondence #1. Davenport, "A Summary of Progress," pp. 13-14.

64. Letter, Charles Davenport to Madison Grant, 3 May 1920: APS B-D27 Grant, Madison #3. See Stoddard, pp. xxix-xxxi, 306-308.

65. Davenport to Billings, 3 May 1903.

66. Davenport to Billings, 3 May 1903.

67. Davenport to Billings, 3 May 1903. MacDowell, p. 19.

68. W. M. Hays, *The American Breeders Association to its Parent, The Association of American Agricultural Colleges and Experiment Stations, Greetings* (circa 1910): Truman. American Breeders' Association, "Minutes," *First Annual Meeting*, 1903, p. 1-2: APS.

69. John H. Noyes, *Essay on Scientific Propagation*, (Oneida, NY: Oneida Community, 1872), section 2, section 15.

70. Author's interview with National Weather Service, 1 October 2002. American Breeders' Association, "Minutes of First Annual Meeting, St. Louis, Missouri," p. 4: ABA. American Breeders' Association, "Constitution and By-Laws of the American Breeders' Association": ABA. American Breeders' Association, "Committees and Their Specific Duties," *Annual Report, American Breeders' Association*, vol. II (1906), p. 11.

71. Charles B. Davenport, secretary, "Report of Committee on Eugenics," *American Breeders Association* (Washington D.C.: American Breeders Association, 1911) vol. VI, pp. 92, 93, 94.

72. Willet M. Hays, "Constructive Eugenics," *The American Breeders Magazine*, Vol. III, No.1 (1912).

73. MacDowell, p. 24. Letter, John Billings to Charles Walcott, 23 December 1903: APS BD27 Cold Spring Harbor Beginnings Correspondence #1. Biography of Andrew Sledd, President of the University of Florida at www.president.ufl.edu. History of Northwestern University Library at www.library.northwestern.edu.

74. Billings to Walcott 23 December, 1903. Letter, Charles Davenport to John Billings, 6 February 1904: APS BD27 Cold Spring Harbor Beginnings Correspondence #2.

75. Billings to Walcott 23 December, 1903. Biography of John Shaw Billings at www.arlingtoncemetery.com.

CHAPTER FOUR

1. Letter, Charles B. Davenport to John S. Billings, 6 February 1904: APS: Cold Spring Harbor Beginnings Correspondence #1. E. Carleton MacDowell, "Charles Benedict Davenport, 1866-1944. A Study of Conflicting Influences," *BIOS* vol. XVII, no. 1, p. 24.

2. MacDowell, p. 24. Carnegie Institution of Washington, *Announcement of Station for Experimental Evolution* (Washington: Carnegie Institution of Washington, 1905), pp. 2-3: CSH: CIW Administrative Files: Dept. of Genetics-Biological Laboratory Plans for Unified Operation.

3. *Announcement of Station*, p. 4.

4. Letter, Charles B. Davenport to Francis Galton, 27 October 1905: APS. Letter, Francis Galton to Charles B. Davenport, 21 November 1905: APS: BD27- Galton, Sir Francis.

5. Charles B. Davenport, "Annual Reports of the Station for Experimental Evolution," *Carnegie Institution Year Book*, (1908) no. 7, p. 90. MacDowell, p. 26.

6. Charles B. Davenport, secretary, "Report of Committee on Eugenics," *American Breeders' Association Annual Report* (1911) vol VI, pp. 92-94. See also Bleecker Van Wagenen, chairman, *Preliminary Report of the Committee of the Eugenic Section of the American Breeder's Association to Study and to Report on the Best Practical Means for Cutting Off the Defective Germ-Plasm in the Human Population*: ABA.

7. Letter, Alexander Graham Bell to Charles B. Davenport, 15 April 1909: APS B:D27 - Alexander Graham Bell #4. Letter, Alexander Graham Bell to Charles B. Davenport, 14 May 1909 APS B:D27 - Alexander Graham Bell #4. Davenport, "Annual Reports of the Station for Experimental Evolution," p. 87.

8. Charles B. Davenport, *Heredity In Relation To Eugenics* (New York: Henry Holt & Company, 1911; reprint, New York: Arno Press Inc., 1972), p. 260. Harry Laughlin, secretary, *Report of the Committee to Study and to Report on the Best Practical Means of Cutting Off the Defective Germ-Plasm in the American Population* (Cold Spring Harbor: Cold Spring Harbor, 1914), p. 16.

9. McDowell, pp. 25-26.

10. Davenport, *Heredity in Relation to Eugenics*, p. 271. Davenport, "Report of Committee on Eugenics," pp. 91, 92.

11. Davenport, "Report of Committee on Eugenics" (1906), pp. 92-93.

12. McDowell, p. 29.

13. Maury Klein, *The Life and Legend of E.H. Harriman*, (Chapel Hill, NC: The University of North Carolina Press, 2000), p. 118, 152, 182-183, 184, 218-219, 357. Letter, William Loeb to C. Hart Merriam, 28 May 1907: APS.

14. Klein, pp. 6, 440-441.
15. "Death of Mrs. Rumsey," *Eugenical News*, vol. XIX (1934), p. 106. McDowell, p. 29. Klein, p. 299.
16. Klein, p. 8. McDowell, p. 29.
17. McDowell, p. 29.
18. Eugenics Record Office, *Official Record of the Gift of the Eugenics Record Office, Cold Spring Harbor, Long Island, New York by Mrs. E.H. Harriman to the Carnegie Institution of Washington and of its Acceptance by the Institution* (Cold Spring Harbor, New York: Eugenics Record Office, 1918), pp. 19, 21: CSH.
19. Letter, Charles B. Davenport to Mrs. E.H. Harriman, 23 May 1910: APS B:D27 Harriman, Mrs. E #1.
20. Davenport to Harriman, 23 May 1910.
21. See Davenport to Harriman, 23 May 1910.
22. Davenport to Harriman, 23 May 1910. Letter, Charles B. Davenport to Mrs. E.H. Harriman, 10 October, 1910: APS B:D27 APS B:D27 Harriman, Mrs. E #1.
23. Davenport to Harriman, 23 May 1910.
24. Davenport to Harriman, 23 May 1910.
25. Letter, Charles B. Davenport to Mrs. E.H. Harriman, 20 July, 1920: APS B:D27 APS B:D27 Harriman, Mrs. E #1.
26. See O.M. Means, *Kirksville, Missouri: Its Business and its Beauties as seen through the Camera* (Journal Print Co, 1900) p. 1-2, 16; see Wallin Directory Company, *Kirksville City Directory*, (Quincy, Illinois: Hoffman Printing Co., 1899), p. 1.
27. P. O. Selby, *History of the First Christian Church (Disciples of Christ) of Kirksville, Missouri* (1964): Truman E-1-1:10. P. O. Selby, *One Hundred Twenty-Three Biographies of Deceased Faculty Members* (Northeast Missouri State Teachers College, 1963), pp. 47-48.
28. Selby, *History of the First Christian*. Interviews with Mrs. Harold McClure, as cited by Frances Janet Hassencahl, "Harry H. Laughlin, 'Expert Eugenics Agent' for the House Committee on Immigration and Naturalization, 1921 to 1931." (Ph. D. diss., Case Western Reserve University, 1970), pp. 45-46. Mark H. Laughlin, *A Reverie: or One Day in a Woman's Life* (Honolulu, HI, n.d.), p. 18-19: Truman E-1-1:10
29. Laughlin, *A Reverie: or One Day in a Woman's Life*. Charles B. Davenport, "Harry Hamilton Laughlin 1880-1943," *Eugenical News* Vol. XXVIII (1943), p. 43. Hassencahl, pp. 42-43. Private Papers of Mrs. George Laughlin as cited by Hassencahl, p. 45. Interview with Mrs. McClure, as cited by Hassencahl, p. 45.
30. Laughlin papers as cited by Hassencahl, pp. 49-50.
31. Interview with Mark Laughlin, cited by Hassencahl, pp. 50-51.
32. Harry H. Laughlin, unpublished manuscript, "World Government: The Structure and Functioning of a Feasible Civil Government of the Earth": Truman B-5-1:10. Harry H. Laughlin, unpublished manuscript, "Chapter II: Text: The Proposed World Constitution": Truman B-5-2B:7. Harry H. Laughlin, unpublished manuscript, "The Principles of Nation-Rating": Truman B-5-1:6. Letter, Harry H. Laughlin to H.G. Wells, 19 February 1921: Truman B-5-4B:12.
33. Letter, Hamilton Fish Armstrong to Harry H. Laughlin, 11 June 1941: Truman C-4-5:9. Letter, Embajada De Colombia to Harry H. Laughlin: 4 June 1941, Truman C-4-5:9. Letter, H.R. Waddell to Harry H. Laughlin, 24 June 1941: Truman C-4-5:9. Letter, Harry H. Laughlin to Ida J. Dacus, 29 May 1941: Truman C-4-5:9. Letter, William Allan to Harry H. Laughlin, 26 June 1941: Truman C-4-5:9. Letter, Mrs. Anthony Conrad Eiser to Harry H. Laughlin, 30 July 1941: Truman C-4-5:9. Letter, Andres Pastoriza to Harry H. Laughlin, 5 June 1941: Truman C-4-5:9. Letter, Paul Popenoe to Harry H. Laughlin, 9 June 1941: Truman C-4-5:9. Letter, Verna B. Grimm to Harry H. Laughlin, 9 June 1941: Truman C-4-5:9. Letter, G. Burke to Harry H. Laughlin, 10 June 1941: Truman C-4-5:9. Letter, W. E. Rendell to Harry H. Laughlin, 5 June 1941: Truman C-4-5:7. Letter, Francisco Castillo Najera to Harry H. Laughlin, 6 June, 1941: Truman C-4-5:7. Letter, Luis Fernandez to Harry H. Laughlin, 10 June 1941: Truman C-4-5:7. Letter, Arturo Lares to Harry H. Laughlin, 4 June 1941: Truman C-4-5:7. Letter, Secretary to Mr. Fosdick to Harry H. Laughlin, 12 June 1941: Truman C-4-5:9. Letter, S. Shepard Jones to Harry H. Laughlin, 12 June 1941: Truman C-4-5:9. Letter, Henry Allen Moe to Harry H. Laughlin, 20 September 1932: Truman C-2-2:11; see also "Conquest by Immigration (Sent to the following)": Truman C-4-3:1.
34. Letter, Harry H. Laughlin to Madison Grant, 16 January 1928: Truman C-2-5:11. Letter, Harry H. Laughlin to Dr. Domingo F. Ramos, 23 September 1927: Truman C-2-5:11. Letter, Harry H. Laughlin to Madison Grant, 26 January 1928: Truman C-2-5:11. Letter, Harry H. Laughlin to Charles B. Davenport, 7 April 1928: Truman C-2-5:11. Letter, G.L.B. to Mr. Carr, 19 January 1928: State Department 59.250.22.33.7 box 6484. Letter, Harry H. Laughlin to Secretary of State Frank B. Kellogg, 23 March 1928: State Department 59.250.22.11.2 box 5502. Letter, Harry H. Laughlin to President Calvin Coolidge, 28 December 1927: State Department 59.250.22.33.7 box 6484. Letter, Husband to

Secretary of State Frank B. Kellogg, 7 April 1928: State Department 59.250.22.11.2 box 5502. Letter, Carr to Secretary of State Frank B. Kellogg, 13 April 1928: State Department 59.250.22.11.2 box 5502. Letter, Carr to Secretary of State Frank B. Kellogg, 12 April 1928: State Department 59.250.22.11.2 box 5502. Letter, Carr to Secretary of State Frank B. Kellogg, 18 April 1928: State Department 59.250.22.11.2 box 5502. Letter, W.H. Williams to Harry H. Laughlin, 8 June 1921: Truman E-2-5:5. Letter, Acting Secretary of the Carnegie Endowment for International Peace to Harry H. Laughlin, 28 October 1919: Truman E-2-5:18.

35. Laughlin papers, cited by Hassencahl, p. 50.
36. Letter, Harry H. Laughlin to Charles B. Davenport, 17 May 1907: CSH Laughlin Correspondence. Letter, Harry H. Laughlin to Charles B. Davenport, 30 May, 1907: CSH Laughlin Correspondence.
37. Hassencahl, p. 54.
38. Davenport to Harriman, 20 July 1910. Davenport to Harriman, 10 October 1910.
39. Davenport to Harriman, 10 October 1910.
40. Harry H. Laughlin, secretary, *Bulletin No. 10B: The Report of the Committee to Study and to Report on the Best Practical Means of Cutting Off the Defective Germ-Plasm in the American Population* (Cold Spring Harbor: Cold Spring Harbor, 1914), p. 145: CSH. Harry H. Laughlin, secretary, *Bulletin No. 10A: The Report of the Committee to Study and to Report on the Best Practical Means of Cutting Off the Defective Germ-Plasm in the American Population* (Cold Spring Harbor: Cold Spring Harbor, 1914), pp. 46-47, 58: CSH.
41. Davenport to Harriman, 20 July, 1910. Harry H. Laughlin, "Report On The Organization and the First Eight Months' Work of the Eugenics Record Office," *American Breeders Magazine*, no. 2, vol. II (1911).
42. Laughlin, "Report On The Organization and the First Eight Months' Work of the Eugenics Record Office."
43. Report on the Organization and the First Eight Months' Work of the ERO, by Laughlin, ABA reprint No. 2 Vol II, 1911 pp. 1-2.
44. Laughlin, "Report On The Organization and the First Eight Months' Work of the Eugenics Record Office."
45. Laughlin, "Report On The Organization and the First Eight Months' Work of the Eugenics Record Office." Carnegie Institution of Washington, *Year Book No. 10* (Washington: Carnegie Institution of Washington, 1912), p. 80. See The Columbia Encyclopedia, 6th ed., s.v. "Huntington's disease."
46. Laughlin, "Report On The Organization and the First Eight Months' Work of the Eugenics

Record Office." Eugenics Record Office, *Report for Six Months Ending March 31, 1911*, CSH, p. 1. *Historical Overview: Development of Public Responsibility for the Mentally Ill in Massachusetts* (article on-line: accessed 19 September 2002); available from www.1856.org. See Charles B. Davenport and David Weeks, *A First Study of Inheritance in Epilepsy: Eugenics Record Office Bulletin No. 4* (Cold Spring Harbor: Cold Spring Harbor, NY, 1911), p. 5: CSH.
47. Davenport and Weeks, p. 2. Eugenics Records Office, "Method for Studying the Hereditary History of Patients as used at the New Jersey State Village for Epileptics, New Jersey State Village for Epileptics Schedules and Forms," circa 1911, p. 6: APS ERO Series I.
48. "Method for Studying the Hereditary History of Patients as used at the New Jersey State Village for Epileptics," pp. 2, 8.
49. Davenport, *Heredity In Relation To Eugenics*, pp. 257-258. See Van Wagenen, p. 4. Also see Paul Popenoe and Roswell Hill Johnson, *Applied Eugenics*, rev. ed. (New York: Macmillan Company, 1935) pp. 396-397 as compared to Frederick Osborn, *Preface to Eugenics* (New York: Harper & Brothers, 1940) p. 14. Also see J. David Smith, *Minds Made Feeble: The Myth and Legacy of the Kallikaks* (Rockville, MD: Aspen Systems Corporation, 1985) pp. 21-36, 83-114.
50. Davenport and Weeks, pp. 2, 19, 29-30. Letter, Charles B. Davenport to Mrs. E. H. Harriman, 18 December 1911: APS B:D27 - Harriman, Mrs. E.H. #3.
51. Davenport and Weeks, pp. 9-10.
52. Davenport and Weeks, p. 1.
53. Davenport and Weeks, p. 30.
54. Laughlin, "Report On The Organization and the First Eight Months' Work of the Eugenics Record Office," pp. 109-110.
55. Laughlin, "Report On The Organization and the First Eight Months' Work of the Eugenics Record Office," p. 110. Also see Albert Edward Wiggam and Stephen S. Visher, "Needed: Faculty Family Allowances," *Eugenics*, Vol. III, No. 12 (December 1930), pp. 445-446. Also see discussion, "The Faculty Birth Rate: Should It Be Increased?," *Eugenics*, Vol. III, No. 12 (December 1930), pp. 458-460.
56. *Official Record of the Gift of the Eugenics Record Office*, pp. 3, 21. Letter, David Starr Jordan to Mrs. E.H. Harriman, 22 July, 1910: APS B:D 27 - Harriman, Mrs. E. #1. Origins of Cold Spring Harbor. Letter, Alexander Graham Bell to Charles B. Davenport, 9 March 1915: APS B:D27 Alexander Graham Bell #7. Letter, Charles B. Davenport to Dr. William H. Welch, 1 March 1915: APS B:D27 Alexander Graham Bell #7. "A County Survey," *Eugenical News* Vol. I (!916), p. 24.

57. Van Wagenen, p. 2. Laughlin, *Bulletin No. 10A*, p. 5.
58. Laughlin, *Bulletin No. 10A*, pp. 5, 6, 12, 17. Dr. Lucien Howe, "Presidential Address of the Eugenics Research Association: The Control of Law of Hereditary Blindness," *Eugenical News*, July 1928, p. 6. See Letter from Lucien Howe to Dr. Best, 4 October 1927: APS Series V.
59. Laughlin, *Bulletin No. 10A*, pp. 7, 8.
60. Laughlin, *Bulletin No. 10A*, pp. 15-16. Davenport, *Heredity In Relation To Eugenics*, p. 221.
61. Laughlin, *Bulletin No. 10A*, p. 15. Van Wagenen, p. 5.
62. Laughlin, *Bulletin No. 10A*, p. 15. Davenport, *Heredity In Relation To Eugenics*, pp. 221-222.
63. Laughlin, *Bulletin No. 10A*, p. 15.
64. Laughlin, *Bulletin No. 10A*, pp. 8, 9.
65. Laughlin, *Bulletin No. 10B*, p. 74, 75. Also see Edwin Black, *The Transfer Agreement*, (Washington, D.C.: Dialog Press, 1999) pp. 4, 26.
66. Laughlin, *Bulletin No. 10B*, pp. 74, 75.
67. Laughlin, *Bulletin No. 10A*, pp. 45-47, 53-56. Davenport, *Heredity In Relation To Eugenics*, p. 259. Van Wagenen, p. 7. Also see The Human Betterment Foundation, *Human Sterilization* (Pasadena: The Human Betterment Foundation, 1929). Also see Popenoe, pp. 150-151. Also see E.S. Gosney and Paul Popenoe, *Sterilization for Human Betterment* (New York: The Macmillan Company, 1929), pp. xv, 21, 31.
68. Laughlin, *Bulletin No. 10A*, pp. 45-46, 55.
69. Laughlin, *Bulletin No. 10A*, pp. 6, 13. Van Wagenen, p. 20. Karl Pearson and Ethel Elderton, *A Second Study of the Influence of Parental Alcoholism on the Physique and Ability of the Offspring* (London: Dulau and Co. Limited, 1910), pp. 39-40.
70. Van Wagenen, p. 13.
71. Laughlin, *Bulletin No. 10A*, p. 9.

CHAPTER FIVE
1. Martin W. Barr, *Mental Defectives* (Philadelphia: Blakiston, 1904; reprint, New York: Arno Press, 1973), p. 195-6. Mark H. Haller, *Eugenics: Hereditarian Attitudes in American Thought* (New Brunswick, New Jersey: Rutgers University Press, 1963), p. 48.
2. "Obituary: Dr. Harry C. Sharp: A Medical Leader," *The New York Times*, 1 November 1940. Elof Axel Carlson, *The Unfit: A History of a Bad Idea* (Cold Spring Harbor, NY: Cold Spring Harbor Laboratory Press, 2001), pp, 207, 208, 224. Dr. A. J. Ochsner, "Surgical Treatment of Habitual Criminals," *Journal of the American Medical Association* vol. XXXIII (1899), p. 867-868.
3. Dr. Harry C. Sharp, "The Severing of the Vasa Deferentia and its Relation to the

Neuropsychopathic Constitution," *New York Medical Journal*, 8 March 1902, p. 413; Dr. Daniel R. Brower, "Medical Aspects of Crime," *Journal of the American Medical Association*, vol. XXXII (1899), pp. 1282- 1287.
4. Sharp, p. 413. Carlson, p. 214.
5. Sharp, p. 412.
6. Sharp, pp. 413-414.
7. "An Act for the Relief of the Poor," 30 January 1824: Indiana Historical Society. Also see Oscar C. McCulloch, "The Tribe of Ishmael: A Study In Social Degradation," *Proceedings of the National Conference of Charities and Correction* (Boston: George H. Ellis, 1888), pp. 154-159.
8. McCulloch, pp. 154, 159.
9. McCulloch, pp. 154, 157-159. Carlson, p. 174.
10. Carlson, pp. 185-186, 188, 190.
11. Thurman B. Rice, "A Chapter In The Early History of Eugenics in Indiana," selected by Paul Popenoe, *Eugenical News* vol. XXXIII No 1-2 (March-June 1948), pp. 24-25.
12. Carlson, pp. 210-211. Rice, p. 27.
13. Carlson, pp. 218-219. Harry H. Laughlin, *Eugenical Sterilization in the United States* (Chicago: Psychopathic Laboratory of the Municipal Court of Chicago, 1922), p. 35.
14. Laughlin, p. 36.
15. Carlson, p. 211.
16. Laughlin, p. 15.
17. Bleecker Van Wagenen, chairman, *Preliminary Report of the Committee of the Eugenic Section of the American Breeder's Association to Study and to Report on the Best Practical Means for Cutting Off the Defective Germ-Plasm in the Human Population*, p. 18: ABA.
18. Laughlin, pp. 40-41.
19. Harry H. Laughlin, secretary, *Bulletin No. 10A: The Report of the Committee to Study and to Report on the Best Practical Means of Cutting Off the Defective Germ-Plasm in the American Population* (Cold Spring Harbor: Cold Spring Harbor, 1914), fold-out on "Sterilization Bills Introduced Into Legislatures, But Which Were Defeated or Have Not Yet Become Laws.": CSH.
20. Laughlin, *Eugenical Sterilization in the United States*, pp. 6,8. Laughlin, *Bulletin 10A*, fold-out on "Analysis of Existing Sterilization Laws, 1913."
21. Laughlin, *Eugenical Sterilization: 1926*, p. 10. Laughlin, *Bulletin 10A*, fold-out on "Analysis of Existing Sterilization Laws, 1913."
22. Laughlin, *Eugenical Sterilization in the United States*, pp. 8-9, 21. Laughlin, *Bulletin 10A*, fold-out on "Analysis of Existing Sterilization Laws, 1913." Laughlin, *Bulletin No. 10A*, fold-out on "Sterilization Bills Introduced Into Legislatures, But Which Were Defeated or Have Not Yet Become Laws."
23. Laughlin, *Eugenical Sterilization in the United*

States, pp. 23-24. Laughlin, *Bulletin 10A*, fold-out on "Analysis of Existing Sterilization Laws, 1913." William A. DeGregorio, *The Complete Book of U.S. Presidents: Third Edition*, (New York: Wing Books, 1991), pp. 416-417, 424-425. Entry number 64927, *The Columbia World of Quotations, 1996* (New York: Bartelby.com, 2001).

24. Laughlin, *Eugenical Sterilization in the United States*, pp. 25-26. Laughlin, *Bulletin 10A*, fold-out on "Analysis of Existing Sterilization Laws, 1913," fold-out continuation.

25. Laughlin, *Bulletin 10A*, fold-out on "Analysis of Existing Sterilization Laws, 1913," fold-out continuation. Van Wagenen, p. 15. Carlson, pp. 216, 226.

26. Van Wagenen, p. 18.

27. Van Wagenen, p. 18.

28. Van Wagenen, p. 18.

29. "Notes on the Early Days of the 'Eugenics Education Society'," unpublished manuscript, p. 11, 13: SA/EUG/B11 Wellcome Library.

30. Overview of Galton's life, at www.mugu.com. Daniel J. Kevles, *In The Name of Eugenics*, (Cambridge, MA: Harvard University Press, 1985), pp. 63-64. Leonard Darwin citation in Michael W. Perry, ed., *Eugenics and Other Evils* (Seattle, WA: Inkling Books, 2000), p. 23. C.W. Saleeby citation in Perry, p. 36. See "The International Eugenics Congress, An Event of Great Importance in the History of Evolution, Has Taken Place," *Journal of the American Medical Association* vol. LIX, no. 7, p. 555. See Dr. Caleb W. Saleeby, "The Discussion of Alcoholism at the Eugenics Congress," *British Journal of Inebriety*, October 1912, pp. 1, 2-3, 5-6. See Dr. Caleb W. Saleeby, "The House of Life: The Mental Deficiency Bill," July 23 1912. See Charles B. Davenport, "A Discussion of the Methods and Results of Dr. Heron's Critique," *Eugenics Record Office Bulletin No. 11: Reply to the Criticism of Recent American Work by Dr. Heron of the Galton Laboratory* (Cold Spring Harbor, NY: Eugenics Record Office, 1914), pp. 23-24. Carnegie Institution of Washington, *Announcement of Station for Experimental Evolution* (Washington: Carnegie Institution of Washington, 1905), pp. 4: APS: Davenport Beginnings of Cold Spring Harbor. The Eugenics Education Society, "Programme," *Problems in Eugenics Vol. II: Report of Proceedings of the First International Eugenical Congress* (Kingsway, W.C.: Eugenics Education Society, 1913), p. 1, 3, 5, 6-13.

31. "Programme," *Problems in Eugenics Vol. II*, p. 5. Jon Alfred Mjøen, "Harmonic and Disharmonic Racecrossing," *Eugenics in Race and State, Vol. II: Scientific Papers of the Second International Congress of Eugenics*, (Baltimore: Wilkins and Wilkins, 1923), pp. 58-60.

32. "London Letter," *Journal of the American Medical Association* vol. LIX (1912), p. 555. "Programme," *Problems in Eugenics Vol. II*, p. 2. Letter, Winston Churchill to unknown recipient, 27 May 1910: PRO- HO 144/1085/193548/1. Letter, William Borland to the Department of State, 25 March, 1912: NA: 59/250/22/10/3-5459 Doc. No. 540.1E1. Letter, Huntington Wilson, Acting Secretary of State, to William Borland: NA: 59/250/22/10/3-5459 Doc. No. 540.1E1. Letter, Philander Chase Knox to Lord Weardale, 28 February 1911: NA: 59/250/22/14/4-5656 Doc. No. 592.7B1/4.

33. Letter, Philander Chase Knox to Mr. Alfred Mitchell Innes, Charge d'affairs of Great Britain, 3 July 1912: NA: 59/250/22/10/3-5459 Doc. No. 540.1A1/2. Letter, Henry L. Stimson to Philander Chase Knox, 20 June 1912: NA: 59/250/22/10/3-5459 Doc. No. 540.1A1/1.

34. Letter, Philander Chase Knox to Governor Phillip L. Goldsborough, 20 June 1912: NA: 59/250/22/10/3-5459 Doc. No. 540.1A1/1. Letter, Philander Chase Knox to Governor Woodrow Wilson, 20 June 1912: NA: 59/250/22/10/3-5459 Doc. No. 540.1A1/1. Letter, Philander Chase Knox to Governor Walter R. Stubbs, 20 June 1912: NA: 59/250/22/10/3-5459 Doc. No. 540.1A1/1. Letter, Philander Chase Knox to Governor James B. McCreary, 20 June 1912: NA: 59/250/22/10/3-5459 Doc. No. 540.1A1/1. Letter, Philander Chase Knox to Dr. L.S. Rowe, President, American Academy of Political and Social Science, 20 June 1912: NA: 59/250/22/10/3-5459 Doc. No. 540.1A1/1. Letter, Philander Chase Knox to Professor H.W. Farnam, President, American Economic Association, 20 June 1912: NA: 59/250/22/10/3-5459 Doc. No. 540.1A1/1. Letter, Philander Chase Knox to Dr. W. W. Keen, President, American Philosophical Society, 20 June 1912: NA: 59/250/22/10/3-5459 Doc. No. 540.1A1/1. Letter, Philander Chase Knox to Dr. Reuben Peterson, President, American Gynecological Society, 20 June 1912: NA: 59/250/22/10/3-5459 Doc. No. 540.1A1/1. Letter, Philander Chase Knox to Dr. W. N. Bullard, President, American Neurological Association, 20 June 1912: NA: 59/250/22/10/3-5459 Doc. No. 540.1A1/1. Letter, Philander Chase Knox to Dr. W. Leslie Carr, President, American Pediatric Society, 20 June 1912: NA: 59/250/22/10/3-5459 Doc. No. 540.1A1/1. Letter, Philander Chase Knox to Dr. John B. Murphy, President, American Medical Association, 20 June 1912: NA: 59/250/22/10/3-5459 Doc. No. 540.1A1/1. Letter, Philander Chase Knox to Dr. Charles E. Bessey, President, American Association for the

454 || WAR AGAINST THE WEAK

Advancement of Science, 20 June 1912: NA: 59/250/22/10/3-5459 Doc. No. 540.1A1/1.
Letter, Philander Chase Knox to Dr. John H. Finley, President, American Social Science Association, 20 June 1912: NA: 59/250/22/10/3-5459 Doc. No. 540.1A1/1.
Letter, Philander Chase Knox to Professor C. E. Seashore, President, American Psychological Association, 20 June 1912: NA: 59/250/22/10/3-5459 Doc. No. 540.1A1/1.

35. Letter, Ira Remsen, President, National Academy of Sciences, to Philander Chase Knox, 24 June 1912: NA: 59/250/22/10/3-5459 Doc. No. 540.1A1/1. Letter, Henry L. Stimson to Philander C. Knox, 8 July 8 1912: NA: 59/250/22/10/3-5459 Doc. No. 540.1A1/1.

36. Joseph Frazier Wall, *Andrew Carnegie* (New York: Oxford University Press, 1970), pp. 644-645.

37. "The International Eugenics Congress." Saleeby, "The Discussion of Alcoholism at the Eugenics Congress," p. 6. Also see Saleeby, "The House of Life: The Mental Deficiency Bill."

38. Saleeby, "The House of Life: The Mental Deficiency Bill."

39. Saleeby, "The Discussion of Alcoholism," p. 6.

40. "The International Eugenics Congress."

41. Charles B. Davenport, *Heredity In Relation To Eugenics* (New York: Henry Holt & Company, 1911; reprint, New York: Arno Press Inc., 1972), p. 241.

42. Davenport, p. 67. "How Heredity Builds Our Lives," *Eugenical News*, Vol. XXVII (1942), p.53.

43. Davenport, pp. 216, 219.

44. Davenport, p. 222.

45. Davenport, p. 1; also see Letter, Charles B. Davenport to Professor V.L. Kellogg, 30 October 1912: APS B:D27 Kellogg, Vernon #3. Davenport, pp. 80-82.

46. Davenport, pp. 255-259.

47. "College Courses in Genetics and Eugenics," *Eugenical News* vol. 1 (1916), pp. 26-27.

48. Carnegie Institution of Washington, "ERO Schedule: Inquiry Into the Nature of Instruction Offered By Schools and Colleges in Eugenics (Not Sex-Hygiene) and Human Heredity": APS: ERO documents, Series X. Letter, Charles B. Davenport to Professor Irving Fisher, 8 February 1916: APS: BD27-Fisher #1.

49. Hamilton Cravens, *The Triumph of Evolution: The Heredity-Environment Controversy, 1900-1941*, (Baltimore: Johns Hopkins University Press, 1988), p. 53.

50. George William Hunter, *A Civic Biology: Presented in Problems* (New York: American Book Company, 1914), p. 263, as cited in Steven Selden, *Inheriting Shame* (New York:

Teachers College Press, 1999), p. 71. Selden, p. 61-69.

51. See Francis Galton, *Inquiries Into Human Faculty And Its Development* (London: JM Dent & Co, 1883), pp. 19-20. See Francis Galton, "On the Anthropometric Laboratory at the late International Health Exhibition," *Journal of the Anthropological Institute*, pp. 205-206, 214-218. James Cattell, "Mental Tests and Measurements," *Mind* (1890), pp. 378-380.

52. Theta H. Wolf, *Alfred Binet* (Chicago: The University of Chicago Press, 1973), pp. 21, 29, 71, 141, 162-165, 172, 177, 179-182, 183-185, 191, 201, 202, 207.

53. The Vineland Training School, "The Vineland Training School- History," at www.vineland.org. Charles B. Davenport and David F. Weeks, *Eugenics Record Office Bulletin No. 4: A First Study of Inheritance in Epilepsy* (Cold Spring Harbor, NY: Eugenic Record Office), pp. 4-5.

54. Henry H. Goddard, *The Kallikak Family: A Study in the Heredity of Feeble-Mindedness* (Vineland, New Jersey: 1913,) pp. vii, 101-110, 116-117.

55. Goddard, pp. 18, 29-30, 103.

56. Goddard, p. 53. Author's interview with James H. Wallace, Jr., director of Photographic Services at the Smithsonian Institution.

57. Goddard, p. 16.

58. Goddard, p. 84.

59. Goddard, p. 109.

60. Goddard, pp. 105-106, 118.

61. Wolf, p. 195. Author's interview with Merriam-Webster Corporation.

62. Letter, Henry H. Goddard to Charles B. Davenport, 25 July 1912, APS B:D27 Davenport - Goddard, Henry H. #4.

63. Henry H. Goddard, "Mental Tests and the Immigrant," *The Journal of Delinquency*, vol. II, no. 5 (September 1917), pp. 243-244. Goddard, *The Kallikaks*, p. 79.

64. Goddard, "Mental Tests and the Immigrant," pp. 249, 266-267.

65. "Mental Differences," *Eugenical News*, vol. 1 (1916), pp. 51-52. "News and Notes," *Eugenical News*, vol. 1 (1916), p. 52.

66. "Measuring Mentality," *Eugenical News*, vol. 1 (1916), p. 59. "The Municipal Psychopathic Clinic," *Eugenical News*, vol. 1 (1916), p. 55.

67. "Negro Efficiency," *Eugenical News*, vol. 1, (1916), p. 79.

68. Arthur H. Estabrook, "National Conference of Charities and Corrections," *Eugenical News*, vol. 1 (1916), pp. 42-43.

69. "The Binet Test in Court," *Eugenical News*, vol. 1 (1916), p. 55.

70. "Record Blank for Point Scale," *Eugenical News*, vol. 1 (1916), p. 56. "Autobiography of Robert Means Yerkes," in Carl Murchison, ed., *History of Psychology in Autobiography* (Worcester, MA:

Clark University Press, 1930), pp. 381-407. "Officers and Committee List of the Eugenics Research Association, January 1927": Truman: ERA Membership Records.

71. Daniel J. Kevles, "Testing the Army's Intelligence: Psychologists and the Military in World War I," *The Journal of American History*, Vol. 55, Issue 3 (Dec., 1968), p. 567-568, 571, 573. Robert M. Yerkes and Clarence S. Yoakum, *Army Mental Tests*, (New York: Henry Holt and Company, 1926), p. 2.

72. Carl C. Brigham, *A Study of American Intelligence* (Princeton, NJ: Princeton University Press, 1923), p. xxii. Examination Alpha, Test 8: Information- cited in Brigham, p. 29 and Diane B. Paul, *Controlling Human Heredity* (Atlantic Highlands, New Jersey: Humanities Press, 1995), p. 66. See United States Historical Census Data Browser at fisher.lib.virginia.edu/census/; Internet, for details on rural population.

73. Brigham, p. 29 and Paul, p. 66.

74. Brigham, pp. 48, 50.

75. Brigham, p. xxii. Raymond E. Fancher, *The Intelligence Men: Makers of the IQ Controversy* (New York: W. W. Norton & Company, 1985), pp. 139, 140.

76. Robert M. Yerkes, *Memoirs of the National Academy of Science*, (Washington D.C.: National Academy of Science, 1921), p. 790-791. Brigham, p. 152.

77. Fancher, pp. 102-103, 140. Columbia Encyclopedia, 6th ed., s.v. "Mental Retardation."

78. "News and Notes," *Eugenical News*, Vol. II (1917), p. 24.

79. Eugenics Research Association, *Active Membership Accession List* (Cold Spring Harbor, NY: Eugenics Research Association, 1922): Truman, ERA Membership Records. Brigham, pp. v-vii, xvii-xviii.

80. Brigham, pp. 174, 178, 180.

81. Brigham, p. 192.

82. Brigham, pp. 182, 210.

83. Nicholas Lemann, *The Big Test: The Secret History of the American Meritocracy*, (New York: Farrar, Straus and Giroux, 1999), p. 30-32.

84. See National Association for the Advancement of Colored People, *Thirty Years of Lynching in the United States, 1889-1918* (New York: NAACP, 1919; reprint, New York: Negro Universities Press, 1969) pp. 45, 70.

85. Kevles, "Testing the Army's Intelligence: Psychologists and the Military in World War I," pp. 576-577, 578.

86. Walter Lippmann, "The Mental Age of Americans," New Republic 32, no. 415 (November 15, 1922). Walter Lippmann, "The Mental Age of Americans," New Republic 32 no. 417 (November 29, 1922). Lewis M.

Terman, "The Great Conspiracy or the Impulse Imperious of Intelligence Testers, Psychoanalyzed and Exposed by Mr. Lippmann," New Republic 33 (December 27, 1922). Also see Ezekiel Cheever, *School Issues* (Baltimore: Warwick & York, Inc., 1924).

87. Henry H. Goddard, "Feeblemindedness: A Question of Definition," *Journal of Psycho-Asthenics*, vol. 33 (1928), p. 224.

88. Goddard, "Feeblemindedness: A Question of Definition," pp. 223, 224.

89. Carl C. Brigham, "Intelligence Tests of Immigrant Groups," *Psychological Review*, Vol. 37 (1929), p. 165.

CHAPTER SIX

1. The Race Betterment Foundation, *Proceedings of the First National Conference on Race Betterment* (Battle Creek, MI: The Race Betterment Foundation, 1914), p. xi. Kellogg Company, "Kellogg's Company History" at www.thekelloggcompany.co.uk. "Race Betterment Foundation and the Eugenics Registry," *Organized Eugenics*, (New Haven, CT: American Eugenics Society), 1931, p. 51. Also see "Brief Notes Made at Conference Held in Sacramento at the Request of the State Board of Control to Consider the Problem of Feeblemindedness, Insanity, and Epilepsy in Relation to Crime, Poverty and Inefficiency," unpublished manuscript, p. 5: California State Archives, Berkeley PD 72/227C: Box 5.

2. *Proceedings, First National Conference on Race Betterment*, pp. 431, 433, 447. Diane B. Paul, *Controlling Human Heredity* (Atlantic Highlands, NJ: Humanities Press International, 1995), p. 9.

3. Charles B. Davenport, "The Importance to the State of Eugenic Investigation," *Proceedings, First National Conference* p. 452.

4. Harry H. Laughlin, "Calculations on the Working Out of a Proposed Program of Sterilization," *Proceedings, First National Conference on Race Betterment*, p. 478.

5. Laughlin, p. 484, 490.

6. Professor Irving Fisher, "A Reply," *Official Proceedings of the Second National Conference on Race Betterment* (Battle Creek, MI: The Race Betterment Foundation, 1915), p. 68.

7. Eugenics Record Office, *First Meeting of the Board of Scientific Directors*, unpublished manuscript, circa December 1912: APS BD27 - Harriman, Mrs. E. H. #1. Johns Hopkins University, *Chronology of the Life of William Henry Welch* at www.medicalarchives.jhmi.edu.

8. *First Meeting of the Board of Scientific Directors.*

9. Letter, Alexander Graham Bell to Charles B. Davenport, 27 December 1912: Truman C-2-3:3.

10. Eugenics Research Association, *Officers and Committee List of the Eugenics Research Association - January 1927* (Cold Spring Harbor, NY: Eugenics Research Association, 1927): Truman, ERA Membership Records. Eugenics Research Association, *Active Membership Accession List* (Cold Spring Harbor, NY: Eugenics Research Association, 1922): Truman, ERA Membership Records.

11. *Active Membership Accession List*. Edwin Katzen-Ellenbogen, "The Detection of a Case of Simulation of Insanity By Means of Association Tests," *Journal of Abnormal Psychology* Vol. VI (1911), p. 19.

12. Madison Grant, *The Passing of the Great Race* (New York: Charles Scribner's Sons, 1936), pp. 50-51, 86, 89. Eugenics Record Office, *Official Record of the Gift of the Eugenics Record Office, Cold Spring Harbor, Long Island, New York by Mrs. E.H. Harriman to the Carnegie Institution of Washington and of its Acceptance by the Institution* (Cold Spring Harbor, New York: Eugenics Record Office, 1918), p. 33: CSH. *Officers and Committee List of the Eugenics Research Association- January 1927*.

13. Lothrop Stoddard, *The Rising Tide of Color* (New York: Charles Scribner's Sons, 1926), pp. 258, 259-260. *Active Membership Accession List*.

14. *Active Membership Accession List*.

15. Edwin Katzen-Ellenbogen, "The Mental Efficiency in Epileptics," *Epilepsia* Vol. 3 (Dec 1912), p. 504. Katzen-Ellenbogen, "The Detection of a Case of Simulation of Insanity By Means of Association Tests." Edwin Katzen-Ellenbogen, "A Critical Essay on Mental Tests in Their Relation to Epilepsy," *Epilepsia* Vol. 4 (1913), p. 130. American Men of Science (1914): NA: RG496/Box 457.

16. NA: RG496/Box 457. "Record of Marriage": NA: RG496/Box 457. Letter, Edwin Katzen-Ellenbogen to 7708 WCG, circa 13 April 1948: NA: RG496/ box 457. "Extract Copy: Review And Recommendations: NA: RG496/Box 457.

17. See "Photo of Edwin Katzen-Ellenbogen" at www.ushmm.org. "A Critical Essay on Mental Tests." Arrest photo of Katzen-Ellenbogen: NA: RG496/Box 457. "Testimony of Karl Hemrick Victor Berthold": NA: RG496/Box 457.

18. "Review of WC Section, Military Affairs Branch." *Active Membership Accession List*.

19. Letter, Olga Heide-Pilat to General Handy, 7 August 1951: NA: RG496/Box 457. See Testimony of Katzen-Ellenbogen: NA: RG496/290/59/14/1-5/Box434.

20. Sworn statement of Walter Hummelsheim: NA: RG496/290/59/14/1-5/Box 444.

21. Timeline of Rockefeller Foundation History at www.rockfound.org.

22. Letter, John D. Rockefeller Jr. to Charles B. Davenport, 27 January 1912: APS: B:D27

Davenport - J.D. Rockefeller. Letter, John D. Rockefeller Jr. to Charles B. Davenport, 27 March 1912: APS: B:D27 Davenport - J.D. Rockefeller. Letter, John D. Rockefeller Jr. to Charles B. Davenport, 2 April 1912: APS: B:D27 Davenport - J.D. Rockefeller. Letter, John D. Rockefeller Jr. to Charles B. Davenport, 8 May 1912: APS: B:D27 Davenport - J.D. Rockefeller. Timeline of Rockefeller Foundation History; see Biography of John D. Rockefeller Jr. at www.brown.edu.

23. Biography of John D. Rockefeller, Jr. Rockefeller to Davenport, 27 January 1912. Rockefeller to Davenport, 27 March 1912. Rockefeller to Davenport, 2 April 1912. Rockefeller to Davenport, 8 May 1912.

24. *First Meeting of the Board of Scientific Directors*.

25. Letter, Charles B. Davenport to Dr. William H. Welch, 1 March 1915: APS B:D27- Harriman, Mrs. E.H. #5. Also see Letter, Charles B. Davenport to Alexander Graham Bell, 5 March 5 1915: APS B:D27 - Alexander Graham Bell #7.

26. Davenport to Welch, 1 March 1915.

27. Letter, Charles B. Davenport to Alexander Graham Bell, 20 March 1915: APS B:D27 - Alexander Graham Bell #7.

28. See Letter, Charles B. Davenport to Mrs. E.H. Harriman, 13 February 1915: APS B:D27 - Harriman, Mrs. E.H. #4. "Conference on the Feebleminded at the Home of Mrs. E.H. Harriman," meeting agenda with notations: APS B:D27 - Harriman, Mrs. E.H. #4.

29. Letter, Robert W. Hebberd to Mrs. E.H. Harriman, 28 October 1913: APS B:D27 - Harriman, Mrs. E.H. #3.

30. "A County Survey," *Eugenical News*, Vol. I (1916) p. 24.

31. Memorandum on Immigration enclosed with letter, Charles B. Davenport to Madison Grant, 6 January 1921: APS B:D27 - Grant, Madison #3.

32. National Committee on Prisons and Prison Labor, *The National Committee on Prisons and Prison Labor - Its Origin, Purpose and Present Activities* (New York: National Committee on Prisons and Prison Labor, 1915) pp. 2-3, 4, 5: APS B:D27 Davenport - Nat'l Committee on Prisons & Prison Labor. "Field Work in a Police Department," *Eugenical News*, Vol. II (1917) p. 21. "Field Workers Appointed," *Eugenical News*, Vol. II (1917), p. 80.

33. "New York State Commission on the Mentally Deficient," *Eugenical News*, Vol. I (1916) pp. 6-7. Letter, Davenport to Harriman, 13 February 1915. "Wanted," *Eugenical News*, Vol. I (1916) p. 48.

34. "Hospital Development Commission," *Eugenical News* Vol. 2 (1917), p. 59.

35. "News and Notes," *Eugenical News* Vol. II

(1917), p. 24. "Field Workers' Returns," *Eugenical News* Vol. I (1916), p. 3. "Field Workers' Returns," *Eugenical News* Vol. I (1916), p. 9. "News and Notes," *Eugenical News* Vol. I (1916), p. 18. "Work of a Field Worker," *Eugenical News* Vol. II (1917), p. 46.

36. Letter, A.G. Smith to C.L. Goodrich, 14 November 1912: APS B:D27 - ABA Committee on Eugenics #2. Letter, C.L. Goodrich to A.G. Smith, 25 November 1912: APS B:D27 - ABA Committee on Eugenics #2. Letter, D.A. Brodie to Charles B. Davenport, 26 November 1912: APS B:D27 - ABA Committee on Eugenics #2. Letter, Charles B. Davenport to D.A. Brodie, 29 November 1912: APS B:D27 - ABA Committee on Eugenics #2.

37. Brodie to Davenport, 26 November 1912. Davenport to Brodie, 29 November 1912.

38. Letter, Charles B. Davenport to George W. Knorr, 3 January 1913: APS B:D27 - ABA Committee on Eugenics #2. Letter, Charles B. Davenport to George W. Knorr, 8 January 1913: APS B:D27 - ABA Committee on Eugenics #2. Letter, Charles B. Davenport to R. DeC. Ward, 8 January 1913: APS B:D27 - ABA Committee on Eugenics #2. Letter, Charles B. Davenport to George W. Knorr, 10 January 1913: APS B:D27 - ABA Committee on Eugenics #2.

39. James Wilson, "Presidential Address: Ninth Annual Meeting," *The American Breeders' Magazine: A Journal of Genetics and Eugenics* Vol. IV (1913), pp. 53, 55, 57.

40. "Foreword," *Eugenical News* Vol. I (1916) p. 1. Truman Library, "Harry H. Laughlin Biography," at www.library.truman.edu. *Official Record of the Gift of the Eugenics Record Office*, p. 33.

41. "Personals," *Eugenical News* Vol. II (1917) p. 12. "Accessions to Archives," *Eugenical News* Vol. II (1917) p. 12. "Voice Inheritance," *Eugenical News* Vol. II (1917) p. 19. "Our Visitors," *Eugenical News* Vol. I (1916) pp. 32-33. "The New Immigration Law," *Eugenical News* Vol. II (1917) p. 22. "Eugenic Legislation," *Eugenical News* Vol. II (1917) p. 29. "Personals," *Eugenical News* Vol. II (1917) p. 71.

42. Letter, Theodore Roosevelt to Charles B. Davenport, 3 January 1913: APS B:D27 Davenport- Roosevelt, Theodore. *What I Think About Eugenics* (n.p., n.d.), Bancroft Library. Dr. Albert Edward Wiggam, as quoted by Thomas F. Gossett, *Race: The History of an Idea in America* (Dallas: Southern Methodist University Press, 1963), p. 403 as cited in Christopher Cerf and Victor Navasky, *The Experts Speak* (New York: Villard Press, 1984), p. 30.

43. Dr. David Heron, "A Criticism of Recent American Work," p. 5, as cited by Charles B. Davenport, "A Discussion of the Methods and Results of Dr. Heron's Critique," *Eugenics Record Office Bulletin No. 11: Reply to the Criticism of Recent American Work by Dr. Heron of the Galton Laboratory* (Cold Spring Harbor, NY: Eugenics Record Office, 1914), p. 3.

44. Heron, pp. 4, 62, as cited by Dr. A.J. Rosanoff, "Mendelism and Neuropathic Heredity: A Reply to Some of Dr. David Heron's Criticisms of Recent American Work," *Eugenics Record Office Bulletin No. 11*, pp. 27, 28: CSH.

45. Heron, p. 67, as cited by Davenport, *Eugenics Record Office Bulletin No. 11*. Heron, p. 30, as cited by David F. Weeks, "Extract from Letter to C.B. Davenport From Dr. David F. Weeks, Superintendent of the New Jersey State Village for Epileptics at Skillman," *Eugenics Record Office Bulletin No. 11*, p. 25.

46. See *Eugenics Record Office Bulletin No. 11*.

47. Davenport, *Eugenics Record Office Bulletin No. 11*, pp. 4-5, 9. Weeks, *Eugenics Record Office Bulletin No. 11*, p. 25. Rosanoff, *Eugenics Record Office Bulletin No. 11*, pp. 35, 36.

48. Davenport, *Eugenics Record Office Bulletin No. 11*, p. 24. Letter, Charles B. Davenport to V.L. Kellogg, 30 October 1912: APS- BD27 Kellogg, Vernon #3.

49. Letter, Charles B. Davenport to Alexander Graham Bell, 25 September 1915: APS B:D27 Alexander Graham Bell #7. Letter, Alexander Graham Bell to Charles B. Davenport, 30 September 1915: APS B:D27 Alexander Graham Bell #7.

50. "Where To Begin," The San Francisco Daily News, 14 October 1915.

51. Letter, Charles B. Davenport to Thomas D. Eliot, 1 November 1915: APS B:D27.

52. Davenport to Eliot, 1 November 1915.

53. Letter, Irving Fisher to Charles B. Davenport, 18 February 1916: APS B:D27 Davenport - Irving Fisher #3.

54. Fisher to Davenport, 18 February 1916.

55. Letter, Charles B. Davenport to Irving Fisher, 25 February 1916: APS B:D27 Davenport - Irving Fisher #3.

56. Draft of letter, Charles B. Davenport to Alexander Graham Bell, n.d.: CSH.

57. Record of telephone call, Alexander Graham Bell to Cold Spring Harbor, 8 April 1916: APS B:D27 Davenport - Bell.

58. Letter, Alexander Graham Bell to Charles B. Davenport, 20 April 1916: APS B:D27 Davenport-Bell.

59. Letter, Alexander Graham Bell to Charles B. Davenport, 18 November 1916: APS B:D27 Davenport- Bell. Letter, Alexander Graham Bell to Charles B. Davenport, 5 January 1917: APS B:D27 Davenport - Bell.

60. *Official Record of the Gift of the Eugenics Record Office*, pp. 21, 24, 25, 28.

61. "Temperament of the Negro," *Eugenical News* Vol. IV (1919) p. 43.

62. "Thalassophilia," *Eugenical News* Vol. V (1920) p. 26.
63. Charles B. Davenport, *Heredity In Relation To Eugenics* (New York: Henry Holt & Company, 1911; reprint, New York: Arno Press Inc., 1972), p. 6.
64. Eugenics Record Office, *Record of Family Traits No. 40688*: MS. COLL. No. 77, ERO, APS Series I. Eugenics Record Office, *Record of Family Traits No. A:0772-1*: MS. COLL. No. 0772, ERO, APS Series I.
65. *Record of Family Traits No. 40688*.
66. Eugenics Record Office, *Family Tree A:0-3a*: MS. COLL. No. 77, ERO, APS Series I, A:01 #6, 1921-1930.
67. Charles B. Davenport and Harry H. Laughlin, *Eugenics Record Office Bulletin No. 13: How To Make A Eugenical Family Study* (Cold Spring Harbor, New York: Eugenics Record Office, 1915) p. 25: CSH.
68. Charts and Measurements: MS. COLL. No. 77, ERO, APS Series I, A:01 #4 Development, 1922-1923. Letter, Brett Ratner to Charles B. Davenport, 15 June 1922: APS Series I, A:01 #4 Development, 1922-1923.
69. Mrs. Anna Wendt Finlayson, *Eugenical Record Office Bulletin No. 15: The Dack Family, A Study in Hereditary Lack of Emotional Control* (Cold Spring Harbor, NY: Eugenical Record Office, 1916) p. 11. Henry H. Goddard, *The Kallikak Family: A Study in the Heredity of Feeble-Mindedness* (Vineland, New Jersey: 1913), p. x. Eugenics Survey of Vermont, "Farm No. 47, Family Name Irving, Mark," *Jamaica Emigrants Questionnaire Interviews*: Vermont PRA-15.
70. Davenport and Laughlin, *Eugenics Record Office Bulletin No. 13*, pp. 4, 28.
71. US Const, Preamble.
72. J. David Smith and K. Ray Nelson, *The Sterilization of Carrie Buck* (Far Hills, NJ: New Horizon Press, 1989) pp. 14, 30.
73. Departments of the Central State Hospital for Epileptic and Feeble-Minded at Petersburg, Virginia, "Official Interrogatories and Papers of Commitment": Emma Buck files, Central Virginia Training Center Archives.
74. Smith, pp. 15-16. "Official Interrogatories and Papers of Commitment."
75. Smith, pp. 1-3, 5-6, 18.
76. "They Told Me I Had To Have An Operation," *Charlottesville (VA) Daily Progress*, 26 February 1980. Author's Interview with former Central Virginia Training Center Superintendent K. Ray Nelson, 14 November 2002.
77. Smith, pp. 17-18.
78. Paul A. Lombardo, "Eugenic Sterilization in Virginia: Aubrey Strode and the Case of Buck v. Bell," (Ph. D. diss, University of Virginia, 1982), pp. 177, 179, 180. "Order of Commitment of Carrie E. Buck," Carrie Buck

vs. Dr. J.H. Bell, 143 Va. 310 pp. 22, 25: Supreme Court of Virginia as cited by Lombardo.
79. A.S. Priddy, *Biennial Report of the State Epileptic Colony* (Lynchburg, VA: State Epileptic Colony, 1923), as cited by Smith, p. 32.
80. Priddy as cited by Smith, p. 32.
81. Priddy as cited by Smith, p. 33.
82. Unnamed Patient Hearing transcript, as cited in The Lynchburg Story, Dir. Stephen Trombley, Prod. Bruce Eadie. Videocassette. Worldview Pictures, 1993. *Poe v. Lynchburg Training School and Hospital*, 518 F. Supp. 789 (W.D. Va. 1981).
83. Lombardo, "Eugenic Sterilization," p. 120. "Writ of Habeas Corpus, filed at the Supreme Court of Appeals of Virginia, November 24, 1917," as cited by Lombardo, "Eugenic Sterilization," p.120.
84. Lombardo, "Eugenic Sterilization," pp. 120-121. "Writ of Habeas Corpus," as cited by Lombardo, "Eugenic Sterilization," p.120.
85. "Deposition of Willie Mallory, December 11, 1917" as cited by Lombardo, "Eugenic Sterilization," p. 122.
86. Lombardo, "Eugenic Sterilization," pp. 124, 126-127.
87. Author's transcription, Letter, George Mallory to A. S. Priddy, 5 November 1917, Virginia State Archive File Drawer #383, Item #2711; also see Lombardo, "Eugenic Sterilization," pp. 127-128; also see Paul A. Lombardo, "Three Generations, No Imbeciles: New Light on *Buck v. Bell*," *New York University Law Review*, Vol. 60 no. 1, pp. 42-43.
88. Letter, A. S. Priddy to George Mallory, 13 November 1917 in "Grounds of Defense, Willie T. Mallory v. A.S. Priddy," February 16, 1918 (Virginia State Archive File Drawers #383, Item #2711) as cited by Lombardo, "Eugenic Sterilization," pp. 128-129. Lombardo, "Eugenic Sterilization," pp. 124-125. W.I. Prichard, "History - Lynchburg Training School and Hospital," *Mental Health in Virginia*, Summer, 1960, as cited in Lombardo, "Eugenic Sterilization," pp. 129-130.
89. Letter, James DeJarnette to John Dickson, 24 October 1947 (Strode Papers, folder 3014A), as cited by Lombardo, "Eugenic Sterilization," p. 132. *Celebration of Dr. J. S. DeJarnette's Fiftieth Anniversary of Continuous Service at the Western State Hospital*, July 21, 1939 (DeJarnette Papers, Western State Hospital, Staunton, Virginia) as cited by Lombardo, "Eugenic Sterilization," p. 132.
90. Harry H. Laughlin, *Eugenical Sterilization in the United States* (Chicago: Psychopathic Laboratory of the Municipal Court of Chicago, 1922) pp. v, 6-50, 446-461. Also see Harry H. Laughlin, *Eugenics Record Office Bulletin No.*

10B: II. The Legal, Legislative, And Administrative Aspects of Sterilization (Cold Spring Harbor, NY: Eugenics Record Office, 1914) pp. 120-131.

91. Letter, A.S. Priddy to Harry H. Laughlin, 14 October 1924: Carrie Buck File, Central Virginia Training Center Archives. Harry H. Laughlin, "Review of the Legal Procedure and Litigation Under the Virginia Sterilization Statute, Which Led to the Decision of the Supreme Court of the United States, Upholding the Statute," *The Legal Status of Eugenical Sterilization* (Washington, DC: Eugenic Record Office of the Carnegie Institute of Washington, 1930), p. 10.

92. Letter, A.S. Priddy to Caroline Wilhelm, 14 March 1924: Carrie Buck File, Central Virginia Training Center Archives. The Lynchburg Story. Letter, Caroline Wilhelm to A.S. Priddy, 5 May 1924: Carrie Buck File, Central Virginia Training Center Archives.

93. Buck v. Bell, pp. 10, 12. Lombardo, "Eugenic Sterilization," p. 183.

94. Strode to Don Preston Peters, 19 July 1939 (Strode Papers, box 30) as cited by Lombardo, "Eugenic Sterilization," p. 183.

95. Priddy to Laughlin, 14 October 1924. Lombardo, "Eugenic Sterilization," pp. 80, 184. Smith, p. 82. Lombardo, "Three Generations," pp. 39, 55. North British and Mercantile Insurance Company, "Fire Insurance Inspection Report, Dee-Whitehead Building": Central Virginia Training Center Archives.

96. Letter, Caroline Wilhelm to A.S. Priddy, 15 October 1924: Carrie Buck File, Central Virginia Training Center Archives.

97. Letter, A.S. Priddy to J. S. DeJarnette, 1 November 1924: Carrie Buck File, Central Virginia Training Center Archives.

98. Priddy to DeJarnette, 1 November 1924.

99. Buck v. Bell, p. 67.

100. Buck v. Bell, p. 67.

101. Harry H. Laughlin, "Analysis of the Hereditary Nature of Carrie Buck," *The Legal Status of Eugenical Sterilization*, pp. 16-17. Priddy to Laughlin, 14 October 1924.

102. Laughlin, "Analysis of the Hereditary Nature of Carrie Buck," pp. 16-17.

103. Laughlin, "Analysis of the Hereditary Nature of Carrie Buck," p. 16.

104. Laughlin, "Analysis of the Hereditary Nature of Carrie Buck," p. 17.

105. Lombardo, "Eugenic Sterilization," pp. 208-210.

106. Harry H. Laughlin, "Opinion of Judge Bennett T. Gordon of the Circuit Court of Amherst County, Virginia," *The Legal Status of Eugenical Sterilization*, pp. 19-21. Smith, pp. 174-175; Lombardo, "Eugenic Sterilization," pp. 210-212.

107. Harry H. Laughlin, "Opinion of Judge Jesse F. West of the Supreme Court of Appeals of Virginia, at Staunton," *The Legal Status of Eugenical Sterilization*, pp. 30-37.

108. State Colony for Epileptics and Feeble-Minded, "Minutes: December 7, 1925,": Central Virginia Training Center Archives.

109. Liva Baker, *The Justice From Beacon Hill: The Life and Times of Oliver Wendell Holmes* (New York: Harper Collins Publishers, 1991), p.3.

110. Baker, pp. 3, 15. Catherine Drinker Bowen, *Yankee From Olympus: Justice Holmes and His Family* (Boston: Little, Brown and Company, 1945), p. 62.

111. David H. Burton, *Oliver Wendell Holmes, Jr.* (Boston: Twayne Publishers, 1980), p. 13.

112. Burton, pp. 28-29. "Overview of 'The Harvard Regiment,'" at www.harvardregiment.org. "Information on The Battle of Antietam," at www.nps.gov. Baker, pp. 97-98.

113. Baker, p.151.

114. Baker, pp. 144, 165-166.

115. Oliver Wendell Holmes, Jr., *The Common Law* (Boston: Little, Brown and Company, 1881; reprint, 1923), p. 1.

116. Baker, pp. 253, 264, 267, 330.

117. U.S. Supreme Court, *Members of the Supreme Court of the United States*. Author's correspondence with Jacques Semmelman, Esq., 14 November 2002 and Lexis-Nexis search by Semmelman, 14 November 2002.

118. Bowen, pp. 372-373, 446. Oliver Wendell Holmes, Jr., "Dissent, Abrams v. US 250 U.S. 616, 624 (1919), *The Mind and Faith of Justice Holmes*, ed. Max Lerner (Garden City, NY: Halcyon House, 1943), p. 312. Oliver Wendell Holmes, Jr. "Dissent, U.S. v. Schwimmer 279 U.S. 644, 653 (1928)," *The Mind and Faith of Justice Holmes*, pp. 327-328. Oliver Wendell Holmes, "For the Court, Schenck vs. U.S. 249 U.S. 47 (1919)," *The Mind and Faith of Justice Holmes*, p. 296.

119. Baker, p. 3.

120. Bowen, p. 187. Holmes, *Common Law*, p. 340.

121. Oliver Wendell Holmes, Jr., "Natural Law," *The Mind and Faith of Justice Holmes*, p. 395.

122. Oliver Wendell Holmes, Jr., "The Soldier's Faith," *The Mind and Faith of Justice Holmes*, pp. 18, 20.

123. Letter from Oliver Wendell Holmes, Jr. to Dean Wigmore, 19 November 1915, as cited by Mark DeWolfe Howe, *Justice Oliver Wendell Holmes: The Shaping Years 1841-1870* (Cambridge, MA: The Belknap Press of Harvard University Press, 1957), p. 25.

124. Letter, Oliver Wendell Holmes, Jr. to Sir Frederick Pollock, 1 February 1920, *Holmes-Pollock Letters: The Correspondence of Mr. Justice Holmes and Sir Frederick Pollock 1874-1932*, ed. Mark DeWolfe Howe (Cambridge, MA:

Harvard University Press, 1942), Vol. II, p. 36.
125. Felix Frankfurter, foreword to *Holmes-Laski Letters Abridged*, ed. by Mark DeWolfe Howe (Clinton, MA: Atheneum, 1963), Vol. I, p. xvi. Letter, Oliver Wendell Holmes, Jr. to Harold J. Laski, 14 June 1922, *Holmes-Laski Letters*, Vol. I, p. 330.
126. Letter, Oliver Wendell Holmes, Jr. to Harold J. Laski, 5 August 1926, *Holmes-Laski Letters*, ed. Mark DeWolfe Howe (Cambridge, MA: Harvard University Press, 1953), Vol. II., p. 862.
127. Holmes to Laski, 21 May 1927, *Holmes-Laski Letters*, p. 946.
128. *Members of the Supreme Court of the United States*. "The People's Attorney," at library.brandeis.edu. See William E. Hellerstein, "Review of The Forgotten Memoir of John Knox," at www.law.uchicago.edu. See *Columbia Encyclopedia*, 6th ed., s.v. "McReynolds, James Clark."
129. Buck v. Bell 274 U.S. 200 (1927).
130. Buck v. Bell 274 U.S. 200 (1927).
131. Buck v. Bell 274 U.S. 200 (1927).
132. Buck v. Bell 274 U.S. 200 (1927).
133. Smith, pp. 16, 179. Lombardo, "Three Generations, No Imbeciles," p. 61.
134. Harry H. Laughlin, *Eugenical Sterilization: 1926; Historical, Legal, and Statistical Review of Eugenical Sterilization in the United States* (New Haven, CT: The American Eugenics Society, 1926), p. 60.
135. Laughlin, *Eugenical Sterilization: 1926*, pp. 21-22, 60. Abraham Myerson et. al., *Eugenical Sterilization: A Reorientation of the Problem* (New York: The Macmillan Company, 1936), p. 10.
136. Human Betterment Foundation, *Legal Status of Eugenical Sterilization* (ca. 1940), Truman D-4-2:11.
137. *Legal Status of Eugenical Sterilization*.
138. E. Carleton MacDowell, "Charles Benedict Davenport, 1866-1944. A Study of Conflicting Influences," *BIOS* Vol. XVII, no. 1, p. 30.

CHAPTER SEVEN
1. See US Const, Amend XIX. *Margaret Sanger: An Autobiography* (W. W. Norton & Company, 1938; New York: Dover Publications, 1971), p.13.
2. Ellen Chesler, *Women of Valor* (New York: Simon & Schuster, 1992), p. 68.
3. Sanger, pp. 86-89, 213-215. Also see Chesler, p. 62. Also see Margaret Sanger, *The Pivot of Civilization* (New York: Brentano's, 1922), p. 29. See Doris Weatherford, *American Women's History*, (New York: Prentice Hall General Reference, 1994), pp. 182-183.
4. Sanger, *An Autobiography*, pp. 90-92. Also see Chesler, p. 63.

5. Sanger, *An Autobiography*, p. 92.
6. Sanger, *An Autobiography*, pp. 92-93, 107-108, 190, 192-209, 292-294. Sanger, *Pivot of Civilization*, pp. 12, 16, 26-27, 272-273. Margaret Sanger, "Address," read at the Thirtieth Annual Meeting luncheon of the Planned Parenthood Federation of America, New York City, 25 October 1950, p. 1: Wellcome Institute, Box 112. David M. Kennedy, *Birth Control in America: The Career of Margaret Sanger*, (New Haven, CT: Yale University Press, 1970), pp. 256-257.
7. Sanger, *Pivot of Civilization*, pp. 14, 18-21 190-192, 194. Sanger, *An Autobiography*, p. 308. See also Sanger, *An Autobiography*, pp. 301-304.
8. See Planned Parenthood Foundation of America, "Our Founder: Margaret Sanger" at www.plannedparenthood.org.
9. Sanger, *Pivot of Civilization*, pp. 101-102. See Julian Huxley, "Towards A Higher Civilization," *Birth Control Review* (December, 1930), p. 344. "Editorial," *Birth Control Review* (March, 1928), p. 73.
10. Sanger, *Pivot of Civilization*, p. 101. Huxley, p. 344.
11. Sanger, *An Autobiography*, pp. 376-377. Margaret Sanger, "A Plan for Peace," *Birth Control Review*, April 1932, pp. 107-108. Margaret Sanger, excerpt from "Racial Betterment," *The Selected Papers of Margaret Sanger: Volume 1:The Woman Rebel, 1900-1928*, edited by Esther Katz (Chicago: University of Illinois Press, 2003), p. 446.
12. Sanger, *Pivot of Civilization*, pp. 104, 108-109, 113-117, 120-121, 123.
13. Sanger, *Pivot of Civilization*, pp. 109, 112, 116. Margaret Sanger, "Is Race Suicide Probable?" *Collier's*, August 15, 1925, p. 25 as selected by Michael W. Perry, ed., *The Pivot of Civilization: In Historical Perspective* (Seattle, WA: Inkling Books, 2001), p. 176.
14. Katz, pp. 333-334. Chesler, pp, 343-344. Margaret Sanger Papers Project, "Notes on Sources," *The National Committee on Federal Legislation for Birth Control 1929-1937* at www.nyu.edu. Henry Pratt Fairchild, *The Melting-Pot Mistake* (Boston: Little, Brown, and Company: 1926), pp. 109-112.
15. See Roswell H. Johnson, "The Eugenic Aspects of Population Theory," *Birth Control Review*, September 1930, pp. 256-258. See Eleanor Dwight Jones, "Practical Race Betterment," *Birth Control Review*, July 1928, pp. 203-204. See American Medicine, "Intelligent or Unintelligent Birth Control?" *Birth Control Review*, May 1919, p. 12. See Sanger, "Address," p. 3. See Perry, p. 176.
16. Victoria C. Woodhull, "The Rapid Multiplication of the Unfit," as selected by Perry, p. 31.

17. Sanger, *Pivot of Civilization*, p. 81.
18. Sanger, *An Autobiography*, p. 11.
19. Sanger, *An Autobiography*, p.29.
20. Sanger, *An Autobiography*, pp. 107-108.
21. Stephen S. Wise, "The Synagogue and Birth Control," *Birth Control Review*, October 1926, pp. 301-302. Margaret Sanger Papers Project, "ABCL Staff, Officers, and Board Members for 1921-1928," *The American Birth Control League 1921-1939* at www.nyu.edu.
22. Sanger, *Pivot of Civilization*, p. 189.
23. Sanger, *Pivot of Civilization*, p. 105.
24. Sanger, *Pivot of Civilization*, p. 108.
25. Sanger, *Pivot of Civilization*, pp. 116-117.
26. Sanger, *Pivot of Civilization*, p. 115.
27. Sanger, *Pivot of Civilization*, p. 123.
28. Sanger, *Pivot of Civilization*, p. 112.
29. Margaret Sanger, *Woman and the New Race* (New York: Brentano's, 1920), Chapter 6.
30. H. G. Wells, introduction to Sanger, *Pivot of Civilization*, p. xvi.
31. "Intelligent or Unintelligent Birth Control?" Also see Sanger, *Woman and the New Race*, Chapter 4.
32. Sanger, *Pivot of Civilization*, p. 104.
33. Sanger, *Pivot of Civilization*, pp. 101-102.
34. Sanger, *Pivot of Civilization*, pp. 277, 282. Also see "Principles and Aims of the American Birth Control League," pamphlet: California State Archives.
35. Sanger, *Woman and the New Race*, Chapter 3.
36. Sanger, *Woman and the New Race*, Chapter 3.
37. Letter, Isabelle Keating to Margaret Sanger, 4 January 1932: Margaret Sanger Papers Project. Letter, Margaret Sanger to Isabelle Keating, 15 January 1932: Margaret Sanger Papers Project.
38. John C. Duvall, "The Purpose of Eugenics," *Birth Control Review*, December 1924, p. 344: California State Archives.
39. Sanger, *Woman and the New Race*, Chapter 7.
40. Sanger, *Woman and the New Race*, Chapter 5.
41. Perry, p. 176.
42. Lothrop Stoddard, *The Rising Tide of Color Against White World Supremacy* (New York: Charles Scribner's Sons, 1926), pp. 303-304.
43. Stoddard, pp. 259-260, 306.
44. Margaret Sanger Papers Project, *The American Birth Control League 1921-1939*. Letter, Margaret Sanger to Henry F. Osborn, 6 October 1921: APS B:D27 Davenport - Sanger, Margaret. See *The American Birth Control League 1921-1939*. "Tentative Program," program of the Sixth International Neo-Malthusian and Birth Control Conference: Truman E-1-1:1.
45. Eugenics Research Association, *Officers and Committee List of the Eugenics Research Association- January 1927* (Cold Spring Harbor, NY: Eugenics Research Association, 1927): Truman, ERA Membership Records. Professor

Irving Fisher, "A Reply," *Official Proceedings of the Second National Conference on Race Betterment* (Battle Creek, MI: The Race Betterment Foundation, 1915).
46. Letter, Sanger to Osborn, 6 October 1921. Letter, Harry H. Laughlin to Irving Fisher, 26 March 1925: Truman E-1-1:1.
47. Margaret Sanger Papers Project, "Staff Members, Officers, Board Members, Chairman and Committee Members," *The National Committee on Federal Legislation for Birth Control 1929-1937* at www.nyu.edu. Margaret Sanger Papers Project, "BCFA Staff, Officers, Board and Committee Members," *The Birth Control Federation of America 1939-1942* at www.nyu.edu. Margaret Sanger Papers Project, "Organization of Council," *The Birth Control Council of America 1937* at www.nyu.edu. Margaret Sanger Papers Project, "BCCRB Staff, Officers, Council Members, and Board Members," *The Birth Control Clinical Research Bureau 1928-1939* at www.nyu.edu. Margaret Sanger Papers Project, faxed list of letters between Margaret Sanger and Henry Pratt Fairchild.
48. Fairchild, pp. 150, 261.
49. *The National Committee on Federal Legislation for Birth Control 1929-1937. The Birth Control Federation of America 1939-1942. The Birth Control Clinical Research Bureau 1928-1939*. American Birth Control League, "World Population Conference," *Eugenical News* Vol. XII (1927), p. 133. Faxed list of letters between Margaret Sanger and Henry Pratt Fairchild.
50. "Tentative Program."
51. Roswell H. Johnson, "Population Control by Immigration," *Birth Control Review*, February 1932, p. 57. "A Plan for Peace," pp. 107-108. Katz, p. 446.
52. See Sanger, *Pivot of Civilization*, pp. 101-102.
53. See Sanger, *Pivot of Civilization*, p. 104. See Margaret Sanger, "An Answer to Mr. Roosevelt," *Birth Control Review*, December 1917, as reprinted in Perry, pp. 156-157.
54. "Eugenics vs. Birth Control," *Eugenical News*, Vol. II (1917), p. 73.
55. Letter, Margaret Sanger to Henry F. Osborn, 6 October 1921: APS B:D27 Davenport-Sanger, Margaret. Letter, Charles B. Davenport to Margaret Sanger, 21 October 1921: APS B:D27 Davenport - Sanger, Margaret.
56. Letter, Charles B. Davenport to Margaret Sanger, 13 February 1925: APS B:D27 Davenport - Sanger, Margaret.
57. Letter, Margaret Sanger to Harry H. Laughlin, 13 March 1925: Truman E-1-1:1. "Tentative Program." Letter, Margaret Sanger to Harry H. Laughlin, 24 March 1925: Truman E-1-1:1. Letter, Harry H. Laughlin to Margaret Sanger, 26 March 1925: Truman E-1-1:1.

58. Margaret Sanger, "Editorial," *The Birth Control Review* Vol. IX, No. 6 (June, 1925), p. 163. See Letter, Paul Popenoe to Madison Grant, 14 April 1928: APS B:D27 Grant, Madison #5. Also see "Birth Control and Eugenics," *Eugenical News* Vol. X (1925), p. 58.

59. "Editorial." "Birth Control and Eugenics." See Paul Popenoe and Roswell Hill Johnson, *Applied Eugenics*, rev. ed. (New York: The MacMillan Company, 1935). Also see "Birth Control and Eugenics," p. 58.

60. "Editorial," pp. 163-164.

61. "Tenth Annual Meeting of the Executive Committee of the Eugenics Research Association," *Eugenical News* Vol. VII (1922), p.89.

62. Letter, Leon F. Whitney to Charles B. Davenport, 3 April 1928: APS B:D27 Davenport - Leon Whitney #1. "Tentative Program." *The National Committee on Federal Legislation for Birth Control 1929-1937. The Birth Control Federation of America 1939-1942. The Birth Control Council of America. The Birth Control Clinical Research Bureau 1928-1939.* Also see Chesler, p. 217.

63. Reverend Albert P. Van Dusen, "Birth Control as Viewed by a Sociologist," *Birth Control Review*, May 1924, p. 133.

64. Duvall, p. 345.

65. Duvall, p. 345. Van Dusen, p. 134.

66. Whitney to Davenport, 3 April 1928.

67. Popenoe to Grant, 14 April 1928.

68. Popenoe to Grant, 14 April 1928.

69. Letter, Madison Grant to Leon F. Whitney, 15 April 1928: APS B:D27 - Grant, Madison #5. Letter, Charles B. Davenport to Madison Grant, 21 April 1928: APS B:D27 - Grant, Madison #5.

70. Davenport to Grant, 21 April 1928. Letter, Charles B. Davenport to Leon F. Whitney, 5 April 1928: APS B:D27 Davenport - Leon Whitney #1.

71. Davenport to Whitney, 5 April 1928.

72. Davenport to Whitney, 5 April 1928.

73. Letter, Henry Pratt Fairchild to Dr. Harry F. Perkins, 9 February 1933: VT PRA-21. See State of Vermont Department of Buildings and General Services, "Content and Historical Significance of Records," *The Papers of the Eugenics Survey of Vermont* at www.bgs.state.vt.us.

74. Letter, Henry Pratt Fairchild to Dr. Harry F. Perkins, 8 March 1933: VT PRA-21.

75. Fairchild to Perkins, 8 March 1933.

76. Letter, Paul Popenoe to Dr. Harry F. Perkins, 16 May 1933: VT PRA-21.

77. Popenoe to Perkins, 16 May 1933.

78. Letter, George Reid Andrews to Members of the Board of Directors, 22 May 1936: VT PRA-21. Letter, Willystine Goodsell to Dr. Harry F. Perkins, 7 June 1936: VT PRA-21.

79. Sanger, "Address," p. 1, 3

80. Sanger, "Address," pp. 3, 4-5.

81. Sanger, "Address," p. 5.

82. Sanger, "Address," pp. 5-6.

83. Sanger, "Address," p. 1. Planned Parenthood Federation of America, Inc., *Books on Planned Parenthood and Related Subjects*, circa 1950: Wellcome Institute, Box 112.

84. Letter, Margaret Sanger to Dr. C.P. Blacker, 5 May 1953: Wellcome Institute Box 112.

CHAPTER EIGHT

1. "Death of Dr. Lucien Howe," *Eugenical News*, Vol. XIV (1929), p. 16. Frank W. Newell, *The American Ophthalmological Society 1864-1989: A Continuation of Wheeler's First Hundred Years* (Rochester, Minnesota: American Ophthalmological Society, 1989), pp. 154-155. "The Howe Laboratory of Ophthalmology," *Eugenical News*, Vol. XI (1926), p 144.

2. "Death of Dr. Lucien Howe." Eugenics Research Association, *Officers and Committee List of the Eugenics Research Association- January 1927* (Cold Spring Harbor, NY: Eugenics Research Association, 1927): Truman, ERA Membership Records. "Report of the Committee on Selective Immigration of the Eugenics Committee of the United States of America," *Eugenical News*, vol. XI (1924), p. 21. See "Eugenics Committee of the United States of America," *Eugenical News*, Vol. X (1923), p. 5.

3. Francis Galton, "Eugenics; Its Definition, Scope and Aims," *Nature*, Vol. 70 No. 1804 (1904), p. 82. Charles B. Davenport, *Eugenics Record Office Bulletin No. 9: State Laws Limiting Marriage Selection Examined in the Light of Eugenics* (Cold Spring Harbor, New York: Eugenics Record Office, 1913), pp. 43-66.

4. Robert Reid Rentoul, *Race Culture: Or, Race Suicide?* (London: The Walter Scott Publishing Co., Ltd., 1906), pp.133-141.

5. Davenport, p. 1. Dr. W.C. Rucker, "More 'Eugenic Laws,'" *The Journal of Heredity*, Vol. VI, No. 5 (May, 1915), pp. 219, 226.

6. Letter, Edward M. Van Cleve of The New York Institute for the Education of the Blind to Lucien Howe, 18 February 1918: APS 77 ERO Series V. Letter, Dr. Harry Best to Lucien Howe, 26 February 1918: APS 77 ERO Series V. Eugenics Record Office, "Cost for the Blind," memorandum, circa 1920: APS 77 ERO Series V. Letter, Harry H. Laughlin to Lucien Howe, 13 November 1920: APS 77 ERO Series I.

7. Lucien Howe, "The Relation of Hereditary Eye Defects to Genetics and Eugenics," *The Journal of Heredity*, Vol. X, No. 8 (November 1919), p. 318. "Abstracts of Papers," *Eugenical News*, Vol. XI (1926), p. 114. "The Blind:

Follow-Up Census Survey," *Eugenical News*, Vol. V (1920), p. 43.

8. Letter, The New Era Printing Company to Harry H. Laughlin, 12 December 1918: APS 77 ERO Series V. "The Blind," *Eugenical News*, Vol. V (1920), p. 42. "Study of Hereditary Blindness," *Eugenical News*, Vol. V (1920), pp. 42-43. "The Blind: Follow-Up Census Survey," *Eugenical News*, Vol. V (1920), p. 43. "Science of Hereditary Blindness," *Eugenical News*, Vol. V (1920), p. 43. "Prevention of Inherited Blindness," *Eugenical News*, Vol. III (1918), p. 64. Letter, Howard J. Banker to Professor George Arps, 6 January 1921: APS 77 ERO Ser. X, HHL Box #3 - Hereditary Blindness Law Research Materials (1921-1928).

9. Eugenics Record Office, *Schedule for Recording First-Hand Pedigree-Data on Hereditary Eye Defect and Blindness*, 1921: APS 77 Series V.

10. *Schedule for Recording First-Hand Pedigree-Data*. Letter, Harry H. Laughlin to Lucien Howe, 30 March 1920: APS 77 Series I: PDR & Correspondence. Eugenics Record Office, *List of School, Etc. For the Blind to Which Questionnaire And Schedule Have Been Sent*, circa 1920: APS 77 Series X: Harry H. Laughlin Box #2 - Hereditary Blindness Corresp #2 1918-1927.

11. *Schedule for Recording First-Hand Pedigree-Data*. Letter, Laughlin to Howe, 30 March 1920. See annotated *List of Fellows of the American Medical Association registered in the Section on Ophthalmology, 1919*: APS 77 ERO Series X.

12. Letter, Banker to Arps, 6 January 1921.

13. Howe, "Relation of Hereditary Eye Defects to Genetics and Eugenics," p. 381. Author's notes on inquiries to Howe Laboratory, American Ophthalmology Society, the National Society for the Prevention of Blindness, the National Eye Institute of the National Institutes of Health, and the Retinitis Pigmentosa Society. Laughlin to Howe, 13 November 1920.

14. Howe, "Relation of Hereditary Eye Defects to Genetics and Eugenics," p. 381.

15. Howe, "Relation of Hereditary Eye Defects to Genetics and Eugenics," pp. 381, 382.

16. Letter, Banker to Arps, 6 January 1921.

17. Letter, Banker to Arps, 6 January 1921.

18. Eugenics Record Office, "Copy - List of Geneticists," circa 1921: APS 77 Series V.

19. Letter, Raymond Pearl to Howard J. Banker, 11 January 1921: APS 77 Series I.

20. Eugenics Record Office, "List of Fellows of the American Medical Association Registered in the Section on Ophthalmology, 1919," circa 1921: APS 77 Series V.

21. *To Amend the Domestic Relations Law, In Relation to Prevention of Hereditary Blindness*, New York (1921), Bill 1597: APS 77 Series X, HHL Box #3, Hereditary Blindness Research Materials (1921-1928).

22. American Medical Association, "Hereditary Eye Defects and Blindness Central Committee," (n.p., circa March 1921): APS 77 ERO Series X.

23. Letter, O.E. Koegel to Dr. Lucien Howe, 7 September 1921: APS 77 Series X. Letter, Leonard W. H. Gibbs to Dr. Lucien Howe, 15 September 1921: APS 77 Series X. Letter, Dr. Hermann M. Biggs to Dr. Lucien Howe, 16 September 1921: APS 77 Series X. Letter, H. S. Birkett to Dr. Lucien Howe, 16 September 1921: APS 77 Series X. Letter, Frank H. Lattin to Dr. Lucien Howe, circa September 1921: APS 77 Series X. Also see Letter, Dr. Lucien Howe to Dr. Harry H. Laughlin, 12 July 1921: APS 77 Series X.

24. Letter, Koegel to Howe, 7 September 1921. Letter, Gibbs to Howe, 15 September 1921. Letter, Biggs to Howe, 16 September 1921. Letter, Birkett to Howe, 16 September 1921. Letter, Lattin to Howe, circa September 1921. Author's notes on APS ERO files.

25. Letter, Lucien Howe to Harry H. Laughlin, 12 January 1922: APS 77 Series I.

26. Howe to Laughlin, 12 January 1922.

27. Letter, Lucien Howe to Harry H. Laughlin, 22 July 1922: APS 77 Series I.

28. Howe to Laughlin, 22 July 1922.

29. Howe to Laughlin, 22 July 1922.

30. Letter, Harry H. Laughlin to Lucien Howe, 7 August 1922: APS 77 Series I. Letter, Lucien Howe to Harry H. Laughlin, 28 August 1922: APS 77 Series I.

31. Letter, Howe to Laughlin, 28 August 1922.

32. Letter, Lucien Howe to Charles B. Davenport and Harry H. Laughlin, 10 February 1923: APS 77 Series I. "Memorandum on the Institutional Cost of the Blind and the Economic Cost of the Blind in the Population at Large", enclosure to Howe to Davenport and Laughlin, 10 February 1923: APS 77 Series I.

33. Howe to Davenport and Laughlin, 10 February 1923.

34. J. P. Chamberlain, "Current Legislation: Eugenics and Limitations of Marriage," *American Bar Association Journal*, 1923, pp. 429-430.

35. *To Amend the Domestic Relations Law, In Relation to Prevention of Hereditary Blindness*, New York (1926), Bill 605: APS 77 Series X, Harry H. Laughlin Box #3, Hereditary Blindness Research Materials (1921-1928). Letter, Lucien Howe to Dr. Best, 4 October 1927: APS 77 Series V. Also see Letter, Edward G. Seibert to Lucien Howe, 15 July 1927: APS 77 Series V.

36. Bill 605. "Abstracts of Papers," p. 114. Consumer Price Index Calculator at www.jsc.nasa.gov. "Memorandum on the General Principle of Bonding Applicants for Marriage License Against the Production of

Offspring Who Would Become Public Charges", memo, ca. 1928: Truman E-1-2:8.
37. Letter, Harry H. Laughlin to Lucien Howe, 30 March 1921: APS 77 Series I. "Memorandum," p 3.
38. Letter, Harry H. Laughlin to Lucien Howe, 4 May 1922: APS 77 Series I. Letter, Lucien Howe to Harry H. Laughlin, 14 April 1922: APS ERO Series X: Hereditary Blindness Corresp.
39. Letter, Harry H. Laughlin to Lucien Howe, 30 December 1922: APS 77 Series I: PDR & Correspondence.
40. Bill 605. "Abstracts of Papers," p. 114.
41. "Memorandum," pp 2, 4.
42. Laughlin to Howe, 30 March 1921.
43. Letter, Harry H. Laughlin to Lucien Howe, 5 December 1922: APS 77 Series: I. Laughlin to Howe, 30 December 1922.
44. Letter, Harry H. Laughlin to Lucien Howe, 9 March 1925: APS 77 Series I: PDR & Correspondence.
45. "Eugenical Responsibility," *Eugenical News*, Vol. XVI (1931), pp. 45, 46.
46. "Eugenical Responsibility," p. 46.
47. "Eugenical Responsibility," pp. 46-47.
48. "Eugenical Responsibility," p. 47.
49. "Death of Dr. Lucien Howe." Newell, p. 155.

CHAPTER NINE
1. Harry H. Laughlin, "Population Schedule for the Census of 1920," *Journal of Heredity*, Vol. X, No. 5 (May 1919), p. 208.
2. U. S. Department of Commerce, *Statistical Directory of State Institutions for the Defective, Dependent, and Delinquent Classes* (Washington, DC: Government Printing Office, 1919), p. 5.
3. See U.S. Department of Commerce, "Report on the Defective, Dependent, and Delinquent Classes," as referenced by University of Wisconsin, "Study Report," *DPLS Catalog of Holdings* at www.wisc.edu. See Documents Service Center, "Decennial Census Information: 1880 Census - Detailed Holdings," at www.columbia.edu. Harry H. Laughlin, "The Socially Inadequate: How Shall We Designate and Sort Them?" *The American Journal of Sociology* Vol. XXVII No. 1 (July 1921) reprinted in *Harry Laughlin Reprints* (Washington, DC: Carnegie Institute of Washington, n.d.), pp. 54, 55-56, 68.
4. Laughlin, "Socially Inadequate," p.68. Letter, Joseph A. Hill to Harry H. Laughlin, 2 January 1918: Truman D-4-5:5.
5. Laughlin, "Socially Inadequate," pp. 57, 62, 67-68. Hill to Laughlin, 2 January 1918.
6. Hill to Laughlin, 2 January 1918. See Harry H. Laughlin, *Statistical Directory of State Institutions for the Defective, Dependent, and Delinquent*

Classes (Washington, DC: Government Printing Office, 1919). Laughlin, "Socially Inadequate," pp. 54, 57-67, 68.
7. Letter, Harry H. Laughlin to Samuel L. Rogers, 20 December 1918: Truman C-4-2:6. Letter, Samuel L. Rogers to Harry H. Laughlin, 23 December 1918: Truman C-4-2:6. See letter, Harry H. Laughlin to William M. Steuart, 22 May 1929: Truman E-2-2:8. Also see letter, Harry H. Laughlin to Irving Fisher, 23 February 1928: Truman D-2-3:22.
8. Laughlin, "Population Schedule," pp. 208-209. Eugenic Research Association, "Meeting of the Executive Committee of the Eugenics Research Association Tuesday, Feb. 18, 1919," Truman ERA Minutes Vol. 2 (n.p., 1919), p.2. See IPUMS-USA, "1910 Sampling Procedures," at www.ipums.umn.edu.
9. Laughlin, "Population Schedule," p. 209. Hill to Laughlin, 2 January 1918.
10. See letter, Julia C. Lathrop to Elizabeth B. Muncey, 17 February 1916: APS Series I A:015-T. See "The Federal Census Schedule," *Eugenical News* Vol. V (1920), pp. 36-37. See "Eugenical Significance of Individual Records," *Eugenical News* Vol. XIII (1928), pp. 8-9. See "A Needed Amendment of the Census Bill: 1929," *Eugenical News* Vol. XIV (1929), pp. 53-55. See letter, Harry H. Laughlin to Lewis W. Douglas, 11 April 1929: Truman C-4-6:17. See letter, Harry H. Laughlin to William M. Steuart, 12 April 1929: Truman E-2-2:8. See letter, Harry H. Laughlin to William M. Steuart, 22 May 1929: Truman E-2-2:8. See "Present Status of the Proposed 'Race-Descent' Item in the Census Population Schedule," *Eugenical News* Vol. XIV (1929), pp. 138-139.
11. Harry H. Laughlin, *Classification Standards to be Followed in Preparing Data for the Schedule "Racial and Diagnostic Records of Inmates of State Institutions,"* (Washington, DC: Government Printing Office, 1922), p. 4: Truman C-4-6:16. J. David Smith, *The Eugenic Assault on America: Scenes in Red, White, and Black* (Fairfax, VA: George Mason University Press, 1993), p. 60.
12. Letter, Walter A. Plecker to Editor, Survey Graphic, 13 March 1925: UVA Library 7284A Powell Papers, Box 56, Folder 1925.7284A.
13. Loyd Thompson and Winfield Scott Downs, ed., *Who's Who in American Medicine 1925* (New York: Who's Who Publications, Inc., 1925), s.v. "Plecker, Walter Ashby." Smith, p. 60. See Joan Charles, *Elizabeth City County, Virginia Will 1800-1859*, (Bowie, MD: Heritage Books, Inc., n.d.). See Blanche Adams Chapman, *Wills and Administration of Elizabeth City County, Virginia 1688-1800* (1980, n.p.). See author's notes on established sources of Elizabeth City County genealogical records.
14. Smith, p. 60. Walter A. Plecker, "The 1930

U.S. Census," n.p., n.d.: Truman D-4-3:12. Letter, Walter A. Plecker to Harry H. Laughlin, 24 November 1928: Truman D-4-3:12.

15. Smith, pp. 60-61. Richard B. Sherman, "'The Last Stand': The Fight for Racial Integrity in Virginia in the 1920's," *The Journal of Southern History*, Vol. 54 Issue 1 (February, 1988), p. 78.

16. Bureau of Vital Statistics, *Eugenics in Relation to the New Family and the Law on Racial Integrity*, (Richmond: Supt. Public Printing, 1924), p. 26.

17. Letter, Walter A. Plecker to Madison Grant, 13 January 1928: Truman D-4-3:12. Letter, Walter A. Plecker to Harry H. Laughlin, 17 November 1930: Truman D-4-3:12.

18. Letter, Walter A. Plecker to Harry H. Laughlin, 25 February 1928: Truman D-4-3:12.

19. Letter, Walter A. Plecker to Undertakers of Virginia, July 1921: Library of Virginia - Richmond. See Letter, Walter A. Plecker to Harry Davis, 4 October 1924: UVA Library 7284A Powell Papers, Box 56, Folder 1924 #2.

20. Plecker to Grant, 13 January 1928.

21. Plecker to Grant, 13 January 1928.

22. Charles B. Davenport, *Eugenics Record Office Bulletin No. 9: State Laws Limiting Marriage Selection Examined in the Light of Eugenics* (Cold Spring Harbor, New York: Eugenics Record Office, 1913), pp. 43-66. Walter A. Plecker, *Virginia's Vanished Race*, (n.p.: July 1947): UVA Library Broadside 1947, Virginia's Vanished Race W.A. Plecker. Arthur Estabrook and Ivan E. McDougle, *Mongrel Virginians: The Win Tribe*, (Baltimore: The Williams & Wilkins Company, 1926), p. 145

23. "Racial Integrity," *Richmond Times-Dispatch*, 18 February 1924. Sherman, p. 81. Philip Reilly, "The Virginia Racial Integrity Act Revisited: The Plecker-Laughlin Correspondence: 1928-1930," *American Journal of Medical Genetics* Vol. 16 (1983), p. 486. Letter, Walter A. Plecker to Harry H. Laughlin, 19 March 1930: Truman D-4-3:12.

24. Smith, pp. 16-17. Sherman, pp. 73-74.

25. The Marcus Garvey and UNIA Papers Project, UCLA "American Series Introduction: Volume VI: September 1924 - December 1927) at www.isop.ucla.edu. "Nuclei," *Eugenical News* Vol. IX (1924), p. 8. "White America," *Eugenical News* Vol. IX (1924), p. 3.

26. John Powell, *A Breach in the Dike: An Analysis of the Sorrells Case Showing the Danger to Racial Integrity from Intermarriage of Whites with So-Called Indians* (Richmond, n.d.), p. 3 as cited by Sherman, p. 81. Sherman, p. 79.

27. Smith, p. 59. Sherman, p. 77. Virginia Department of Health, "The New Virginia Law to Preserve Racial Integrity," *Virginia Health Bulletin*, Vol. XVI Extra No. 2 (March 1924), p. 4.

28. John Powell and E.S. Cox, "Is White America to Become a Negroid Nation?" *Richmond Times-Dispatch*, 22 July 1923. "Editorial: Racial Integrity," *Richmond Times-Dispatch*, 22 July 1923.

29. "Powell Asks Law Guarding Racial Purity," *Richmond Times-Dispatch*, 13 February 1924.

30. "Racial Integrity," *Richmond Times-Dispatch*, 18 February 1924.

31. Sherman, pp. 77-78. Letter, Walter A. Plecker to Reverend Wendell White, 10 May 1924: UVA Library 7284A Powell Papers, Box 56, Folder 1924 #1. Plecker to Davis, 4 October 1924.

32. "Against Miscegenation," *Eugenical News*, Vol. IX (1924), p. 48. Virginia Department of Health, "The New Virginia Law to Preserve Racial Integrity." Smith, p. 20.

33. Virginia Department of Health, "Instructions to Local Registrars and Other Agents In Administration of the Law," *Virginia Health Bulletin*, Vol. XVI Extra No. 1 (March 1924), pp.1, 2, attached sample record card.

34. "The New Virginia Law to Preserve Racial Integrity," pp. 1, 2, 4.

35. Plecker to Davis, 4 October 1924.

36. Letter, Walter A. Plecker to Mrs. Robert H. Cheatham and Mrs. Mary Gildon, 30 April 1924: UVA Library 7284A Powell Papers, Box 56, Folder 1924 #1.

37. Plecker to Cheatham and Gildon, 30 April 1924.

38. Plecker to Cheatham and Gildon, 30 April 1924.

39. Letter, Walter A. Plecker to W. H. Clark, 29 July 1924: UVA Library 7284A Powell Papers, Box 56, Folder 1924 #1.

40. Plecker to Clark, 29 July 1924.

41. Letter, Walter A. Plecker to Pal S. Beverly, 12 October 1929: Truman D-4-3:12.

42. Plecker to Beverly, 12 October 1929.

43. Plecker to Beverly, 12 October 1929.

44. Letter, Walter A. Plecker to Mascott Hamilton, 10 October 1930: Truman D-4-3:12.

45. Plecker to Hamilton, 10 October 1930.

46. Plecker to Hamilton, 10 October 1930.

47. Letter, Walter A. Plecker to Mrs. Frank C. Clark, 1 May 1930: Truman D-4-3:12.

48. Plecker to Mrs. Frank C. Clark, 1 May 1930.

49. Letter, Walter A. Plecker to Harry H. Laughlin, 24 May 1929: Truman D-4-3:12.

50. Smith,pp. 66-67.

51. Plecker to Laughlin, 24 November 1928.

52. Letter, Walter A. Plecker to Dr. H. V. Fitzgerald, County School Board, 11 July 1940: UVA Library 7284A Powell Papers, Box 56, Folder 1940 #4. Plecker to Laughlin, 17 November 1930. Smith, p. 67.

53. Plecker to Fitzgerald, 11 July 1940.

54. Smith, pp. 93-94.

55. Bureau of Vital Statistics, State Board of Health, *The New Family and Race Improvement*, (Richmond, VA: Bureau of Vital Statistics, 1925), p. 3. "Report of the Bureau of Vital Statistics State Board of Health to the Governor of Virginia," *Virginia Health Bulletin* Vol. XIX, No. 1, (January 1927), pp. 9-10.

56. *Eugenics in Relation to the New Family*, pp. 6-7.

57. Walter A. Plecker, "Virginia's Effort to Preserve Racial Integrity," *A Decade of Progress in Eugenics: Scientific Papers of the Third International Congress of Eugenics* (Baltimore: The Williams & Wilkins Company, 1934), p. 105. "The Twenty Second Annual Meeting of the Eugenics Research Association: Virginia's Methods of Research in Racial Integrity," *Eugenical News*, Vol. XX (1935), p. 25. Letter, Walter A. Plecker to Harry H. Laughlin, 12 December 1928: Truman D-4-3:12. Letter, Harry H. Laughlin to Walter A. Plecker, 18 May 1929: Truman D-4-3:12. Letter, Harry H. Laughlin to Walter A. Plecker, 17 March 1930: Truman D-4-3:12. Plecker to Laughlin, 29 March 1930. Walter A. Plecker, "Race Mixture and the Next Census," *Eugenics* Vol. II No. 3 (March 1929), p. 1.

58. Walter A. Plecker, "Virginia's Attempt to Adjust the Color Problem," *Journal of American Public Health* (1925), pp. 111, 114.

59. Plecker, "The 1930 U.S. Census."

60. Letter, John Powell to George H. Roberts, 28 February 1925: UVA Library 7284A Powell Papers, Box 56, Folder 1925. Plecker to White, 10 May 1924.

61. Powell to Roberts, 28 February 1925.

62. Plecker to Davis, 4 October 1924. Plecker to Laughlin, 12 December 1928. Plecker to White, 10 May 1924.

63. Plecker to Laughlin, 24 May 1929. See letter, Walter A. Plecker to Harry H. Laughlin, 26 June 1928: Truman D-4-3:12. See Wisconsin (1929), Bill 409 S: Truman D-4-3:12. See chart attached to letter, Walter A. Plecker to Madison Grant, 18 June 1931: Truman D-4-3:12 as compared to Davenport pp. 43-66. Letter, Harry H. Laughlin to Walter A. Plecker, 22 November 1928: Truman D-4-3:12. Laughlin to Plecker, 18 May 1929.

64. Letter, Walter A. Plecker to A. H. Crismond, 21 August 1940: UVA Library 7284A Powell Papers, Box 56, Folder 1940 #4.

65. Letter, Walter A. Plecker to Luke M. Smith, 3 September 1940: UVA Library 7284A Powell Papers, Box 56, Folder 1940 #4. Letter, Walter A. Plecker to Samuel L. Adams, 11 December 1924: UVA Library 7284A Powell Papers, Box 56, Folder 1924 #2. *Eugenics in Relation to the New Family*, p. 22.

66. Stuart E. Brown, Jr., Lorraine F. Myers and Eileen M. Chappel, *Pocahontas' Descendants* (Baltimore, MD: The Pocahontas Foundation, 1985), pp. iii, 347, 409, 410, 421, 426, 442.

67. See University of Wisconsin, "Help Files-Genealogical Research Using U.S. Census Data," at www.uwm.edu. See U.S. Department of Commerce, "1860 Census: Instructions to the Marshals"at www.ipums.umn.edu as compared to U.S. Department of Commerce, "1870 Census: Instructions to the Marshals" at www.ipums.umn.edu as compared to U.S. Department of Commerce, "1880 Census: Instructions to the Marshals" at www.ipums.umn.edu as compared to U.S. Department of Commerce, "1890 Census: Instructions to the Marshals" at www.ipums.umn.edu as compared to U.S. Department of Commerce, "1900 Census: Instructions to the Marshals" at www.ipums.umn.edu as compared to U.S. Department of Commerce, "1910 Census: Instructions to the Marshals" at www.ipums.umn.edu. See "The Most Prolific People in the United States," *Eugenical News* Vol. XXIII (1938), pp. 29-31. See "American Indians Made Citizens," *Eugenical News* Vol. IX (1924), p. 73. See Library of Congress, "Today in History," at memory.loc.gov.

68. Plecker to Davis, 4 October 1924. Department of Commerce, "U.S. Census Estimate of Indians in Virginia, Vol. III, 1920 Population," n.d.: UVA Library 7284A Powell Papers, Box 56, Folder 1921. See Plecker to Adams, 11 December 1924.

69. Plecker to Davis, 4 October 1924.

70. *Reclaiming our Heritage*, prod. Sharon Bryant, Virginia Foundation for the Humanities, 1997, videocassette. Estabrook and McDougle, p. 145.

71. Plecker to Davis, 4 October 1924. "Abstracts of Papers at 1934 ERA Meeting," *Eugenical News*, Vol. XX (1935), p. 25.

72. Plecker to W. H. Clark, 29 July 1924.

73. Letter, Walter A. Plecker to A. P. Bohannon, 3 May 1938: UVA Library 7284A Powell Papers, Box 56, Folder 1938 #4.

74. Letter, Walter A. Plecker to R. S. Major, 28 August 1942: UVA Library 7284A Powell Papers, Box 56, Folder 1942 #4.

75. Letter, Walter A. Plecker to William E. Bradby, 2 February 1942: UVA Library 7284A Powell Papers, Box 56, Folder 1942 #4.

76. Estabrook and McDougle, p. 15. See Plecker to Davis, 4 October 1924.

77. "Mongrel Virginians," *Eugenical News* Vol. XI (1926), p. 70. The Wilkins & Wilkins Company, "Mongrel Virginians," (Baltimore: The Williams & Wilkins Company, circa 1926): Truman E-2-2. Estabrook and McDougle, pp. 3, 8. See Estabrook and McDougle, pp. 13, 177.

78. Estabrook and McDougle, pp. 13,14.

79. Estabrook and McDougle, pp. 199-200.

80. Estabrook and McDougle, p. 201.
81. Estabrook and McDougle, p. 202.
82. Smith, pp. 86, 87. "Mongrel Virginians," *Eugenical News*, p. 70.
83. Smith, pp. 95, 96. "Petition For Mandamus," *Sorrells v. A.T. Shields, Clerk* (1924): UVA Library 7284A Powell Papers, Box 56, Folder 1924 #2.
84. Smith, p. 98.
85. "Petition For Mandamus." Letter, Leon H. Bazile to John Powell, 26 November 1924: UVA Library 7284A Powell Papers, Box 56, Folder 1924 #2.
86. Letter, Bazile to Powell, 26 November 1924. Smith, p. 75. Sherman, pp. 81, 85.
87. Letter, Walter A. Plecker to the Virginia Department of Health, 27 May 1946: UVA Library 7284A Powell Papers, Box 56, Folder 1944-46. Letter, Walter A. Plecker to Dr. I. C. Riggin, 27 May 1946: UVA Library 7284A Powell Papers, Box 56, Folder 1944-46. See Plecker, *Virginia's Vanished Race*.

CHAPTER TEN
1. Letter, John C. Merriam to Charles B. Davenport, 20 June 1923: CIW Genetics: Eugenics Record Office Misc. Correspondence 2 of 2. See Fourth Report of the Committee on Selective Immigration of the American Eugenics Society, memorandum, circa July 1929: Truman C-4-4:6.
2. US Department of Commerce, *Historical Statistics of the United States, Colonial Times to 1970*, (Washington DC: US Department of Commerce, 1976).
3. US Department of Commerce.
4. The National Park Service, "Statue of Liberty: Museum Exhibits," at www.nps.gov. Emma Lazarus, "The New Colossus."
5. Margo J. Anderson, *The American Census* (New Haven, CT: Yale University Press, 1988), pp. 132, 133. Paul Burnett, "The Red Scare," at www.law.umkc.edu.
6. Anderson, pp. 134, 139.
7. Anti-Defamation League, "Extremism in America: Ku Klux Klan," at www.adl.org. Burnett. John Higham, *Strangers in the Land* (New Brunswick, NJ: Rutgers University Press, 1955), pp 264-265. William M. Tuttle, Jr., *Race Riot: Chicago in the Red Summer of 1919* (New York, NY: Atheneum, 1970), pp 16-20, 22-23. Anderson, p. 133.
8. Robert DeC. Ward, "Our Immigration Laws From the Viewpoint of National Eugenics," *National Geographic*, January 1912: Truman C-4-2:7. Letter, Irving Fisher to Charles B. Davenport, 2 March 1912: APS BD27 Fisher #7. Letter, Charles B. Davenport to Irving Fisher, 4 March 1912: APS BD27 Fisher #7.

9. *Biographical Directory of the U. S. Congress*, s.v. Albert Johnson at bioguide.congress.gov. City of Hiawatha Home Page at www.cityofhiawatha.org. City of Atchison Visitor Information at www.atchisonkansas.net.
10. "Course Outline of Eugenics," 1933, p. 14: Truman E-2-2:17. Harry H. Laughlin, "A Bill," n.p., circa 1917: Truman C-2-4:5. Letter, Prescott F. Hall to Charles B. Davenport, 1 October 1920: APS B:D 27.
11. Prescott F. Hall, "Immigration Restrictions and World Eugenics," *Journal of Heredity* Vol. X, No. 3 (March, 1919), p. 126.
12. Letters, Albert Johnson to Madison Grant, 19 March 1924 and Madison Grant to Albert Johnson, 12 December 1923 as cited by Hassencahl, p. 209. Letter, Madison Grant to Charles B. Davenport, 24 January 1921:APS B:D 27 - Grant, Madison #4. Letter, Madison Grant to John C. Merriam, 26 November 1924:APS B:D 27 - Grant, Madison #4. Letter, Charles B. Davenport to Madison Grant, 17 March 1921:APS B:D 27 - Grant, Madison #4. Letter, Madison Grant to Charles B. Davenport, 29 January 1921:APS B:D 27 - Grant, Madison #4. Letter, Charles B. Davenport to Madison Grant, 27 November 1920:APS B:D 27 - Grant, Madison #4. Letter, Charles B. Davenport to Madison Grant, 6 January 1921:APS B:D 27 - Grant, Madison #4. Martha Ragsdale, "The National Origins Plan of Immigration Restriction," (Nashville: Vanderbilt University) unpublished manuscript, p. 140: PRA #18.
13. See Harry H. Laughlin, *Biological Aspects of Immigration*: CSHL: Harry H. Laughlin Reprints.
14. Laughlin, pp. 4-5, 6, 13, 18.
15. Laughlin, pp. 3, 4, 21.
16. Laughlin, pp. 4, 5.
17. Laughlin, pp. 23-26.
18. Memorandum, Harry H. Laughlin to John C. Merriam, 17 October 1922: Truman C-4-5:6. "Sample of Schedule used in the Melting Pot Survey," attachment to Harry H. Laughlin, "Definite Proposal for a New and More Thorough Study of Crime Among Aliens and the Descendants of Recent Immigrants in American Criminalistic Institutions," circa 1921: C-4-5:10. Letter, Harry H. Laughlin to Albert Johnson, 30 December 1924: Truman C-2-4:5.
19. "Immigration Limits for the Year Ending July 1, 1922," *Eugenical News* Vol. VI (1921). Robert DeCourcey Ward, "Immigration and the Three Per Cent Restrictive Law," *Journal of Heredity* Vol. XII No. 7 (August-September 1921), pp. 319-325. "First Report of the Committee on Selective Immigration of the Eugenics Committee of the United States of America,"

memorandum, circa 1924: Truman C-4-4:3. Harry H. Laughlin, "Scientific Investigations by the Committee on Immigration and Naturalization of the House of Representatives: Abstract of Studies Made for the Committee," memorandum, circa May 1922: Truman C-2-4:5.

20. Harry H. Laughlin, *Classification Standards to be Followed in Preparing Data for the Schedule "Racial and Diagnostic Records of Inmates of State Institutions,"* (Washington, DC: Government Printing Office, 1922), pp. 4, 7: Truman C-4-6:16. Letter, Harry H. Laughlin to Albert Johnson, circa 1922: Truman C-2-4:5.

21. "Report of Harry H. Laughlin for the Ten Months September 1, 1923 - June 30, 1924," memorandum, circa 1924, p. 3: C-2-3:3. "Report of Harry H. Laughlin for the Year Ending September 1st, 1923," Memorandum circa 1923: Truman C-2-5:15. House Committee on Immigration and Naturalization, *Statement of Dr. Harry H. Laughlin,* 67th Cong., #3rd sess., 21 November 1922, p. 734.

22. House Committee, *Statement of Dr. Harry H. Laughlin,* 21 November 1922, pp. 756, 760.

23. House Committee, *Statement of Dr. Harry H. Laughlin,* 21 November 1922, pp. 725, 752, 759.

24. "Biological Research in Immigration," memorandum, circa 1920: Truman C-2-4:5. Letter, Harry H. Laughlin to Frank Babbott, 18 February 1922: Truman C-4-3:5. Letter, Harry H. Laughlin to Frank Babbott, 9 July 1925: Truman C-4-3:5. Davenport to Grant, 17 March 1921. Harry H. Laughlin, *Immigration and Conquest: A Report of The Special Committee on Immigration and Naturalization of the Chamber of Commerce of the State of New York* (n.p., 1939), p. 8.

25. "First Report of the Committee on Selective Immigration of the Eugenics Committee of the United States of America," draft copy, circa 1924: Truman C-4-4:3. "Eugenics Committee of the United States of America," memorandum, circa 1924: APS 576.06 AM3 AES-Eugenics Committee of USA Documents.

26. "First Report of the Committee on Selective Immigration," p. 2.

27. "First Report of the Committee on Selective Immigration," pp. 2, 3, 4.

28. "Annual Meeting of the Eugenics Research Association," *Eugenical News* Vol. VIII (1923), p. 53. Eugenics Research Association, *Active Membership Accession List* (Cold Spring Harbor, NY: Eugenics Research Association, 1922): Truman, ERA Membership Records. US Department of Labor, "Portraits: James J. Davis," at www.dol.gov. Merriam to Davenport, 20 June 1923.

29. Merriam to Davenport, 20 June 1923.

30. Merriam to Davenport, 20 June 1923. Letter, Harry H. Laughlin to Charles B. Davenport, 26 November 1923: Truman C-2-6:17.

31. Merriam to Davenport, 20 June 1923.

32. Memorandum, Charles B. Davenport to Harry H. Laughlin, 26 June 1923: Truman C-4-3:9.

33. Certificate of appointment: Mark Laughlin Collection as cited by Hassencahl, p. 191. See Harry H. Laughlin, "Report of Harry H. Laughlin for the Year Ending September 1st, 1923."

34. Laughlin to Johnson, circa 1922, p. 9.

35. "Personals," *Eugenical News* Vol. VIII (1923), p. 94. Letter, Harry H. Laughlin to Dr. Albert Govaerts, 17 March 1923: Truman C-4-6:19. Letter, Harry H. Laughlin to Charles B. Davenport, 1 October 1923: Truman D-2-6:17.

36. Laughlin to Davenport, 1 October 1923. Letter, Harry. H. Laughlin to Charles B. Davenport, 22 November 1923: Truman C-2-6:17. "Report of Harry H. Laughlin for the Ten Months September 1, 1923 - June 30, 1924." Letter, Harry H. Laughlin to Judge Harry Olson, 12 October 1923: Truman D-2-3:6.

37. Laughlin to Olson, 12 October 1923.

38. "Report of Harry H. Laughlin for the Ten Months September 1, 1923 - June 30, 1924," p. 2. "Dr. Albert Govaerts of Belgium," *Eugenical News* Vol. VII (1922), p. 64. Laughlin to Davenport, 22 November 1923. Laughlin to Davenport, 26 November 1923.

39. "Report of Harry H. Laughlin for the Ten Months September 1, 1923 - June 30, 1924," p. 3. Laughlin to Davenport, 22 November 1923.

40. "Report of Harry H. Laughlin for the Ten Months September 1, 1923 - June 30, 1924," p. 2. Laughlin to Davenport, 22 November 1923. Laughlin to Davenport, 26 November 1923.

41. "Report of Harry H. Laughlin for the Ten Months September 1, 1923 - June 30, 1924," pp. 2, 3. Laughlin to Davenport, 22 November 1923. Laughlin to Davenport, 26 November 1923. Laughlin to Davenport, 1 October 1923.

42. Laughlin to Davenport, 26 November 1923.

43. Laughlin to Davenport, 26 November 1923. Harry H. Laughlin, "Interdepartmental Authority," memorandum circa December 1923: Truman D-4-3:13.

44. Laughlin, "Interdepartmental Authority."

45. "Report of Harry H. Laughlin for the Ten Months September 1, 1923 - June 30, 1924," pp. 1, 6. Laughlin to Davenport, 1 October 1923.

46. "Report of Harry H. Laughlin for the Year Ending September 1st, 1923." Laughlin to Davenport, 22 November 1923.

47. Laughlin to Davenport, 22 November 1923. Letter, Charles B. Davenport to Harry H. Laughlin, 21 December 1923: Truman C-2-6:17.

48. Laughlin to Davenport, 26 November 1923.
49. Laughlin to Davenport, 22 November 1923.
50. Davenport to Laughlin, 21 December 1923.
51. "Report of Harry H. Laughlin for the Ten Months September 1, 1923 - June 30, 1924," p. 1. "Report of the Committee on Selective Immigration of the Eugenics Committee of the United States of America," *Eugenical News* Vol. IX (1924), pp. 21-24.
52. "Secretary Davis on Immigration," *Eugenical News* Vol. IX (1924), p. 37.
53. House Committee on Immigration and Naturalization, *Statement of Dr. Harry H. Laughlin*, 68th Cong., 1st sess., 8 March 1924, pp. 1279, 1281, 1283, 1294, 1295.
54. House Committee, *Statement of Dr. Harry H. Laughlin*, 8 March 1924, pp. 1311, 1322, 1323, 1340.
55. House Committee, *Statement of Dr. Harry H. Laughlin*, 8 March 1924, p. 1300.
56. "Memorandum and Outline of Tentative Working Agreement Between the Carnegie Institution of Washington and the State Department of the Federal Government in Reference to Collaboration in the Collection of First-Hand Data on Immigration at its Sources," memorandum, circa June 1924: Truman C-4-3:9. Letter, W. M. Gilbert to Harry H. Laughlin, 11 September 1924: Truman C-4-3:9.
57. "Memorandum and Outline of Tentative Working Agreement Between the Carnegie Institution of Washington and the State Department." "Report of Harry H. Laughlin for the Ten Months September 1, 1923 - June 30, 1924," p. 4. "The Several Filterings of the Immigrant Stream Directed Toward the United States," attachment to Letter, Harry H. Laughlin to Albert Johnson, 30 December 1924: Truman C-2-4:5.
58. "Memorandum and Outline of Tentative Working Agreement Between the Carnegie Institution of Washington and the State Department" p. 3. See Ezekiel Cheever, *School Issues* (Baltimore: Warwick & York, Inc., 1924): CIW Genetics: Eugenics Record Office Misc. Correspondence 2 of 2.
59. See Cheever.
60. Cheever, pp 28-29.
61. Cheever, p 38.
62. Cheever, pp 41, 42-43.
63. Cheever, p 44.
64. Cheever, p 19.
65. Letter, Charles B. Davenport to Lewellys F. Barker, 18 April 1924: CIW Genetics: Eugenics Record Office Misc. Correspondence 2 of 2. Letter, Robert DeC. Ward to Harry H. Laughlin, 20 March 1924: Truman C-4-1:8. Letter, Harry H. Laughlin to Robert DeC. Ward, 1 April 1924: Truman C-4-1:8

66. Davenport to Barker, 18 April 1924.
67. US Department of Justice, "Immigration Act of May 26, 1924 (43 Statutes-at-Large 153)," at www.ins.usdoj.gov. Ragsdale, p. 17. Anderson, p. 146.
68. Anderson, pp. 147, 149. Ragsdale, p. 42.
69. Alfred P. Schultz, *Race or Mongrel* (Boston: L. C. Page and Company, 1908) as cited by Ragsdale, p. 11.
70. Ragsdale, p. 35.
71. See Ragsdale, p. 41. Ragsdale, pp. 41-42.
72. Ragsdale, p. 42.
73. Ragsdale, p. 43.
74. Ragsdale, pp. 41, 45, 46, 48, 49. Fourth Report of Committee on Selective Immigration, p. 6.
75. Letter, Harry H. Laughlin to Frank L. Babbott, 31 January 1927: Truman C-4-3:5.
76. Anderson, p. 149. "Immigration Act of May 26, 1924," pp. 422-423. See Fourth Report of Committee on Selective Immigration.
77. Robert DeC. Ward, "Higher Mental and Physical Standards for Immigrants", reprinted from *The Scientific Monthly*, Vol. IX (1924) p 539: Truman C-4-1:8. See Fourth Report of Committee on Selective Immigration, pp. 20, 28-30. Draft copy, "Immigration Service," (n.d), p. 2: Truman C-2-4:5.
78. US Department of Justice, "Immigration and Nationality Act of June 27, 1952 (INA) (66 Statutes-at-Large 163)," at www.ins.usdoj.gov. US Department of Justice, "Immigration and Nationality Act" at www.ins.usdoj.gov.

CHAPTER ELEVEN

1. See Robert Reid Rentoul, *Race Culture; Or Race Suicide?* (London: The Walter Scott Publishing Co., Ltd., 1906), pp. 4-5, 19-22. See Richard A. Soloway, *Demography and Degeneration: Eugenics and the Declining Birthrate in Twentieth-Century Britain*, (Chapel Hill, NC: The University of North Carolina Press, 1990), pp. 2-4.
2. Pauline M.H. Mazumdar, *Eugenics, Human Genetics and Human Failings* (London: Routledge, 1992), pp. 72-80, 89, 125, 143. Arthur H. Estabrook and Charles B. Davenport, *The Nam Family: A Study in Cacogenics* (Cold Spring Harbor, NY: Cold Spring Harbor Press, 1912), p. 1. "The Eugenics Record Office," *Eugenical News*, Vol. I (1916), p. 2. Charles B. Davenport, "First Report of Station for Experimental Evolution Under Department of Experimental Biology," *Carnegie Institution of Washington Year Book No. 3 1904* (Washington, DC: Carnegie Institution of Washington, 1905), pp. 22, 23, 33-34. American Breeders' Association, "Minutes of First Annual Meeting: St. Louis, Missouri: December 29th and 30th, 1903" memorandum circa 1904, pp. 1-3. Francis Galton, *Memories of*

my Life, (London: Methuen & Co., 1908), pp. 310, 320-321. See Francis Galton, "Eugenics; Its Definitions, Scope and Aims": University College London, Galton Papers, 138/9.

3. Rentoul, pp. 164, 165. Author's interview with Indiana State Library, 9 December 2002. Rentoul, p. i, xiv.

4. Francis Galton, "Eugenics: Its Definition, Scope and Aims," (paper read at a Meeting of the Sociological Society, 16 May 1904): UCL Galton Papers 138/9. Rentoul, p.164. Also see "An Easy Way of Sterilizing Degenerates," *The British Medical Journal,* 13 August 1904, pp. 346-347.

5. Rentoul, pp. i, 17-22, 24-25, 109-110, 133-142.

6. Rentoul, pp. 10, 44, 101, 155.

7. Rentoul, p. 133.

8. Rentoul, pp. 31-32.

9. Lady Georgina Chambers, "Notes on the Early Days of the 'Eugenics Education Society,'" pp 2, 3: Wellcome SA/EUG/B-11. Mazumdar, pp. 24, 25, 27, 29, 30. Letter, Leonard Darwin to David Starr Jordan, 1 January 1914: Hoover Institution Archives, Horder, Box 60, Folder 52. Also see Phyllis Grosskurth, *Havelock Ellis, A Biography* (London: Allen Lane, 1980), p. 412n.

10. Rentoul, p. 169. Letter, C.S. Tromp to R. Chalmers, 14 September 1906: PRO HO 45/10341/139871. "The Isle of Lundy," at www.lundy.org.uk.

11. Francis Galton, *Restrictions in Marriage* (American Journal of Sociology, 1906), p. 3. Francis Galton, *Memories of My Life* (London: Methuen & Co., 1908), p. 310. Major Leonard Darwin, "First Steps Towards Eugenic Reform," *Eugenics Review,* Vol. 4 (ca. April 1912), pp 34-35 as selected in G. K. Chesterton, *Eugenics and Other Evils,* edited by and including additional articles selected by Michael W. Perry (Seattle, WA: Inkling Press, 2000), pp 144-145.

12. "Eugenical Sterilization in England," *Eugenical News* Vol. X (1925), pp. 134-135. Letter, Hugh MacEwen to Sir George Newman, 12 August 1930: PRO MH79/291. Letter, A. Neville to A.S. Moshinsky, 20 February 1937: PRO MH79/291.

13. Soloway, pp. 74-75.

14. "Notes on the Early Days," p. 33.

15. "Notes on the Early Days," pp. 3, 6-7. "A Large Family" and "A Decadent Family", Admissions forms for Sandlebridge Boarding Special School: UCL, Galton Papers, 138/8. "Notes on the Early Days of the 'Eugenics Education Society,'" pp. 4, 9. Dr. Caleb W. Saleeby, "The House of Life: The Mental Deficiency Bill," July 23 1912.

16. Letter, Sybil Gotto to Francis Galton, 11 December 1909: UCL, Galton Papers, 240/7.

"Eugenics: Prof. Karl Pearson on its Methods," *The Standard,* 3 January 1910. See "Notes on the Early Days," p. 32.

17. Carnegie Institution of Washington, *Announcement of Station for Experimental Evolution* (Washington: Carnegie Institution of Washington, 1905), p. 4: APS: Davenport Beginnings of Cold Spring Harbor. The Eugenics Education Society, "Programme," *Problems in Eugenics Vol. II: Report of Proceedings of the First International Eugenical Congress* (Kingsway, W.C.: Eugenics Education Society, 1913), pp. 3, 5, 6-13.

18. "Programme," *Problems in Eugenics Vol. II,* p. 2.

19. Saleeby, "The House of Life: The Mental Deficiency Bill."

20. Saleeby, "The Discussion of Alcoholism," p. 6. Richard Allen Soloway, *Birth Control and the Population Question in England, 1877-1930* (Chapel Hill, NC: University of North Carolina Press, 1982), p. 17. Rentoul, p. i. "Notes on the Early Days," pp. 4, 9.

21. "The International Eugenics Congress." Saleeby, "The Discussion of Alcoholism at the Eugenics Congress," p. 6. Saleeby, "The House of Life: The Mental Deficiency Bill."

22. Grotto to Galton, 11 December 1909.

23. Michael Warren, *A Chronology of State Medicine, Public Health, Welfare and Related Services in Britain: 1066 - 1999.*

24. Mazumdar, pp. 22-23. Daniel J. Kevles, *In The Name of Eugenics,* (Cambridge, MA: Harvard University Press, 1985), p. 98.

25. Lord Riddell, "Sterilization of the Unfit: A Paper for the Medico-Legal Society," memorandum, circa February 1929, p. 17: PRO MH 58/103.

26. Mazumdar, pp. 23-24. Kevles, p. 98.

27. "Editorial Notes", *Eugenics Review* Vol. 2 (October 1910), pp. 163-164. Letter, Winston Churchill to unknown recipient, 27 May 1910: PRO HO 144/1085/193548/1.

28. Darwin, "First Steps Towards Eugenic Reform."

29. Darwin, "First Steps Towards Eugenic Reform." Riddell, p. 17.

30. Caleb Saleeby, *The Progress of Eugenics* (London: Cassell, 1914), p. 181, as selected by Perry, p. 133.

31. Darwin, "First Steps Towards Eugenic Reform." "The Mental Deficiency Act, 1913," *Eugenics Review* Vol. 5 (Apr. 1913- Jan. 1914), p. 290, as selected by Perry, p. 148.

32. Saleeby, "The House of Life."

33. Darwin, "First Steps Towards Eugenic Reform."

34. Saleeby, "The House of Life."

35. Saleeby, "The House of Life."

36. "Mental Deficiency Bill," *Eugenics Review* Vol. 4 (circa January 1913), p. 420, as selected by

Perry, p. 146. R. Langdon-Down, "The Mental Deficiency Bill," *Eugenics Review* Vol. 5 (circa April 1913-January 1914), pp. 166-167, as selected by Perry, p. 147.

37. "The Mental Deficiency Act, 1913," p. 148. Eugenics Society, "The Sterilization of Mental Defectives," draft of leaflet, circa 1929: PRO MH 58/104A.

38. Saleeby, *The Progress of Eugenics*, pp. 188-189, as selected by Perry, p. 134.

39. "The Mental Deficiency Act, 1913," p. 290, as selected by Perry, p. 148. "The Mental Deficiency Act," *Eugenics Review* Vol. 9 (April 1917 - January 1918), p. 263 as cited by Perry, pp. 148-149.

40. See Harry H. Laughlin, *Eugenical Sterilization in the United States* (Chicago: Psychopathic Laboratory of the Municipal Court of Chicago, 1922).

41. Galton, *Memories of my Life*, pp. 293-294, 320-321. "Galton Laboratory for National Eugenics and the Biometric Laboratory," *Organized Eugenics*, (New Haven, CT: American Eugenics Society), 1931, p. 37. Soloway, Demography and Degenration, p. 163. C. P. Blacker, *Eugenics: Galton and After* (Westport, CT: Hyperion Press, Inc.), p. 237. Mazumdar, pp. 82, 85.

42. Letter, Francis Benedict to Karl Pearson, 20 November 1920: UCL, Pearson Papers, 653/2. Letter, Francis Benedict to Karl Pearson, 13 December 1920: UCL, Pearson Papers, 653/2.

43. Mazumdar, pp. 77, 85-87, 289, 328.

44. Mazumdar, p. 72. Report of the Committee appointed to consider the Eugenic Aspect of Poor Law Reform, "Section I: The Eugenic Principle in Poor Law Administration," *Eugenic Review* Vol. 2 (1910-1911) pp. 167-177 as cited by Mazumdar, p. 72. Eugenics Education Society, *Third Annual Report* (1911), p. 18 as cited by Mazumdar, pp. 71-72.

45. Mazumdar, pp. 71-72, 133-135, 205-207. MacNicol, p 429.

46. Mazumdar, pp. 72, 73. "Metropolitan Relieving Officers' Association: Eugenics and the Poor Law," *The Poor-Law Officers' Journal* 26 September 1913 p. 1217. "Life and Scenes in London #1: 'Bethnal Green,'" *The Nineteenth Century* (June 1924) as cited by Casebook: Jack The Ripper at www.casebook.org.

47. "Metropolitan Relieving Officers' Association: Eugenics and the Poor Law."

48. Mazumdar, pp. 109-121, 124, 125. See "Rothamsted," at nolimits.nmw.ac.uk.

49. Mazumdar, pp. 125, 126, 137, 142, 294.

50. Letter, Harry H. Laughlin to Harry Olson, 12 October 1923: Truman D-2-3:6. Letter, Harry H. Laughlin to Charles B. Davenport, 22 November 1923: Truman C-2-6:17.

51. Laughlin to Olson, 12 October 1923.

52. Generally see David Starr Jordan, *War and the Breed: The Relation of War to the Downfall of Nations* (1915).

53. Jordan.

54. Jordan. *Eugenics Review*, Vol. 6, No. 3 (October 1914), pp. 197-198 as cited by Soloway, p. 141.

55. Letter, Cora Hodson to Elton Mayo, 27 June 1927: Eugenics Society Paper C210, as cited by Mazumdar, pp. 127-128.

56. Mazumdar, p. 133.

57. Mazumdar, p. 137. Letter, Cora Hodson to Sir Walter Moley Fletcher, 15 September 1927: PRO FDI/1734.

58. "Population studies in Edinburgh," *Eugenics Review* Vol. 18 (1926-27), pp. 227-230 as cited by Mazumdar p. 137.

59. Letter, Cora Hodson to Sir Walter Moley Fletcher, 15 September 1927. Letter, Cora Hodson to Miss C. H. Paterson, 8 February 1926: Wellcome Box 112. Mazumdar, pp. 133-137, 142. See Daniel Kevles, *In The Name of Eugenics* (Cambridge: Harvard University Press, 1985), pp. 100-101. See John MacNicol, "The Voluntary Sterilization Campaign in Britain, 1918-39," *The Journal of the History of Sexuality*, Vol. 2., No. 3 (1992), p. 429.

60. Caleb Saleeby, "Two Decades of Eugenics," *The Sociological Review* 16 (July 1924), pp. 251-253 as cited by Perry, p. 135. "History and Survey of the Eugenics Movement, Committee on the," *Organized Eugenics*, p. 17.

61. Letter, Cora Hodson to Irving Fisher, 17 June 1925: Truman C-2-5:6. Letter, Field Secretary to Cora Hodson, 29 June 1925: Truman C-2-5:6.

62. Letter, Cora Hodson to S. Wayne Evans, 9 June 1931: Wellcome SA/EUG/E-1.

63. Letter, Paul M. Kinsie to Harry H. Laughlin, 28 March 1928: Truman C-2-5:6.

64. Eugenics Education Society, "Minutes of Proceedings at A Meeting held at The Rooms of the Royal Society, Burlington House, London, W. on Tuesday, January 29th, 1924": Truman D_5-2:13. Harry H. Laughlin, "Eugenics in America," *Eugenics Review* April 1925.

65. Laughlin, "Eugenics in America."

66. Cora Hodson, "Draft of Letter to 'The Times.'"

67. "Segregation versus Sterilization," *Eugenical News* vol. X (1925), pp. 2-3.

68. C. P. Blacker, *Eugenics: Galton and After* (Cambridge, MA: Harvard University Press 1952 reprinted by Westport, CT: Hyperion Press, Inc., 1987), p. 203. "The Sterilization of Mental Defectives."

69. Mazumdar, pp. 197, 198. MacNicol pp 428, 429.

70. Ministry of Health, "Existing Position in U.K.": PRO MH 58/104A.

71. Letter, Ellen Askwith to Neville Chamberlain,

16 February 1929: PRO MH 58/103. Letter, Sir Bernard Mallet to Neville Chamberlain, 18 February 1929: PRO MH 58/103 98826.

72. Letter, Frederick. J. Willis to Leonard Darwin, 8 July 1927: PRO MH 51/547. Frederick J. Willis, "Sterilization Bill," draft attached to letter, 8 July 1927: PRO MH 51/547. Bernard Mallet, "Draft of Sterilization Bill," circa 1929: PRO MH 51/547.

73. Letter, Lord Riddell to Neville Chamberlain, 27 April 1929: PRO MH 58/103. Riddell, pp. 1, 9, 17, 20.

74. Riddell, p. 21.

75. Mazumdar, p. 204. "Committee for Legalizing Eugenic Sterilization," circa 1929, pp. 16, 28-29: PRO MH 58/103 98826.

76. *The Columbia Encyclopedia*, 6th ed., s.v. "Great Britain." Riddell, pp. 9, 10, 11.

77. Soloway, Demography and Degeneration, pp. 163, 381 footnote 3. C. P. Blacker, "Eugenics In Prospect and Retrospect," *The Galton Lecture, 1945* (Hamish Hamilton Medical Books, n.d.), p. 18.

78. Letter, Cora Hodson to Ernst Rüdin, 24 July 1930: Eugenic Society Papers C300 as cited by Mazumdar, p. 205. Ernst Rüdin, "Psychiatrische Indikation zur Sterilisierung," *Das kommende Geschlecht* 5 (1929), pp. 1-19: Eugenics Society Papers C300 as cited by Mazumdar p. 206. Mazumdar pp 205, 309 footnote 21.

79. Letter, Cora Hodson to S. Wayne Evans, 11 June 1930: Wellcome SA/EUG/E-1.

80. Letter, Cora Hodson to Charles B. Davenport, 15 February 1930: APS B:D 27 - IFEO 1930 #1. See Letter, Cora Hodson to Charles B. Davenport, 25 March 1930: APS B:D 27 - IFEO 1930 #1. See Letter, Charles B. Davenport to Cora Hodson, 31 March 1930: APS B:D 27 - IFEO 1930 #1. See Letter, Charles B. Davenport to Cora Hodson, 13 May 1930: APS B:D 27 - IFEO 1930 #1. See Letter, Charles B. Davenport to Cora Hodson, 13 June 1930: APS B:D 27 - IFEO 1930 #1.

81. Hodson to Davenport, 25 March 1930. Davenport to Hodson, 31 March 1930. Letter, Cora Hodson to Sir Walter Moley Fletcher, 10 April 1930: APS B:D 27 - IFEO 1930 #1.

82. Davenport to Hodson, 13 May 1930.

83. Davenport to Hodson, 13 June 1930. "1930 Meeting of International Federation of Eugenic Organization: Programme and Time Table": Truman C-2-4:3.

84. Eric Donaldson, "Operations on Mentally Deficient Patients in the Poor Law Hospital," pp. 1,4: PRO MH 79/291.

85. Donaldson, p.2.

86. Donaldson, p.3.

87. Donaldson, pp. 2-3.

88. Donaldson, pp. 2, 3, 4. Letter, Eric Donaldson to Hugh MacEwen, 9 August 1930: PRO MH 79/291. MacEwen to Newman, 12 August 1930.

89. Donaldson, p. 1. Letter, Laurence Brock to R. H. H. Keenlyside, 1 August 1930: PRO MH 79/291.

90. Letter, Lionel L. Westrope to the Ministry of Health, 14 October 1930: PRO MH 79/291. "Ambulance Notes," *L.N.E.R. Magazine* Vol. 29 No. 5 (May, 1939).

91. MacNicol, pp. 431, 432. Mazumdar. pp. 211, 212. See *Casti Connubii: Encyclical of Pope Pius XI on Christian Marriages.*

92. *Casti Connubii*, section 68.

93. *Casti Connubii*, section 70.

94. *Casti Connubii*, sections 63, 64. Exodus 20:13 NIV Study Bible.

95. Eugenics Society, *Sterilization of Mental Defectives*, n.p., n.d.: Wellcome SA/EUG/N-32. "Should the Unfit be Sterilized?" newspaper clipping, n.p., n.d.: Wellcome SA/EUG/N-33. *Committee for Legalizing Sterilization*, p. 16.

96. MacNicol, pp. 429, 435. Eugenics Society, *Annual Report 1931-32*, n.p., circa 1932, p. 6: Wellcome SA/EUG/A-24.

97. MacNicol, p 429. Mazumdar pp. 211, 212.

98. Letter, Eugenics Society to Michael Pease, 17 August 1931: Wellcome Box 112. "Committee for Legalising Sterilization", pp. 15-17. Mazumdar, p. 206.

99. Letter, British Embassy to Sir John Simon, 17 November 1938: Wellcome Box 112. Brock Committee, "Summary of Principal Recommendation," p. 1: PRO MH 51/210. Mazumdar, p. 203. Brock Committee, "Section 86: The Problem of the Carrier": PRO MH 51/210.

100. Mazumdar, pp. 210-211.

101. Blacker, pp 303-304. "Population and Its Control," *Eugenical News*, Vol. XX (1935), p. 100. "Publication of the State Law: Part I: Given out in Berlin - 25th July 1933, No. 86" circa 1933: Wellcome Library Box 112.

CHAPTER TWELVE

1. Harry Laughlin, secretary, *Bulletin #10A: Report of the Committee to Study and to Report on the Best Practical Means of Cutting Off the Defective Germ-Plasm in the American Population: I. The Scope of the Committee's Work* (Cold Spring Harbor, NY: Cold Spring Harbor, 1914), pp. 1, 9.

2. See *Problems in Eugenics Vol. II: Report of Proceedings of the First International Eugenical Congress* (Kingsway, W.C.: Eugenics Education Society, 1913). Caleb W. Saleeby, "The House of Life: The Mental Deficiency Bill," 22 July 1912. See Laughlin, *Bulletin #10A*. See Harry H. Laughlin, secretary, *Bulletin #10B: : Report of*

the Committee to Study and to Report on the Best Practical Means of Cutting Off the Defective Germ-Plasm in the American Population: II. The Legal, Legislative and Administrative Aspects of Sterilization (Cold Spring Harbor, NY: Cold Spring Harbor, 1914).

3. The Eugenics Education Society, "Programme," *Problems in Eugenics Vol. II: Report of Proceedings of the First International Eugenical Congress* (Kingsway, W.C.: Eugenics Education Society, 1913), p. 3. "History of the International Organisation of Eugenics," memorandum circa November 1923, pp. 1-12: Truman C-2-1:2.

4. "History of the International Organisation of Eugenics," p. 3.

5. Letter, Charles B. Davenport to Madison Grant, 10 October 1919: APS B:D27 - Grant, Madison #1. Letter, Charles B. Davenport to Madison Grant, 2 April 1920: APS B:D27 - Grant, Madison #3. Letter, Madison Grant to Charles B. Davenport, 7 April 1920: APS B:D27 - Grant, Madison #3. Letter, Madison Grant to Charles B. Davenport, 13 April 1920: APS B:D27 - Grant, Madison #3. Letter, Alvey A. Adee to Charles B. Davenport, 5 February 1921: NA 59/250/22/10/3/2620. Letter, C. C. Kimble to Envoy Extraordinary & Minister Plenipotentiary of the United States of America, 17 March 1921: NA 59/250/22/10/3/5459. Letter, Charles S. Hartman to Sr. Dr. Dn. N. Clemente Ponce, 7 June 1921: NA 59/250/22/10/3/5459.

6. Grant to Davenport, 13 April 1920. Davenport to Grant, 2 April 1920. Lothrop Stoddard, *The Rising Tide of Color* (New York: Charles Scribner's Sons, 1926), p. i. "Second Eugenics Congress," *Eugenical News* Vol. VI (1921) p. 65. American Museum of Natural History, "Timeline" at www.amnh.org. Second International Congress of Eugenics, *Eugenics, Genetics and the Family: Volume I: Scientific Papers* (Baltimore: Williams & Wilkins Company, 1923), p. i. Harry H. Laughlin, *The Second International Exhibition of Eugenics* (Baltimore: Williams & Wilkins Co., 1923), p. 13.

7. *Eugenics, Genetics and the Family: Volume I: Scientific Papers*, p. ii. "Second Eugenics Congress," p. 64. Laughlin, *The Second International Exhibition of Eugenics*, p. 16. Herny Fairfield Osborn, "Address of Welcome," *Eugenics, Genetics and the Family: Volume I: Scientific Papers*, pp. 1, 3.

8. Osborn, p. 2.

9. *Eugenics, Genetics and the Family: Volume I: Scientific Papers*, pp. iii-v.

10. Letter, Hermann Lundborg to Charles B. Davenport, 29 August 1921: APS B:D27. Also See Letter, Charles B. Davenport to Hermann Lundborg, 24 October 1921: APS B:D27. Also

See Letter, Hermann Lundborg to Charles B. Davenport, 28 November 1921: APS B:D27.

11. Charles B. Davenport, "Research in Eugenics," *Eugenics, Genetics and the Family: Volume I: Scientific Papers*, p. 20.

12. Laughlin, *The Second International Exhibition of Eugenics*, pp. 13, 33, 36, 152-153. "II. International Congress of Eugenics," *Eugenical News*, Vol. VI (1921), p. 28. Princeton University, "Dodge-Osborn Hall," at etc.princeton.edu.

13. Arthur H. Estabrook, "The Second International Eugenics Congress," speech given to the Indiana Academy of Science, 2 December 1921: Truman E-2-4:9.

14. "Resolution Passed by the Executive Session of the Second International Congress, September 27, 1921," memorandum, n.d.: Wellcome SA/EUG/E11. "The International Eugenics Commission," *Eugenical News* Vol. VI (1921), p. 67. Letter, Charles B. Davenport to Dr. Erwin Baur, 30 March 1923: APS B:D27 Davenport & Erwin Baur. See "Meeting of International Commission at Lund - 1923" article, n.p., n.d.: Truman C-4-6:19.

15. "International Commission of Eugenics," *Eugenical News* Vol. VII (1922) p. 117. "History of the International Organisation of Eugenics," pp. 5-6, 7.

16. "Personals," *Eugenical News* Vol. VIII (1923), p. 94. Letter, John C. Merriam to Charles B. Davenport, 20 June 1923: CIW Genetics: Eugenics Record Office Misc. 2 of 2. "Minutes of the Meetings of the International Commission of Eugenics Held in the Rooms of the Medical Faculty of the University of Lund: Saturday, September 1st and Monday, September 3rd, 1923," memorandum: Truman C-2-1:2.

17. "History of the International Organisation of Eugenics," pp. 9, 12. "Report of Sub-Committee on Ultimate Program to be Developed by the Eugenics Society of the United States of America," *Eugenical News*, Vol. VIII (1923), p. 73. "Eugenics in India," *Eugenical News* Vol. VII (1922), p. 2. "Eugenics in Japan," *Eugenical News* Vol. VII (1922), p. 104.

18. "Meeting of International Commission at Lund - 1923," n.p. n.d. article: Truman C-4-6:19. Letter, Charles B. Davenport to Dr. Erwin Baur, 30 March 1923: APS B:D27 - Davenport & Erwin Baur. Letter, Leonard Darwin to Herman Lundborg, 21 November 1925: APS B:D27. Dr. Timothy Holian, "The German Hyperinflation of 1923: A Seventy-Fifth Anniversary Retrospective," at www.mwsc.edu. "Resolution Passed by the Executive Session of the Second International Eugenics Congress," p. 3.

19. "The International Commission of Eugenics," minutes of 14 July 1925 meeting: Truman C-2-5:6. See "Eugenics in the University of Padua," *Eugenical News* Vol. X (1925), p. 164. See "Eugenical Sterilization in Denmark," *Eugenical News* Vol. X (1925), p. 178. See "The International Federation of Eugenic Organizations," *Eugenical News* Vol. XI (1926), p. 100. See "Immigration to Norway," *Eugenical News* Vol. XI (1926), p. 139.

20. "The International Commission of Eugenics."

21. "Memorandum of the Objects of the International Federation of the Eugenics Organizations," circa September 1928: APS B:D27 - IFEO 1928 #2.

22. "Seventh Meeting of the International Commission of Eugenics," *Eugenical News* Vol. X (1925), p. 117. "International Federation of Eugenic Organizations," *Eugenical News* Vol. XII (1927), p. 153. "Memorandum of the Objects of the International Federation of the Eugenics Organizations."

23. Bent Sigurd Hansen, "Something Rotten in the State of Denmark: Eugenics and the Ascent of the Welfare State," in *Eugenics and the Welfare State*, edited by Gunnar Broberg and Nils Roll-Hansen (East Lansing, MI: Michigan State University Press, 1996), pp. 44, 51. See Nils Roll-Hansen, "Conclusion: Scandinavian Eugenics in the International Context," in Broberg and Roll-Hansen, p. 268. See William H. Schneider, "The Eugenics Movement in France, 1890-1940," in *The Wellborn Science*, edited by Mark B. Adams (New York: Oxford University Press, 1990), pp. 80-83. "Pan's Plans," *Eugenical News*, Vol. IX (1924), p. 80.

24. "Belgium Society of Eugenics," *Eugenical News* Vol. V (1920), p. 63. "Revue D'Eugenique," *Eugenical News* Vol. VI (1921), p. 43. "Societe Belge D'Eugenique," *Eugenical News* Vol. V (1920), p. 54.

25. "Revue D'Eugenique," p. 43. "Societe Belge D'Eugenique," p. 54. "Belgium Society of Eugenics," p. 63. "Foreign Notes," *Eugenical News* Vol. VI (1921), p. 72.

26. "Belgian Eugenics Society," *Eugenical News*, Vol. VII (1922), p. 14. "Dr. Albert Govaerts of Belgium," *Eugenical News*, Vol. VII (1922), p. 64. "The New Belgian Eugenics Office," *Eugenical News*, Vol. VII (1922), p. 92. "National Office of Eugenics in Belgium," *Eugenical News*, Vol. VII (1922), p. 120. "The Hereditary Factor in the Etiology of Tuberculosis," *Eugenical News*, Vol. VIII (1923), p. 32.

27. "Prenuptial Examinations in Belgium, Luxemburg and Germany," *Eugenical News*, Vol. XII (1927), p. 114. "The New Belgian Eugenics Office," p. 92. Letter, Harry H. Laughlin to Dr. Albert Govaerts, 18 July 1923: Truman C-4-6:19. Letter, Harry H. Laughlin to Charles B. Davenport, 22 November 1923: Truman C-2-6:14. Letter, Harry H. Laughlin to Charles B. Davenport, 26 November 1923: Truman C-2-6:17.

28. Angus McLaren, *Our Own Master Race: Eugenics in Canada, 1885 - 1945* (Toronto, Ontario: McClelland & Stewart, Inc.), pp. 43, 47, 107, 181 f74. National Council of Women, *13th Report* (Toronto, Ontario: Johnstone, 1907), pp. 56, 58 as cited by McLaren, p. 38. R. W. Bruce Smith, "Mental Sanitation," *Canada Lancet*, Vol. 41 (1907-1908), p. 976 as cited by McLaren, p. 42. F. McKevley Bell, "Social Maladies," *Queen's Quarterly* Vol. 16 (1908-09), p. 52 as cited by McLaren, p. 52.

29. McLaren, pp. 42, 125, 159-160. See Brian L. Ross, "An Unusual Defeat: The Manitoba Controversy over Eugenical Sterilization in 1933," unpublished paper, Institute for the History and Philosophy of Science and Technology, University of Toronto, 1981, as cited by McLaren, p. 196 f8. "Eugenical Sterilization in Canada," *Eugenical News*, Vol. XIII (1928), p. 47. Timothy J. Christian, "The Mentally Ill and Human Rights in Alberta," unpublished University of Alberta paper, n.d., pp. 13-20, 25-29 as cited by McLaren p. 100.

30. Terry L. Chapman, "The Early Eugenics Movement in Western Canada," *Alberta History* Vol. 25 (1977), pp. 9-17 as cited by McLaren, pp. 90-91. Statutes of the Province of British Columbia, 1933, "An Act Respecting Sexual Sterilization," ch. 59, 7 April 1933 as cited by McLaren, p. 91. M. Stewart, "Some Aspects of Eugenical Sterilization in British Columbia with Special Reference to Patients Sterilized from Essondale Provincial Hospital since 1935," Provincial Archives of British Columbia, Provincial Secretary, Mental Health Services, GR 542, box 14, "Sterilization" as cited by McLaren, p. 160.

31. See "Dr. A. Forel's Views," *Eugenical News*, Vol. XI (1926). Véronique Mottier, "Narratives of National Identity: Sexuality, Race, and the Swiss 'Dream of Order,'" paper presented at the European Consortium for Political Research Annual Joint Sessions, Workshop: The Political Uses of Narrative, at Mannheim 26-31 March 1999, pp. 11, 13. "The Julius Klaus Fund," *Eugenical News*, Vol. VIII (1923), p. 36.

32. Mottier, p. 11. "The Julius Klaus Fund," p. 36. "Accessions to Archives of the Eugenics Record Office, January, 1924," *Eugenical News*, Vol. IX (1924), p. 19. "Julius Klaus-Stiftung," *Eugenical News*, Vol. X (1925) pp. 139-140. "Meeting of International Commission," *Eugenical News*, Vol. VIII (1923), p. 116.

33. Mottier, pp. 14, 15, 16. "Dr. A. Forel's Views," p. 134. "Sterilization in Switzerland," *Eugenical News*, Vol. XI (1926), p. 91. "New Sterilization

Statutes," *Eugenical News,* Vol XIV (1929), p.
63. Robert N. Proctor, *Racial Hygiene: Medicine
Under the Nazis* (Cambridge, MA: Harvard
University Press, 1988), p. 97.
34. "Eugenics in Denmark," *Eugenical News,* Vol.
VIII (1923), pp. 6. Hansen, pp. 9, 19. "Eugenics
in Denmark," *Eugenical News,* Vol. X (1925), p.
81. "Danish-Sterilization Law," *Eugenical News,*
Vol. XIV (1929), pp. 122-124. "Eugenical
Registration in Denmark," *Eugenical News,* Vol.
XV (1930), p. 100. Hilda von Hellmer Wullen,
"Eugenics in Other Lands," *Journal of Heredity*
Vol. XXVIII No. 8 (August 1937), p. 274.
Letter, Cora Hodson to Sören Hansen, 28 June
1928: APS B:D27.
35. "Eugenical Sterilization in Denmark," *Eugenical
News* Vol. XXII (1927), p. 178. Letter, Harry H.
Laughlin to Harry Olson, 15 November 1927:
Truman D-2-2:16. "Danish-Sterlization Law,"
p. 122.
36. Rockefeller Foundation, "University of
Copenhagen - Institute of Human Genetics,"
June 1939 Appraisal, pp. 3, 5: RF RG 1.2 Ser
713 Box 2 Folder 15. Letter, Tage Kemp to the
Rockefeller Foundation, 17 November 1932:
RF RG 1.2 Ser 713 Box 2 Folder 15.
37. "University of Copenhagen - Institute of
Human Genetics," p. 1.
38. "Race Hygiene in Scandinavia," *Eugenical News,*
Vol. IV (1919), p. 88. Osborn, p. 1. See
"Photograph of Jon Alfred Mjøen in Library,"
at www.amphilsoc.org. "Part I: The American
Eugenics Society, Inc.: B. Early History and
Development," *Organized Eugenics,* January
1931, p. 3.
39. "Race Hygiene in Scandinavia," *Eugenical News,*
Vol. IV (1919), p. 88. "Der Nordske Race,"
Eugenical News, Vol. V (1920), p. 2. "Personal
Notes," *Eugenical News,* Vol. VII (1922), p. 113.
"Notes and News," *Eugenical News,* Vol. VIII
(1923), p. 88. "Eugenical Activities in the
Different Countries: V. Eugenics in Norway,"
Eugenical News, Vol. X (1925), pp. 55-57.
"Immigration to Norway," *Eugenical News,* Vol.
XI (1926), p. 139. "American Lecture Tour of
Dr. Mjoen." *Eugenical News,* Vol. XII (1927), p.
24. "Galton Society," *Eugenical News,* Vol. XII
(1927), p. 54. "Dr. Mjoen's Lectures," *Eugenical
News,* Vol. XII (1927), pp. 139-140. See
"Photograph of Jon Alfred Mjøen and Leon
Whitney," at www.amphilsoc.org. Klaus
Hansen, "The Norwegian Sterilization Law of
1934 and its Practical Results," *Eugenical News,*
Vol. XXI (1936), p. 129. Nils Roll-Hansen,
"Norwegian Eugenics: Sterilization as Social
Reform," in Broberg and Roll-Hansen, pp. 176,
178.
40. "Race-Biology in Sweden," *Eugenical News,* Vol.
VII (1922), p. 121. Letter, Charles B.
Davenport to Herman Lundborg, 2 June 1923:

APS B:D27. Letter, Herman Lundborg to
Charles B. Davenport, 9 February 1926: APS
B:D27. Letter, Charles B. Davenport to
Herman Lundborg, 1 March 1926: APS B:D27.
Letter, Herman Lundborg to Charles B.
Davenport, 24 April 1926: APS B:D27. Letter,
Charles B. Davenport to Herman Lundborg, 1
October 1928: APS B:D27. Gunnar Broberg
and Mattias Tydén, "Eugenics in Sweden:
Efficient Care," in Broberg and Roll-Hansen,
pp. 102-103, 109-110. Kungl. Maj:t, *Lag om
sterilisering av vissa sinnessjuka, sinneslöa eller
andra som lida av rubbad själsverksamhet,* Svensk
författningssamling no. 1934/171 [Sterilization
Act of 1934] (Stockholm: P.A. Norstedt &
Sönners förlag, 1934) as cited by Broberg and
Tydén, p. 103. Kungl. Maj:t, *Lag om sterilisering,*
Svensk författningssamling no. 1941/282
[Sterilization Act of 1941] (Stockholm: P.A.
Norstedt & Sönners förlag, 1941), as cited by
Broberg and Tydén, p. 108. *Sveriges Offentliga
Statistik: Allmän hälso- och sjukvärd* [Annual
Reports on Health Published by the Swedish
Central Bureau of Statistics] (Stockholm:
Statistiska centralbyrän, 1935-1976) as cited by
Broberg and Tydén, pp. 108-109.
41. "Sterilization Law in Finland," *Eugenical News,*
Vol. XXIII (1938), p. 47. "Race-Hygiene in
Roumania," *Eugenical News,* Vol. XI (1926), p.
136. "The Italian Society of Genetics and
Eugenics," *Eugenical News,* Vol. X (1925), p. 13.
"Eugenical Activities in the Different
Countries," *Eugenical News,* Vol. X (1925), pp.
49-51. "Eugenical Efforts in Hungary,"
Eugenical News, Vol. XVI (1931), pp. 172-173.
"Vienna Society of Eugenics," *Eugenical News,*
Vol. X (1925), p. 152. von Hellmer Wullen, p.
271. "International Congress of Genetics"
Rockefeller Foundation memorandum, 4 June
1931: RF RG 1.1 Ser 100 Box 40 Folder 365.
"University of Copenhagen - Prof. Thomsen"
Rockefeller Foundation memorandum, 28
September 1934: RF RG 1.1 Ser 713 Box 2
Folder 15. Radiogram to Gregg, 13 Mat 1932:
RF RG 1.1 Ser 717 Box 10 Folder 63. Paul
Weindling, *Health, Race and German Politics
Between National Unification and Nazism 1870 -
1945* (Cambridge: Cambridge University Press,
1989), p. 468.
42. *A Decade of Progress in Eugenics: Scientific Papers
of the Third International Congress of Eugenics*
(Baltimore: The Williams & Wilkins Company,
1934), p. i. See Paul Popenoe and Roswell Hill
Johnson, *Applied Eugenics,* rev. ed. (New York:
The MacMillan Company, 1935). See Charles
B. Davenport, *Heredity in Relation to Eugenics*
(New York: Henry Holt and Company, 1911:
Reprint, New York: Arno Press & The New
York Times, 1972). See E.S. Gosney and Paul
Popenoe, *Sterilization for Human Betterment*

(New York: The MacMillan Company, 1929). See Harry H. Laughlin, *Eugenical Sterilization: 1926: Historical, Legal, and Statistical Review of Eugenical Sterilization in the United States* (Lancaster, PA: Lancaster Press, 1926). See Harry H. Laughlin, *Immigration and Conquest* (New York: The Special Committee on Immigration and Naturalization of the Chamber of Commerce of the State of New York, 1939). See "Forward," *Eugenical News*, Vol. I (1916), p. 1. "College Courses in Genetics and Eugenics," *Eugenical News* vol. 1 (1916), pp. 26-27.

43. D. V. Glass, "Population Policies and Their Objectives," *Eugenical News*, Vol. XXVII (1942), p. 8. Schneider, pp. 78, 79. "Actual Aspect of the Problem of Eugenical Sterilization in France," *Eugenical News*, Vol. XXI (1936), p. 105. Historical Sample of the Netherlands, "Sources- Population Registers," at www.iisg.nl. Nancy Leys Stepan, *"The Hour of Eugenics": Race, Gender, and Nation in Latin America* (Ithaca, NY: Cornell University Press, 1991), p. 126. von Hellmer Wullen, p. 274. "Sterilization Law in Finland," p. 47.

CHAPTER THIRTEEN

1. Bleecker Van Wagenen, Chairman, *Preliminary Report of the Committee of the Eugenic Section of the American Breeder's Association to Study and to Report on the Best Practical Means for Cutting Off the Defective Germ-Plasm in the Human Population*, p. 5: ABA.

2. "The Richardson Lethal Chamber (patented) for the Painless Extinction of Lower Animal Life," undated pamphlet: UCD Special Collections.

3. Robert W. Chambers, *The King in Yellow* (F. Tennyson Neely, 1895), p. 9. Arnold White, *Efficiency and Empire* (London: Methuen and Co., 1901), pp. 116-117 as cited by Dan Stone, *Breeding Superman: Nietzsche, Race and Eugenics in Edwardian and Interwar Britain* (Liverpool: Liverpool University Press, 2002), p. 125.

4. H.G. Wells, *A Modern Utopia* (New York: Charles Scribner's Sons, 1905), p. 129 as cited by Michael W. Perry, editor, *The Pivot of Civilization: In Historical Perspective* (Seattle: Inkling Books, 2001), p. 36. Eden Paul, "Eugenics, Birth-Control, and Socialism" in *Population and Birth Control*, edited by Eden and Cedar Paul (New York: Critic and Guide, 1917), pp. 144-146 as cited by Perry, p. 108.

5. Robert Reid Rentoul, *Race Culture; or, Race Suicide?* (London: The Walter Scott Publishing Company, Ltd., 1906), pp. 178, 179.

6. Article, *Daily Express*, 4 March 1910 as cited by Stone, p. 127. Stone, p. 128.

7. C.W. Wilson as quoted in article, *Birmingham Post*, 4 February 1910, as cited by Stone, p. 127. Arnold White, *The Views of 'Vanoc': An Englishman's Outlook* (London: Kegan, Paul, Trench, Trübner, and Co., 1910), pp. 282-283 as cited by Stone, p. 126.

8. A.F. Tredgold, "Eugenics and the Future Progress of Man," *Eugenics Review* Vol. III (1911), p. 100 as cited by Stone, p. 126. Caleb Saleeby, *The Methods of Race Regeneration* (London: Cassell & Co., 1911), pp. 46-47, as cited by Stone, p. 126. *Proceedings of the First National Conference on Race Betterment*, (Race Betterment Foundation, 1914), p. 477.

9. Martin A. Elks, "The 'Lethal Chamber': Further Evidence for the Euthanasia Option," *Mental Retardation*, Vol. 31 No. 4 (August 1993), p. 203. A. F. Tredgold, *A Textbook of Mental Deficiency (Amentia)*, 2d ed. (New York: William Wood, 1915), p. 455 as cited by Elks, p. 203. A. F. Tredgold, *A Textbook of Mental Deficiency (Amentia)*, 6th ed. (New York: William Wood, 1937), pp. 517-518 as cited by Elks, p. 203. A. F. Tredgold, *A Textbook of Mental Deficiency (Amentia)*, 7th ed. (New York: William Wood, 1977), p. 491 as cited by Elks, pp. 203-204.

10. W. Duncan McKim, *Heredity and Human Progress* (New York: G. P. Putnam's Sons, 1900), pp. 120, 168 as cited by Mark H. Haller, *Eugenics: Hereditarian Attitudes in American Thought* (New Brunswick, NJ: Rutgers University Press, 1963), p. 42. McKim, pp. 192-193, as cited by Russell Hollander, "Euthanasia and Mental Retardation: Suggesting the Unthinkable," *Mental Retardation* Vol. 27 No. 2 (April 1989), p. 58.

11. A. Johnson, "Report of Committee on Colonies for Segregation of Defectives," *Proceedings of the National Conference on Charities and Corrections, 1903* (Fred J. Heer, 1903), p. 249 as cited by Hollander, pp. 58-59. E.R. Johnstone, "President's Address," *Journal of Psycho-Asthenics*, Vol. 8 (1904), pp. 65-70 as cited by Hollander, pp. 55, 58. Haller, p. 207 f. f. 5.

12. Milroy, "Discussion and Minutes," *Journal of Psycho-Asthenics*, Vol. 10 (1906), p. 224 as cited by Hollander, p. 58. Rentoul, p. 178.

13. E.B. Sherlock, *The Feeble-minded: A Guide to Study and Practice* (New York: Macmillan, 1911), p. 267 as cited by Elks, p. 202.

14. Henry H. Goddard, *The Kallikak Family: A Study in the Heredity of Feeble-Mindedness* (Vineland, New Jersey: 1913), pp. 101, 105, 106-108.

15. Harry H. Laughlin, secretary, *Bulletin No. 10A: The Report of the Committee to Study and to Report on the Best Practical Means of Cutting Off the Defective Germ-Plasm in the American Population* (Cold Spring Harbor: Cold Spring Harbor, 1914), pp. 46, 55: CSH.

16. William J. Robinson, *Eugenics, Marriage, and*

Birth Control (New York: The Critic and Guide Company, 1917), p. 74. Margaret Sanger, *The Pivot of Civilization* (New York: Brentano's, 1922), pp. 100-101.

17. *Proceedings of the First National Conference on Race Betterment*, (Race Betterment Foundation, 1914), pp. 502, 503.

18. Paul Popenoe and Roswell Hill Johnson, *Applied Eugenics* (New York: Macmillan, 1918), p. 184.

19. Eugenics Record Office, *Official Record of the Gift of the Eugenics Record Office, Cold Spring Harbor, Long Island, New York by Mrs. E.H. Harriman to the Carnegie Institution of Washington and of its Acceptance by the Institution* (Cold Spring Harbor, New York: Eugenics Record Office, 1918), p. 33: CSH. "American Eugenical Society, Inc.: B. Early History and Development," *Organized Eugenics*, January 1931, pp. 3-4, 7. Madison Grant, *The Passing of the Great Race* (New York: Charles Scribner's Sons, 1936), p. 49.

20. Martin S. Pernick, *The Black Stork: Eugenics and the Death of "Defective" Babies in American Medicine and Motion Pictures Since 1915* (New York: Oxford University Press, 1996), pp. 3-4. "Jury Clears, Yet Condemns Dr. Haiselden," *The Chicago Daily Tribune*, 20 November 1915.

21. "Jury Clears, Yet Condemns Dr. Haiselden."

22. "Jury Clears, Yet Condemns Dr. Haiselden."

23. Pernick, pp. 9, 10. "Jury Clears, Yet Condemns Dr. Haiselden." *Chicago Tribune*, 20 December 1915, p. 7 as cited by Pernick, p. 41. *Chicago American*, 3 December 1915, p. 6 as cited by Pernick, p. 41.

24. "Jury Clears, Yet Condemns Dr. Haiselden."

25. Pernick, p. 5. "Jury Clears, Yet Condemns Dr. Haiselden."

26. "Jury Clears, Yet Condemns Dr. Haiselden."

27. "Jury Clears, Yet Condemns Dr. Haiselden." *New York Times*, 18 November 1915 as cited by Pernick, pp. 7-8. *New York Times*, 24 December 1915 as cited by Pernick, pp. 7-8. *Chicago Herald*, 23 December 1915, as cited by Pernick, pp. 7-8.

28. "Baby Dies; Physician Upheld," *Chicago Daily Tribune*, 18 November 1915 as cited by Pernick, figure 3. "Was the Doctor Right? Some Independent Opinions," *The Independent*, 3 January 1916.

29. *Chicago Tribune*, 20 December 1915, p. 7 as cited by Pernick, p. 41. *Chicago American*, 3 December 1915, p. 6 as cited by Pernick, p. 41.

30. Pernick, pp. 4, 5, 87. *Chicago Herald*, 24 July 1917, p. 14 as cited by Pernick, p. 87. *New York Times*, 13 November 1917, p. 12 as cited by Pernick, p. 87.

31. "Was the Doctor Right?"

32. *Chicago American*, 23 November, 1915 through 30 December, 1915, as cited by Pernick, p. 4. *Chicago Herald*, 27 July, 1916 as cited by

Pernick, p. 10. Harry J. Haiselden, "Regarding the Meter Baby of Chicago," *Medical Review of Reviews*, 23 (Oct 1917), p. 697 as quoted in author's interview with Martin Pernick, 22 December 2002.

33. See Neal Black, *Animal Health: A Century of Progress*, Chapter 4. See BBC, "Cattle TB Threatens Farmers," 27 June 2002.

34. State of Illinois Board of Administration, *Volume II: Biennial Reports of the State Charitable Institutions: October 1, 1914 to September 30, 1916* (State of Illinois, 1917), p. 695. "Superintendent Leonard's Report to Board of Administration," *Institution Quarterly* Vol. 7 (1916), pp. 117-118.

35. Elks, p. 201. "Regarding the Meter Baby of Chicago," and *Chicago Examiner*, 25 July 1917, p. 6 as quoted in author's interview with Martin Pernick, 22 December 2002.

36. "The Report of Judge Scully's Committee of Three Woman," *Institution Quarterly*, Vol. 7 (1916), p. 113. "Superintendent Leonard's Report to Board of Administration," p. 117.

37. *Biennial Reports of the State Charitable Institutions: October 1, 1914 to September 30, 1916*, pp. 678, 682, 686. Patrick Almond Curtis, "Eugenic Reformers, Cultural Perceptions of Dependent Populations, and the Care of the Feebleminded in Illinois: 1909-1920," (Ph. D. diss., University of Illinois at Chicago, 1983), p. 89. See Paul Popenoe and Roswell Hill Johnson, *Applied Eugenics* rev. ed. (New York: Macmillan, 1935), pp. 90, 92-93, 94. *Eugenics, Genetics and the Family: Volume I: Scientific Papers of the Second International Congress*, (Baltimore: Williams & Wilkins Co., 1923), pp. 178-181. "Dr. Albert Govaerts of Belgium," *Eugenical News*, Vol. VII (1922), p. 64.

38. Charles B. Davenport, *Heredity In Relation To Eugenics* (New York: Henry Holt & Company, 1911; reprint, New York: Arno Press Inc., 1972), pp. 164-165. Popenoe and Johnson, 1918 as cited by Elks, p. 205.

39. Charles Henderson, "The Relation of Philanthropy to Social Order and Progress," *National Conference of Charities and Corrections: Proceedings of the Twenty-Sixth Annual Session* (Cincinnati: May, 1899), p. 4 as cited by Curtis, pp. 53, 55. "Propagation of the Unfit," *Institution Quarterly*, 1 (May 1910), p. 35 as cited by Curtis, pp. 68-69.

40. Charles B. Davenport, "Presidential Address," *A Decade of Progress in Eugenics: Scientific Papers of the Third International Congress of Eugenics* (Baltimore: The Williams & Wilkins Company, 1934), p. 21.

41. *Chicago American*, 24 November 1915, p. 2 as cited by Pernick, p. 84.

42. Illinois Department of Human Services, *A Brief History of the Lincoln Developmental Center*, p. 1.

Biennial Reports of the State Charitable Institutions: October 1, 1914 to September 30, 1916, pp. 679, 681, 686. Also see K. Charlie Lakin, "Demographic Studies of Residential Facilities for the Mentally Retarded: A Historical Review of Methodologies & Findings," University of Minnesota Department of Psychoeducational Studies.

43. Harrison L. Harley, "Observations on the Operation of the Illinois Commitment Law for the Feeble-Minded," *Institution Quarterly* Vol. 8 (1917), p. 97.

44. *Biennial Reports of the State Charitable Institutions: October 1, 1914 to September 30, 1916*, pp. 682, 686, 687. See Harley, p. 97.

45. *Biennial Reports of the State Charitable Institutions: October 1, 1914 to September 30, 1916*, pp. 677, 678, 679, 683. Martin W. Barr, *Mental Defectives: Their History, Treatment and Training* (Philadelphia: P. Blakiston's Son & Co., 1904 reprint New York: Arno Press, 1973), pp. 195-196. "The Municipal Psychopathic Clinic," *Eugenical News*, Vol. I (1916), p. 55. Harry H. Laughlin, *Eugenical Sterilization in the United States* (Chicago: Psychopathic Laboratory of the Municipal Court of Chicago, 1922), p. i. "Tenth Annual Business Meeting of the Eugenics Research Association, Cold Spring Harbor, June 10, 1922," *Eugenical News*, Vol. VII (1922), p. 91. Tredgold, 2d ed., p. 455 as cited by Elks, p. 203.

46. Curtis, pp. 78, 80-81, 148.

47. Testimony of Dr. David Braddock, *Boudreau v. Ryan*, Northern District of Illinois 00 C 5392 (2001). See K. Charlie Lakin, "Demographic Studies of Residential Facilities for the Mentally Retarded," *Developmental Disabilities Project on Residential Services and Community Adjustment Project Report No. 3* (University of Minnestoa Department of Psychoeducational Studies), circa 1979, pp. 88, 89.

48. Pernick, pp. 144, 151. *The Black Stork*, dir. Leopold and Theodore Wharton, 1917 as cited by Pernick, Figures 11, 16, 17, 22.

49. Advertisement, "The Black Stork," *Chicago Herald*, 1 April 1917, p. 7 as cited by Pernick, Figure 7. Advertisement, "The Black Stork," *Motography*, 14 April 1917, p. 2 as cited by Pernick, Figure 5. *Exhibitor's Trade Review*, 14 February 1917, p. 850 as cited by Pernick, p. 88.

50. Pernick, pp. 157-158.

51. War Department, "General Orders, No. 62." Nevada State Library and Archives, "An Outline of Capital Punishment in Nevada," at dmla.clan.lib.nv.us. Florida Corrections Commission, "Execution Methods Used by States: Executions in the U.S. - A Brief History," at www.fcc.state.fl.us. Popenoe and Johnson 1918, p. 184. Popenoe and Johnson, rev. ed., p. 135. Davenport, p. 63.

52. Carnegie Institution of Washington, *Announcement of Station for Experimental Evolution* (Washington: Carnegie Institution of Washington, 1905), pp. 2-3: CSH: CIW Administrative Files: Dept. of Genetics-Biological Laboratory Plans for Unified Operation. RAC 1.1/717/10/64 as cited by Paul J. Weindling, "From Philanthropy to International Science Policy: Rockefeller Funding of Biomedical Sciences in Germany 1920-1940," in Nicolass A. Rupke, ed., *Science, Politics and the Public Good: Essays in Honor of Margaret Gowing* (New York: Macmillan Press, 1988), p. 132. RAC 1.1/717/20/187 as cited by Weindling, p. 132.

53. "A Biological Court: Treating the Cause," *Eugenical News* Vol. IX (1924), p. 92.

54. See William L. Shirer, *The Rise and Fall of the Third Reich: A History of Nazi Germany* (New York: Simon and Schuster, 1960), pp. 29-31. See Adolf Hitler, *Mein Kampf*, trans. Ralph Manheim, (Boston: Houghton Mifflin Company, 1943), pp. 394-397, 400-405. Paul J. Weindling, *Health, Race and German Politics Between National Unification and Nazism, 1870-1945* (Cambridge: Cambridge University Press, 1989), pp. 308-310.

55. Elof Axel Carlson, *The Unfit: A History of a Bad Idea* (Cold Spring Harbor, NY: Cold Spring Harbor Laboratory Press, 2001), p. 323. Benno Müller-Hill, *Murderous Science*, G. R. Fraser, trans. (Cold Spring Harbor, NY: Cold Spring Harbor Laboratory Press: 1998), p. 121. Erwin Baur, Eugen Fischer, Fritz Lenz, *Human Heredity*, trans. Eden and Cedar Paul (New York: The Macmillan Company, 1931), pp. 442, 590, 593, 594.

56. *Autobiography of Leon F. Whitney*, unpublished manuscript circa 1973, p. 205: APS Manuscript Collection.

57. Grant, p. 49.

58. Whitney, p. 204, 205.

59. Whitney, p. 205.

CHAPTER FOURTEEN

1. Gustav Boeters, "Die Unfruchtbarmachung der geistig Minderwertigen," *Zwickauer Tageblatt* Sonderabdruck (Special Edition), n.d.: Bundesarchiv Berlin, R86, Akte 2374, Blatt 5. Gustav Boeters, "Die Unfruchtbarmachung der geistig Minderwertigen," *Wissenschaftliche Beilage der Leipziger Lehrerzeitung*, Nr. 28 (August 1924): Bundesarchiv Berlin, R86, Akte 2374, Blatt. 214. Robert N. Proctor, *Racial Hygiene: Medicine Under the Nazis* (Cambridge, MA: Harvard University Press, 1988), pp. 98, 360. Paul J. Weindling, *Health, Race and German Politics Between National Unification and Nazism, 1870-1945* (Cambridge: Cambridge

University Press, 1989), pp. 389-393, 450, 471, 526.
2. Weindling, p. 69.
3. Weindling, pp. 75, 76. Fritz Lenz, "Eugenics in Germany," Paul Popenoe, trans., *Journal of Heredity* Vol. XV No. 5 (May, 1924), pp. 223, 224.
4. "The Richardson Lethal Chamber (patented) for the Painless Extinction of Lower Animal Life," undated pamphlet: UCD Special Collections. Weindling, pp. 69, 77, 123. Lenz, p. 223. Proctor, p. 15.
5. Proctor, pp. 14-16. Letter, Alfred Ploetz to G. Hauptmann, 2 April 1897 as cited by Weindling, p. 127. Letter, Alfred Ploetz to G. Hauptmann, 11 May 1901 as cited by Weindling, p. 128. Letter, Alfred Ploetz to Ernst Rüdin as cited by Weindling, p. 128. Weindling, p. 129.
6. Proctor, pp. 17, 20-21. Weindling, pp. 141-142, Lenz, p. 225. "Eugenicists in Germany in 1946," *Eugenical News* Vol. XXXI (1946), p. 21.
7. Weindling, p. 394. Elof Axel Carlson, *The Unfit: A History of a Bad Idea* (Cold Spring Harbor, NY: Cold Spring Harbor Laboratory Press, 2001), pp. 323-324. Robert Jay Lifton, *The Nazi Doctors: Medical Killing and the Psychology of Genocide* (New York: Basic Books, 1986), p. 46.
8. Carnegie Institution of Washington, *Announcement of Station for Experimental Evolution* (Washington: Carnegie Institution of Washington, 1905), p. 4: CSH: CIW Administrative Files: Dept. of Genetics-Biological Laboratory Plans for Unified Operation. Letter, Dr. H. Iltis to Charles B. Davenport, 11 October, 1907: APS B:D27. Letter, Charles B. Davenport to Eugen Fischer, 15 August 1908: APS B:D27 Fischer. Letter, Charles B. Davenport to Eugen Fischer, 9 November 1908: APS B:D27. Letter, Charles B. Davenport to Eugen Fischer, 12 January 1909: APS B:D27 Fischer. Letter, Eugen Fischer to Charles B. Davenport, 22 December 1908: APS B:D27.
9. *Announcement of Station for Experimental Evolution*, p. 4. Davenport to Fischer, 15 August 1908. Weindling, pp. 143, 237. Benno Müller-Hill, *Murderous Science*, G. R. Fraser, trans. (Cold Spring Harbor, NY: Cold Spring Harbor Laboratory Press: 1998), pp. 123, 228 f. 167. See Charles B. Davenport, *Race Crossing In Jamaica*, (Washington: Carnegie Institute of Washington, 1929). See *State Laws Limiting Marriage Selection*, Eugenics Record Office (Cold Spring Harbor: Cold Spring Harbor, 1913), pp. 31-36. See Charles B. Davenport, "Heredity and Race Eugenics," p. 10: APS: B-D27.
10. Lady Georgina Chambers, "Notes on the Early Days of the 'Eugenics Education Society,'" p. 11: Wellcome SA/EUG/B-11.

11. Weindling, pp. 141-147.
12. Weindling, pp. 145, 151.
13. Lenz, pp. 225-226. Weindling, pp. 152-153.
14. The Eugenics Education Society, "Programme," *Problems in Eugenics Vol. II: Report of Proceedings of the First International Eugenical Congress* (Kingsway, W.C.: Eugenics Education Society, 1913), pp. 2, 3. "History of the International Organisation of Eugenics," memorandum circa November 1923, pp. 1-12: Truman C-2-1:2. Weindling, pp. 152-153.
15. See Géza von Hoffmann, *Die Rassenhygiene in den Vereinigten Staaten von Nordamerika* (Munich: J.F. Lehmanns Verlag, 1913).
16. Von Hoffmann, p. 14. Lenz, p. 226.
17. Letter, Géza von Hoffmann to Harry H. Laughlin, 27 December 1913: Truman D-5-4:7. Letter, Géza von Hoffmann to Harry H. Laughlin: 26 May 1914: Truman D-5-4:7
18. Letter, Hoffmann to Laughlin, 27 December 1913. Letter, Hoffmann to Laughlin, 26 May 1914.
19. "German Progress in Genetics," *Journal of Heredity* Vol. V No. 6 (June 1914), p. 243. Dr. von Stradonitz, "Bismarck's Heredity," *Journal of Heredity* Vol. V No. 6 (June 1914), p. 254. Alfred Ploetz, "The First and Last Child," *Journal of Heredity* Vol. V No. 6 (June 1914), p. 268. "International Genetics Conference," *Journal of Heredity* Vol. V No. 7 (July 1914), p. 300. C. Fruwirth, "New Publications: Handbuch der landwirtschaftlichen Pflanzenzüchtung," *Journal of Heredity* Vol. V No. 7 (July 1914), p. 304. Eugen Fischer, "Racial Hybridization," *Journal of Heredity* Vol. V No. 10 (October 1914), pp. 465-467.
20. KCET/Los Angeles, The British Broadcasting Corporation, and The Imperial War Museum of London, "Interactive Timeline," *The Great War and the Shaping of the 20th Century* at www.pbs.org. Lenz, p. 226. Weindling, p. 314. See Madison Grant, *The Passing of the Great Race* (New York: Charles Scribner's Sons, 1936). See letter, Erwin Baur to Charles B. Davenport, 24 November 1920: APS B:D27 - Davenport & Baur. See Deborah E. Lipstadt, *Beyond Belief* (New York: The Free Press, 1986), pp. 8-9.
21. "Babies- More, Fewer or None," *Eugenical News*, Vol. II (1917), p. 31.
22. "Interactive Timeline," *The Great War and the Shaping of the 20th Century*. William L. Shirer, *The Rise and Fall of the Third Reich: A History of Nazi Germany* (New York: Simon and Schuster, 1960), pp. 52-57.
23. Harry H. Laughlin, "National Eugenics in Germany," *Eugenics Review*, January 1921, reprinted in *Harry Laughlin Reprints* (Washington, DC: Carnegie Institute of Washington, n.d.). "Eugenical Research Association," *Eugenical News*, Vol. V (1920), p.

44. "Eugenics Research Association - Eighth Annual Meeting," *Eugenical News*, Vol. V (1920), p. 52. "National Eugenics in Germany," *Eugenical News*, Vol. V (1920), pp. 55-56.

24. Letter, Baur to Davenport, 24 November 1920. Letter, Charles B. Davenport to Erwin Baur, 16 December 1920: APS B:D27 - Baur, Erwin. "Notes and News," *Eugenical News*, Vol. VI (1921), p. 8. Letter, Charles B. Davenport to Eugen Fischer, 20 May 1921: APS B:D27 - Fischer. Letter, Charles B. Davenport to Agnes Bluhm, 30 August 1921: APS B:D27 - Bluhm, Agnes.

25. Shirer, p. 51. U.S. Holocaust Museum, "Blacks During the Holocaust," at www.ushmm.org. Adolf Hitler, *Mein Kampf*, trans. Ralph Manheim, (Boston: Houghton Mifflin Company, 1943), Volume I, Chapter XI, p. 325.

26. Baur to Davenport, 24 November 1920. Letter, Charles B. Davenport to Erwin Baur, 30 March 1923: APS B:D27 - Baur, Erwin. "Hygiene Congress Abandoned," *Eugenical News* Vol. VI (1921), p. 32.

27. Baur to Davenport, 24 November 1920. Letter, Charles B. Davenport to Herman Lundborg, 24 October 1921: APS B:D27. "Minutes of the Meetings of the International Commission of Eugenics," memorandum circa 1923: Truman C-2-1:2. Davenport to Bluhm, 30 August 1921. See letter, Agnes Bluhm to Charles B. Davenport, 17 October 1921: APS B:D27 - Bluhm, Agnes. See letter, Agnes Bluhm to Charles B. Davenport, 24 November 1921: APS B:D27 - Bluhm, Agnes. See letter, Richard Crane to Charles Evans Hughes, 7 September 1921: NA 59/250/22/10/3/5459. See letter, Alvey A. Adee to C. C. Little, 21 September 1921: NA 59/250/22/10/3/5459. See letter, American Minister in Nicaragua to Charles Evans Hughes, 13 June 1921: NA 59/250/22/10/3/5459. See letter, Benjamin Jefferson to Don Mariano Zelaya B., 12 April 1921: NA 59/250/22/10/3/5459. See letter, Mariano Zalaya B. to Benjamin Jefferson, 15 April 1921: NA 59/250/22/10/3/5459. See letter, C. C. Little to the Envoy Extraordinary & Minister Plenipotentiary of the United State of America, 17 March 1921: NA 59/250/22/10/3/5459. Letter, Charles Hartman to Sr. Dr. Dn. N. Clemente Ponce, 7 June 1921: NA 59/250/22/10/3/5459. Letter, Walker Smith to Charles Evans Hughes, 29 June 1921: NA 59/250/22/10/3/5459.

28. Shirer, p. 61. French Diplomatic Archives, "Chronology: 1918-1939: Inter-War Period," at www.france.diplomatie.fr. Missouri Western College, "The German Hyperinflation of 1923: A Seventy-Fifth Anniversary Retrospective," at www.mwsc.edu.

29. Davenport to Baur, 30 March 1923. Letter,

Charles B. Davenport to Cora Hodson, 17 December 1925: APS B:D27 - Davenport: International Federation of Eugenic Orgs. See letter, Cora Hodson to Charles B. Davenport, 26 November 1925: APS B:D27 - Davenport: International Federation of Eugenic Orgs. See letter, Cora Hodson to Charles B. Davenport, 30 December 1925: APS B:D27 - Davenport: International Federation of Eugenic Orgs. See letter, Charles B. Davenport to Herman Lundborg, 21 November 1925: APS B:D27 - Davenport: International Federation of Eugenic Orgs. See letter, Herman Lundborg to Leonard Darwin, 8 December 1925: APS B:D27 - Davenport: International Federation of Eugenic Orgs. See letter, Charles B. Davenport to Cora Hodson, 22 January 1926: APS B:D27 - Davenport: International Federation of Eugenic Orgs. See letter, Charles B. Davenport to Herman Lundborg, 21 November 1925: APS B:D27 - Davenport: International Federation of Eugenic Orgs.

30. Davenport to Baur, 30 March 1923.

31. Letter, Fritz Lenz to Charles Davenport, 8 August 1923: APS B:D27 - Lenz, F.

32. Adam Smith [George J.W. Goodman] *Paper Money* (New York: Summit Books, 1981), pp. 57-62. Robert L. Hetzel, "German Monetary History in the First Half of the Twentieth Century," *Economic Quarterly* Vol. 88/1 (Winter 2002).

33. Shirer, pp. 68-79. Ian Kershaw, *Hitler: 1889 - 1936: Hubris* (New York: W. W. Norton & Company, 1998), p. 240. Müller-Hill, p. 121.

34. Shirer, p. 84. Hitler, Volume I, Chapter XI, pp. 302-327. Hitler, Volume II, Chapter XIV, pp. 654-655.

35. Lifton, p. 31. See Annette Horn, review of Aenne Baeumer-Schleinkofer, *Nazi Biology and Schools*, at web.uct.ac.za. See University of California, Berkeley Campus, "Biography of Ernst Haeckel (1934-1919)," at www.ucmp.berkeley.edu.

36. Kershaw, pp. 240, 241-242. Elof Axel Carlson, *The Unfit: A History of a Bad Idea* (Cold Spring Harbor, NY: Cold Spring Harbor Laboratory Press, 2001), p. 323. Müller-Hill, pp. 8, 120. Weindling, pp. 330, 397.

37. Letter, Erwin Baur to Charles B. Davenport, 24 November 1920: APS B:D27 - Davenport & Baur.

38. Baur to Davenport, 24 November 1920.

39. Baur to Davenport, 24 November 1920.

40. Baur to Davenport, 24 November 1920. See Harvard University, "Blakeslee, Albert Francis, 1874-1954. Papers of Albert F. Blakeslee, 1912-1960: A Guide" at oasis.Harvard.edu.

41. Davenport to Baur, 30 March 1923.

42. Lifton, p. 23. Lenz to Davenport, 8 August 1923.

43. Lenz to Davenport, 8 August 1923.
44. Fritz Lenz, *Menschliche Auslese und Rassenhygiene*, vol. II of Erwin Baur, Eugen Fischer, and Fritz Lenz, *Grundriss der menschlichen Erblichkeitslehre und Rassenhygiene* (Munich: J. F. Lehmanns Verlag, 1923), p. 147 as cited by Lifton, p. 23.
45. See APS B:D27 - Fisher # 1. See APS B:D27 - Fisher #2. See APS B:D27 - Fisher #3. See APS B:D27 - Fisher #4. See APS B:D27 - Fisher #5. See APS B:D27 - Fisher #6.
46. Erwin Baur, Eugen Fischer, and Fritz Lenz, *Menschliche Erblichkeitslehre*, vol I. of Baur, Fischer, and Lenz, pp. 299-305. Lenz, *Menschliche Auslese und Rassenhygiene*, pp. 233-237.
47. Erwin Baur, Eugen Fischer and Fritz Lenz, *Human Heredity*, 3rd Ed., trans. Eden & Cedar Paul (New York: The MacMillan Company, 1931), pp. 202, 204-206, 208, 311, 390 f. 2, 429 f. 2, 429 f. 3, 628-629, 638, 666-671, 680-681. Lenz, *Menschliche Auslese und Rassenhygiene*, p. 126.
48. Davenport to Baur, 30 March 1923. "Human Genetics and Eugenics," *Eugenical News*, Vol. VIII (1923), p. 47. "Human Selection," *Eugenical News*, Vol. VIII (1923), pp. 96-97. "Heredity and Eugenics: A Review," *The Journal of Heredity*, Vol. XIV, No. 7 (October 1923), p. 336. "Human Genetics and Eugenics," *Eugenical News*, Vol. VIII (1923), p. 47. "Human Selection," *Eugenical News*, Vol. VIII (1923), pp. 96-97. "A Compendium of Eugenics," *Eugenical News*, Vol. VII (1922), p. 97.
49. Proctor, pp. 26-27, 203, 344 f. 57, 344 f3 59. Weindling, p. 314. See Madison Grant, *Der Untergang der grossen Rasse: Die Rassen als Grundlage der Geschichte Europas* (Berlin: J. F. Lehmanns Verlag, 1925). See Von Hoffmann.
50. Weindling, p. 311. Carlson, pp. 323. Müller-Hill, p. 121. Gary D. Stark, *Entrepreneurs of Ideology* (Chapel Hill, NC: The University of North Carolina Press, 1981), pp. 170, 279.
51. Hitler, Volume I, Chapter X., p. 255.
52. Hitler, Volume II, Chapter II, pp. 403-404.
53. Hitler, Volume II, Chapter II, p. 402. Hitler, Volume II, Chapter II, pp. 404-405.
54. Hitler, Volume I, Chapter XI, p. 285.
55. Grant, p. 16.
56. Hitler, Volume II, Chapter II, pp. 388-389, 390.
57. Grant, p. 17.
58. Hitler, Volume I, Chapter XI, p. 286.
59. Hitler, Volume II, Chapter III, pp. 439-440.
60. Hitler, Volume I, Chapter II, p. 29. Hitler, Volume II, Chapter III, pp. 439-440. Hitler, Volume I, Chapter XI, p. 286.
61. Otto Wagener, *Hitler: Memoirs of a Confidant*, trans. Henry Ashby Turner (Yale University Press, 1987), pp. 145-146.
62. Richard Breiting, *Secret Conversations with Hitler*, edit. Edouard Calic, trans. Richard Barry, (New York: The John Day Company, 1968), p. 81.
63. Norman Cameron and R.H. Stevens, trans. *Hitler's Table Talk: 1941-1944: His Private Conversations*, (New York City: Enigma Books, 2000), pp. 670, 675.
64. Lifton, pp. 46-48. Hitler, Volume II Chapter XV, p. 679.
65. Shirer, pp. 3-5, 170-184. "Delegates Urge Wider Practice of Sterilization," *Richmond Times-Dispatch*, 16 January 1934.
66. *Autobiography of Leon F. Whitney*, unpublished manuscript circa 1973, p. 205: APS Manuscript Collection.
67. See APS B:D27 - Davenport - Ernst Rüdin. See APS B:D27 - Davenport - Baur, Erwin. See APS B:D27 - Fisher # 1. See APS B:D27 - Fisher #2. See APS B:D27 - Fisher #3. See APS B:D27 - Fisher #4. See APS B:D27 - Fisher #5. See APS B:D27 - Fisher #6. See Universitatsarchiv Münster- Nachlass Verschuer, Nr. 4.
68. Zentralarchiv der Deutschen Demokratischen Republik, Potsdam: Reichsinnenministerium 10160, Film 23063 as cited by Müller-Hill, p. 34. See Proctor, p. 106. Stephen Trombley, *The Right to Reproduce: A History of Coercive Sterilization* (London: Weidenfeld & Nicolson, 1988), p. 117.
69. Human Betterment Foundation, "Report to the Board of Directors of the Human Betterment Foundation for the Year Ending February 12, 1936.": Bailey/Howe Library: Perkins Papers.

CHAPTER FIFTEEN

1. See "Notes on the Second Conference," n.p., 27 September 1929: APS B:D27 - Davenport - Gini, Corrado #2. See "The Meeting of the International Federation of Eugenic Organizations," *Eugenical News*, Vol. XIV (1929), pp. 153-157.
2. "Notes on the Second Conference."
3. "Notes on the Second Conference." "The Meeting of the International Federation of Eugenic Organizations," pp. 154, 156.
4. "Notes on the Second Conference." "The Meeting of the International Federation of Eugenic Organizations," p. 156.
5. The Meeting of the International Federation of Eugenic Organizations," p. 155.
6. "Notes on the Second Conference." "The Meeting of the International Federation of Eugenic Organizations," pp. 156, 157.
7. See *Eugenical News*, Vol. VIII (1923).
8. "Eugenical Notes," *Eugenical News* Vol. VIII (1923), p. 22. "Archiv Für Rassen Und Gesellschafts-Biologie," *Eugenical News* Vol. IX (1924), p. 51. "Notes and News," *Eugenical News* Vol. IX (1924), p. 64. "Archiv Fuer

Rassen- U. Gesellschafts-Biologie," *Eugenical News* Vol. X (1925), p. 31. "Archiv Fuer Rassen- U. Gesellschafts-Biologie," *Eugenical News* Vol. X (1925), p. 88. "Archiv Fuer Rassen- Und Gesellschafts-Biologie," *Eugenical News* Vol. X (1925), p. 152. "Archiv Fuer Rassen- U. Gesellschafts-Biologie," *Eugenical News* Vol. XI (1926), p. 9. "Archiv Fuer Rassen- U. Gesellschafts-Biologie," *Eugenical News* Vol. XI (1926), p. 41. "Archiv Fuer Rassen- U. Gesellschafts-Biologie," *Eugenical News* Vol. XI (1926), p. 92. "Archiv Fuer Rassen- U. Gesellschafts-Biologie," *Eugenical News* Vol. XI (1926), p. 134. "Archiv Fuer Rassen- U. Gesellschafts-Biologie," *Eugenical News* Vol. XII (1927), p. 31. "Current Periodicals," *Eugenical News* Vol. XII (1927), p. 64. "Archiv Fuer Rassen- U. Gesellschafts-Biologie," *Eugenical News* Vol. XII (1927), p. 91. "Archiv Fuer Rassen- U. Gesellschafts-Biologie," *Eugenical News* Vol. XII (1927), p. 180. "Archiv Fuer Rassen- U. Gesellschafts-Biologie," *Eugenical News* Vol. XIII (1928), p. 32. "Current Periodicals," *Eugenical News* Vol. XIII (1928), p. 72. "Current Periodicals," *Eugenical News* Vol. XIII (1928), p. 104. "Current Periodicals," *Eugenical News* Vol. XIII (1928), p. 162. "Current Periodicals," *Eugenical News* Vol. XIV (1929), p. 32. "Current Periodicals," *Eugenical News* Vol. XIV (1929), p. 48. "Archiv Fuer Rassen- Und Gesellschafts-Biologie," *Eugenical News* Vol. XIV (1929), p. 88. "Archiv Fuer Rassen- Und Gesellschafts-Biologie," *Eugenical News* Vol. XIV (1929), p. 126. "Archiv Für Rassen- Und Gesellschafts-Biologie," *Eugenical News* Vol. XV (1930), p. 16. "Archiv Fur Rassen- Und Gesellschafts-Biologie," *Eugenical News* Vol. XV (1930), p. 88. "Archiv Für Rassen-u. Gesellschafts-Biologie," *Eugenical News* Vol. XV (1930), p. 132. "Archiv Für Rassen- Und Gesellschafts-Biologie," *Eugenical News* Vol. XVI (1931), p. 184. "Archiv Für Rassen- Und Gesellschafts-Biologie," *Eugenical News* Vol. XVII (1932), p. 30.

9. Fritz Lenz, "Eugenics in Germany," trans. Paul Popenoe, *Journal of Heredity* Vol. XV No. 5 (May, 1924), pp. 223-231. "Race Hygiene," *Eugenical News*, Vol. IX (1924), p. 86.

10. "Berlin (From Our Regular Correspondent)," *Journal of the American Medical Association*, Vol. 82, No. 21 (May 1924), pp. 1709, 1710.

11. "Are the Gifted Families in America Maintaining Themselves?" *Eugenical News* Vol. XI (1926), pp. 2-4.

12. "Anthropology Iconography," *Eugenical News* Vol. XI (1926), p. 144. Paul J. Weindling, *Health, Race and German Politics Between National Unification and Nazism, 1870-1945* (Cambridge: Cambridge University Press, 1989), pp. 310-311.

13. "Current Periodicals," *Eugenical News* Vol. XII (1927), p. 64.

14. "Races of Central Europe," *Eugenical News* Vol. IX (1924), p. 34. "Archiv Fuer Rassen- Und Gesellschafts-Biologie," *Eugenical News* Vol. X (1925), p. 152. "Archiv F. Rassen-U. Gesellschafts-Biologie," *Eugenical News* Vol. XII (1927), p. 180. "Noses and Ears," *Eugenical News* Vol. XIV (1929), p. 55.

15. Nobel Museum, "The Foundation of the Kaiser Wilhelm Institute for Medical Research," at www.nobel.se. Nobel Museum, "The Nobel Prize in Chemistry 1918" at www.nobel.se. Nobel Museum, "The Nobel Prize in Chemistry 1936" at www.nobel.se. Nobel Museum, "The Nobel Prize in Physics 1918" at www.nobel.se. Nobel Museum, "The Nobel Prize in Physics 1921" at www.nobel.se. Memorandum, Charles B. Davenport to Cora Hodson, 27 October 1928: APS B:D27 - IFEO 1928 #2.

16. Matthias M. Weber, "Psychiatric Research and Science Policy in Germany. The History of the Deutsche Forschungsanstalt für Psychiatrie (German Institute for Psychiatric Research) in Munich from 1917 to 1945," *History of Psychiatry* xi (2000), p. 239. See Paul J. Weindling, "From Philanthropy to International Science Policy: Rockefeller Funding of Biomedical Sciences in Germany 1920-1940," in Nicolaas A. Rupke, ed., *Science, Politics and the Public Good: Essays in Honor of Margaret Gowing* (New York: Macmillan Press, 1988), p. 131.

17. Angus Rae, "Osler Vindicated: The Ghost of Flexner Laid to Rest," *Canadian Medical Association Journal* 164 (13) 26 June 2001, p. 1860. Weindling, "From Philanthropy to International Science Policy: Rockefeller Funding of Biomedical Sciences in Germany 1920-1940," p. 121. See Flexner, Abraham *Medical Education in Europe: A Report to the Carnegie Foundation* (1912). See Nancy Rockefeller, "The Flexner Report in Context," at www.library.ucsf.edu.

18. See Abraham Flexner, *Prostitution in Europe* (New York: The Century Company, 1914). Weindling, "From Philanthropy to International Science Policy: Rockefeller Funding of Biomedical Sciences in Germany 1920-1940," pp. 121, 123. Kristie Macrakis, *Surviving the Swastika: Scientific Research in Nazi Germany* (New York: Oxford University Press, 1993), pp. 18-22. Paul J. Weindling, *Health, Race and German Politics Between National Unification and Nazism, 1870-1945* (Cambridge: Cambridge University Press, 1989), pp. 324-325.

19. Weindling, "From Philanthropy to International Science Policy," pp. 123, 124-125, 127, 128. Weindling, *Health, Race and German*

Politics Between National Unification and Nazism,
1870-1945, p. 335.
20. Rockefeller Archives, "History," Vol. 15, p.
3794 as cited by Weindling, "From
Philanthropy to International Science Policy,"
pp. 124-125, 127.
21. Weindling, "From Philanthropy to
International Science Policy," pp. 126-127.
Letter, Fritz Haber to Friedrich Schmidt-Ott, 6
March 1923: BAB R 73, Akte 217 (Notgemein-
schaft der deutschen Wissenschaft - now:
Deutsche Forschungsgemeinschaft). Letter
from E. Uhlenbruch to Friedrich Schmidt-Ott,
22 March 1923: BAB R 73, Akte 217
(Notgemeinschaft der deutschen Wissenschaft -
now: Deutsche Forschungsgemeinschaft).
22. "VII. Bericht über die Deutsche
Forschungsanstalt für Psychiatrie (Kaiser-
Wilhelm-Institut) in München zur
Stiftungsratssitzung am 5. Februar 1927,"
Zeitschrift für die gesamte Neurologie und
Psychiatrie, p. 344: BAB R 1501, Akte 126 789,
Blatt 148-150. Author's communication with
Paul Weindling, 23 January 2003.
23. Weindling, *Health, Race and German Politics*
Between National Unification and Nazism, 1870-
1945, p. 336. Weber, pp. 250, 251. Robert N.
Proctor, *Racial Hygiene: Medicine Under the*
Nazis (Cambridge, MA: Harvard University
Press, 1988), p. 112.
24. Matthais M. Weber, *Ernst Rüdin: Eine Kritische*
Biographie (Berlin: J. Springer-Verlag, 1993).
Proctor, p. 17. Also see Weindling, *Health, Race*
and German Politics Between National Unification
and Nazism, 1870-1945, pp. 72, 150, 185-186.
25. Weindling, *Health, Race and German Politics*
Between National Unification and Nazism, 1870-
1945, pp. 384, 385.
26. "The German Genetic Association," *Journal of*
Heredity, Vol. XIII No. 5 (May 1922), p. 200.
"Notes and News," *Eugenical News* Vol. IX
(1924), p. 56. "Heredity of Insanity," *Eugenical*
News Vol. IX (1924), p. 83. "The Genealogical
Section of the Psychiatric Institute of Munich,"
Eugenical News Vol. X (1925), p. 118. "Berlin
(From Our Regular Correspondent)," *Journal of*
the American Medical Association Vol. 94 No. 3
(Dec. 1929), p. 201.
27. "Meeting of International Federation of
Eugenic Organizations," *Eugenical News* Vol.
XIII (1928), pp. 129, 131. "Membership and
Organization of the International Federation of
Eugenic Organizations," *Eugenical News* Vol.
XV (1930), p. 168. "The International
Federation of Eugenics Organizations,"
Eugenical News Vol. XVIII (1933), p. 16.
28. "Fifth International Congress of Genetics,"
Eugenical News Vol. XII (1927), p. 152.
29. "Fifth International Congress of Genetics," p.
152. Weindling, *Health, Race and German Politics*

Between National Unification and Nazism, 1870-
1945, p. 435.
30. "Fifth International Congress of Genetics," p.
150, 152. Fifth International Congress,
"Program," p. 4: APS B:D27 - International
Congress of Genetics 5th and 6th. William E.
Seildeman, "Science and Inhumanity: The
Kaiser-Wilhelm/Max Planck Society," *If Not*
Now an e-journal Vol. 2 (Winter 2000), at
www.soec.at. Invitation to Charles B.
Davenport from Fifth International Congress
of Genetics: APS B:D27 - International
Congress of Genetics 5th and 6th. Weindling,
Health, Race and German Politics Between
National Unification and Nazism, 1870-1945, p.
436.
31. Fifth International Congress of Genetics," pp.
150, 152. Fifth International Congress,
"Program," pp. 9-11, 22-23.
32. "Fifth International Congress of Genetics," pp.
150, 152. "Program," pp. 12-19. Fifth
International Congress of Genetics, "List of the
Papers Announced at the Congress," n.p., n.d.,
pp. 7-13: APS B:D27 - International Congress
of Genetics 5th and 6th. Invitation to Charles B.
Davenport from Fifth International Congress
of Genetics.
33. Letter, Charles B. Davenport to Eugen Fischer,
5 October 1926: APS B:D27 - Fischer. "Fifth
International Congress of Genetics," p. 152.
Letter, Charles B. Davenport to Hermann
Muckermann, 6 October 1928: APS B:D27 -
Davenport - Muckermann, Dr. H. Letter,
Charles B. Davenport to Eugen Fischer, 3
October, 1928 and attached letter, Charles B.
Davenport to Eugen Fischer: APS B:D27 -
IFEO 1928 #2. Letter, Charles B. Davenport to
Eugen Fischer, 4 December 1928: APS B:D27 -
Fischer.
34. Hans-Walter Schmuhl, *Hirnforschung und*
Krankenmord, Das Kaiser-Wilhelm-Institut für
Hirnforschung 1937-1945, (Berlin: 2000).
"Cécile and Oskar Vogt: On the Occasion of
her 75th and his 80th Birthday," *Neurology* Vol. 1
No. 3 (May-June 1951), pp. 183, 184. Max
Delbrück Center for Molecular Medicine,
"History" at www.mdc-berlin.de. Tage Kemp,
"To The Rockefeller Foundation: Report of vis-
its to various Institutes, Laboratories etc. for
Human Genetics in Europe." (July-October
1934), pp. 59-60: RF 1.2/713/2/16. Letter,
Norma S. Thompson to Adolf von Harnack, 24
May 1929: RF 1.1 717 10 64. Letter, George J.
Beal to Adolf von Harnack, 4 June 1929: RF 1.1
717 10 64. Review of recommendation on the
Kaiser Wilhelm Institute for Brain Research, 22
May 1929: RF 1.1 717 10 64. Review of appro-
priations to the Kaiser Wilhelm Institute for
Brain Research, 9 May 1932: RF 1.1 717 10 64.
35. Letter, Charles B. Davenport to Wickliffe

Draper, 23 February 1926: APS B:D27 - Davenport - W. P. Draper #1. Letter, Charles B. Davenport to Wickliffe Draper, 15 March 1926: APS B:D 27 - W.P. Draper #1.

36. Davenport to Draper, 23 February 1926. Davenport to Draper, 15 March 1926. Letter, Charles B. Davenport to Cora Hodson, 18 March 1926: APS B:D27 - Davenport: Int'l Fed of Eugenic Orgs.

37. Charles B. Davenport and Morris Steggerda, *Race Crossing in Jamaica* (Washington: Carnegie Institution of Washington, 1929), pp. 3, 4, 9.

38. See Edwin Black, *IBM and the Holocaust* (New York: Crown Publishers, 2001), Chapter II.

39. Generally see Black, especially Chapters II, III, IV and V.

40. Hermann Krüger, "Das Hollerith-Lochkarten-Verfahren im Fursorgewesen," *Hollerith Nachrichten* 47 (March 1935), pp. 615, 618. "Secret Report: PW Intelligence Bulletin No. 2/57," April 25, 1945, p. 1: CSDIC. Davenport and Steggerda, p. 4. See Black, Chapter II. See "Report of the Advisory Committee on the Eugenics Record Office," circa 1935: Truman C-2-2:2.

41. "List of data for Columns of Hollerith Cards," memorandum circa 1926: APS B:D27 - Davenport - Draper Fund for Race Crossing #2. Letter, Charles B. Davenport to Morris Steggerda, 8 April 1927: APS B:D27 - April #3 Davenport - Steggerda.

42. Generally see Black, especially Chapters IV, V, VII and VIII.

43. See Davenport and Steggerda. Letter, Charles B. Davenport to Eugen Fischer, 17 February 1927: APS B:D27 - Fischer. Letter, Charles B. Davenport to Cora Hodson, 23 February 1927: APS B:D27 -Davenport: Int'l Federation of Eugenic Orgs. Letter, Charles B. Davenport to Hermann Lundborg, 17 May 1928: APS B:D27.

44. Letter, Charles B. Davenport to Henry L. Bolley, 14 November 1928: APS B:D27 - Committee on Race Crossing. Letter, Charles B. Davenport to Bennet Allen, 14 November 1928: APS B:D27 - Committee on Race Crossing. Letter, Charles B. Davenport to Trevor Kincaid, 14 November 1928: APS B:D27 - IFEO Committee on Race Crossing #3. Letter, Charles B. Davenport to W.E. Bryan, 14 November 1928: APS B:D27 - Committee on Race Crossing. Letter, J.P. Anderson to Charles B. Davenport, 20 March 1929: APS B:D27 - Committee on Race Crossing. Letter, Raymond Bellamy to Charles B. Davenport, 17 January 1928: APS B:D27 - Committee on Race Crossing.

45. Letter, Bellamy to Davenport, 17 January 1928. Letter, W. E. Bryan to Charles B. Davenport, 10 January 1929: APS B:D27 - Committee on

Race Crossing. Letter, J.S. Blitch to Charles B. Davenport, 10 January 1929: APS B:D27 - Committee on Race Crossing. Letter, B.M. Allen to Charles B. Davenport, 5 December 1929: APS B:D27 - Committee on Race Crossing. Letter, Henry Bolley to Charles B. Davenport, 21 November 1928: APS B:D27 - Committee on Race Crossing.

46. "Form Letter" circa February 1929: APS B:D27 - IFEO Committee on Race Crossing #3. Letter, Charles B. Davenport to E.A. Arce, 28 February 1929: APS B:D27 - IFEO Committee on Race Crossing #3.

47. Letter, Halfdan Bryn to Charles B. Davenport, 4 April 1929: APS B:D27 - IFEO Committee on Race Crossing #3. Letter, Charles B. Davenport to Halfdan Bryn, 19 April 1929: APS B:D27 - IFEO Committee on Race Crossing #3. Letter, V. Bunak to Charles B. Davenport, 20 March 1929: APS B:D27 - IFEO Committee on Race Crossing #3. Letter, Charles B. Davenport to V. Bunak, 18 April 1929: APS B:D27 - IFEO Committee on Race Crossing #3. Letter, G. Arnold to Charles B. Davenport, 4 April 1929: APS B:D27 - IFEO Committee on Race Crossing #3. Letter, Charles B. Davenport to G. Arnold, 10 May 1929: APS B:D27 - IFEO Committee on Race Crossing #3. Letter, A. de Assis to Charles B. Davenport, 28 March 1929: APS B:D27 - IFEO Committee on Race Crossing #3. Letter, Davidson Black to Charles B. Davenport, 1 April 1927: APS B:D27 - IFEO Committee on Race Crossing #3. Letter, H. J. V. Bijlmer to Charles B. Davenport, 24 April 1929: APS B:D27 - IFEO Committee on Race Crossing #3. Letter, Charles B. Davenport to H. J. V. Bijlmer, 27 June 1929: APS B:D27 - IFEO Committee on Race Crossing #3. Letter, Charles B. Davenport to F. W. Caine, 8 July 1929: APS B:D27 - IFEO Committee on Race Crossing #3. Letter, E. Lucas Bridges to Charles B. Davenport, 15 May 1929: APS B:D27 - IFEO Committee on Race Crossing #3. Letter, Charles B. Davenport to E. Lucas Bridges, 9 July 1929: APS B:D27 - IFEO Committee on Race Crossing #3. Letter, H. Beroot to Charles B. Davenport, 9 March 1929: APS B:D27 - IFEO Committee on Race Crossing #3.

48. Letter, Prescott Childs to Charles B. Davenport, 30 April 1929: APS B:D27 - IFEO Committee on Race Crossing #3. Letter, C.C. Hanson to Charles B. Davenport, 8 May 1929: APS B:D27 - IFEO Committee on Race Crossing #3. Letter, Charles H. Albrecht, 14 May 1929: APS B:D27 - IFEO Committee on Race Crossing #3. Letter, Harry E. Carlson to Charles B. Davenport, 29 May 1929: APS B:D27 - IFEO Committee on Race Crossing

#3. Letter, Charles B. Davenport to the American Consul in Magallanes, Chile: APS B:D27 - IFEO Committee on Race Crossing #3. Letter, Lewis V. Boyle to Charles B. Davenport, 2 May 1929: APS B:D27 - IFEO Committee on Race Crossing #3.

49. Letter, Eugen Fischer to Charles B. Davenport, 19 July 1929: APS B:D27 - Fischer. "The Meeting of the International Federation of Eugenic Organizations," p. 156.

50. "The Meeting of the International Federation of Eugenic Organizations," p. 154.

51. "The Meeting of the International Federation of Eugenic Organizations," pp. 155, 157.

52. Letter, Charles B. Davenport to Eugen Fischer, 2 December 1929: APS B:D27. Letter, Charles B. Davenport to Alfred Ploetz, 1 October 1932: APS B:D27 - Ploetz, Alfred.

53. Letter, Eugen Fischer to Charles B. Davenport, 22 December 1929: APS B:D27. Letter, Charles B. Davenport to Eugen Fischer, 3 February 1930. Charles B. Davenport and Eugen Fischer, "Studies on Human Race Crossings," memorandum circa 1930: APS B:D27.

54. Letter, F. Schmidt-Ott to Edmund E. Day, 20 September 1929: RF 1.1 717 20 187.

55. Schmidt-Ott to Day, 20 September 1929.

56. "Progress Report: Grant to Notgemeinschaft for Anthropological Studies of the Population of Germany," 10 June 1933: RF 1.1 717 20 187. "Action RF 29137- Anthropological Investigation of the German People," memorandum of 2 October 1933: RF 1.1 717 20 187. Letter, Norma S. Thompson to Dr. F. Schmidt-Ott, 14 November 1929: RF 1.1 717 20 187. Letter, Edmund E. Day to Dr. F. Schmidt-Ott, 27 November 1929: RF 1.1 717 20 187. Letter, George J. Beal to R. Letort, 6 December 1929: RF 1.1 717 20 187. Letter, R. Letort to George J. Beal, 6 January 1930: RF 1.1 717 20 187. Letter, Dr. F. Schmidt-Ott to Edmund E. Day, 5 September 1929: RF 1.1 717 20 187.

57. "Archiv Für Rassen-u. Gesellschafts Biologie," Eugenical News, Vol. XV (1930), p. 152. "Jews in West Africa," Eugenical News, Vol. XV (1930), pp. 142-143. "Books on Human Heredity," Eugenical News, Vol. XV (1930), p. 143. Gary D. Stark, Entrepreneurs of Ideology (Chapel Hill, NC: The University of North Carolina Press, 1981), p. 223. "Ninth Meeting of the International Federation of Eugenic Organizations," Eugenical News, Vol. XV (1930), p. 162.

58. Ernst Rüdin, "Hereditary Transmission of Mental Diseases," Eugenical News Vol. XV (1930), pp. 171-174.

59. "Hereditary Transmission of Mental Diseases," pp. 172, 174.

60. Memorandum from D.P. O'Brien to Alan Gregg, 10 November 1933: RF 1.1 717 9 46.

61. "Memo for Officer's Action: Forschungsanstalt fur Psychiatrie," memorandum circa December 1933: RF 1.1 717 9 56. Letter, Thomas B. Appleget to Benson Y. Landis, 23 February 1934: RF 1.1 717 9 56. Memorandum, D.P. O'Brien to Alan Gregg, 27 November 1934: RF 1.1 717 9 56. "University of Copenhagen - Institute of Human Genetics," memorandum, June 1939: RF 1.1 Ser 713A Box 2 Folder 15. Letter, Tage Kemp to The Rockefeller Foundation, 17 November 1932: RF 1.2 Ser 713 Box 2 Folder 15.

62. "From HAS' diary: June 4, 1931," inter-office correspondence: RF 1.1 Ser 100 Box 40 Folder 365.

63. Erwin Baur, Eugen Fischer and Fritz Lenz, Human Heredity, 3rd Ed., trans. Eden & Cedar Paul (New York: The MacMillan Company, 1931), pp. 677, 680, 681.

64. "Heredity and Eugenics," Eugenical News Vol. XVI (1931), pp. 220-221.

65. See Kershaw, pp. 336-337.

66. "Hitler and Race Pride," Eugenical News Vol. XVII (1932), pp. 60-61.

67. Radiogram to Alan Gregg, 13 May 1932: RF 1.1 Ser 7171 Box 10 Folder 63. Gerald Jonas, The Circuit Riders (New York: W.W. Norton & Company, 1989), p. 111.

68. "Eugenics in the service of public welfare: Report of the proceedings of a committee convened by the Prussian State Health Council on 2 July 1932," Veröffentlichungen aus dem Gebiete der Medizinalverwaltung Vol. XXXVIII, part 5, p. 98 as cited by Müller-Hill, p. 30. R.B. Goldschmidt, Im Wandel das Bleibende: Mein Lebensweg (Hamburg and Berlin, 1963), p. 264 as cited by Müller-Hill, p. 30.

69. A Decade of Progress in Eugenics: Scientific Papers of the Third International Congress of Eugenics (Baltimore: The Williams and Wilkins Company, 1934), pp. i, xi, 17. Letter, Charles B. Davenport to Eugen Fischer, 28 January 1932: APS B:D27.

70. K. Holler, "The Nordic Movement in Germany," Eugenical News Vol. XVII (1932), pp. 117, 119.

71. Edwin Black, The Transfer Agreement (Washington: Dialog Press, 1999), p. 3. William L. Shirer, The Rise and Fall of the Third Reich: A History of Nazi Germany (New York: Simon and Schuster, 1960), p. 173. See Shirer, pp. 159, 172.

72. Black, The Transfer Agreement, p. 3. Shirer, pp. 4-5.

73. The Newseum, "Holocaust: The Untold Story," at www.newseum.org. Generally see Deborah E. Lipstadt, Beyond Belief (New York: The Free Press, 1986). Generally see Black, The Transfer Agreement.

74. Shirer, pp. 196, 430.
75. Black, *The Transfer Agreement,* pp. 177-185. See "German Sterilization Progress," *Eugenical News,* Vol. XIX (1934), p. 38. See "Sterilization in Germany," *Eugenical News,* Vol. XX (1935), p. 13. See "Applied Eugenics in Germany," *Eugenical News,* Vol. XX (1935), p. 100.
76. Raul Hilberg, *The Destruction of the European Jews* (New York: Harper Colophon Books, 1961), p. 45. Shirer, pp. 201-202, 221-224. "Jewish Refugees from Germany," *Eugenical News* Vol. XIX (1934), p. 44. Yad Vashem: The Holocaust Martyr's and Heroes' Remembrance Authority, "March 22: Dachau camp established," *Chronology of the Holocaust 1933-1936* at www.yad-vashem.org.il. Also see Black, *The Transfer Agreement,* pp. 177-179.
77. "Eugenical Sterilization In Germany," *Eugenical News* Vol. XVIII (1933), pp. 91-93. Weber, p. 251.
78. "Eugenical Sterilization In Germany," p. 91. "Human Sterilization in Germany and the United States," *Journal of the American Medical Association* Vol. 102 No. 18, p. 1501.
79. "Berlin: From Our Regular Correspondent," *Journal of the American Medical Association,* Vol. 103 No. 13 (August 1935), p. 1051. Proctor, p. 103.
80. C.P. Blacker, "Eugenics in Germany," 8 August 1933, pp. 4,5: Wellcome Box 112.
81. "The New Format," *Eugenical News* Vol. XVII (1932), p. 16. "Eugenical Sterilization In Germany," pp. 89, 91-93.
82. "Eugenical Sterilization In Germany," p. 90.
83. "Eugenical Sterilization In Germany," p. 90.
84. Ernst Rüdin, "Eugenic Sterilization: An Urgent Need," *Birth Control Review* April 1933, pp. 102, 103-104.
85. "Berlin: From Our Regular Correspondent," *Journal of the American Medical Association,* Vol. 101 No. 11 (Sept. 9, 1933), pp. 866-867.
86. "Berlin: From Our Regular Correspondent," *Journal of the American Medical Association,* Vol. 100 No. 23 (June 10, 1933), p. 1877.
87. "Race-Culture in Germany," *Eugenical News* Vol. XVIII (1933), p. 111.
88. *Haushaltsplan der Kaiser Wilhelm Instituts für Anthropologie,* (1933): R 36, Akte 1366 Deutscher Gemeindetag. Weindling, *Health, Race and German Politics Between National Unification and Nazism, 1870-1945,* p. 315. Proctor, p. 40
89. Letter, H. J. Muller to Robert A. Lambert, 7 June 1933: RF RG 1.1 Ser 717 Box 10 Folder 64.
90. Muller to Lambert, 7 June 1933.
91. Muller to Lambert, 7 June 1933.
92. Weindling, "From Philanthropy to International Science Policy," p. 132.
93. W.W. Peter, "Germany's Sterilization Program," *American Journal of Public Health,* No. 3 Vol. 24 (March 1934)
94. "German Children Face Sterilization," *New York Times,* 5 January 1934: Truman C-2-7:3. "Berlin: From our Regular Correspondent," *Journal of the American Medical Association* Vol. 103 No. 13 (Sept 1935), p. 1051. Proctor p. 106.
95. "The Eleventh Federation Meeting," *Eugenical News* Vol. XIX (1934), p. 107. Lipstadt, pp. 13-15.
96. "Nazis Insist Reich Be 'Race-Minded,'" *New York Times,* 7 January 1934: Truman C-2-7:3. "Question of Admitting German Refugees Under Bond Studied by Labor Department," *New York Times,* 20 January 1934: Truman C-2-7:3. "Catholics Exempt on Sterilization," *New York Times,* 31 December 1933: Truman C-2-7:2. "German Children Face Sterilization." "Jewish Immigrants Aided," undated article: Truman C-2-7:3. "11,060 Loans Made to Jews on Farms," *New York Times,* undated: Truman C-2-7:3.
97. "Nazis Insist Reich Be 'Race-Minded.'" Letter, Harry H. Laughlin to Madison Grant, 13 January 1934: Truman D-2-5:5.
98. Wilhelm Frick, "German Population and Race Politics," *Eugenical News* Vol. XIX (1934), pp. 33-38. "A French View," *Eugenical News* Vol. XIX (1934), p. 39. "Eugenics in Germany," *Eugenical News* Vol. XIX (1934), pp. 40-41. "Jewish Refugees from Germany," p. 44.
99. "Notes," *Eugenical News,* Vol. XIX (1934), p. 60.
100. Carl Hammesfahr, "The Mother of Nations," in "Eugenical Propaganda in Germany," *Eugenical News,* Vol. XIX (1934), p. 45.
101. Masthead, *Eugenical News,* Vol. XVIII (1933), p. 14. Masthead, *Eugenical News,* Vol. XIX (1934), p. 16. Masthead, *Eugenical News,* Vol. XX (1935), p. 98.
102. "A Letter from Dr. Ploetz," *Eugenical News,* Vol. XIX (1934), p. 129. "Jewish Physicians in Berlin," *Eugenical News,* Vol. XIX (1934), p. 126.
103. Letter, Thomas B. Appleget to George K. Strode, 23 February 1938: R 1.1 717 9 56. Letter, Bruce Bliven to the Rockefeller Foundation, 20 December 1933: RF 1.1 717 9 56. Letter, Benson Y. Landis to the Rockefeller Foundation, 20 February 1934: RF 1.1 717 9 56. Letter, Walther Spielmeyer to D.P. O'Brien, 4 November 1933: RF 1.1 717 9 46.
104. Letter, Thomas B. Appleget to Bruce Bliven, 10 January 1934: RF 1.1. 717 9 56. Letter, Thomas B. Appleget to Bruce Bliven, 31 January 1934: RF 1.1 717 9 56.
105. Landis to the Rockefeller Foundation, 20 February 1934.
106. Landis to the Rockefeller Foundation, 20 February 1934. Letter, Thomas B. Appleget to

Raymond Fosdick, 7 February 1934: RF1.1 717 9 56.

107. Appleget to Strode, 23 February 1934. Memorandum, 13 March 1934: RF 1.1 717 9 56. Letter, George K. Strode to Thomas B. Appleget, 6 March 1934: RF 1.1 717 9 56.

108. Memorandum from D.P. O'Brien to Alan Gregg, 12 December 1934: RF 1.1 717 9 56. Memorandum from D.P. O'Brien to Alan Gregg, 27 November 1934: RF 1.1 717 9 56. Letter, Walther Spielmeyer to Alan Gregg, 22 October 1934: RF 1.1 717 9 56.

109. Memorandum, 28 August 1934: RF 1.1 717 13 123. Weindling, ""From Philanthropy to International Science Policy," p. 133.

110. "Archiv Fuer Rassen- Und Gesellschatfs-Biologie," *Eugenical News*, Vol. XXI (1926), p. 184. Kristie Macrakis, *Surviving the Swastika*, (New York: Oxford University Press, 1993), p. 114.

111. J.W. Schotte, "To L.H. La Motte: Confidential Report on Our Dealings with War Ministries of Europe," circa spring 1940, p. 1: NA RG 60. Also see Black, *IBM and the Holocaust*, Chapter VIII.

112. Memorandum, W.D. Jones to Thomas J. Watson, 10 January 1934: IBM Files. Also see Black, *IBM and the Holocaust*, pp. 81-84.

113. *Denkschrift zur Einweihung der neuen Arbeitsstätte der Deutschen Hollerith Maschinen Gesellschaft m.b.H in Berlin-Lichterfelde,* January 8 1934, pp. 39-40: USHMM Library. Also see Black, *IBM and the Holocaust*, p. 82.

114. *Denkschrift zur Einweihung der neuen Arbeitsstätte der Deutschen Hollerith Maschinen Gesellschaft m.b.H in Berlin-Lichterfelde,* January 8 1934, pp. 39-40. Also see Black, *IBM and the Holocaust*, p. 82.

115. Memorandum, Jones to Watson, 10 January 1934. Letter, Thomas J. Watson to Willy Heidinger, 26 February 1934: IBM Files. Also see Black, *IBM and the Holocaust*, pp. 85-86.

116. Edgar Schultze, "Die verfeinerte Auswertung statistischer Zusammenhänge mit Hilfe des Hollerith-Lochkartenverfahrens," *Hollerith Nachrichten* 40 (August 1934), pp. 505-517. Also see Black, *IBM and the Holocaust*, p. 92.

117. Hermann Krüger, "Das Hollerith-Lochkarten-Verfahren im Führsorgewesen," *Hollerith Nachrichten* 47 (March 1935), pp. 615, 618. Also see Black, *IBM and the Holocaust*, pp. 94-95. *Hollerith Nachrichten* 45 (January 1935), p. 588. Also see Black, *IBM and the Holocaust*, p. 95.

118. Friedrich Zahn, "Fortbildung der moderne Bevölkerungsstatistik durch erbbiologische Bestandsaufnahmen," *Allgemeines Statistisches Archiv* 27 (1937/38), pp. 194-195. Also see Black, *IBM and the Holocaust*, p. 96.

119. Hilberg, pp. 46-47.

120. Black, *IBM and the Holocaust*, pp. 109-111.

121. Hilberg, pp. 46, 48.

122. Generally see Black, *IBM and the Holocaust.*

123. "Reich Adopts Swastika As Nation's Flag; Hitler's Reply To 'Insult,'" *New York Times* 16 September 1935.

124. Harry H. Laughlin, "Jewish Studies: Outline of a Plan for the Determination of Jewish Racial Quota-Fulfillments Among Both Institutional and College Jews in the United States," memorandum circa 1924: Truman C-2-4:5.

125. Stefan Kühl, *Die Internationale der Rassisten: Aufstieg und Niedergang der internationalen Bewegung für Eugenik und Rassenhygiene im zwanzigsten Jahrhundert* (Frankfurt/Main: Campus, 1997), p. 138. Letter, Dr. E. Rodenwaldt to Carl Schneider, 11 December 1935: B-1523/5 UH.

126. Letter, Carl Schneider to Dr. Gampp, 10 June 1936: B-1523/6 Universitätarchiv Heidelberg. Letter, Carl Schneider to Harry H. Laughlin, 16 May 1936: Truman E-1-3:8. Letter, Harry H. Laughlin to Carl Schneider, 11 August 1936: Truman E-1-3:8.

127. Proctor, pp. 188-191. Lifton, pp. 64-66, 71-75.

128. Shirer, pp. 322-357, 428-455. Eugen Kogon, *Theory and Practice of Hell* (New York: Berkley Books, 1980), pp. 48-49. Lipstadt, pp. 121-125, 143.

129. Letter, Waldemar Kaempffert to Eugenical News, 15 October 1935: Truman D-2-3:5. Letter, J. H. Landman to Harry H. Laughlin, 13 September 1935: Truman D-2-3:5. Letter, Walter Landauer to Charles B. Davenport, 29 February 1935.

130. See "Berlin: From Our Regular Correspondent," *Journal of the American Medical Association*, Vol. 106 (Nov 1936), p. 1582. See "Berlin: From Our Regular Correspondent," *Journal of the American Medical Association*, Vol. 113 No. 24 (Nov 1939), p. 2163. See "Berlin: From Our Regular Correspondent," *Journal of the American Medical Association*, Vol. 112 No. 19 (April 1939), p. 1981.

131. Seeley G. Mudd Manuscript Library, "The Raymond Blaine Fosdick Papers," at libweb.princeton.edu. Weindling, "From Philanthropy to International Science Policy," pp. 133, 134, 135. Macrakis, p. 114. Letter, Hans Bauer to Kaiser Wilhelm-Institut Für Biologie, 18 August 1935: BDA I. Abt., Rep 1A, Nr. 1052/6 (Foreign and International Affairs). Letter, Fritz von Wettstein to the Generalverwaltung der Kaiser Wilhelm-Gesellschaftzur Förderung der Wissenschaften, 10 June 1936: BDA I. Abt., Rep 1A, Nr. 1054/2 (Foreign and International Affairs). Letter, G. Gottschewski to Herr Reichs-und Preussischen Minister für Wissenschaft, Erziehung und Volksbildung: BDA I. Abt., Rep 1A, Nr. 1054/2

(Foreign and International Affairs). Fritz von Wettstein, "An das Reichsministerium für Wissenschaft, Erziehung und Volksbildung," 27 July 1938: BDA I. Abt., Rep 1A, Nr. 1061/3 (Foreign and International Affairs).

132. Letter, Raymond Fosdick to Selskar M. Gunn, 6 June 1939: RF 1.1 717 16 150.

133. "U.S. Eugenist Hails Nazi Racial Policy," *New York Times*, 29 August 1935: Truman D-2-3:5. See Letter, Kaempffert to Eugenical News, 15 October 1935. See Letter, Landman to Laughlin, 13 September 1935.

134. "U.S. Eugenist Hails Nazi Racial Policy." "The German Racial Policy," *Eugenical News* Vol. XXI (1936), p. 25.

135. C.M. Goethe, "Patriotism and Racial Standards," *Eugenical News* Vol. XXI (1936), pp. 65-66.

136. Marie E. Kopp, "Sterilization" in *American Eugenics: Being the Proceedings at the Annual Meeting and Round Table Conferences of the American Eugenics Society* (New York) 7 May 1936, pp. 56, 57, 61, 64. Kopp, "The German Program of Marriage Promotion through State Loan," *Eugenical News* Vol. XXI (1936), pp. 121-129.

137. Kopp, "Sterilization," p. 65.

138. Harry H. Laughlin, "Eugenics In Germany: Motion Picture Showing How Germany is Presenting and Attacking Her Problems in Applied Eugenics," *Eugenical News*, Vol. XXII - (1937), pp. 65-66. Letter, Harry H. Laughlin to Wickliffe Draper, 9 December 1938: Truman D-2-3:14. Letter, Harry H. Laughlin to Wickliffe Draper, 15 March 1937: Truman D-2-3:14. *Zeil und Weg* Vol. 7 (1937), No. 14, p. 361. See Martin S. Pernick, *The Black Stork: Eugenics and the Death of "Defective" Babies in American Medicine and Motion Pictures Since 1915* (New York: Oxford University Press, 1996), pp. 164-166.

139. Laughlin, p. 66. Pernick, p. 165.

140. Shirer, pp. 430-434. Martin Gilbert, *The Holocaust* (New York: Holt, Rinehart and Winston, 1985), pp. 69-70. Weindling, *Health, Race and German Politics Between National Unification and Nazism, 1870-1945*, pp. 502-503. Generally see Proctor. Generally see Lifton.

141. Weindling, *Health, Race and German Politics Between National Unification and Nazism, 1870-1945*, p. 515. Proctor, pp. 112-114. Zentral-archiv der Deutschen Demokratischen Republik, Potsdam: Reichsinnenministerium 10160, Film 23063 as cited by Müller-Hill, p. 34.

142. Müller-Hill, pp. 137-139, 164. Tage U.H. Ellinger, "On the Breeding of Aryans," *Journal of Heredity* Vol. XXXIII (April 1942), p. 141. Mary Mills, "Propaganda and Children during

the Hitler Years," at www.nizkor.org. Pernick, p. 165. Proctor, pp. 196-197.

143. Müller-Hill, p. 13.

144. Henry Friedlander, *The Origins of Nazi Genocide* (Chapel Hill, NC: The University of Carolina Press, 1995) pp. 21-22, 81.

145. Stephen Trombley, *The Right to Reproduce: A History of Coercive Sterilization* (London: Weidenfeld & Nicolson, 1988), p. 116. Nancy L. Gallagher, *Breeding Better Vermonters* (Hanover, NH: University of New England Press, 1999), p. 140.

146. Lifton, pp. 62-79. Proctor, pp. 188-194. Müller-Hill, p. 15.

147. Lothrop Stoddard, *Into the Darkness* (Newport Beach, CA: Noontide Press, 2000), p. 187.

148. Stoddard, pp. 187-188.

149. Stoddard, p. 189.

150. Stoddard, p. 201, 205.

151. Stoddard, p. xi, 192, 196.

152. Lifton, p. 31.

CHAPTER SIXTEEN

1. Brig. Gen. Eric F. Wood, "Inspection of German Concentration Camp for Political Prisoners Located at Buchenwald on the North Edge of Weimar," report of 16 April 1945: PRO FO 371 151185. Direct examination of Katzen-Ellenbogen, *U.S. v. Josias Prince Zu Waldeck et. al.*, Case No. 000-50-9, p. 4220: NA RG 496/451. Direct examination of Rous, *U.S. v. Josias Prince Zu Waldeck et. al.*, p. 1626.

2. Direct examination of Horn, *U.S. v. Josias Prince Zu Waldeck et. al.*, pp. 896-897, 1624-1626, 1627, 1639, 1640, 1642. Cross-examination of Katzen-Ellenbogen, *U.S. v. Josias Prince Zu Waldeck et. al.*, p. 4366. Testimony of Karl Hemrick Victor Berthold, 17 January 1947, p. 16: NA RG 496/290/59/14/1-5 Box # 444. Physical measurements of defendants: NA RG 496/290/59/14/1-5 Box #442. Wood, p. 2. Direct examination of Katzen-Ellenbogen, pp. 4216-4217.

3. Eugen Kogon, *The Theory and Practice of Hell* (New York: Berkley Books, 1980), p. 210. Direct examination of Katzen-Ellenbogen, pp. 4224-4225, 4243-4245, 4261-4262, 4291. Cross-examination of Katzen-Ellenbogen, pp. 4344, 4383. Testimony of Berthold, pp. 16-17. Statement of Walter Hummelsheim, p. 2: NA RG 496/290/59/14/1-5 Box #444

4. Direct examination of Rous, pp. 1625-1627. Cross-examination of Rous, *U.S. v. Josias Prince Zu Waldeck et. al.*, pp. 1640-1643. See Wood, p. 2.

5. Direct examination of Katzen-Ellenbogen, pp. 4241, 4295, 4300. Re-direct examination of Katzen-Ellenbogen, *U.S. v. Josias Prince Zu Waldeck et. al.*, p. 4386. Wood, p. 2. Cross-examination of Katzen-Ellenbogen, p. 4366.

See photograph of Edwin Katzen-Ellenbogen: NA RG 496/290/59/14/1-5 Box #434. Direct examination of Siebeneichler, *U.S. v. Josias Prince Zu Waldeck et. al.*, pp. 2320-2321. Direct examination of Challe, *U.S. v. Josias Prince Zu Waldeck et. al.*, pp. 438-439. Statement of Edwin Katzen-Ellenbogen, *U.S. v. Josias Prince Zu Waldeck et. al.*, p. 4401. Memorandum, Leon Alexander to A.H. Rosenfeld, Jr., 28 January 1947: NA RG 496/290/59/14/1-5 Box #444. Statement, Edwin Katzen-Ellenbogen to the 7708 War Crime Group, 13 April 1948: NA RG 496/457. "Buchenwald Doctor Lists Atrocities," *Stars and Stripes*, 15 May 1947: NA RG 496/290/59/14/1-5 Box 445.

6. Ben A. Smith, "Review of the War Crimes Section, Military Affairs Branch, Judge Advocate Division, Headquarters, European Command, ATO 403, U.S. Army," Report of 31 July 1950, pp. 1,2, 6: NA RG 496/290/59/14/5-7 Box #457. "The Guilty," *Life Magazine*, 25 August 1947. Sentences, *U.S. v. Josias Prince Zu Waldeck et. al.*, p. 5713. Letter, "Owner of a copy of Bourgemeister Madonna" to Edwin Katzen-Ellenbogen, 31 August 1947: NA RG 496/457. Letter, Evelyn Kranz to "Edwin Katzenellenbogen," 15 August 1947: NA RG 496/457. Letter, Olga Pilat to the High Military Court at Dachau, 20 September 1947: NA RG 496/457. Letter, Mieczyslaw Lurczynski to Board of Review, 20 April 1948: NA RG 496/457. Signed testimonial of A. Simonart, 15 March 1950: NA RG 496/457. Letter, W. R. Graham to Board of Review, 5 April 1950: NA RG 496/457.

7. Eugenics Research Association, *Active Membership Accession List* (Cold Spring Harbor, NY: Eugenics Research Association, 1922): Truman, ERA Membership Records. Direct examination of Katzen-Ellenbogen, pp. 4198-4199.

8. *American Men of Science.* (New York: Bowker, 1910 edition), s.v. "KatzenEllenbogen." Letter, Graham to Board of Review, 5 April 1950. Commonwealth of Massachusetts, "Record of Marriage," 30 March 1939: NA RG 496/457. Direct testimony of Katzen-Ellenbogen, pp. 4193-4194. *The Jewish Encyclopedia*, s.v. "Katzenellenbogen."

9. H. Hoefle, "Medical Report on Katzenellenbogen, Edwin," 30 January 1950, p. 2: NA RG 496/457. *American Men of Science.* Direct examination of Katzen-Ellenbogen, pp. 4194-4197. "Record of Marriage." Supreme Judicial Court of Massachusetts, "Edward Peter Pierce: A Memorial," at www.massreports.com. Sworn deposition of "Dr. Edwin Marie Katzen-Ellenbogen": NA RG 496/290/59/14/1-5 Box #435.

10. Direct testimony of Katzen-Ellenbogen, pp. 4197-4199. Edwin Katzen-Ellenbogen, "The Detection of a Case of Simulation of Insanity By Means of Association Tests," *Journal of Abnormal Psychology* Vol. VI (1911), p. 19. See Massachusetts Department of Correction, "People Executed by Electrocution in Massachusetts," at www.state.ma.us.

11. Direct testimony of Katzen-Ellenbogen, pp. 4198-4199. "A Critical Essay on Mental Tests in Their Relation to Epilepsy," *Epilepsia* Vol. 4 (1913), pp. 130, 140. Edwin Katzen-Ellenbogen, "The Mental Efficiency in Epileptics," *Epilepsia* Vol. 3 (Dec 1912), p. 504.

12. Direct examination of Katzen-Ellenbogen, pp. 4198-4199.

13. Direct examination of Katzen-Ellenbogen, pp. 4199, 4200. Edwin Katzen-Ellenbogen, "A Critical Essay on Mental Tests in Their Relation to Epilepsy," p. 130. *Active Membership Accession List.*

14. Direct examination of Katzen-Ellenbogen, p. 4200. Letter, Pilat to the High American Military Court, 20 September 1947. Letter, Olga Heide-Pilat to General Handy, 7 August 1951: NA RG 496/457. Letter, "Edwin Katzen Ellenbogen" to the Prison Director of Landsberg, 7 January 1950: NA RG 496/457. Letter, "Owner of the Bourgemeister Madonna" to Katzen-Ellenbogen, 31 August 1947: NA RG 496/457.

15. Letter, "Owner of a copy of Bourgemeister Madonna" to Katzen-Ellenbogen, 31 August 1947. Letter, Krantz to "Katzenellenbogen," 15 August 1947.

16. Letter, Pilat to "The High American Military Court," 20 September 1947. Deposition of Edwin Katzen-Ellenbogen, 13 April 1948, p. 3. Letter, Olga Pilat to the Chief Commander of EUCOM, 5 June 1951: NA RG 496/457. Affadavit, Axel and Jan Helge Heide, 20 February 1950: NA RG 496/457.

17. Direct examination of Katzen-Ellenbogen, p. 4200.

18. Direct examination of Katzen-Ellenbogen, pp. 4200, 4201-4202.

19. Direct examination of Katzen-Ellenbogen, pp. 4203-4205.

20. Direct examination of Katzen-Ellenbogen, pp. 4204, 4205.

21. David A. Hackett, ed. and trans., *The Buchenwald Report* (Boulder, CO: Westview Press, 1995), pp. 112-115.

22. Kogon, pp. 155, 157, 172. Opening statement of the prosecution, *United States of America v. Karl Brandt et al.* at www.ess.uwe.ac.uk. Testimony of M. Dubost, "The Trial of German Major War Criminals, Forty-Fifth Day: Tuesday 29th January 1946," at www.nizkor.org. *The Buchenwald Report*, p. 79.

Buchenwald Camp: The Report of a Parliamentary Delegation, April 1945, p. 6: NA RG 496/290/59/14/1-5 Box #440. Kogon, p. 172.

23. "Statement by a Jewish-Christian Prisoner," p. 5: PRO FO 371/21757. *Buchenwald Camp: The Report of a Parliamentary Delegation*, p. 6. "Buchenwald Atrocities," at www.scrapbook-pages.com. *The Buchenwald Report*, pp. 238-239. See Johannes Tuchel, *Die Inspektion der Konzentrations-Lager: Das System des Terrors: 1938-1945*, p. 100.

24. *The Buchenwald Report*, p. 113. Wood, pp. 2, 3, 4. *Buchenwald Camp: The Report of a Parliamentary Delegation*, p. 6. "George Vanier: Canadian diplomat reports his experience," 27 April 1945 at www.nizkor.org. "The Trial of German Major War Criminals, Forty-Fifth Day: Tuesday 29th January 1946." Cross-examination of Biermann, *U.S. v. Josias Prince Zu Waldeck et. al.*, p. 549. Cross-examination of Wilhelm, *U.S. v. Josias Prince Zu Waldeck et. al.*, p. 4447. Memorandum to Commanding General, Third U.S. Army, 25 May 1945 at www.nizkor.org. Direct examination of Sitte, *U.S. v. Josias Prince Zu Waldeck et. al.*, pp. 365-366.

25. *Nuremberg Military Tribunal* Volume V, p. 973.

26. *Buchenwald Camp: The Report of a Parliamentary Delegation*, p. 7.

27. PW Intelligence Bulletin No. 2/20 December 1944 at www.lib.uconn.edu/DoddCenter. Deposition of Isaak Egon Oschshorn, 5 September 1945 at www.lib.uconn.edu/DoddCenter. Direct examination of Katzen-Ellenbogen, p. 4206. Deposition of Erich Kather, 9 May 1947: NA RG 496/290/59/14/1-5 Box #435. Direct examination of Horn, pp. 905-906. Direct examination of Katzen-Ellenbogen, p. 4284, 4294-4296. Direct examination of Siebeneichler, pp. 2318-2319. Cross-examination of Katzen-Ellenbogen, p. 4371. Photograph of Katzen-Ellenbogen.

28. Direct examination of Horn, pp. 898-899. Direct examination of Katzen-Ellenbogen, pp. 4207-4210, 4214, 4237-4238, 4244. Affidavit of August Bender, 3 April 1948: NA RG 496/457. Affidavit of August Bender, 20 February 1950: NA RG 496/290/59/14/5-7 Box # 457.

29. Direct examination of Katzen-Ellenbogen, pp. 4209, 4214, 4241, 4244, 4245. Cross-examination of Katzen-Ellenbogen, pp. 4371, 4386. Direct examination of Siebeneichler, p. 2320. Statement of Hummelsheim, p. 2. Direct examination of Horn, p. 912. Testimony of Berthold, p. 17.

30. Direct examination of Katzen-Ellenbogen, pp. 4238-4239.

31. Direct examination of Katzen-Ellenbogen, pp. 4238-4239.

32. Direct examination of Katzen-Ellenbogen, pp. 4240-4241. *The Buchenwald Report*, p. 42.

33. Direct examination of Katzen-Ellenbogen, p. 4239.

34. Direct examination of Katzen-Ellenbogen, pp. 4225, 4226, 4273-4275. Statement of Hummelsheim, p. 2. Direct examination of Horn, pp. 897-898, 907-911. Cross-examination of Katzen-Ellenbogen, p. 4355. Smith, p. 2. Recross examination of Kogon, *U.S. v. Josias Prince Zu Waldeck et. al.*, p. 946.

35. Cross-examination of Kogon, *U.S. v. Josias Prince Zu Waldeck et. al.*, p. 939. Redirect examination of Horn, *U.S. v. Josias Prince Zu Waldeck et. al.*, p. 915.

36. Direct examination of Edwin Katzen-Ellenbogen, p. 4223.

37. Direct examination of Challe, p. 438. Cross-examination of Challe, *U.S. v. Josias Prince Zu Waldeck et. al.*, p. 439.

38. Smith, p. 3.

39. Cross-examination of Katzen-Ellenbogen, pp. 4361-4632.

40. Jens-Christian Wagner, *Das KZ Mittelbau-Dora* (Wallstein Verlag, 2001) p. 53. United States Holocaust Museum, "Dora-Mittelbau," at www.ushmm.org. *The Buchenwald Report*, pp. 219-220.

41. Cross-examination of Katzen-Ellenbogen, pp. 4340-4341, 4342-4343.

42. Cross-examination of Katzen-Ellenbogen, p. 4343.

43. Statement of Hummelsheim, p. 2. Direct examination of Katzen-Ellenbogen, pp. 4278-4281.

44. Kogon, pp. ix, x. Opening statement of the prosecution, *United States of America v. Karl Brandt et al.* Re-direct examination of Kogon, *U.S. v. Josias Prince Zu Waldeck et. al.*, p. 948. Generally see Eugen Kogon et. al., *Terror und Gewaltkriminalität: Herausforderung für den Rechtsstaat: Diskussionsprotokoll Reihe Hessen-forum*. Generally see Eugen Kogon, *Die Stunde der Ingenieure: technolog. Intelligenz u. Politik*.

45. Direct testimony of Katzen-Ellenbogen, pp. 4232, 4300. Kogon, *The Theory and Practice of Hell*, pp. 229-230. "Extracts from the Affidavit of Waldemar Hoven 24 October 1946, Concerning the Killing of Inmates by Phenol and Other Means," at www.mazal.org. Cross-examination of Katzen-Ellenbogen, pp. 4344-4345, 4352. Re-direct of Katzen-Ellebogen, p. 4387. Statement of Katzen-Ellenbogen, p. 4397.

46. Direct examination of Katzen-Ellenbogen, p. 4245. Cross-examination of Katzen-Ellenbogen, pp. 4344-4345, 4352. *The Buchenwald*, pp. 225, 319-320.

47. Cross-examination of Katzen-Ellenbogen, p. 4352.

48. Cross-examination of Katzen-Ellenbogen, pp. 4352-4353.

49. Robert Jay Lifton, *The Nazi Doctors: Medical Killing and the Psychology of Genocide* (New York: Basic Books, 1986), p. 362. Author's communication with Dr. Harry Stein.
50. Direct examination of Katzen-Ellenbogen, pp. 4228, 4243-4244. Author's interview with Dr. Harry Stein, 20 January 2003.
51. Letter, Andre Simonart to Edwin Katzen-Ellenbogen, 2 May 1948: NA RG 496/457. Letter, Andre Simonart to Edwin Katzen-Ellenbogen, 18 May 1947: NA RG 496/290/59/14/1-5 Box #434. Affidavit of August Bender, 20 February 1950. Direct examination of Katzen-Ellenbogen, p. 4214. Direct examination of Siebeneichler, pp. 2323-2324. Statement of Edwin Katzen-Ellenbogen, p. 4401. List of exhibits for Katzen-Ellenbogen trial: NA RG 496/290/59/14/1-5 Box #439. Direct examination of Horn, p. 900.
52. Direct examination of Sitte, p. 363. Author's interview with Dr. Harry Stein. "RuSHA Case: Introduction," at www.mazal.org. Direction examination of Horn, p. 897. See "A. Forced Germanization of Enemy Nationals," at www.mazal.org.
53. Direct examination of Horn, p. 897.
54. Direct examination of Katzen-Ellenbogen, p. 4271.
55. Cross-examination of Katzen-Ellenbogen, p. 4328.
56. Direct examination of Katzen-Ellenbogen, p. 4302.
57. Redirect of Kogon, p. 943.
58. Cross-examination of Katzen-Ellenbogen, p. 4349.
59. Cross-examination of Katzen-Ellenbogen, p. 4373.
60. Cross-examination of Katzen-Ellenbogen, pp. 4373-4374.
61. Cross-examination of Katzen-Ellenbogen, pp. 4393-4394.
62. Cross-examination of Katzen-Ellenbogen, p. 4400.
63. Statement of Katzen-Ellenbogen, pp. 4400-4401.
64. Smith, p. 1. "The Guilty."
65. Opening statement of the prosecution, *U.S. v. Josias Prince Zu Waldeck et. al.*, p. 43. "Review and Recommendations of the Deputy Judge Advocate for War Crimes," p. 60: NA RG 496/290/59/14/1-5 Box #433.
66. Smith, pp. 1-2.
67. "Appraisal of Petition for Clemency," 25 March 1953: NA RG 496/457. Smith, pp. 2, 3. Statement, Katzen-Ellenbogen to the 7708 War Crime Group. Letter, "Edwin K. Ellenbogen" to the Two Representatives of the Inspector General's Office, 1 November 1949: NA RG 496/457. Letter, "Edwin K. Ellenbogen" to General Thomas T. Handy, 18 February 1950:

NA RG 496/457. Letter, "E. K. Ellenbogen" to General Thomas T. Handy, 2 April 1950: NA RG 496/290/59/14/5-7 Box #457. Letter, "Edwin Katzen Ellenbogen" to the Prison Director, 7 January 1950: NA RG 496/457. Letter, "Edwin Katzen Ellenbogen," to Commander-in-Chief, United States Army, Europe, 10 March 1953: NA RG 495/457. Internal Route Slip, Subject Edwin Katzen-Ellenbogen, 25 March 1953: NA RG 496/457.

CHAPTER SEVENTEEN
1. Lucette Matalon Lagnado and Sheila Cohn Dekel, *Children of the Flames* (New York: William Morrow and Company, Inc., 1991), p. 47. Miklos Nyiszli, *Auschwitz: A Doctor's Eyewitness Account* (New York: Arcade Publishing, 1993), pp. 17-19.
2. Nyiszli, p. 19. Robert Jay Lifton, *The Nazi Doctors: Medical Killing and the Psychology of Genocide* (New York: Basic Books, 1986), p. 164.
3. Nyiszli, pp. 19, 52-54. Lifton, p. 166. Rudolf Hoess, *Commandant of Auschwitz* (New York: Popular Library, 1959), pp. 176-177.
4. Lifton, p. 165 Edwin Black, *IBM and the Holocaust* (New York: Crown Publishers, 2001), pp. 352-353.
5. Generally see Yisrael Gutman and Michael Berenbaum, ed., *Anatomy of the Auschwitz Death Camp* (Indianapolis: Indiana University Press, 1994). See Gerald Reitlinger, *The Final Solution* (New York: A.S. Barnes & Company, Inc., 1961), pp. 113-117. Also see Hoess, p. 183.
6. Lifton, pp. 169, 269-271, 360-363. Nyiszli, pp. 102, 103, 134-137.
7. Benno Müller-Hill, *Murderous Science*, G. R. Fraser, trans. (Cold Spring Harbor, NY: Cold Spring Harbor Laboratory Press: 1998), p. 127. Benno Müller-Hill, "The Blood from Auschwitz and the Silence of the Scholars," *History and Philosophy of the Life Sciences* Vol. 21 (1999), p. 332.
8. Henry Friedlander, *The Origins of Nazi Genocide* (Chapel Hill, NC: The University of Carolina Press, 1995), p. 14. Müller-Hill, *Murderous Science* pp. 216-217. "Notes and News," *Eugenical News* vol. X (1925), p. 24. Paul J. Weindling, *Health, Race and German Politics Between National Unification and Nazism, 1870-1945* (Cambridge: Cambridge University Press, 1989), p. 436.
9. Robert N. Proctor, *Racial Hygiene: Medicine Under the Nazis* (Cambridge, MA: Harvard University Press, 1988), pp. 42-43, 81, 188. Black, *IBM and the Holocaust* (New York: Crown Publishers, 2001), p. 109. "Reich Health Leader," *Who's Who in Nazi Germany* (London: Wiederfield and Nicols, 1982).
10. Proctor, pp. 104-105.

11. Black, pp. 93-95. "Verschuer's Institute," *Eugenical News* Vol. XX (1935), pp. 59-60. Posner and Ware, pp. 11-12. Arthur L. Caplan, ed., *When Medicine Went Mad* (Totowa, NJ: Humana Press, 1992), p. 28. Lifton, p. 27. Weindling, pp. 526-530, 553. "Berlin: From Our Regular Correspondent," *Journal of the American Medical Association*, Vol. 102 No. 8 (24 February 1934), p. 631. "Berlin: From Our Regular Correspondent," *Journal of the American Medical Association*, Vol. 104 No. 22 (1 June 1933), p. 2011. "Berlin: From Our Regular Correspondent," *Journal of the American Medical Association*, Vol. 104 No. 23 (8 June 1935), p. 2110. "Berlin: From Our Regular Correspondent," *Journal of the American Medical Association*, Vol. 103 No. 13 (30 September 1936), p. 1052.

12. Müller-Hill, *Murderous Science*, p. 12. Proctor, p. 81. Posner and Ware, p. 11.

13. Proctor, p. 259. Müller-Hill, "The Blood from Auschwitz," pp. 333, 334.

14. "Notes and News," p. 24. "Archiv Fuer Rassen-Und Gesellschafts-Biologie," *Eugenical News* Vol. XI (1926), p. 9. "Archiv f. Rassen-U. Gesellschafts-Biologie," *Eugenical News* Vol. XII (1927), p. 180. "Meeting of International Federation of Eugenic Organizations," *Eugenical News* Vol. XIII (1928), p. 129. See "Archiv Für Rassen-Und Gesellschafts-Biologie," *Eugenical News* Vol. XIII (1928), p. 32. See "Current Periodicals," *Eugenical News* Vol. XIII (1928), p. 162.

15. "Berlin: From our Regular Correspondent," *Journal of the American Medical Association* Vol. 102 No. 1 (6 January 1934), p. 57. "The Inheritance of Tuberculosis," *Journal of Heredity* (January 1934), p. 26. "Marriage Consultation Centers in Central Europe," *Eugenical News* Vol. XIX (1934), p. 78. "Berlin: From our Regular Correspondent," *Journal of the American Medical Association* Vol. 103 No. 6 (1 September 1934), p. 767. "New German Etymology for Eugenics," *Eugenical News* Vol. XIX (1934), p. 125. "German Eugenics, 1934," *Eugenical News* Vol. XIX (1934), pp. 141-142.

16. "Verschuer's Institute," pp. 59-60.

17. Letter, C.H. Danforth to Otmar von Verschuer, 15 April 1936: Universitätsarchiv Münster: Nachlass Verschuer Nr. 4. Letter, Henry H. Goddard to Otmar von Verschuer, 7 July 1936: Universitätsarchiv Münster: Nachlass Verschuer Nr. 4, Letter, Paul Popenoe to Otmar von Verschuer, 16 July 1936: Universitätsarchiv Münster: Nachlass Verschuer Nr. 4. Letter, E.S. Gosney to Otmar von Verschuer, 24 May 1937: Universitätsarchiv Münster: Nachlass Verschuer Nr. 4. Letter, E.S. Gosney to Otmar von Verschuer, 15 June 1937: Universitätsarchiv Münster: Nachlass Verschuer Nr. 4. Letter, E.S. Gosney to Otmar

von Verschuer, 26 June 1937: Universitätsarchiv Münster: Nachlass Verschuer Nr. 4.

18. Letter, Harry H. Laughlin to Otmar von Verschuer, 16 June 1936: Universitätsarchiv Münster: Nachlass Verschuer Nr. 4. Letter, Harry H. Laughlin to Otmar von Verschuer, 23 July 1936: Universitätsarchiv Münster: Nachlass Verschuer Nr. 4.

19. Letter, Otmar von Verschuer to Harry H. Laughlin, 2 September 1936: Universitätsarchiv Münster: Nachlass Verschuer Nr. 4.

20. "Verschuer's Institute of Genetics and Race Hygiene," *Eugenical News* Vol. XX (1935), p. 104. "Berlin: From Our Regular Correspondent," *Journal of the American Medical Association* Vol. 103 No. 13 (30 September 1935), pp. 1052-1053. "Berlin: From Our Regular Correspondent," *Journal of the American Medical Association* Vol. 106 No. 4 (23 January, 1935), p. 308.

21. Proctor, pp. 196-197. "The Hereditary Aspect of Pathology," *Eugenical News* Vol. XXI (1936), pp. 21-22.

22. Letter, C.M. Goethe to Otmar von Verschuer, 15 April 1937: Universitätsarchiv Münster: Nachlass Verschuer Nr. 4. See letter, C.M. Goethe to Otmar von Verschuer, 23 December 1937: Universitätsarchiv Münster: Nachlass Verschuer Nr. 4. See letter, C.M. Goethe to Otmar von Verschuer, 26 February 1938: Universitätsarchiv Münster: Nachlass Verschuer Nr. 4. See letter, C.M. Goethe to Otmar von Verschuer, 22 November 1938: Universitätsarchiv Münster: Nachlass Verschuer Nr. 4.

23. Goethe to von Verschuer, 23 December 1937.

24. Goethe to von Verschuer, 26 February 1938. Goethe to von Verschuer, 22 November 1938. Martin Gilbert, *The Holocaust* (New York: Holt, Rinehart and Winston, 1985), pp. 64-65.

25. Letter, C.H. Danforth to Otmar von Verschuer, 28 February 1939: Universitätsarchiv Münster: Nachlass Verschuer Nr. 4.

26. Letter, Charles B. Davenport to Otmar von Verschuer, 15 December 1937: Universitätsarchiv Münster: Nachlass Verschuer Nr. 4. Letter, Otmar von Verschuer to Charles B. Davenport, 5 January 1938: Universitätsarchiv Münster: Nachlass Verschuer Nr. 4.

27. Weindling, pp. 559, 561. Gerald L. Posner and John Ware, *Mengele: The Complete Story* (New York: Cooper Square Press, 1986), pp. 9, 10-11. Lifton, p. 339.

28. Posner and Ware, pp. 12-13. Müller-Hill, *Murderous Science*, pp. 38, 128. Also see RG242, *RuSHA roll #D5462*, beginning Fr. 2892.

29. Posner and Ware, pp. 12-13. Müller-Hill, *Murderous Science*, pp. 38, 128. Müller-Hill, "The Blood from Auschwitz," pp. 336-337.

30. Müller-Hill, *Murderous Science*, pp. 128, 163.
31. *RuSHA roll #D5462*. Posner and Ware, p. 14. Lifton, p. 340. Universitäts-Institut Erbbiologie und Rassenhygiene, memorandum of 30 September 1938: BAK: R 73 Akte 15342.
32. Letter, rector of the Johann Wolfgang Goethe-Universität to the Reichsminister für Wissenschaft, Erziehung und Volksbildung, 17 December 1938: BAK: R 4901 Akte 3016 Blatt 2-6. Letter, Otmar von Verschuer to Reichsminister für Wissenschaft, Erziehung und Volksbildung, 29 November 1938: BAK: R 4901 Akte 3016 Blatt 2-6. Letter, Dr. H. Schade to Reichsminister für Wissenschaft, Erziehung und Volksbildung, 17 December 1938: BAK: R 4901 Akte 3016 Blatt 2-6. Letter, rector of the Johann Wolfgang Goethe-Universität to the Reichsminister für Wissenschaft, Erziehung und Volksbildung, 16 December 1938: BAK: R 4901 Akte 3016 Blatt 2-6. Letter, Otmar von Verschuer to the rector of the Johann Wolfgang Goethe-Universität, 8 December 1938: BAK: R 4901 Akte 3016 Blatt 2-6. Report, Reichsminister für Wissenschaft, Erziehung und Volksbildung, 25 July 1939, pp. 184-187: BAK: R 4901 Akte 3016 Blatt 2-6.
33. Posner and Ware, p. 16. "Universitäts-Institut für Erbbiologie und Rassenhygiene Frankfurt a.M. Gartenstr.140 (Deutschland)," memorandum circa January 1939: Universitätsarchiv Münster: Nachlass Verschuer Nr. 4. "Geburtstagsliste," circa 1940: Universitätsarchiv Münster: Nachlass Verschuer, Nr. 4.
34. Posner and Ware, pp. 14-15, 328 f. 19, 328 f. 20.
35. Posner and Ware, pp. 16-17. Müller-Hill, *Murderous Science*, p. 226 f. 144.
36. Weindling, pp. 557-558. Gilbert, pp. 154, 176.
37. Müller-Hill, pp. 17, 48-49, 53-54.
38. Weindling, pp. 557-558.
39. Proctor, pp. 43-44. Otmar von Verschuer, *Leitfaden der Rassenhygiene* (Leipzig, 1941), p. 127 as cited by Proctor, p. 211. Müller-Hill, *Murderous Science*, p. 226 f 144. "Geburtstagsliste."
40. Müller-Hill, *Murderous Science*, p. 226 f. 144.
41. "Wanted, Photographs of Twins," *Journal of Heredity* (Oct. 1918), p. 262. Columbia Encyclopedia, 6th ed., s.v. "multiple birth."
42. Francis Galton, "The History of Twins, as a Criterion of the Relative Powers of Nature and Nurture," *Journal of the Anthropological Institute*, (1875), pp. 391, 392-393.
43. "'Probably Mendelian' and 'Clearly Hereditary' traits," lantern slide circa 1921: Truman Lantern Slides, Brown Box 835. Charles B. Davenport, *Heredity In Relation To Eugenics* (New York: Henry Holt & Company, 1911; reprint, New York: Arno Press Inc., 1972), p. 180. Erwin Baur, Eugen Fischer and Fritz Lenz,

Human Heredity, 3rd Ed., trans. Eden & Cedar Paul (New York: The MacMillan Company, 1931), pp. 554-555.
44. "Index," *Eugenical News* Vol. I (1916), p. 93. "Notes on Genetics," *Eugenical News* Vol. I (1916), p. 67. "Sex of Twins," *Eugenical News* Vol. II (1917), p. 5. "Special Data Wanted," *Eugenical News* Vol. II (1917), p. 54.
45. "Wanted, Photographs of Twins," p. 262. "Six Hundred Twins Already Discovered," *Journal of Heredity* (May 1919), p. 210.
46. Eugenics Record Office, "Schedule No. 10: Schedule for the Study of Twins," circa 1918: APS Series I: PDR and Correspondence. "A Strain Producing Multiple Births," *Journal of Heredity* (Nov. 1919), p. 382. "Notes and News," *Eugenical News*, Vol. VI (1921), p. 32.
47. "The Pendleton Twins," *Eugenical News* Vol. III (1918), p. 77. "Twins in Russia," *Eugenical News* Vol. IV (1919), p. 52.
48. "II-2-a Racial differences in twin frequency," undated lecture notes: APS B:D27 Davenport-Twins #4. "Twins in the Census," *Eugenical News* Vol. IV (1920), p. 16. Charles B. Davenport and Morris Steggerda, *Race Crossing in Jamaica* (Washington: Carnegie Institution of Washington, 1929), pp. 445-452.
49. "Diagnosing Twin Pregnancy With Stethoscope," *Eugenical News* Vol. VII (1922), p. 48. "Similar Tumors in Twins," *Eugenical News* Vol. VII (1922), p. 71.
50. "How to Tell Identical Twins," *Eugenical News* Vol. XI (1926), p. 41. H. J. Muller, "The Determination of Twin Identity," *Journal of Heredity* Vol. XVII No. 6 (June 1926), p. 195. Paul Popenoe, "Twins Reared Apart," *Journal of Heredity* Vol. XIII No. 3 (March 1922), p. 142. See "A New Study of Twins (A Review)," *Journal of Heredity* Vol. XV No. 4 (April 1924), p. 165. See "Causes of Twinning," *Journal of Heredity* Vol. XIV No. 8 (November 1923), p. 370. See "Eugenic Studies in Scandinavia," *Eugenical News* Vol. VIII (1923), p. 52. See "Fact and Fiction," *Eugenical News* Vol. VIII (1923), p. 71. See "A Father of Twins," *Eugenical News* Vol. IX (1924), p. 3. See "Parallel Behavior," *Eugenical News* Vol. X (1925), p. 46.
51. Paul Popenoe and Roswell Hill Johnson, *Applied Eugenics* rev. ed. (New York: Macmillan, 1935), p. 6. Baur, Fischer, and Lenz, pp. 554-555, 557, 590-591.
52. R.A. Fisher, "New Data on the Genesis of Twins," in *Eugenics, Genetics and the Family: Volume I: Scientific Papers of the Second International Congress of Eugenics* (Baltimore: Williams & Wilkins Co., 1923), p. 195. "Booth 2-Physical Anthropometry," *Eugenical News* Vol. XVII (1932), p. 142.
53. Alfred Gordon, "The Problems of Heredity and

Eugenics," *Eugenical News* Vol. XX (1935), pp. 50, 52-53. "Identical Twins," *Eugenical News* Vol. XIX (1934), p. 134.

54. Letter, John William Draper to Charles B. Davenport, 30 December 1924: APS B:D27: Davenport - John William Draper #3. Letter, Charles B. Davenport to John William Draper, 9 January 1925: APS B:D27: Davenport - John William Draper #3.

55. "Notes and News," *Eugenical News* Vol. XII (1927), p. 164. "Twins With Tuberculosis (Rev. of Diehl and Verschuer, *Zwillingstuberkulose*)," *Journal of Heredity* Vol. XXVIII No. 3 (March 1937), pp. 91-96. "On the Inheritance of Darwin's Tubercle," *Eugenical News* Vol. XX (1935), pp. 3-4. See "Eugenics in Germany," *Eugenical News* Vol. XIX (1934), p. 42.

56. "Berlin: From Our Regular Correspondent," *Journal of the American Medical Association* Vol. 106 No. 4 (23 January 1936), p. 308.

57. Weindling, p. 555. "The Hereditary Aspect of Pathology," *Eugenical News* Vol. XXI (1936), pp. 21-22. "Verschuer's Institute," *Eugenical News* Vol. XXI (1936), pp. 59-60.

58. Paul Popenoe, "Twins and Criminals," *Journal of Heredity* Vol. XXVII No. 10 (October 1936), pp. 388-390.

59. "Universitäts-Institut für Erbbiologie und Rassenhygiene Frankfurt a.M. Gartenstr. 140 (Deutschland)."

60. "Bericht über die im Jahre 1938 bisher durchgeführten und für die nächste Zeit geplanten Forschungen," 30 September 1938: BAK: R 73 Akte 15342.

61. *Ziel und Weg* Vol. 14, No. 9 (1939), p. 449.

62. Tage U.H. Ellinger, "On the Breeding of Aryans," *Journal of Heredity* Vol. XXXIII No. 4 (April 1942), p. 141.

63. Ellinger, pp. 141-142.

64. *Ziel und Weg* Vol. 14, No. 9 (1939), p. 449. "Wehrmachtsauftragsnummer: S 4891-5378," undated report: BAK: R 73 Akte 15342, Blatt 64.

65. Müller-Hill, *Murderous Science*, p. 76. "Bericht über die Fortführung der erbpsychologischen Forschung," 14 March 1944: BAK: R 73 Akte 15342, Blatt 66.

66. "Wehrmachtsauftragsnummer: S 4891-5378." Posner and Ware, p.

67. Lagnado and Dekel, p. 47.

68. Lagnado and Dekel, p. 51.

69. Lagnado and Dekel, p. 45.

70. Lagnado and Dekel, p. 47.

71. Lagnado and Dekel, p. 55.

72. Lagnado and Dekel, p. 31.

73. Lagnado and Dekel, p. 36.

74. Posner and Ware, pp. 36, 43. Lagnado and Dekel, p. 42. Nyiszli, p. 60.

75. Nyiszli, pp. 19, 23, 56-58.

76. Nyiszli, pp. 39-40.

77. Nyiszli, pp. 31-32, 58.

78. Posner and Ware, p. 27. Lifton, pp. 164-165, 342-343. Nyiszli, pp. 132-133.

79. Lifton, p. 342. Nyiszli , p. 57. Posner and Ware, pp. 48, 49. Testimony of Vera Alexander at the Jerusalem tribunal of Josef Mengele, February 1985 as cited by Helena Kubica, "The Crimes of Josef Mengele," in Gutman and Berenbaum, p. 324. Lagnado and Dekel, pp. 58-59, 64, 67.

80. Lifton, p. 343. Posner and Ware, p. 45. Gisella Perl, *I Was a Doctor in Auschwitz* (New York: International Universities Press, 1948), pp. 110-111 as cited by Posner and Ware, pp. 46-47.

81. Nyiszli, p. 58. Lagnado and Dekel, pp. 66, 70-71. Posner and Ware, pp. 36-37, 40. Testimony of unnamed survivor at the Jerusalem tribunal of Josef Mengele, February 1985as cited by Posner and Ware, pp. 37-38.

82. Testimony of Vera Alexander as cited in Gutman and Berenbaum, p. 324.

83. Lifton, p. 351.

84. Nyiszli, pp. 59, 61, 132.

85. Lifton, p. 355. Lagnado and Dekel, p. 61. Gilbert, p. 687.

86. Lifton, p. 360. Nyiszli, pp. 56, 57, 102-103. Hermann Langbein, Menschen in Auschwitz (Vienna: Europaverlag, 1972) as cited by Lifton, p. 362. Lagnado and Dekel, p. 65.

87. Lagnado and Dekel, pp. 59-60.

88. Nyiszli, p. 61.

89. Nyiszli, p. 11. Gilbert, p. 756.

90. "Bericht über die Fortführung der erbpsychologischen Forschung."

91. Nyiszli, p. 63.

92. Nyiszli, pp. 175-181.

93. Nyiszli, p. 65.

94. "Hereditary Eye Defects," *Eugenical News* Vol. XVIII (1933), pp. 43-44. See "Twins: Like and Unlike," *Eugenical News* Vol. XXII (1937), pp. 119-120. See "11. Myopia in Identical Twins," *Eugenical News* Vol. XXII (1937), p. 64.

95. "Supplement to Schedule for the Study of Twins," attachment to letter, Harry F. Perkins to Harry H. Laughlin, 29 May 1936: Truman D-2-4:2.

96. Weindling, p. 560.

97. Testimony of Jancu Vexler before Judge Horst von Glasenapp, 13 March 1973 as cited by Kubica in Gutman and Berenbaum, p. 326. "Israel Visions of Hell: Pursuing the 'Angel of Death'" *Time Magazine*, 18 February 1985.

98. Lifton, p. 362.

99. Karin Magnussen, "Bericht über die Durchrihrung der Arbeiten zur Erforsonung der Erbbedingtheit der Entwicklung der Augenfarbe als Grundlage für Rassen-und Abstammungsuntersuchungen," 15 March 1944: BAK: R 73 Akte 15342 Blatt 68.

100. Müller-Hill, *Murderous Science*, pp. 171-172.

101. "Vernehmung: Vorgeladen erscheint Frau Dr. Karin Magnussen, geb. 9.2.08 in Bremen, wohnhaft Bremen, Hagenanerstrasse 7," 25 May 1949: Max Planck Archive Abt. I, Rep. 3, Nr. 26.

102. Also see Major Leonard Darwin, "First Steps Towards Eugenic Reform," *Eugenics Review*, Vol. 4 (ca. April 1912), pp 34-35 as selected in G. K. Chesterton, *Eugenics and Other Evils*, edited by and including additional articles selected by Michael W. Perry (Seattle, WA: Inkling Press, 2000), pp 144-145. Also see "Berlin: From Our Regular Correspondent," *Journal of the American Medical Association* Vol. 104 No. 23 (8 June 1935), p. 2110.

103. Charles B. Davenport, "Research in Eugenics," in *Eugenics, Genetics and the Family*, p. 25.

104. "Fifth International Congress of Genetics," *Eugenical News* Vol. XII (1927), p. 152. "Radiogram to Gregg, 13 May 1932: RF 1.1 Ser 7171 Box 10 Folder 63.

105. Tracy B. Kittredge, "Progress Report: Grant to Notgemeinschaft for Anthropological Studies of the Population of Germany," 10 June 1933: RF 1.1 717 20 187. Letter, Norma S. Thompson to Schmidt-Ott, 14 November 1929: RF 1.1 717 20 187.

106. "Genetics: Blood Groups of Twins," *Eugenical News* Vol. XX (1935), p. 41.

107. Paul J. Weindling, "From Philanthropy to International Science Policy: Rockefeller Funding of Biomedical Sciences in Germany 1920-1940," in Nicolaas A. Rupke, ed., *Science, Politics and the Public Good: Essays in Honor of Margaret Gowing* (New York: Macmillan Press, 1988), pp. 129, 133, 135. Letter, Raymond B. Fosdick to Selskar M. Gunn, 6 June 1939: RF 1.1 717 16 150.

108. Weindling, *Health, Race and German Politics Between National Unification and Nazism, 1870-1945*, pp. 560-561.

109. Müller-Hill, *Murderous Science*, p. 76. "Wehrmachtsauftragsnummer: S 4891-5378."

110. Unidentified survivor as cited by Posner and Ware, p. 37. Lagnado and Dekel, pp. 62, 66.

111. Lagnado and Dekel, p. 37. Müller-Hill, *Murderous Science*, p. 128.

112. Posner and Ware, p. 52.

113. Müller-Hill, *Murderous Science*, p. 169.

114. Müller-Hill, *Murderous Science*, pp. 169-170.

115. David A. Hackett, ed. and trans., *The Buchenwald Report* (Boulder, CO: Westview Press, 1995), pp. 72, 369-370.

116. See Joachim C. Fest, *The Face of the Third Reich: Portraits of the Nazi Leadership* (New York: Pantheon Books, 1970), p. 379 f. 44. See Heinz Höhne, *The Order of the Death's Head: The Story of Hitler's SS* (New York: Coward-McCann, 1970), p. 157. See author's communication with Harry W. Mazal, 10 February 2003.

117. H. Nachtsheim, "Bericht über die im Halbjahr 1943/44 im Auftrage des Reichsforschungsrates durchgeführten Untersuchungen zur vergleichenden und experimentellen Erbpathologie," 15 March 1944: BAK: R 73, Akte 15342, Blatt 61-63.

118. Nachtsheim, "Bericht."

119. "Wehrmachtsauftragsnummer: SS 4891-5377," undated report: BAK: R 73, Akte 15342, Blatt 65.

120. Michael Shevell, "Racial Hygiene, Active Euthanasia, and Julius Hallervorden," *Neurology* 42 (November 1992), pp. 2214, 2216. Michael Shevell, "Reply from the Author," *Neurology* 43 (July 1993), p. 1453. Peter S. Harper, "Naming of Syndromes and Unethical Activities: The Case of Hallervorden and Spatz," *Lancet* 348 (1996), p. 1224. Müller-Hill, *Murderous Science*, p. 244. Jürgen Peiffer, "Neuropathology in the Third Reich," *Brain Pathology* 1 (1991), p. 127.

121. Shevell, pp. 2216-2217.

122. Letter, Thomas B. Appleget to Frederick Strauss, 14 April 1936: RF 1.1 717 9 58. Weindling, "From Philanthropy to International Science Policy," p. 133. Letter, Dr. Telschow to Dr. Hoffmann, 27 July 1938: Max Planck Archive I. Abt., Rep 1A, Nr. 1061/2. Kristie Macrakis, *Surviving the Swastika: Scientific Research in Nazi Germany* (New York: Oxford University Press, 1993), pp. 85-86.

123. "Natural Sciences-Program and Policy: Past Program and Proposed Future Program," pp. 79-80: RF RG 3.1 915 1 6. Diane B. Paul, "The Rockefeller Foundation and the Origins of Behavior Genetics," in Keith R. Benson, Jane Maienschein and Ronald Rainger, eds., *The Expansion of American Biology* (New Brunswick, NJ: Rutgers University Press, 1991), p. 266.

124. "Natural Sciences," p. 80.

125. Lagnado and Dekel, pp. 89, 92.

CHAPTER EIGHTEEN

1. Gerald L. Posner and John Ware, *Mengele: The Complete Story* (New York: Cooper Square Press, 2000), pp. 57-64, 91, 93, 94. Office of Special Investigations, "United Nations War Crimes List, Number 8" as cited by Posner and Ware, p. 63. "Universitäts-Institut für Erbbiologie und Rassenhygiene Frankfurt a.M. Gartenstr.140 (Deutschland)," memorandum circa January 1939: Universitätsarchiv Münster: Nachlass Verschuer Nr. 4.

2. Otmar von Verschuer, "Bevölkerungs- und Rassenfrage in Europa," *Europäischer Wissenschaftsdienst*, 1 (1944), pp. 11-14 as cited by Benno Müller-Hill, "The Blood from Auschwitz and the Silence of the Scholars," *History and Philosophy of the Life Sciences* Vol. 21 (1999), p. 335. Müller-Hill, p. 348.

496 || WAR AGAINST THE WEAK

3. Max Planck Archiv: Nachlass Nachtsheim as cited by Müller-Hill, p. 348.
4. Generally see James P. O'Donnell, *The Bunker: The History of the Reich Chancellery Group* (New York: Bantam Books, 1978).
5. "Vertriebene Wissenschaft: Deutsche Gelehrte, die ins Ausland gingen," *Die Neue Zeitung* 15 April 1946. "Kunst und Kultur in Kürze," *Die Neue Zeitung* 3 May 1946. Müller-Hill, p. 350. Letter, Otmar von Verschuer to Paul Popenoe, 28 August, 1946: Universitätsarchiv Münster: Nachlass Verschuer, Nr. 4.
6. "Personalien v. Verschuer.": Max Planck Archiv: II Abt. Rep. 0001A as cited by Müller-Hill pp. 350-351.
7. "Personalien v. Verschuer." as cited by Müller-Hill pp. 350-351.
8. "Personalien v. Verschuer." as cited by Müller-Hill p. 351.
9. "Personalien v. Verschuer." as cited by Müller-Hill p. 351. Müller-Hill, p. 351.
10. Letter, Paul Popenoe to Otmar von Verschuer, 25 July 1946: Universitätsarchiv Münster: Nachlass Verschuer, Nr. 4. Letter, Otmar von Verschuer to Paul Popenoe, (sic) 31 September 1946: Universitätsarchiv Münster: Nachlass Verschuer, Nr. 4.
11. Verschuer to Popenoe, 28 August, 1946.
12. Von Verschuer to Popenoe, (sic) 31 September 1946.
13. Letter, Paul Popenoe to Otmar von Verschuer, 7 November 1946: Universitätsarchiv Münster: Nachlass Verschuer, Nr. 4. Letter, Fritz Lenz to Otmar von Verschuer, 14 September 1946: Universitätsarchiv Münster: Nachlass Verschuer, Nr. 4.
14. Letter, Paul Popenoe to Otmar von Verschuer, 8 February 1947: Universitätsarchiv Münster: Nachlass Verschuer, Nr. 4. Letter, Otmar von Verschuer to Paul Popenoe, 26 February 1947: Universitätsarchiv Münster: Nachlass Verschuer, Nr. 4. Letter, Paul Popenoe to Otmar von Verschuer, 12 May 1947: Universitätsarchiv Münster: Nachlass Verschuer, Nr. 4. See letter, Otmar von Verschuer to Paul Popenoe, 20 December 1946: Universitätsarchiv Münster: Nachlass Verschuer, Nr. 4.
15. Letter, Paul Popenoe to Otmar von Verschuer, 29 October 1947: Universitätsarchiv Münster: Nachlass Verschuer, Nr. 4. Letter, C.M. Goethe to Otmar von Verschuer, 16 April 1948: Universitätsarchiv Münster: Nachlass Verschuer, Nr. 4. See letter, Otmar von Verschuer to Lee R. Dice, 18 September 1947: Universitätsarchiv Münster: Nachlass Verschuer, Nr. 4. See letter, Otmar von Verschuer to Lee R. Dice, 12 January 1948: Universitätsarchiv Münster: Nachlass Verschuer, Nr. 4. See letter, Lee R. Dice to Otmar von Verschuer, 27 February 1948:

Universitätsarchiv Münster: Nachlass Verschuer, Nr. 4. See letter, Otmar von Verschuer to Lee R. Dice, 21 April 1948: Universitätsarchiv Münster: Nachlass Verschuer, Nr. 4. See letter, Otmar von Verschuer to the "Charles Fremont Dight Institute for the Study of Human Genetics," 12 August 1948: Universitätsarchiv Münster: Nachlass Verschuer, Nr. 4.
16. Müller-Hill, pp. 351-352. Letter, von Lewinski to Heubner 23 December 1946 as cited by Müller-Hill, pp. 351-352. "Abschrift. Denkschrift betreffend Herrn Prof. Dr. med. Otmar Frhr. v. Verschuer," as cited by Müller-Hill, pp. 352-353.
17. Müller-Hill, p. 355. American Society of Human Genetics, "ASHG Past Presidents, 1948-2004," at www.faseb.org.
18. Müller-Hill, pp. 354-355. Generally see Otmar von Verschuer and E. Kober, *Die Frage der erblichen Disposition zum Krebs* (Mainz: Verlag der Akademie der Wissenschaft und der Literatur, 1956).
19. Author's interview with Dr. Kurt Hirschhorn, 12 February 2003.
20. "Personalien v. Verschuer." as cited by Müller-Hill p. 351. Müller-Hill, pp. 351, 354, 355.
21. "Universitäts-Institut für Erbbiologie und Rassenhygiene Frankfurt a.M. Gartenstr.140 (Deutschland)," memorandum circa January 1939: Universitätsarchiv Münster: Nachlass Verschuer Nr. 4. Nachlass von Verschuer as cited by Müller-Hill, p. 348. Karin Magnussen, "Bericht über die Durchrihrung der Arbeiten zur Erforsonung der Erbbedingtheit der Entwicklung der Augenfarbe als Grundlage für Rassen-und Abstammungsuntersuchungen," 15 March 1944: BAK: R 73 Akte 15342 Blatt 68.
22. "Medical Experiments of the Holocaust and Nazi Medicine," at www.remember.org. Peter Tyson, "The Experiments," at www.pbs.org.
23. Leo Alexander as quoted by Barry Siegel, "Can Evil Beget Good? Nazi Data: A Dilemma for Science," *Los Angeles Times*, 30 October 1988.
24. *Nuremberg Military Tribunal, Green Book, Volume I*, p. 421. Anti-Defamation League, "Library Won't Be Named for Nazi Scientist, Air Force Assures ADL," press release of 14 February 2003.
25. "Library Won't Be Named for Nazi Scientist, Air Force Assures ADL." "Objections lead OSU to remove picture of Nazi," *Daily Illini* 29 October 1993. Cable News Network, "U.S. News Briefs," 24 September 1995.
26. New Mexico Museum of Space History, "International Space Hall of Fame: Hubertus Strughold," at www.spacefame.org. Author's interview with New Mexico Museum of Space History, 13 February 2003.
27. Michael Shevell, "Racial Hygiene, Active

Euthanasia, and Julius Hallervorden," *Neurology* 42 (November 1992), pp. 2214. National Institute of Neurological Disorders and Stroke, "The First Scientific Workshop of Hallervorden-Spatz Syndrome," at www.ninds.nih.gov.
28. Rubin I. Kuzniecky and Bradley K. Evans, "Julius Hallervorden: To the Editor," *Neurology* 43 (July 1993), p. 1452. Peter S. Harper, "Naming of Syndromes and Unethical Activities: The Case of Hallervorden and Spatz," *Lancet* 348 (1996), pp. 1224, 1225.
29. NBIA Disorders Association, "Welcome to the NBIA Disorders Association Homepage," at www.hssa.org. Author's interview with Patricia Wood, 12 February 2003. NBIA Disorders Association, "NBIA Frequently Asked Questions," at www.hssa.org.
30. "The First Scientific Workshop of Hallervorden-Spatz Syndrome." National Institute of Health, "NINDS Hallervorden-Spatz Disease Information Page," at www.ninds.nih.gov.
31. "Brain sections to be buried?" *Nature* Vol. 339 (15 June 1989), p. 498. Wolfgang Neugebauer and Georg Stacher, "Nazi Child 'Euthanasia' in Vienna and the Scientific Exploitation of Its Victims before and after 1945," *Digestive Diseases* 17 (1999), pp. 282, 283. Dr. William E. Seidelman, "Medicine and Murder in the Third Reich," *Dimensions: A Journal of Holocaust Studies* Vol. 13 No. 1 (1999).
32. Jürgen Peiffer, "Neuropathology in the Third Reich: Memorial to those Victims of National-Socialist Atrocities in Germany who were Used by Medical Science," *Brain Pathology* 1 (1991), pp. 125-126.
33. Peiffer, p. 129.

CHAPTER NINETEEN
1. E. Carleton MacDowell, "Charles Benedict Davenport, 1866-1944. A Study of Conflicting Influences," *BIOS* vol. XVII, no. 1, pp. 4, 8, 33, 34, 36.
2. MacDowell, no. 1, p. 36.
3. Charles B. Davenport, *Heredity In Relation To Eugenics* (New York: Henry Holt & Company, 1911; reprint, New York: Arno Press Inc., 1972), pp. 253-254.
4. "Personals," *Eugenical News* Vol. I (1916), p. 65. MacDowell, p. 36. See McGill University, "Symptoms [of Polio]," at sprojects.mmi.mcgill.ca.
5. MacDowell, p. 34. Letter, Charles B. Davenport to W.M. Gilbert, 29 June 1934: CSH: CIW Charles Davenport Corresp. 1933-34. Letter, A.F. Blakeslee to Charles B. Davenport, 5 July 1934: CSH: CIW Charles Davenport Corresp. 1933-34. Letter, Charles B. Daven-

port to John C. Merriam, 29 June 1934: CSH: CIW Charles Davenport Corresp. 1933-34.
6. Letter, Charles B. Davenport to A.F. Blakeslee, 19 July 1934: CSH: CIW Charles Davenport Corresp. 1933-34. MacDowell, p. 34. See Charles B. Davenport, "Harry Hamilton Laughlin: 1880-1943," *Eugenical News* Vol. XXVIII (1943), p. 43.
7. MacDowell, p. 37. See The Whaling Museum Society, Inc., "Second Annual Report," 29 July 1944.
8. MacDowell, p. 37.
9. See C.B. Davenport and A.J. Rosanoff, "Reply to the Criticism of Recent American Work by Dr. Heron of the Galton Laboratory," *Eugenics Record Office Bulletin No. 11* (Cold Spring Harbor, NY: Eugenics Record Office, 1914). See Ezekiel Cheever, *School Issues* (Baltimore: Warwick & York, Inc., 1924): CIW Genetics: Eugenics Record Office Misc. Correspondence 2 of 2.
10. Letter, Charles B. Davenport to Harry H. Laughlin, 16 April 1928 as cited by Frances Janet Hassencahl, "Harry H. Laughlin, 'Expert Eugenics Agent' for the House Committee on Immigration and Naturalization, 1921 to 1931." (Ph. D. diss., Case Western Reserve University, 1970), p. 329.
11. Davenport to Laughlin, 16 April 1928, as cited by Hassencahl, p. 329.
12. Hassencahl, pp. 330-331. A.V. Kidder, "Memorandum for Dr. Merriam re Meeting of advisory committee on Eugenics Record Office," circa February 1929: Truman C-2-3:3.
13. Kidder.
14. Contract regarding *Eugenical News*, 20 November 1938: Truman C-2-3:3. Harry H. Laughlin, "Memoranda on Origin of the Eugenical News and the Relation of the Eugenics Record Office and the Eugenics Research Association in its Publication," 11 September 1934: Truman C-2-2:2.
15. Laughlin, "Memoranda on Origin of the Eugenical News."
16. Laughlin, "Memoranda on Origin of the Eugenical News."
17. Letter, Harry H. Laughlin to A.V. Kidder, 30 October 1934: Truman C-2-2:2. Letter, Harry H. Laughlin to A.V. Kidder, 5 November 1934: Truman C-2-2:2.
18. Letter, A.V. Kidder to Harry H. Laughlin, 1 November 1934: Truman C-2-2:2. Letter, A.V. Kidder to Harry H. Laughlin, 3 November 1934: Truman C-2-2:2.
19. "Report of the advisory committee on the Eugenics Record Office," circa June 1935, pp. 1, 6-7: Truman C-2-2:2.
20. "Report of the advisory committee," pp. 2, 3.
21. "Report of the advisory committee," pp. 2, 3.
22. "Report of the advisory committee," pp. 3-5, 6.

23. "Report of the advisory committee," p. 6.
24. "Report of the advisory committee," p. 1. Letter, L.C. Dunn to John C. Merriam, 3 July 1935: Truman C-2-2:2.
25. Dunn to Merriam, 3 July 1935.
26. Dunn to Merriam, 3 July 1935.
27. Generally see David S. Wyman, *Paper Walls* (Amherst, MA: University of Massachusetts Press, 1968). Also see letter, Vannevar Bush to Harry H. Laughlin, 22 March 1939: Truman C-4-3:1.
28. Harry H. Laughlin, "Eugenics in Germany," *Eugenical News* Vol. XXII (1937), pp. 65-66. "A New German Eugenical Quarterly," *Eugenical News* Vol. XXII (1937), p. 88. "The Twenty-Fifth Annual Meeting of the Eugenics Research Association," *Eugenical News* Vol. XXII (1937), pp. 66-67. "Who Knows the Answer?" *Eugenical News* Vol. XXII (1937), pp. 48-49. See Martin S. Pernick, *The Black Stork: Eugenics and the Death of "Defective" Babies in American Medicine and Motion Pictures Since 1915* (New York: Oxford University Press, 1996), pp. 164-166.
29. Martin Gilbert, *The Holocaust* (New York: Holt, Rinehart and Winston, 1985), pp. 64-65. Rudolph M. Binder, "Germany's Population Policy," *Eugenical News* Vol. XXIII (1938), pp. 113-116. Lucy Dawidowicz, *The War Against The Jews, 1933-1945* (New York: Bantam Books, 1975), pp. 128-129.
30. Letter, Vannevar Bush to Harry H. Laughlin, 4 January 1939: Truman D-2-3:13. Bush to Laughlin, 22 March 1939.
31. Postcard, William A. Elam to Harry H. Laughlin, 9 July 1939: Truman C-4-3:1.
32. Harry H. Laughlin, *Immigration and Conquest: A Report of The Special Committee on Immigration and Naturalization of the Chamber of Commerce of the State of New York* (New York: The Special Committee on Immigration and Naturalization of the Chamber of Commerce of the State of New York, 1939), pp. i, 8, 9.
33. Laughlin, *Immigration and Conquest*, p. 20. "Conquest by Immigration (Sent to the following)," circa 1939: Truman C-4-3:1.
34. Contract regarding *Eugenical News*, 20 November 1938. Letter, Vannevar Bush to Harry H. Laughlin, 4 May 1939 as cited by Kevles, p. 199. See *Eugenical News* Vol. XXIII (1938) as compared to *Eugenical News* Vol. XXIV (1939). See Carnegie Institution of Washington, "Memorandum Regarding Mail Addressed to the Genetics Record Office (formerly Eugenics Record Office," circa 1940: CSH GRO Correspondence 1940. See letter, Harry H. Laughlin to John B. Trevor, 18 November 1939: Truman C-2-4:10. See "Independence of the Eugenics Record Office," memorandum circa 1940: Truman C-2-4:10.

See "Tentative Plan for a Laboratory of National Eugenics at Cold Spring Harbor, Long Island," memorandum circa 1940: Truman C-2-4:10. See "Private Notes for Senator Walcott on the Eugenics Record Office," memorandum circa 1940: Truman C-2-4:10.
35. Davenport, "Harry Hamilton Laughlin: 1880-1943," pp. 42, 43. Hassencahl, p. 356. Truman Library, "Harry H. Laughlin biography," at library.truman.edu.
36. Hassencahl, pp. 65-66. See Daniel J. Kevles, *In The Name of Eugenics*, (Cambridge, MA: Harvard University Press, 1985), p. 199.
37. See Harry H. Laughlin, secretary, *Bulletin No. 10A: The Report of the Committee to Study and to Report on the Best Practical Means of Cutting Off the Defective Germ-Plasm in the American Population* (Cold Spring Harbor: Cold Spring Harbor, 1914), pp. 12, 13, 16, 17, 18, 25, 45-47.
38. "Memorandum Regarding Mail Addressed to the Genetics Record Office (formerly Eugenics Record Office)."
39. "Memorandum Regarding Mail Addressed to the Genetics Record Office (formerly Eugenics Record Office)." Letter, Elizabeth M. Howe to Charles B. Davenport, 19 February 1940: CSH GRO Correspondence 1940. Letter, Albert F. Blakeslee to Elizabeth M. Howe, 23 February 1940: CSH GRO Correspondence 1940.
40. Letter, Albert F. Blakeslee to H. C. Coryell, 12 March 1940: CSH GRO Correspondence 1940. See letter, Albert F. Blakeslee to Owen Hannant, 2 March 1940: CSH GRO Correspondence 1940. See letter, Albert F. Blakeslee to Mrs. W.D. Dobson, 2 March 1940: CSH GRO Correspondence 1940. See letter, Albert F. Blakeslee to H. Earl Trobaugh, 6 April 1940: CSH GRO Correspondence 1940. See letter, "Secretary to the Director" to Hiram H. Maxim, 30 March 1953: CSH GRO Correspondence 1947-54. See Laughlin to Trevor, 18 November 1939. See "Independence of the Eugenics Record Office."
41. Letter, Jane Arnold Betts to the Eugenics Record Office, 29 February 1944: CSH GRO Correspondence 1942-45. Letter, M. Demerec to Jane Arnold Betts, 7 March 1944: CSH GRO Correspondence 1942-45. See letter, Blakeslee to Hannant, 2 March 1940. See letter, Blakeslee to Dobson, 2 March 1940. See letter, Blakeslee to Trobaugh, 6 April 1940. See letter, M. Demerec to Abraham Sohulman, 29 April 1946: CSH GRO Correspondence 1942-45. See letter, M. Demerec to Mrs. James A. Bruun, 2 May 1953: CSH GRO Correspondence 1947-54. See letter, "Secretary to the Director" to Mrs. Fred Gillum, 26 January 1954: CSH GRO Correspondence 1947-54. See letter, Agnes C. Fisher to Minnie J. Williams, 24 April 1959:

CSH GRO Correspondence 1955-62. See letter, "Secretary to Director" to Donald Clough McCaffree, 4 January 1963: CSH GRO Correspondence 1963-65. See letter, Agnes C. Fisher to Frank Merrick Semans, 6 January 1966: CSH GRO Correspondence 1966-69. See letter, "Secretary to the Director" to Elsie Van Guilder, 12 January 1966: CSH GRO Correspondence 1966-69. See letter, "Secretary to the Director" to Mrs. Merle Roberts, 12 January 1967: CSH GRO Correspondence 1966-69. See letter, "Secretary to the Director" to Mrs. Eugene W. Coling, 28 March 1967: CSH GRO Correspondence 1966-69. See letter, Agnes C. Fisher to Mrs. Harold F. Zink, 3 March 1969: CSH GRO Correspondence 1966-69. See letter, Agnes C. Fisher to Mrs. L. C. Strong, Jr., 3 June 1969: CSH GRO Correspondence 1966-69. See letter, A.D. Hershey to William Hartley, 20 October 1970: CSH GRO Correspondence 1970-77. See letter, Agnes Fisher to E. Taylor Campbell, 14 September 1976: CSH GRO Correspondence 1970-77. See letter, Agnes C. Fisher to Noel C. Stevenson, 2 February 1977: CSH GRO Correspondence 1970-77. See letter, Agnes Fisher to Dr. Sheldon C. Reed, 30 March 1977: CSH GRO Correspondence 1970-77.

42. Letter, L.R. Dice to M. Demerec, 8 May 1947: GRO CSH GRO Correspondence 1947-54. See letter, M. Demerec to Sheldon Reed, 29 September 1947: CSH GRO Correspondence 1947-54. Letter, M. Demerec to Dewey G. Steele, 15 July 1955: CSH GRO Correspondence 1955-62.

43. Letter, Demerec to Reed, 29 September 1947.

44. Letter, Sheldon Reed to M. Demerec, 8 October 1947: CSH GRO Correspondence 1947-54. Letter, Sheldon Reed to M. Demerec, 5 December 1947: CSH GRO Correspondence 1947-54.

45. Letter, Sheldon C. Reed to M. Demerec, 19 April 1948: CSH GRO Correspondence 1947-54. Letter, Sheldon C. Reed to M. Demerec, 22 January 1948: CSH GRO Correspondence 1947-54. Letter, Claude F. Rogers to Russell F. Barnes, 25 June 1948: CSH GRO Correspondence 1947-54. "Secretary to the Director" to Maxim, 30 March 1953. Letter, M. Demerec to John Fall, 11 August 1948: CSH GRO Correspondence 1947-54. Letter, M. Demerec to Dewey G. Steele, 15 July 1955: CSH GRO Correspondence 1955-62.

46. Letter, Mrs. James A. Bruun to the Eugenics Record Office, circa May 1953: CSH GRO Correspondence 1947-54. Letter, Clifford Frazier to the Eugenics Record Office, 19 February 1952: CSH GRO Correspondence 1947-54. Letter, Minnie J. Williams to Eugenics Record Office, 24 March 1959: CSH

GRO Correspondence 1955-62. Letter, Elsie Van Guilder to American Breeders' Association Eugenics Section, 9 January 1966: CSH GRO Correspondence 1966-69. Letter, E. Taylor Campbell to Eugenics Record Office, 11 September 1976: CSH GRO Correspondence 1970-77.

47. Also see CSH GRO Correspondence Files.

48. Jonas Robitscher, ed. *Eugenic Sterilization* (Springfield, IL: Charles C. Thomas, 1973), p. 123.

49. Letter, Joseph Juhan to Charles Davenport, circa December 1976: CSH GRO Correspondence 1970-77.

50. Letter, Agnes C. Fisher to Joseph Juhan, 13 December 1976: CSH GRO Correspondence 1970-77.

51. "ACLU Says Legal Action Certain," *Richmond Times-Dispatch*, 6 April 1980. Memorandum, Carol Donovan to Suzanne Lynn and Janet Benshoof, 16 June 1980: American Civil Liberties Union Foundation. "Important Notice to All Persons Who Were Residents of Lynchburg Training School and Hospital," circa 1981: ACLU - Sterilization Settlement. American Civil Liberties Union, "Virginia Apologizes for Forced Sterilizations," at archive.aclu.org. Mark R. Warner, "Statement of Governor Mark R. Warner: On the 75th Anniversary of the Buck v. Bell Decision," at www.governor.state.va.us. "Virginia Governor Apologizes for Eugenics Law," *USA Today* 2 May 2002. "Oregon Apologizes for Sterilizations," *Associated Press* 2 December 2002. Tim Smith, "Hodges Apologizes for Sterilizations," *The Greenville News* 8 January 2003. "South Carolina Governor Apologizes for Forced Sterilizations," *The Wall Street Journal* 9 January 2003. "State Issues Apology for Policy of Sterilization," *Los Angeles Times* 12 March 2003.

52. North Carolina General Assembly, "Statute Chapter 35: Sterilization Procedures" at www.ncga.state.nc.us. See D. Anthony D'Esposo, "Recent Developments in Post-Partum Sterilization," *Eugenical News* Vol. XXXVI (1951), pp. 59-61. See "Notes and News Relating to Eugenics: Overpopulation in Puerto Rico," *Eugenical News* Vol. XXVIII (1943), pp. 29-30. See "Notes and News Relating to Eugenics: Puerto Rico's Population Problem," *Eugenical News* Vol. XXX (1945), pp. 61-62. See "Sterilization Outside the United States," *Eugenical News* Vol. XXXI (1946), p. 11. Generally see United States General Accounting Office, "Report B-164031 (5)," 27 February 1978.

53. United States General Accounting Office, "Report B-164031 (5): Enclosure I," 27 February 1978.

54. *Loving v. Virginia*, 388 U.S. 1 (1967).

55. *Loving v. Virginia*, 388 U.S. 1 (1967).
56. *Loving v. Virginia*, 388 U.S. 1 (1967).
57. "Alabama Repeals Century-Old Ban on Interracial Marriages," *Cable News Network* 8 November 2000.
58. See "Memorandum for Dr. Merriam re Meeting of advisory committee on Eugenics Record Office."
59. *Buck v. Bell*, 274 U.S. 200 (1927).
60. "Nazi Retribution Widened by Eden," *New York Times* 18 December 1942. "Allies Describe Outrages on Jews," *New York Times* 20 December 1942.
61. Hebrew Union College, "An Inventory to the Raphaël Lemkin Papers: Biographical Sketch," at www.huc.edu. Raphaël Lemkin, *Axis Rule in Occupied Europe: Analysis, Proposals for Redress* (Washington DC: Carnegie Endowment for International Peace, 1944), pp. xi, xiv, xv.
62. "The Moscow Conference; October 1943," at www.yale.edu/lawweb/avalon. "Nuremberg Trials Final Report Appendix D: Control Council Law No. 10," at www.yale.edu/lawweb/avalon.
63. Lemkin, *Axis Rule in Occupied Europe*, pp. xv, 79-95. Raphaël Lemkin, "Genocide - A Modern Crime," *Free World* Vol. 4 (April 1945), pp. 39-43.
64. Rauschning, as cited by Lemkin, *Axis Rule in Occupied Europe*, Chapter IX, Section II, f. 29.
65. Lemkin, "Genocide - A Modern Crime." "An Inventory to the Raphaël Lemkin Papers."
66. "Nuremberg Trial Proceedings Vol. 1: London Agreement of August 8th 1945," at www.yale.edu/lawweb/avalon. "Trial of German Major War Criminals Nuremberg: First Day: Tuesday, 20th November, 1945 (Part 1 of 10)," at www.nizkor.org. Jewish Virtual Library, "Hans Frank," at www.us-israel.org. Jewish Virtual Library, "Julius Streicher," at www.us-israel.org. Jewish Virtual Library, "Hans Fritzche," at www.us-israel.org. Mazal Library, "Trials of War Cirminals before the Nuernberg Military Tribunals under Control Council Law No. 10," at www.mazal.org.
67. United Nations Resolution 95 (I), "Affirmation of the Principles of International Law Recognized by the Charter of the Nürnberg Tribunal.": United Nations Archives. United Nations Resolution 96 (I), "The Crime of Genocide.": United Nations Archives.
68. Office of the High Commissioner for Human Rights, "Convention on the Prevention and Punishment of the Crime of Genocide," at www.unhchr.ch.
69. Office of the High Commissioner for Human Rights. See *Nuremberg Military Tribunal, Green Book, Volume IV*, p. 626.
70. *Nuremberg Military Tribunal, Green Book, Volume IV*, pp. 609-614, 617.
71. *Nuremberg Military Tribunal, Green Book, Volume IV*, p. 610.
72. *Nuremberg Military Tribunal, Green Book, Volume V*, pp. 3-4.
73. *Nuremberg Military Tribunal, Green Book, Volume IV*, pp. 611-612.
74. *Nuremberg Military Tribunal, Green Book, Volume IV*, pp. 613, 1121-1127.
75. *Nuremberg Military Tribunal, Green Book, Volume IV*, pp. 675, 680, 681-682, 1101.
76. *Nuremberg Military Tribunal, Green Book, Volume IV*, p. 682. Memorial Lidice, "History," at www.lidice-memorial.cz.
77. *Nuremberg Military Tribunal, Green Book, Volume IV*, pp. 613, 614.
78. *Nuremberg Military Tribunal, Green Book, Volume IV*, pp. 614, 616.
79. *Nuremberg Military Tribunal, Green Book, Volume IV*, p. 617.
80. "Protocol of the Wannsee Conference," 20 January 1942 at www.ghwk.de.
81. *Nuremberg Military Tribunal, Green Book, Volume IV*, p. 1180.
82. *Nuremberg Military Tribunal, Green Book, Volume IV*, pp. 1158, 1159.
83. *Nuremberg Military Tribunal, Green Book, Volume IV*, p. 1159.
84. *Nuremberg Military Tribunal, Green Book, Volume IV*, p. 1159.
85. *Nuremberg Military Tribunal, Green Book, Volume V*, p. 166.

CHAPTER TWENTY
1. American Breeders' Association, "Minutes," *First Annual Meeting*, 1903, pp. 1-2: APS. American Breeders' Association, "Committees and Their Specific Duties," *Annual Report, American Breeders' Association*, vol. II (1906), p. 11. E. Carlton MacDowell, "Charles Benedict Davenport, 1866-1944: A Study of Conflicting Influences," *BIOS* vol. XVII no. 1, pp. 23-24. American Genetic Association, "Overview," at lsvl.la.asu.edu/aga.
2. Robert C. Olby, "Horticulture: The Font for the Baptism of Genetics," *Nature Reviews* Vol. 1 (October 2000), pp. 65, 68. See letter, William Bateson to Adam Sedgewick, 4 May 1905.
3. University of Cambridge Department of Genetics, "A Brief History of the Department," at www.gen.cam.ac.uk.
4. American Genetic Association, "Overview." Eugenics Research Association, *Active Membership Accession List* (Cold Spring Harbor, NY: Eugenics Research Association, 1922): Truman ERA Membership Records. "College Courses in Genetics and Eugenics," *Eugenical News* vol. 1 (1916), pp. 26-27. See *Eugenical News* Vol. XV (1930) as compared to *Eugenical News* Vol. XIV (1929).

5. Edmund W. Sinnott and L.C. Dunn, *Principles of Genetics: An Elementary Text, with Problems* (New York: McGraw-Hill, 1925), p. 406.

6. Letter, L.C. Dunn to John C. Merriam, 3 July 1935: Truman C-2-2:2.

7. "Report of the Committee on Human Heredity," memorandum attached to letter, Laurence H. Snyder to Harry H. Laughlin, 27 September 1937: Truman D-2-3:21. Laurence H. Synder, "Presidential Address: Present Trends in the Study of Human Inheritance," *Eugenical News* Vol. XXIII (1938), p. 61.

8. "Report of the Committee on Human Heredity."

9. "Report of the Committee on Human Heredity."

10. Synder, p. 61.

11. Snyder, p. 61.

12. Tage U.H. Ellinger, "On the Breeding of Aryans," *Journal of Heredity* Vol. XXXIII (April 1942), pp. 141-142.

13. Ellinger, pp. 141-142.

14. Ellinger, pp. 141-142.

15. Ellinger, p. 141.

16. Ellinger, p. 142.

17. Ellinger, p. 142.

18. Ellinger, p. 142.

19. Ellinger, pp. 142-143.

20. Ellinger, p. 143.

21. "Eugenics After the War," *Eugenical News* Vol. XXVIII (1943), p. 11.

22. "Eugenics After the War," p. 11.

23. "Eugenics in 1952," *Eugenical News* Vol. XXVIII (1943), p. 12.

24. "Eugenics in 1952," p. 13.

25. "Maternity, the Family and the Future," *Eugenical News* Vol. XXVIII (1943), p. 22.

26. "Notes and News Relating to Eugenics," *Eugenical News* Vol. XXIX (1944), p. 33. "Discussion: Eugenics After the War," *Eugenical News* Vol. XXX (1945), p. 23.

27. "Eugenics in England," *Eugenical News* Vol. XXX (1945), pp. 34, 36.

28. "Eugenics and Modern Life: Retrospect and Prospect," *Eugenical News* Vol. XXXI (1946), p. 33.

29. "Eugenics and Modern Life: Retrospect and Prospect," pp. 33, 34-35.

30. Report, Tage Kemp to the Rockefeller Foundation, 17 November 1932: RF RG 1.2 Ser 713 Box 2 Folder 15.

31. Letter, Daniel P. O'Brien to Tage Kemp, 29 June 1934: RF RG 1.2 Ser 713 Box 2 Folder 15.

32. O'Brien to Kemp, 29 June 1934. Tage Kemp, "To the Rockefeller Foundation: Report of Visits to Various Institutes, Laboratories etc. for Human Genetics in Europe," circa December 1934, pp. 54-55.: RF RG 1.2 Ser 713 Box 2 Folder 16.

33. Kemp, pp. 59-61, 62-63.

34. Kemp, p. 57.

35. See "University of Copenhagen - Institute of Human Genetics," memorandum circa June 1939: RF RG 1.2 Ser 713 Box 2 Folder 15.

36. "To The Rockefeller Foundation," memorandum circa 1935: RF RG 1.2 Ser 713A Box 2 Folder 17. See "To the Ministry for Public Education," memorandum circa February 1935: RF RG 1.2 Ser 713A Box 2 Folder 17.

37. "University of Copenhagen - Institute of Human Genetics." "To The Rockefeller Foundation." "News and Notes Relating to Eugenics," *Eugenical News* Vol. XXXII (1947), p. 30.

38. Frederick Osborn, "History of the American Eugenics Society," unpublished draft of 20 January 1971, p. 15: APS: AES Records - Osborn Papers #2 - History of the AES. Wake Forest University, "Centennial: Wake Forest 'Firsts' in Medicine," at www.wfubmc.edu. "The Twenty-Sixth Annual Meeting of the Eugenics Research Association, June 2 1938," *Eugenical News* Vol. XXIII (1938), p. 72. Manson Meads, *The Miracle on Hawthorne Hill* (Winston-Salem, NC: Wake Forest University, 1988), p. 32. William Allan, "The Relationship of Eugenics to Public Health," *Eugenical News* Vol. XXI (1936), p. 74. See William Allan, "The Inheritance of the Shaking Palsy," *Eugenical News* Vol. XX (1935), p. 72.

39. "Forsyth in the Forefront," *Winston-Salem Journal.* "Lifting the Curtain on a Shameful Era," *Winston-Salem Journal.*

40. C. Nash Herndon, "Human Resources From the Viewpoint of Medical Genetics," *Eugenical News* Vol. XXXV (1950), pp. 6-8. "Lifting the Curtain On a Shameful Era." See "Against Their Will: North Carolina's Sterilization Program."

41. Osborn, "History of the American Eugenics Society," pp. 16-17.

42. Osborn, "History of the American Eugenics Society," pp. 4, 17. American Philosophical Society, "Frederick Henry Osborn Papers," at www.amphilsoc.org. "The Twenty-Sixth Annual Meeting of the Eugenics Research Association, June 2 1938." "B. Early History and Development," *Organized Eugenics,* (New Haven, CT: American Eugenics Society), 1931, p. 3.

43. Osborn, "History of the American Eugenics Society," p. 17.

44. Frederick Osborn, "Population and The Progress of Civilization", presented at Hunter College, 22 December 1959: APS: AES Records: Osborn Papers #9.

45. Letter, Frank Lorimer to Frederick Osborn. 1 October 1959: APS: AES Records - Osborn Papers #2 - Letters on Eugenics.

46. Frederick Osborn, "Draft Prepared for the

Directors' Meeting, April 23rd: Eugenics: Retrospect and Prospect," draft of 26 March 1959: APS: AES Records - Osborn Papers.

47. Letter, Bruce Wallace to Frederick Osborn, 11 April 1961: APS: AES Records - Osborn Papers #2 - Letters on Eugenics.

48. Letter, Sheldon C. Reed to Harry L. Shapiro, 15 May 1961: APS: AES Records.

49. Letter, Frederick Osborn to P.S. Barrows, 25 August 1965: APS: AES Records - Osborn Papers #2 - Letters on Eugenics. Letter, Frederick Osborn to Frank Lorimer, 17 September 1959: APS. Letter, Gordon Allen to Frederick Osborn, 23 May 1961: APS: AES Records. Frederick Osborn, "A Program of Eugenics," undated paper circa May 1961: APS.

50. Osborn to P.S. Barrows, 25 August 1965.

51. Letter, Frederick Osborn to Paul Popenoe, 25 March 1965: APS: AES Records - Osborn Papers #2 - Letters on Eugenics. Letter, Frederick Osborn to Sheldon E. Hermanson, 12 April 1965: APS: AES Records - Osborn Papers #2 - Letters on Eugenics. See Osborn, "History of the American Eugenics Society," p. 20.

52. Letter, Frederick Osborn to Alexander Robertson, 11 October 1967: APS: AES Records - Osborn Papers #2 - Letters on Eugenics. Letter, Dick Hoefnagel to Carnegie Institute of Washington, 5 February 1968: CSH GRO Correspondence 1966-69.

53. "#4519: Human Betterment League of North Carolina Records: Inventory," at www.lib.unc.edu. AIM25, "Penrose, Correspondence," at www.aim25.ac.uk. University College of London, "UCL & Galton,"at collection.ucl.ac.uk. "The Annals of Human Genetics Homepage,"at www.gene.ucl.ac.uk. "The Galton Institute: Annual Report, 1989.": Wellcome SA/EUG/A 95.

54. Osborn, "History of the American Eugenics Society," p. 9. Author's interview with Jay Olshansky, 10 February 2003. "American Eugenics Society," at www.amphilsoc.org.

55. See letter, B. C. Lake to Sewall Wright, 2 June 1954: APS: AES Records. See "American Genetic Association,"at lsvl.la.asu.edu/aga.

56. See Planned Parenthood Federation of America, Inc., "About Us," at www.plannedparenthood.org. See John R. Weeks, "Vignettes of PAA History: Milbank, Princeton and the War,"at www.pop.psu.edu. See "American Eugenics Society." See "Population Council - Fiftieth Anniversary - Officers," at www.popcouncil.org.

57. James D. Watson, "President of Cold Spring Harbor Laboratory," at www.cshl.org. See "Celebrating 50 years of DNA," at www.dna50.org.

CHAPTER TWENTY-ONE

1. "Test-tube Baby Pioneer Backs Human Cloning," News.Telegraph.Co.UK, 6 September 2002. "Scientists: Cloned Sheep Dolly has 'Old' DNA," Cable News Network, 27 May 1999. "World: Asia-Pacific Giant Pandas Follow Dolly," British Broadcasting Corporation, 20 July 1998. "Scientists Await Birth of First Cloned Endangered Species," Cable News Network, 5 January 2001. "Researchers Clone Pigs," Cable News Network, 16 August 2000. "Carbon Copy Cat Cloned," Nature, 14 February 2002. Texas A&M, "Texas A&M Clones First Cat," press release of 14 February 2002. "OWU Professor Clones Mastodon Genes From Intestinal DNA," Ohio Wesleyan University On-Line.

2. "Clones in the Real World," AP Wire Service, 4 March 2001. "Clonaid: Baby 'Clone' Returns Home," Cable News Network 1 January 2003. World Future Society, "Futurist Update," January 2000 at www.wfs.org. "Doctor Refuses to Identify Mother, Give Evidence of Provocative Claim," SiliconValley.com 27 November 2002. "Revealed: Couple Try to Have First Human Clone Baby," Sunday Herald, 21 July 2002.

3. Dorothy Wertz, "DNA Forensics: Professional and Patient Attitudes Internationally," at hgm2002.hgu.mrc.ac.uk. Federal Bureau of Investigation, "Mission Statement and Background," at www.fbi.gov. Federal Bureau of Investigation, "FBI Laboratory: Forensic Systems," at www.fbi.gov. Attorney General Transcript, "News Conference- DNA Initiatives," 4 March 2002 as cited by Electronic Privacy Information Center and Privacy International, "Privacy and Human Rights 2002," circa 2002, p. 30.

4. "DNA Database 'Should Include Every Citizen,'" NewScientist.com 12 September 2002. 2nd International DNA users' Conference for Investigative Officers, "Minutes: Forensic DNA Analysis in China," at www.Interpol.int. "The National DNA Data Bank of Canada," brochure circa 2002. 1st International DNA users' Conference for Investigative Officers, "Austrian National DNA Database," circa November 1999 at www.Interpol.int. 1st International DNA users' Conference for Investigative Officers, "The DNA Database in The Netherlands," circa November 1999 at www.Interpol.int. "Bundestag Establishes Legal Footing for Genetic Database," DE News 24 June 1998. "Privacy and Human Rights 2002," p. 109. "In Louisiana, Debate Over a DNA Dragnet," The Christian Science Monitor, 21 February 2003.

5. "Fingerprint Fear," NewScientist.com, 2 May 2001. R. Chen, P.S. Rabinovitch, D.A. Crispin, M.J. Edmond, K.M. Koprowicz, M.P. Bronner

and T.A. Brentnall, "DNA Fingerprinting Abnormalities can Distinguish Ulcerative Colitis Patients with Dysplasia and Cancer from Those Who Are Dysplasia/Cancer-Free," *American Journal of Pathology*, 162 (2) (February 2003), pp. 665-672.

6. See 1st International DNA Users' Conference for Investigative Officers, "Minutes," circa 1999 at www.Interpol.int. See 2nd International DNA Users' Conference for Investigative Officers, "Minutes," circa 2001 at www.Interpol.int.

7. Armed Forces Department of Pathology, "The Department of Defense DNA Registry," at www.afip.org. Douglas J. Gillert, "Who Are You? DNA Registry Knows," at www.defenselink.mil.

8. Connecticut Department of Social Services, "DSS's Biometric ID Project," at www.dss.state.ct.us. Also see Gregory H. Smith, "Securing our Personal Genome," (master's thesis, Indiana University, 2003).

9. See National Cancer Institute, "Understanding Gene Testing," at press2.nci.nih.gov.

10. Generally see Francis Galton, *Finger Prints* (New York: Da Capo Press, 1965).

11. MIB Group, Inc., "About Us" at www.mib.com. Author's interviews with MIB Group, Inc. officials, 28 February 2003. Author's investigation of MIB Group, Inc., February-March 2003.

12. Author's investigation of MIB Group, Inc., 28 February 2003. Also see author's interviews with MIB Group, Inc. officials, 28 February 2003.

13. See American Academy of Actuaries, "Genetic Information and Medical Expense Insurance," monograph of June 2000, pp. 2,3.

14. American Academy of Actuaries, p. 2.

15. American Academy of Actuaries, pp. 2-3, 27-30.

16. American Academy of Actuaries, "Issue Brief: The Use of Genetic Information in Disability Income and Long-Term Care Insurance," briefing paper of 2002, p. 7.

17. "Health Insurance Companies Accused of Genetic Bias," *British Broadcasting Corporation*, 11 December 1998. Ann Deering, "Risk Reporter: Genetic Discrimination," *Risk Management Magazine*.

18. "Genetic Data 'Insurance Fear,'" *British Broadcasting Corporation*, 27 November 2000. "The Price of Having the Wrong Genes," *British Broadcasting Corporation*, 22 January 2001. "Britain Moves to Ban Insurance Genetic Tests," *Washington Post*, 30 April 2001.

19. Association of British Insurers, "Genetic Testing: Background," at www.abi.org.uk. "Insurance Firm Admits Using Genetic Screening," *The London Times*, 8 February 2001. "The Price of Having the Wrong Genes." "Insurance Companies Announce Genetic Testing Halt," *The Scientist*, 30 October 2001.

Author's interview with Norwich Union, 28 February 2003 and 5 March 2003.

20. "'Moratorium on Genetic Data Use," *British Broadcasting Corporation*, 23 October 2001. Author's interview with Norwich Union, 28 February 2003 and 5 March 2003.

21. Trudo Lemmens, "Selective Justice, Genetic Discrimination, and Insurance: Should We Single Out Genes in Our Laws?" *McGill Law Journal* 45 347 (2000), pp. 353-354. Author's interview with Industrial Alliance, 3 March 2003.

22. Author's interview with Industrial Alliance, 3 March 2003.

23. "UK Forum for Genetics and Insurance," at www.ukfgi.org.uk. International Actuarial Association, "3. Astin News / Nouvelles D'Astin," *IAA Bulletin AAI No. 36* (2002). MIB Group, Inc., "Special Section: Genetics," at www.knowledgedigest.com.

24. Office of Technology Assessment, "Genetic Monitoring and Screening in the Workplace 44–45 (1990); "Are Your Genes Right for Your Job?" 3 Cal Law 25, 27 (May 1983) as cited in Privacy International, "Privacy and Human Rights 2002, and International Survey of Privacy Laws and Developments," (2000), pp. 82–83. Office of Technology Assessment, "The Role of Genetic Testing in the Prevention of Occupational Disease," (April 1983), p. 35. American Management Association as cited by The National Workrights Institute, "Genetic Discrimination in the Workplace Fact Sheet," at www.workrights.org.

25. "A Bloody Mess At One Federal Lab," *U.S. News & World Report* 23 June 1997. Author's interview with Lawrence Berkeley Laboratory, 3 March 2003.

26. "Court Declares Right to Genetic Privacy," *U.S. News & World Report* 16 February 1998. "Privacy and Human Rights 2002," p. 83. Author's interview with Lawrence Berkeley Laboratory, 3 March 2003.

27. Burlington Northern Santa Fe Railroad, "BNSF and EEOC Settle Genetic Testing Case Under Americans with Disabilities Act," press release of 8 May 2002. Burlington Northern Santa Fe Railroad, "BNSF Ends DNA Testing For Carpal Tunnel Syndrome," press release of 12 February 2001. Author's interview with Burlington Northern Santa Fe Railroad, 27 February 2003.

28. Author's interview with Burlington Northern Santa Fe Railroad, 27 February 2003. Author's interview with Athena Diagnostics, 28 February 2003.

29. "China is Thwarted by Jobs Ruling," *The Guardian*, 1 October 2000.

30. "DNA Database 'Should Include Every Citizen.'" "Privacy and Human Rights 2002," p.

30. "DNA Tests for All Will Cut Crime, Says Pioneer," *News.Telegraph.Co.UK* 19 February 2001.

31. "Agreement Between the Minister of Health and Social Security and Islensk refdagreining ehf. in Connection With the Issuing of a License to Create and Operate a Health Sector Database (HSD)," at www.mannvernd.is. "Profiling an Entire Nation," *Associated Press* 18 February 1999. Mannvernd, "Opts Out From Icelandic Health Sector Database," at www.mannvernd.is. Mannvernd, "Status of Lawsuits Against the Health Sector Database Act and Related Matters," at www.mannvernd.is. International Business Machines and decODE Genetics, "decODE and IBM Form Strategic Alliance to Deliver Technology Solution for Applying Genetics to Drug Discovery," press release of 23 January 2003.

32. Jean Pierre Sørensen, "IBM Life Sciences-Nordic: Pulling Together The Pieces," Powerpoint Presentation circa 2003.

33. Mark D. Uehling, "Decoding Estonia," *Bio IT World* 10 February 2003 as cited by www.genomics.ee/index.php?lang=eng&nid=14 0; Internet.

34. "Banking on Genes," *The Scientist*, 4 December 2000. Autogen Limited, "ASX Announcement," press release of 17 November 2000.

35. Author's communication with Autogen Limited, 3 March 2003. Autogen Limited, "ASX Announcement," press release of 2 March 1999.

36. The Wellcome Trust, "UK Biobank: A Study of Genes, Environment and Health," at www.wellcome.ac.uk. "Phones Download DNA," *Nature*, 26 February 2003. "Report of the Bioethics Advisory Committee the Israel Academy of Sciences and Humanities," at stwww.weizmann.ac.il.

37. Author's communication with Congresswoman Louise M. Slaughter (D-NY), March 2003.

38. Bionet, "Who Owns Your Genes?" at www.bionetonline.org. School of Health and Related Research, "Genetic Testing in the Workplace," at www.shef.ac.uk.

39. Author's communication with Congresswoman Louise M. Slaughter (D-NY), March 2003.

40. "Perfect Features: Science May Pave Way For Designer Babies," *ABCNews.com*, 26 December 2002. "A Way to Choose a Baby's Gender," *Los Angeles Times*, 3 March 2003.

41. "Couple 'Choose' To Have Deaf Baby," *British Broadcasting Corporation*, 8 April 2002. "Babies, Deaf by Design," *The Australian*, 16 April 2002.

42. "Stupidity Should Be Cured, Says DNA Discoverer," *New Scientist.com*, 28 February 2003.

43. "Stupidity Should Be Cured, Says DNA Discoverer."

44. Lee M. Silver, *Remaking Eden* (New York: Avon Books, 1997), pp. 4-8.

45. "Details of Hybrid Clone Revealed," *British Broadcasting Corporation*, 18 June 1999. "Frankenfish Wiping Out Wild Salmon," *Associated Newspapers Ltd.*, 19 September 1999. "Genetically Modified Monkey Could Be Key to Curing Some Diseases," *Cable News Network*, 18 January 2001. "Scientists Want A Life," AP Wire Services, 21 November 2002. "Scientists Planning to Make New Form of Life," *Washington Post* 21 November 2002. Author's communication with The Center for the Advancement of Genomics, March 2003.

Major Sources

ARCHIVAL REPOSITORIES

Original papers and documents were accessed at several dozen archival repositories, record collections and unprocessed files in storage. The challenging range of repositories spanned the gamut from governmental and organizational archives to corporate and private files. Many key records are held by the special collections and manuscript departments of libraries, such as the Laughlin Papers in the Special Collections of Pickler Memorial Library at Truman University. I estimate there are some five hundred key and niche repositories of eugenic information in the United States and just as many overseas. Most of them are listed below, but space precludes a complete roster.

UNITED STATES

American Breeders Association Files (ABA)	*Maryland*
American Civil Liberties Union Files	*Richmond*
American Genetics Association Files	*Maryland*
American Heritage Center	*Laramie, WY*
American Philosophical Society (APS)	*Philadelphia*
Charles B. Davenport Papers	
American Eugenics Society Records	
Leon F. Whitney Collection	
Eugenics Record Office Records	
California Institute of Technology Archives	*Pasadena, CA*
Ezra Gosney / Human Betterment Foundation Papers	
California State Archive	*Sacramento*
Carnegie Institute of Washington (CIW)	*Washington, DC*
Central Virginia Training Center Files	*Lynchburg, VA*
Chicago Tribune Newspaper Morgue	*Chicago*
Cold Spring Harbor Archive (CSH)	*Cold Spring Harbor*
Cook County Circuit Court Archives	*Chicago*
Hoover Institute Archives	*Stanford, CA*
Indiana Historical Society	*Indianapolis*
Indiana State Archives	*Indianapolis*
Indiana State Library	*Indianapolis*
Margaret Sanger Papers Project at NYU	*New York*
Monacan National Tribal Archives Files	*Madison Heights, VA*
National Archives (NA)	*College Park, MD*
RG 29 Bureau of the Census	
RG 40 Commerce	
RG 43 Conference Commissions and Expositions	
RG 59 State Department	
RG 60 Department of Justice	
RG 242 Captured German Records	
RG 238 War Crimes Records	
RG 330 Department of Defense	
Pickler Memorial Library, Truman State University (Truman)	*Kirksville, MO*
Harry H. Laughlin Papers.	

Planned Parenthood Foundation   *New York*
Records of the Montgomery County Courthouse   *Christiansburg, VA*
Richmond Times-Dispatch Newspaper Morgue   *Richmond*
Rockefeller Foundation Archives (RAC)   *Sleepy Hollow, NY*
    RG 1.1 Projects
    RG 1.2 Projects
    RG 3.1 Administration, Program and Policy
    RG 6.1 Field Officers
    RG 10 Fellowship Recorder Cards
Smith College   *North Hampton, MA*
    Sophia Smith Collection
Tamiment-Wagner Labor Archives Archive   *New York*
University of California at Berkeley Archive   *Berkeley, CA*
    71/3C William E. Ritter Papers
    72/227C Berkeley PD
    C-B 403 August Vollmer Papers
    C-B 927 Robert H. Lowie Papers
    CU-23 UCB Department of Anthropology
University of California at Davis   *Davis, CA*
Vermont Public Records (VT-PRA)   *Middlesex, VT*
    Eugenics Survey of Vermont and the Vermont Commission
       on County Life

UNITED KINGDOM
House of Lords Records Office   *London*
Public Records Office (PRO)   *London*
    Colonial Office
    Department of Education
    Department of Technical Co-Operation, Ministry of
       Overseas Development
    Dominions Office
    Foreign Office
    General Register Office
    Home Office
    Medical Research Council
    Ministry of Health
University College of London (UCL)   *London*
    Galton Papers
    Pearson Papers
    Penrose Papers
Wellcome Library   *London*
    SA-EUG Eugenics Society
    PP-MCS Marie Stopes Papers
    GC-088 Rockefeller Papers

GERMANY
Buchenwald Archiv   *Weimar*
Bundesarchiv Berlin (BAB)   *Berlin*
    NS 2 Rasse- und Siedlungshauptamt
    NS 5 Deutsche Arbeitsfront
    R 2 Reichsfinanzministerium
    R 3 Reichsministerium für Rüstung und Kriegsproduction
    R 7 Reichswirtschaftsministerium
    R 36 Deutscher Gemeindetag
    R 86 Reichsgesundheitsamt
    R 1501 Reichsministerium des Inneren
    R 1509 Reichssippenamt
    R 4901 Reichsministerium für Wissenschaft,
       Erziehung und Volksbildung

Bundesarchiv Koblenz (BAK)   *Koblenz*
  R 73 Notgemeinschaft der deutschen Wissenschaft
Max Planck Archiv   *Berlin*
  I. Abt., Rep. 1A, Nr. 762-781 Presse
  I. Abt., Rep. 1A, Nr. 1050-1065 Auslands- und
    internationale Angelegenheiten
  I. Abt., Rep. 1A, Nr. 1076-1086 Besuche durch
    ausländische Gelehrte
  I. Abt., Rep. 1A, Nr. 1094 Rockefeller Foundation
  I. Abt., Rep. 1A, Nr. 2443-2451 Kaiser-Wilhelm-Institut
    für Psychiatrie
  I. Abt., Rep. 3, Nr. 4 Jahresberichte des Kaiser-Wilhelm-
    Instituts für Anthropologie, menschliche Erblehre
    und Eugenik
  I. Abt., Rep. 3, Nr. 23 International Federation of
    Eugenic Organisations
  I. Abt., Rep. 3, Nr. 26 Kaiser-Wilhelm-Institut für
    Anthropologie, menschlich Erblehre und Eugenik
  V. a Abt., Rep. 16 Verschuer
  Max-Planck-Institut für Psychiatrie, München (Deutsche   Munich
    Forschungsanstalt für Psychiatry) Historisches Archiv
    der Klinik
  GDA (ehemalige Genealogisch-Demographische Abteilung)
Universitätsarchiv Heidelberg   *Heidelberg*
  B-1523/3-7 Ehrepromotionen
  H-III-869/2 Akten der medizinischen Fakultät
Universitätsarchiv Münster   *Münster*
  Nachlass Verschuer, Nr. 4

LIBRARIES

Libraries are crucial to research on eugenics because so much information resides in period secondary sources. In addition, each library maintains its own unique and often precious collection of obscure literature and local materials. Sometimes the most valuable materials are found in small community libraries. I estimate there are hundreds of libraries in the United States, and just as many overseas, containing important secondary materials. Most of the libraries we accessed are listed below, but space precludes a complete roster.

UNITED STATES

Alderman Library, University of Virginia	*Charlottesville, VA*
Bailey/Howe Library, University of Vermont	*Burlington, VT*
Bancroft Library, University of California at Berkeley	*Berkeley, CA*
Bobst Library, New York University	*New York*
Boston Public Library	*Boston*
California Institute of Technology Library	*Pasadena*
Carnegie Library, Cold Spring Harbor Laboratory	*Cold Spring Harbor*
Charles C. Sherrod Library, East Tennessee State University	*Johnson City, TN*
Chicago Historical Society Research Center	*Chicago*
Chicago Public Library	*Chicago*
Clapp Library, Occidental College	*Los Angeles*
Dag Hammarskjöld Library, United Nations	*New York*
Dahlgren Memorial Library, Georgetown University	*Washington, DC*
Enoch Pratt Free Library	*Baltimore*
Fairfax County Public Library	*Fairfax, VA*
Fenwick Library, George Mason University	*Fairfax, VA*
Gelman Library, George Washington University	*Washington, DC*
History Office and Library, Immigration and Naturalization	*Washington, DC*
Hodges Library, University of Tennessee at Knoxville	*Knoxville*
Howard-Tilton Memorial Library, Tulane University	*New Orleans*
Illinois State Historical Library	*Springfield, IL*

Indiana State Library	*Indianapolis*
John Crerar Library, University of Chicago	*Chicago*
Kellogg-Hubbard Library	*Montpelier, VT*
Kuhn Library, University of Maryland, Baltimore County	*Baltimore*
Lane Medical Library, Stanford University Medical School	*Stanford, CA*
Lauinger Memorial Library, Georgetown University	*Washington, DC*
Lehman Social Sciences Library, Columbia University	*New York*
Library of Congress	*Washington, DC*
Library of the American Philosophical Society	*Philadelphia*
Library of the American University	*Washington, DC*
Library of the University of the District of Columbia	*Washington, DC*
Library of Virginia	*Richmond*
Library, U.S. Holocaust Memorial Museum	*Washington, DC*
Library, YIVO Institute for Jewish Research	*New York*
Macdonald DeWitt Library, Ulster County Community College	*Stone Ridge, NY*
McCormick Library, Planned Parenthood Foundation	*New York*
McKeldin Library, University of Maryland College Park	*College Park, MD*
Memorial Library, University of Wisconsin–Madison	*Madison, WI*
Merriam Library, California State University	*Chico, CA*
Montgomery College Library	*Rockville, MD*
Montgomery County Public Libraries	*Rockville, MD*
Mullen Library, Catholic University of America	*Washington, DC*
National Library of Medicine, National Institutes of Health	*Bethesda, MD*
New York Academy of Medicine Library	*New York*
New York Public Library	*New York*
Newman Library, Virginia Polytechnic Institute	*Blacksburg, VA*
Orange Public Library	*Orange, NJ*
Pickler Memorial Library, Truman State University	*Kirksville, MO*
Princeton University Library	*Princeton, NJ*
Schlesinger Library, Harvard University	*Cambridge, MA*
Science, Industry & Business Library, New York Public Library	*New York*
Sheridan Libraries, Johns Hopkins University	*Baltimore*
Smith Memorial Library, Indiana Historical Society	*Indianapolis*
Washington College of Law Library, American University	*Washington, DC*
Washington Research Library Consortium	*Upper Marlboro, MD*

CANADA

Osler History of Medicine Library, McGill University	*Montreal, QC*
McLennan-Redpath Library, McGill University	*Montreal, QC*

FRANCE

Bibliothèque Nationale de France	*Paris*

GERMANY

Bibliothek des Archivs zur Geschichte der Max-Planck-Gesellschaft	*Berlin*
Bibliothek für Geschichte der Medizin, Freie Universität Berlin	*Berlin*
Bibliothek des Otto-Suhr-Institutes für Politikwissenschaft, Freie Universität	*Berlin*
Bibliothek des Zentrums für Antisemitismusforschung, Technische Universität	*Berlin*
Staatsbibliothek Berlin	*Berlin*

UNITED KINGDOM

Bodleian Library, Oxford University	*London*
British Library	*London*
Library of the Public Record Office	*London*
Library of the University College of London	*London*
Wellcome Library	*London*

JOURNAL, NEWSPAPERS AND MEDIA

Scores of publications and media outlets were consulted, both as sources of period materials and for topical information. These covered a spectrum, from obscure professional and medical journals, to Nazi-era scientific and political media, to the eugenics media, to contemporary publications and news organizations. In some cases, every issue of a publication was surveyed for as many as forty years; *Eugenical News* is an example. In other instances, we studied select editions. Many of the publications and media outlets we surveyed are listed below, but space precludes a complete roster.

JOURNALS
American Bar Association Journal
American Journal of Medical Genetics
American Journal of Pathology
American Journal of Public Health
American Journal of Sociology
Bio IT World
BIOS
Birth Control Review
Brain Pathology
British Journal of Inebriety
The British Medical Journal
Canadian Medical Association Journal
Digestive Diseases
Dimensions
Epilepsia
History and Philosophy of the Life Sciences
History of Psychiatry
IAA Bulletin AAI
Institutional Quarterly
Journal of Abnormal Psychology
Journal of American History
Journal of American Public Health
Journal of Contemporary Health Law and Policy
Journal of Delinquency
Journal of Psycho-Asthenics
Journal of Southern History
Journal of the American Medical Association
Journal of the Anthropological Institute
Journal of the History of Sexuality
Journal of the History of Biology
Lancet
McGill Law Journal
Mental Retardation
National Geographic
Nature
Nature Reviews
Neurology
New York Medical Journal
New York University Law Review
Osiris
Psychological Review
The Standard

EUGENIC MEDIA
American Breeders' Magazine
Eugenical News
Eugenics
Eugenics Quarterly
Eugenics Review
Journal of Heredity

GERMAN MEDIA

Abhandlungen aus dem Gebiete der Sexualforschung
Allgemeines Statistisches Archiv Bevölkerungsfragen
Archiv für Rassen- und Gesellschaftsbiologie
Der Erbarzt
Deutsches Ärtzeblatt
Die Neue Zeitung
Fortschritte der Erbpathologie, Rassenhygiene und ihrer Grenzgebiete
Hollerith Nachrichten
Neues Volk
Rassenpolitische Auslands-Korrespondenz
Schleswig-Holsteinische Hochschullblätter
Sexual-Probleme, Zeitschrift für Sexualwissenschaft und Sexualpolik
Völkischer Beobachter
Volk und Rasse
Zeitschrift für Morphologie und Anthropologie Festschrift
Zeitschrift für Rassenkunde
Zeitschrift für Sexualwissenschaft
Ziel und Weg

NEWSPAPERS, MAGAZINES, WIRE SERVICES AND OTHER MEDIA

Associated Press
Atlantic Monthly
The Australian
British Broadcasting Corporation
Cable News Network
Chicago Tribune
Christian Science Monitor
Economic Quarterly
Free World
The Guardian
The Independent
London Times
Los Angeles Times
Mind
New Republic
New York Times
The Pedagogical Seminary
The Poor-Law Officers' Journal Reuters
Richmond Times-Dispatch
Risk Management Magazine
San Francisco Daily News
Scientist
Time Magazine
U.S. News & World Report
Washington Post
Winston-Salem Journal

UNPUBLISHED MANUSCRIPTS

Numerous university dissertations, theses and other unpublished manuscripts and monographs were consulted. Some of the salient ones are listed below.

American Academy of Actuaries. "Genetic Information and Medical Expense Insurance." June 2000.
Curtis, Patrick Almond. "Eugenic Reformers, Cultural Perceptions of Dependent Populations, and the Care of the Feebleminded in Illinois: 1909-1920." Ph. D. diss., University of Illinois at Chicago, 1983.
Hassencahl, Francis Janet. "Harry H. Laughlin, 'Expert Eugenics Agent' for the House Committee on Immigration and Naturalization." Ph. D. diss., Case Western Reserve University, 1970.

Lombardo, Paul A. "Eugenic Sterilization in Virginia: Aubrey Strode and the Case of Buck v. Bell." Ph. D. diss, University of Virginia, 1982.
Mehler, Barry. "A History of the American Eugenics Society, 1921-1940." Ph. D. diss., University of Illinois, 1988.
Mottier, Véronique. "Narratives of National Identity: Sexuality, Race, and the Swiss 'Dream of Order.'" Paper presented at the European Consortium for Political Research Annual Joint Sessions, Workshop: The Political Uses of Narrative, at Mannheim 26-31 March 1999.
Smith, Gregory H. "Securing our Personal Genome." Forthcoming Master's thesis, Indiana University, 2003.

DOCUMENTARIES

Film documentaries, including independently produced videos, provide an excellent source of eyewitness testimony and visual insight. Some of the salient videos utilized are listed below.

Baron, Saskia and Paul Sen, director and Dunja Noack, producer. Science and the Swastika. Videocassette. The History Channel, 2001.
Bryant, Sharon, producer. Reclaiming our Heritage. Virginia Foundation for the Humanities, 1997, videocassette.
Trombley, Stephen, director and Brucie Eadier, producer. The Lynchburg Story. Videocassette. Worldview Pictures, 1993.
Blumenstein, Rob, producer-director. History's Mysteries: Hitler's Perfect Children. The History Channel, 2000.

MAJOR JOURNAL ARTICLES

I consulted numerous scholarly articles of great value. Some of the salient articles are listed below.

"Cécile and Oskar Vogt: On the Occasion of her 75th and his 80th Birthday." Neurology Vol. 1 No. 3 (May-June 1951).
Elks, Martin A. "The 'Lethal Chamber': Further Evidence for the Euthanasia Option." Mental Retardation, Vol. 31 No. 4 (August 1993).
Kevles, Daniel J. "Testing the Army's Intelligence: Psychologists and the Military in World War I." The Journal of American History, Vol. 55, Issue 3 (Dec., 1968).
Lakin, K. Charlie. "Demographic Studies of Residential Facilities for the Mentally Retarded." Developmental Disabilities Project on Residential Services and Community Adjustment Project Report No. 3. University of Minnesota Department of Psychoeducational Studies, circa 1979.
Lemkin, Raphaël. "Genocide- A Modern Crime." Free World Vol. 4 (April 1945).
Lemmens, Trudo. "Selective Justice, Genetic Discrimination, and Insurance: Should We Single Out Genes in Our Laws?" McGill Law Journal 45 347 (2000).
Lombardo, Paul A. "Medicine, Eugenics, and the Supreme Court: From Coercive Sterilization to Reproductive Freedom." The Journal of Contemporary Health Law and Policy Volume 13 (1996).
Lombardo, Paul A. "Three Generations, No Imbeciles: New Light on Buck v. Bell." New York University Law Review, Vol. 60 no. 1.
MacDowell, E. Carlton. "Charles Benedict Davenport, 1866-1944: A Study of Conflicting Influences." BIOS vol. XVII no. 1.
MacNicol, John. "The Voluntary Sterilization Campaign in Britain, 1918-39." The Journal of the History of Sexuality, Vol. 2., No. 3 (1992).
Müller-Hill, Benno. "The Blood from Auschwitz and the Silence of the Scholars." History and Philosophy of the Life Sciences Vol. 21 (1999).
Neugebauer, Wolfgang and Georg Stacher. "Nazi Child 'Euthanasia' in Vienna and the Scientific Exploitation of Its Victims before and after 1945." Digestive Diseases 17 (1999).
Peiffer, Jürgen. "Neuropathology in the Third Reich: Memorial to those Victims of National-Socialist Atrocities in Germany who were Used by Medical Science." Brain Pathology 1 (1991).
Reilly, Philip. "The Virginia Racial Integrity Act Revisited: The Plecker-Laughlin Correspondence: 1928-1930." American Journal of Medical Genetics Vol. 16 (1983).
Sherman, Richard B. "'The Last Stand': The Fight for Racial Integrity in Virginia in the 1920's." The Journal of Southern History, Vol. 54 Issue 1 (February, 1988).
Shevell, Michael. "Racial Hygiene, Active Euthanasia, and Julius Hallervorden." Neurology 42 (November 1992) and "Reply from the Author." Neurology 43 (July 1993).

512 || WAR AGAINST THE WEAK

Seidelman, William E. "Medicine and Murder in the Third Reich." *Dimensions: A Journal of Holocaust Studies* Vol. 13 No. 1 (1999).
Weber, Matthias M. "Psychiatric Research and Science Policy in Germany. The History of the Deutsche Forschungsanstalt für Psychiatrie (German Institute for Psychiatric Research) in Munich from 1917 to 1945." *History of Psychiatry* xi (2000).

BIBLIOGRAPHY

Literally hundreds of books were consulted, from period eugenic literature to scholarly works on a range of topics. It would be impossible to list them all. However, a few hundred of the salient volumes are listed below. Listing books here by no means suggests they are reliable; many of these books were consulted because of their period insights rather than their accuracy. This list, then, includes both the most credible and the least credible.

Adams, Mark B. *The Wellborn Science.* New York: Oxford University Press, 1990.
Aly, Götz, Peter Chroust, and Christian Pross. *Cleansing the Fatherland: Nazi Medicine and Racial Hygiene.* Trans. Belinda Cooper. Baltimore: Johns Hopkins University Press, 1994.
Anderson, Margo J. *The American Census.* New Haven, CT: Yale University Press, 1988.
Astor, Gerald. *The "Last" Nazi: The Life and Times of Dr. Joseph Mengele.* New York: Donald I. Fine, Inc., 1985.
Baker, Liva. *The Justice From Beacon Hill: The Life and Times of Oliver Wendell Holmes.* New York: Harper Collins Publishers, 1991.
Bannister, Robert C. *Social Darwinism: Science and Myth in Anglo-American Social Thought.* Philadelphia: Temple University Press, 1979.
Barr, Martin W. *Mental Defectives: Their History, Treatment and Training.* Philadelphia: P. Blakiston's Son & Co., 1904; New York: Arno Press, 1973.
Baur, Erwin, Eugen Fischer and Fritz Lenz. *Human Heredity.* 3rd Ed. Trans. Eden & Cedar Paul. New York: The MacMillan Company, 1931.
Baur, Erwin, Eugen Fischer, and Fritz Lenz. *Grundriss der menschlichen Erblichkeitslehre und Rassenhygiene.* Munich: J. F. Lehmanns Verlag, 1921.
Baur, Erwin, Eugen Fischer, and Fritz Lenz. *Grundriss der menschlichen Erblichkeitslehre und Rassenhygiene,* 2nd ed. Munich: J. F. Lehmanns Verlag, 1923.
Benson, Keith R., Jane Maienschein, and Ronald Rainger, editors. *Expansion of American Biology.* New Brunswick, NJ: Rutgers University Press, 1991.
Biesold, Horst. *Crying Hands: Eugenics and Deaf People in Nazi Germany.* Trans. William Sayers. Washington DC: Gallaudet University Press, 1988.
Black, Edwin. *IBM and the Holocaust: The Strategic Alliance Between Nazi Germany and America's Most Powerful Corporation.* New York: Crown Publishers, 2001.
Black, Edwin. *The Transfer Agreement: The Dramatic Story of the Pact Between the Third Reich and Jewish Palestine.* Washington: Dialog Press, 1999; New York: Carol & Graf, 2001.
Blacker, C. P. "Eugenics In Prospect and Retrospect," In *The Galton Lecture, 1945.* Hamish Hamilton Medical Books, n.d.
Blacker, C. P. *Eugenics: Galton and After.* Cambridge, MA: Harvard University Press 1952; Westport, CT: Hyperion Press, Inc., 1987.
Bock, Gisela, *Zwangssterilisation im Nationalsozialismus. Studien zur Rassenpolitik und Frauenpolitik.* Opladen: Westdeutscher Verlag, 1986.
Bohstedt, John. *Riots and Community Politics in England and Wales 1790-1810.* Cambridge: Harvard University Press, 1983.
Bowen, Catherine Drinker. *Yankee From Olympus: Justice Holmes and His Family.* Boston: Little, Brown and Company, 1945.
Breiting, Richard. *Secret Conversations with Hitler.* Edit. Edouard Calic. Trans. Richard Barry. New York: The John Day Company, 1968.
Brigham, Carl C. *A Study of American Intelligence.* Princeton, NJ: Princeton University Press, 1923.
Brown, Stuart E. Jr., Lorraine F. Myers and Eileen M. Chappel. *Pocahontas' Descendants.* Baltimore, MD: The Pocahontas Foundation, 1985.
Browning, Christopher R. *The Path to Genocide: Essays on Launching the Final Solution.* Cambridge: Cambridge University Press, 1992.
Bruehl, Charles P. *Birth Control and Eugenics: In the Light of Fundamental Ethical Principles.* New York: Joseph P. Wagner, Inc., 1928.
Burleigh, Michael. *Death and Deliverance: 'Euthanasia' in Germany 1900-1945.* Cambridge: Cambridge University Press, 1994.

Burton, David H. *Oliver Wendell Holmes, Jr.* Boston: Twayne Publishers, 1980.

Burton, David H., editor. *Progressive Masks: Letters of Oliver Wendell Holmes, Jr., and Franklin Ford.* Newark, DE. University of Delaware Press, 1982.

Cameron, Norman and R.H. Stevens, trans. *Hitler's Table Talk: 1941-1944: His Private Conversations.* New York City: Enigma Books, 2000.

Caplan, Arthur L., ed. *When Medicine Went Mad.* Totowa, NJ: Humana Press, 1992.

Carlson, Elof Axel. *The Unfit: A History of a Bad Idea.* Cold Spring Harbor, NY: Cold Spring Harbor Laboratory Press, 2001.

Chambers, Robert W. *The King in Yellow.* F. Tennyson Neely, 1895.

Chapman, Blanche Adams. *Wills and Administration of Elizabeth City County, Virginia 1688-1800.* N.p., 1980.

Charles, Joan. *Elizabeth City County, Virginia Will 1800-1859.* Bowie, MD: Heritage Books, Inc., n.d.

Chesler, Ellen. *Women of Valor.* New York: Simon & Schuster, 1992.

Chesterton, G. K. *Eugenics and Other Evils.* Ed. comp. Michael W. Perry. Seattle, WA: Inkling Press, 2000.

Cravens, Hamilton. *The Triumph of Evolution: The Heredity-Environment Controversy, 1900-1941.* Baltimore: Johns Hopkins University Press, 1988.

Darwin, Charles. *The Origin of the Species.* 6th ed. New York: D. Appleton & Co., 1881.

Darwin, Charles. *The Variation of Animals and Plants under Domestication.* London: John Murray, 1868; New York: D. Appleton & Co., 1883.

Davenport, Charles B. and Morris Steggerda. *Race Crossing in Jamaica.* Washington: Carnegie Institution of Washington, 1929.

Davenport, Charles B. *Heredity In Relation To Eugenics.* New York: Henry Holt & Company, 1911; New York: Arno Press Inc., 1972.

Dawidowicz, Lucy. *The War Against The Jews, 1933-1945.* New York: Bantam Books, 1975.

Deichmann, Ute. *Biologists Under Hitler.* Trans. Thomas Dunlap. Cambridge, MA: Harvard University Press, 1996.

Dowbiggin, Ian. *A Merciful End: The Euthanasia Movement in Modern America.* Oxford: Oxford University Press, 2003.

Dugdale, Richard L. *The Jukes.* New York: Putnam, 1910.

Durance, Jill and William Shamblin, ed. *Appalachian Ways.* Washington D.C.: The Appalachian Regional Commission, 1976.

Estabrook, Arthur H. and Charles B. Davenport. *The Nam Family: A Study in Cacogenics.* Cold Spring Harbor, NY: Cold Spring Harbor Press, 1912.

Estabrook, Arthur H. and Ivan E. McDougle. *Mongrel Virginians: The Win Tribe.* Baltimore: The Williams & Wilkins Company, 1926.

Eugenics Education Society, *Problems in Eugenics, Vols. I and II: Report of Proceedings of the First International Eugenics Congress.* Kingsway, UK: Eugenics Education Society, 1913.

Evers, Alf. *Woodstock: History of an American Town.* Woodstock, New York: The Overlook Press, 1987.

Fairchild, Henry Pratt. *The Melting-Pot Mistake.* Boston: Little, Brown, and Company, 1926.

Fancher, Raymond E. *The Intelligence Men: Makers of the IQ Controversy.* New York: W. W. Norton & Company, 1985.

Fest, Joachim C. *The Face of the Third Reich: Portraits of the Nazi Leadership.* New York: Pantheon Books, 1970.

Flexner, Abraham. *Medical Education in Europe: A Report to the Carnegie Foundation.* n.p. 1912.

Flexner, Abraham. *Prostitution in Europe.* New York: The Century Company, 1914.

Fosdick, Raymond. *Story of the Rockefeller Foundation.* New York: Harper, 1952.

Friedlander, Henry. *The Origins of Nazi Genocide.* Chapel Hill, NC: The University of Carolina Press, 1995.

Gallagher, Nancy L. *Breeding Better Vermonters.* Hanover, NH: University of New England Press, 1999.

Galton, Francis. *Finger Prints.* New York: Da Capo Press, 1965.

Galton, Francis. *Hereditary Genius: An Inquiry Into Its Laws And Consequences.* 2nd ed. London: Macmillan & Co., 1892; London: Watts & Co., 1950.

Galton, Francis. *Inquiries Into Human Faculty And Its Development.* London: JM Dent & Co., 1883.

Galton, Francis. *Memories of My Life.* London: Methuen & Co., 1908.

Galton, Francis. *Natural Inheritance.* London: Macmillan & Co., 1889.

Gilbert, Martin. *The Holocaust.* New York: Holt, Rinehart and Winston, 1985.

Goddard, Henry H. *Feeble-mindedness: Its Causes and Consequences.* New York: Arno Press, 1973.

Goddard, Henry H. *The Kallikak Family: A Study in the Heredity of Feeble-Mindedness.* Vineland, NJ: 1913.

Gordon, Harold J., Jr. *Hitler and the Beer Hall Putsch.* Princeton, NJ: Princeton University Press, 1972.

Gosney, E.S. and Paul Popenoe, *Sterilization for Human Betterment.* New York: The MacMillan Company, 1929.

Gould, Stephen Jay. *The Mismeasure of Man.* Rev. ed. New York: Norton, 1996.
Grant, Madison. *Der Untergang der grossen Rasse: Die Rassen als Grundlage der Geschichte Europas.* Berlin: J. F. Lehmanns Verlag, 1925.
Grant, Madison. *The Passing of the Great Race.* New York: Charles Scribner's Sons, 1936.
Greenwood, James. *The Seven Curses of London.* London: S. Rivers and Co., 1869.
Grosskurth, Phyllis. *Havelock Ellis, A Biography.* London: Allen Lane, 1980.
Broberg, Gunnar and Nils Roll-Hansen, ed. *Eugenics and the Welfare State.* East Lansing, MI: Michigan State University Press, 1996.
Gutman, Yisrael and Michael Berenbaum, ed. *Anatomy of the Auschwitz Death Camp.* Indianapolis: Indiana University Press, 1994.
Hackett, David A., ed. and trans. *The Buchenwald Report.* Boulder, CO: Westview Press, 1995.
Haller, Mark H. *Eugenics: Hereditarian Attitudes in American Thought.* New Brunswick, NJ: Rutgers University Press, 1963.
Hazlitt, Henry. *The Conquest of Poverty.* New Rochelle, NY: Arlington House, 1973.
Higham, John. *Strangers in the Land.* New Brunswick, NJ: Rutgers University Press, 1955.
Hilberg, Raul. *The Destruction of the European Jews.* New York: Harper Colophon Books, 1961.
Hitler, Adolf. *Mein Kampf.* Trans. Ralph Manheim. Boston: Houghton Mifflin Company, 1943.
Hoess, Rudolf. *Commandant of Auschwitz.* New York: Popular Library, 1959.
Höhne, Heins. *The Order of the Death's Head: The Story of Hitler's SS.* New York: Coward-McCann, 1970.
Holmes, Oliver Wendell Jr. *The Common Law.* Boston: Little, Brown and Company, 1881; 1923.
Holmes, Oliver Wendell Jr. *Collected Legal Papers.* New York. Harcourt, Brace and Co., 1921.
Howe, Mark DeWolfe, ed. *Holmes-Laski Letters.* Cambridge, MA: Harvard University Press, 1953.
Howe, Mark DeWolfe, ed. *Holmes-Laski Letters Abridged.* Clinton, MA: Atheneum, 1963.
Howe, Mark DeWolfe, ed. *Holmes-Pollock Letters: The Correspondence of Mr. Justice Holmes and Sir Frederick Pollock 1874-1932.* Cambridge, MA: Harvard University Press, 1942.
Howe, Mark DeWolfe. *Justice Oliver Wendell Holmes: The Shaping Years 1841-1870.* Cambridge, MA: The Belknap Press of Harvard University Press, 1957.
International Congress of Eugenics. *A Decade of Progress in Eugenics: Scientific Papers of the Third International Congress of Eugenics.* Baltimore: The Williams and Wilkins Company, 1934.
International Congress of Eugenics. *Eugenics, Genetics and the Family: Volume I: Scientific Papers of the Second International Congress.* Baltimore: Williams & Wilkins Co., 1923.
International Military Tribunal. *Nuremberg Military Tribunal, Blue Book.*
International Military Tribunal. *Nuremberg Military Tribunal, Green Book.*
Jennings, H.S. *The Biological Basis of Human Nature.* New York: W.W. Norton & Company, 1930.
Jonas, Gerald. *The Circuit Riders.* New York: W.W. Norton & Company, 1989.
Jordan, David Starr. *War and the Breed: The Relation of War to the Downfall of Nations.* Boston: The Beacon Press, 1915.
Kater, Michael H. *Doctors Under Hitler.* Chapel Hill, NC: University of North Carolina Press, 1989.
Katz, Esther. *The Selected Papers of Margaret Sanger: Volume 1:The Woman Rebel, 1900-1928.* Chicago: University of Illinois Press, 2003.
Kennedy, David M. *Birth Control in America: The Career of Margaret Sanger.* New Haven, CT: Yale University Press, 1970.
Kershaw, Ian. *Hitler: 1889 - 1936: Hubris.* New York: W. W. Norton & Company, 1998.
Kevles, Daniel J. *In The Name of Eugenics.* Cambridge, MA: Harvard University Press, 1985.
Klein, Maury. *The Life and Legend of E.H. Harriman.* Chapel Hill, NC: The University of North Carolina Press, 2000.
Kogon, Eugen. *The Theory and Practice of Hell.* New York: Berkley Books, 1980.
Krauß, Hans. *Die Grundgedanken der Erbkunde und Rassenhygiene in Frage und Antwort.* München: Verlag der Ärztlichen Rundschau O. Gmelin, 1935.
Krieger, Heinrich. *Das Rassenrecht in den Vereinigten Staaten.* Berlin: Junker und Dünnhaupt Verlag, 1936.
Kruta, V. and V. Orel, "Johann Gregor Mendel." In *Dictionary of Scientific Biography.* New York: Scribner's, 1970-1980.
Kühl, Stefan. *Die Internationale der Rassisten: Aufstieg und Niedergang der internationalen Bewegung für Eugenik und Rassenhygiene im zwanzigsten Jahrhundert.* Frankfurt/Main: Campus, 1997.
Kühl, Stefan. *The Nazi Connection: Eugenics, American Racism, and German National Socialism.* New York: Oxford University Press, 1994.
Lagnado, Lucette Matalon and Sheila Cohn Dekel. *Children of the Flames.* New York: William Morrow and Company, Inc., 1991.
Lapon, Lenny. *Mass Murderers in White Coats: Psychiatric Genocide in Nazi Germany and the United States.* Springfield, MA: Psychiatric Genocide Research Institute, 1986.

Laughlin, Harry H. *Eugenical Sterilization in the United States.* Chicago: Psychopathic Laboratory of the Municipal Court of Chicago, 1922.

Laughlin, Harry H. *Eugenical Sterilization: 1926: Historical, Legal, and Statistical Review of Eugenical Sterilization in the United States.* Lancaster, PA: Lancaster Press, 1926.

Laughlin, Harry H. *Immigration and Conquest: A Report of The Special Committee on Immigration and Naturalization of the Chamber of Commerce of the State of New York.* New York: The Special Committee on Immigration and Naturalization of the Chamber of Commerce of the State of New York, 1939.

Laughlin, Harry H. *The Legal Status of Eugenical Sterilization.* Washington, DC: Eugenic Record Office of the Carnegie Institute of Washington, 1930.

Laughlin, Harry H. *The Second International Exhibition of Eugenics.* Baltimore: Williams & Wilkins Co., 1923.

Laughlin, Harry H. *Statistical Directory of State Institutions for the Defective, Dependent, and Delinquent Classes.* Washington, DC: Government Printing Office, 1919.

Lemann, Nicholas. *The Big Test: The Secret History of the American Meritocracy.* New York: Farrar, Straus and Giroux, 1999.

Lemkin, Raphaël. *Axis Rule in Occupied Europe: Analysis, Proposals for Redress.* Washington DC: Carnegie Endowment for International Peace, 1944.

Leonard, E. M. *The Early History of English Poor Relief.* Cambridge: Cambridge University Press, 1900; London: Frank Cass & Co., 1965.

Lifton, Robert Jay. *The Nazi Doctors: Medical Killing and the Psychology of Genocide.* New York: Basic Books, 1986.

Lipstadt, Deborah E. *Beyond Belief.* New York: The Free Press, 1986.

Lookstein, Haskel. *Were We Our Brother's Keepers? The Public Response of American Jews to the Holocaust, 1938-1944.* New York: Vintage Books, 1985.

Macrakis, Kristie. *Surviving the Swastika: Scientific Research in Nazi Germany.* New York: Oxford University Press, 1993.

Malthus, Thomas R. *An Essay on the Principle of Population.* Comp. Donald Winch. Cambridge: Cambridge University Press, 1992.

Marks, Paula Mitchell. *In a Barren Land: American Indian Dispossession and Survival.* New York: William Morrow and Company, Inc., 1998.

Max Lerner, ed. *The Mind and Faith of Justice Holmes.* Garden City, NY: Halcyon House, 1943.

Mazumdar, Pauline M.H. *Eugenics, Human Genetics and Human Failings.* London: Routledge, 1992.

McCann, Carole R. *Birth Control Politics in the United States, 1916-1945.* Ithaca, New York: Cornell University Press, 1994.

McCulloch, Oscar C. "The Tribe of Ishmael: A Study In Social Degradation." In *Proceedings of the National Conference of Charities and Correction.* Boston: George H. Ellis, 1888.

McLaren, Angus. *Our Own Master Race: Eugenics in Canada, 1885 - 1945.* Toronto, Ontario: McClelland & Stewart, Inc.

McWilliams, Carey. *North From Mexico: The Spanish-Speaking People of the United States.* New York: Greenwood Press, 1968.

Meads, Manson. *The Miracle on Hawthorne Hill.* Winston-Salem, NC: Wake Forest University, 1988.

Means, O.M. *Kirksville, Missouri: Its Business and its Beauties as seen through the Camera.* Journal Print Co, 1900.

Miller, Marvin D. *Terminating the "Socially Inadequate": The American Eugenicists and the German Race Hygenists California to Cold Spring Harbor, Long Island to Germany.* Commack, NY: Malamud-Rose, 1996.

Müller, Filip. *Eyewitness Auschwitz: Three Years in the Gas Chambers.* Chicago: Ivan R. Dee, 1979.

Müller-Hill, Benno. *Murderous Science.* Trans. G. R. Fraser. Cold Spring Harbor, NY: Cold Spring Harbor Laboratory Press: 1998.

Müller-Hill, Benno. *Tödliche Wissenschaft: die Aussonderung von Juden, Zigeunern und Geisteskranken, 1933-1945.* Hamburg: Rowohlt, 1984.

Myerson, Abraham, James B. Ayer, Tracy J. Putnam, Clyde E. Keller and Leo Alexander. *Eugenical Sterilization: A Reorientation of the Problem.* New York: The Macmillan Company, 1936.

Nash, Gary B. *Red, White, and Black: The Peoples of Early America.* Englewood Cliffs, NJ: Prentice-Hall, Inc., 1974.

National Association for the Advancement of Colored People. *Thirty Years of Lynching in the United States, 1889-1918.* New York: NAACP, 1919; reprint, New York: Negro Universities Press, 1969.

Newell, Frank W. *The American Ophthalmological Society 1864-1989: A Continuation of Wheeler's First Hundred Years.* Rochester, Minnesota: American Ophthalmological Society, 1989.

Noyes, John H. *Essay on Scientific Propagation.* Oneida, NY: Oneida Community, 1872.

Nyiszli, Miklos. *Auschwitz: A Doctor's Eyewitness Account.* New York: Arcade Publishing, 1993.
O'Donnell, James P. *The Bunker: The History of the Reich Chancellery Group.* New York: Bantam Books, 1978.
Orel, Vitezslav. *Gregor Mendel: The First Geneticist.* Oxford: Oxford University Press, 1996.
Osborn, Frederick. *Preface to Eugenics.* New York: Harper & Brothers, 1940.
Paul, Diane B. "The Rockefeller Foundation and the Origins of Behavior Genetics." In *The Expansion of American Biology,* edited by Keith R. Benson, Jane Maienschein and Ronald Rainger. New Brunswick, NJ: Rutgers University Press, 1991.
Paul, Diane B. *Controlling Human Heredity.* Atlantic Highlands, NJ: Humanities Press International, Inc., 1995.
Pearson, Karl and Ethel Elderton. *A Second Study of the Influence of Parental Alcoholism on the Physique and Ability of the Offspring.* London: Dulau and Co. Limited, 1910.
Pearson, Karl. *The Life, Letters, and Labours of Francis Galton.* Cambridge: Cambridge at the University Press, 1930.
Pernick, Martin S. *The Black Stork: Eugenics and the Death of "Defective" Babies in American Medicine and Motion Pictures Since 1915.* New York: Oxford University Press, 1996.
Popenoe, Paul and Roswell Hill Johnson. *Applied Eugenics.* Rev. ed. New York: Macmillan Company, 1935.
Posner, Gerald L. and John Ware. *Mengele: The Complete Story.* New York: Cooper Square Press, 1986.
Priddy, A.S. *Biennial Report of the State Epileptic Colony.* Lynchburg, VA: State Epileptic Colony, 1923.
Proctor, Robert N. *Racial Hygiene: Medicine Under the Nazis.* Cambridge, MA: Harvard University Press, 1988.
Race Betterment Foundation. *Official Proceedings of the Second National Conference on Race Betterment.* Battle Creek, MI: The Race Betterment Foundation, 1915.
Race Betterment Foundation. *Proceedings of the First National Conference on Race Betterment.* Battle Creek, MI: The Race Betterment Foundation, 1914.
Race Betterment Foundation. *Proceedings of the Third Race Betterment Conference.* Battle Creek, MI: Race Betterment Foundation, 1928.
Reeder, Carolyn and Jack Reeder. *Shenandoah Heritage: The Story of the People Before the Park.* Washington D.C.: The Potomac Appalachian Trail Club, 1978.
Reilly, Philip R. *The Surgical Solution: A History of Involuntary Sterilization in the United States.* Baltimore: Johns Hopkins University Press, 1991.
Reitlinger, Gerald. *The Final Solution.* New York: A.S. Barnes & Company, Inc., 1961.
Rentoul, Robert Reid. *Proposed Sterilization of Certain Mental and Physical Degenerates: An Appeal to Asylum Managers and Others.* London: The Walter Scott Publishing Co., 1903.
Rentoul, Robert Reid. *Race Culture: Or, Race Suicide?* London: The Walter Scott Publishing Co., Ltd., 1906.
Robinson, William J. *Eugenics, Marriage, and Birth Control.* New York: The Critic and Guide Company, 1917.
Robitscher, Jonas, ed. *Eugenic Sterilization.* Springfield, IL: Charles C. Thomas, 1973.
Rosenbaum, Ron. *Explaining Hitler: The Search for the Origins Of His Evil.* New York. Random House, 1998.
Rüdin, Ernst, ed. *Rassenhygiene im völkischen Staat: Tatsachen und Richtlinien.* Munich: J.F. Lehmann, 1934.
Rudwick, Elliott M. *Race Riot at East St. Louis, July 2 1917.* Carbondale, IL. Southern Illinois University Press, 1964.
Sanger, Margaret. *Margaret Sanger: An Autobiography.* New York: W. W. Norton & Company, 1938; Dover Publications, Inc., 1971.
Sanger, Margaret. *The Pivot of Civilization.* New York: Brentano's, 1922.
Sanger, Margaret. *The Pivot of Civilization: In Historical Perspective.* Ed. comp. Michael W. Perry. Seattle, WA: Inkling Books, 2001.
Sanger, Margaret. *Woman and the New Race.* New York: Brentano's, 1920.
Schmuhl, Hans-Walter. *Hirnforschung und Krankenmord, Das Kaiser-Wilhelm-Institut für Hirnforschung 1937-1945.* Berlin: 2000.
Schneider, William H. *Quality and Quantity: The Quest for Biological Regeneration in Twentieth-Century France.* Cambridge: Cambridge University Press, 1990.
Schneider, William H., ed. *Rockefeller Philanthropy & Modern Biomedicine: International Initiatives from World War I to the Cold War.* Bloomington, IN: Indiana University Press, 2002.
Selby, P. O. *History of the First Christian Church (Disciples of Christ) of Kirksville, Missouri.* 1964.
Selby, P. O. *One Hundred Twenty-Three Biographies of Deceased Faculty Members.* Northeast Missouri State Teachers College, 1963.
Selden, Steven. *Inheriting Shame.* New York: Teachers College Press, 1999.
Shirer, William L. *The Rise and Fall of the Third Reich: A History of Nazi Germany.* New York: Simon and Schuster, 1960.
Silver, Lee M. *Remaking Eden.* New York: Avon Books, 1997.
Slack, Paul. *The English Poor Law 1531-1782.* London: Macmillan Education Ltd., 1990.
Smith, Adam [George J.W. Goodman]. *Paper Money.* New York: Summit Books, 1981.

Smith, J. David and K. Ray Nelson. *The Sterilization of Carrie Buck*. Far Hills, NJ: New Horizon Press, 1989.
Smith, J. David. *The Eugenic Assault on America: Scenes in Red, White, and Black*. Fairfax, VA: George Mason University Press, 1993.
Smith, J. David. *Minds Made Feeble: The Myth and Legacy of the Kallikaks*. Rockville, MD: Aspen Systems Corporation, 1985.
Soloway, Richard Allen. *Birth Control and the Population Question in England, 1877-1930*. Chapel Hill, NC: University of North Carolina Press, 1982.
Soloway, Richard Allen. *Demography and Degeneration: Eugenics and the Declining Birthrate in Twentieth-Century Britain*. Chapel Hill, NC: The University of North Carolina Press, 1990.
Spencer, Herbert. *The Principles of Biology*. New York: D. Appleton and Company, 1884.
Spencer, Herbert. *Social Statics*. New York: Robert Schalkenback Foundation, 1970.
Stark, Gary D. *Entrepreneurs of Ideology*. Chapel Hill, NC: The University of North Carolina Press, 1981.
State of Illinois Board of Administration. *Volume II: Biennial Reports of the State Charitable Institutions: October 1, 1914 to September 30, 1916*. State of Illinois, 1917.
Stepan, Nancy Leys. *"The Hour of Eugenics": Race, Gender, and Nation in Latin America*. Ithaca, NY: Cornell University Press, 1991.
Stoddard, Lothrop. *Into the Darkness: An Uncensored Report from Inside the Third Reich at War*. Newport Beach, CA: The Noontide Press, 2000.
Stoddard, Lothrop. *The Rising Tide of Color Against White World Supremacy*. New York: Charles Scribner's Sons, 1926.
Stone, Dan. *Breeding Superman: Nietzsche, Race and Eugenics in Edwardian and Interwar Britain*. Liverpool: Liverpool University Press, 2002.
The Eugenics Education Society. *Problems in Eugenics Vol. II: Report of Proceedings of the First International Eugenical Congress*. Kingsway, W.C.: Eugenics Education Society, 1913.
The Human Betterment Foundation. *Human Sterilization*. Pasadena: The Human Betterment Foundation, 1929.
Thompson, Loyd and Winfield Scott Downs, ed. *Who's Who in American Medicine 1925*. New York: Who's Who Publications, Inc., 1925.
Trombley, Stephen. *The Right to Reproduce: A History of Coercive Sterilization*. London: Weidenfeld & Nicolson, 1988.
Tuttle, William M., Jr, *Race Riot: Chicago in the Red Summer of 1919*. New York, NY: Atheneum, 1970.
Van Pelt, Robert Jan. *The Case for Auschwitz: Evidence from the Irving Trial*. Bloomington, IN: Indiana University Press, 2002.
von Hoffmann, Géza. *Die Rassenhygiene in den Vereinigten Staaten von Nordamerika*. Munich: J.F. Lehmanns Verlag, 1913.
von Verschuer, Otmar and E. Kober. *Die Frage der erblichen Disposition zum Krebs*. Mainz: Verlag der Akademie der Wissenschaft und der Literatur, 1956.
Wagener, Otto. *Hitler: Memoirs of a Confidant*. Trans. Henry Ashby Turner. Yale University Press, 1987.
Wagner, Jens-Christian. *Das KZ Mittelbau-Dora*. Wallstein Verlag, 2001.
Wall, Joseph Frazier. *Andrew Carnegie*. New York: Oxford University Press, 1970.
Wallin Directory Company. *Kirksville City Directory*. Quincy, Illinois: Hoffman Printing Co., 1899.
Weatherford, Doris. *American Women's History*. New York: Prentice Hall General Reference, 1994.
Weber, Matthias M. *Ernst Rüdin: Eine kritische Biographie*. Berlin: J. Springer-Verlag, 1993.
Weindling, Paul J. "From Philanthropy to International Science Policy: Rockefeller Funding of Biomedical Sciences in Germany 1920-1940." In Nicolass A. Rupke, ed., *Science, Politics and the Public Good: Essays in Honor of Margaret Gowing*. New York: Macmillan Press, 1988.
Weindling, Paul J. *Epidemics and Genocide in Eastern Europe: 1890-1945*. New York: Oxford University Press, 2000.
Weindling, Paul J. *Health, Race and German Politics Between National Unification and Nazism, 1870-1940*. Cambridge: Cambridge University Press, 1989.
Weismann, August. *Essays Upon Heredity and Kindred Biological Problems*. Oxford: Clarendon Press, 1889.
Wistrich, Robert S. *Who's Who in Nazi Germany*. London: Wiedenfield and Nicolson, 1982.
Wolf, Theta H. *Alfred Binet*. Chicago: The University of Chicago Press, 1973.
Wyman, David S. *Paper Walls*. Amherst, MA: University of Massachusetts Press, 1968.
Yerkes, Robert M. and Clarence S. Yoakum. *Army Mental Tests*. New York: Henry Holt and Company, 1926.
Yerkes, Robert M. *Memoirs of the National Academy of Science*. Washington D.C.: National Academy of Science, 1921.
Yerkes, Robert M. "Autobiography of Robert Mearns Yerkes." in Carl Murchison, ed., *History of Psychology in Autobiography*. Worcester, MA: Clark University Press, 1930.
Zangwill, Israel. "The Melting Pot: Drama in Four Acts." New York: The Macmillan Company, 1909; reprint, 1919.

SEARCH ENGINES

Modern research cannot be efficiently undertaken without the use of Internet search engines. I have listed here some of the engines we employed to search across the worldwide web as well as institutional databases.

American Book Exchange	www.abe.com
Google	www.google.com
Lexis-Nexis	www.lexis-nexis.com
Proquest	www.proquest.com
Worldcat	www.oclc.org/worldcat

INTERNET SOURCES

While Internet research is essential to contemporary historical investigation, I discovered that virtually nothing on the web dedicated to eugenics was reliable, including some websites operated by respected academic entities. At the same time, I found certain noneugenic sites extremely valuable for their background and contextual information, especially when the site was an official organizational or governmental site. Hence while I consulted and searched through hundreds, perhaps thousands of websites, only a precious few of the most reliable are listed below.

American Philosophical Society	www.amphilsoc.org
Anti-Defamation League	www.adl.org
Avalon Project	www.yale.edu/lawweb/avalon
BioMed Central / PubMed	www.biomedcentral.com
British Broadcasting Corporation	www.bbc.co.uk
Dodd Research Center	www.lib.uconn.edu/DoddCenter
Galton.org	www.galton.org
Jewish Virtual Library	www.us-israel.org
Mazal Library	www.mazal.org
National Library of Medicine, NIH	www.nlm.nih.gov
Nizkor Project	www.nizkor.org
PreventGenocide.org	www.preventgenocide.org/lemkin
Public Broadcasting Service	www.pbs.org
Remember.org	www.remember.org
ScrapbookPages.com	www.scrapbookpages.com
The Scientist	www.thescientist.com
U.S. Department of Justice	www.doj.gov
U.S. House of Representatives	www.house.gov
U.S. Senate	www.senate.gov
Wellcome Library	www.wellcome.ac.uk
Yad Vashem	www.yad-vashem.org.il

Index

ABA. *See* American Breeders Association
Abel, Wolfgang, 316
abortion opponents, 127
ACLU. *See* American Civil Liberties Union
Adkins, Reable, 179
Africa
    Jews in, 295
    mixed race individuals, 263, 265–66
African-Americans
    amount of white blood, 79, 83, 165, 166
    death rates, 163
    definitions of, 165, 166
    "Free Issue" Negroes, 170, 171
    intelligence test scores, 79, 81, 83–84
    lynchings, 23, 84
    multiple birth frequency, 350
    music of, 105, 165, 180
    segregation, 22, 171, 172–73
    six-fingered boys, 97
    slaves, 21
    soldiers in World War I, 186
    stereotypes of, 38, 210
    sterilizations, 5, 421
    See also race
AGA. *See* American Genetic Association
Agriculture, Department of, 47, 97–98, 208
Air Force, U.S., 382
Alabama, ban on interracial marriages, 401
Alberta, Canada, sterilization law, 242
Albrecht, Charles, 293
alcoholism, 27, 58, 285, 299
Algonquin Indians, 176
Allan, William, 421
Allen, Bennet, 292
Allport, Frank, 151
Alma College, 75
AMA. *See* American Medical Association

American Academy of Actuaries, 432–33
American Bar Association Journal, 154
American Birth Control League, 132, 133, 138–40
    See also birth control movement
American Breeders Association (ABA), 291
    Agriculture Department and, 97–98
    establishment, 38–39
    Eugenics Committee, 39, 44, 411
    Eugenics Section
        committee on elimination of defective germ plasm, 57–61, 69–70, 73, 235
        English eugenics and, 210, 224
        Eugenics Record Office and, 47–48, 98
        formation, 44
        meetings, 97
        support for euthanasia, 247, 250–51
    meetings, 44, 45, 51, 97
    See also American Genetic Association
American Breeders Magazine, 40, 412
    See also Journal of Heredity
American Civil Liberties Union (ACLU), 399, 400
American Committee on Sterilization, 217
American eugenics
    comparison to English movement, 208, 219
    evolution of
        decline of, 385, 411, 413–14, 425
        defections from, 412–13
        development of field, 21–25, 65
        postwar changes, 417–18
    global ambitions, 61, 185, 235
    goals, 7, 21, 29, 88, 107
    influence
        on American thinkers, 99

American Committee on Sterilization, influence (*continued*)
dominance of international meetings, 213, 235, 237, 239, 264
in England, 207, 208–10, 212, 224–26
in Europe, 213, 243, 244, 245
in foreign countries, 240
in Germany, 258–59, 261, 264–65, 277, 312, 408–9
on Hitler, 259, 274–76, 277, 298
on social reform movements, 125
lack of scientific foundation, 99–100, 106–7, 414, 423
organizations, 87, 140, 219
reactions to
criticism of, 99–104
press coverage of plans, 101–4
public opposition to proposals, 87
ridicule of, 387
relations with birth control movement, 137–42
relations with German eugenics movement, 270–72, 277
American influence, 258–59, 261, 264–65, 277, 312, 408–9
American support for Nazi eugenics, 7, 277, 297, 300–301, 304–5, 313, 314–16, 317–18, 340–44, 388, 392–93, 414–17
common belief in Nordic supremacy, 266, 297
German dominance, 286, 294, 299
influence of Germans in United States, 281
joint projects, 288, 303–4
opposition to Nazi eugenics, 313, 417, 418
partnership, 280–83, 342
postwar communications, 377–79
reactions to Nazi atrocities, 411, 423–24
during Weimar Republic, 266
research bodies, 89–90, 219
See also Carnegie Institution, Station for Experimental Evolution (Cold Spring Harbor); Davenport, Charles; Laughlin, Harry Hamilton
American Eugenics Society (AES)
finances, 142
founding, 137, 244
goals, 424
leaders, 137–38

members, 163, 377
name change, 425
postwar changes, 417–18, 422–25
presidents, 259, 418, 422–24
proposed alliance with birth control movement, 138–40
publications, 138–40, 174, 394, 424
relations with British organizations, 229–30
relations with Carnegie Institution, 421
relations with German eugenics movement, 313, 377, 378
Ultimate Program, 239
during World War II, 422
American Genetic Association (AGA), 219, 349, 412, 425
See also American Breeders Association
American Hebrew, 306–7
American Indians
bans on interracial marriages, 146
census statistics, 176–77
citizenship, 177
in colonial Virginia, 168, 176
interracial marriages, 146, 165, 168, 176, 177–79
music of, 180
reservations, 156
seen as unfit, 179–80
sterilizations, 5, 400
Virginia racial laws and, 176–81
American Jewish Congress, 129
American Journal of Public Health, 304
American Management Association, 437
American Medical Association (AMA), 72
Section on Ophthalmology, 147–48, 149–51, 153
American Medical Society, 63
American Museum of Natural History, 236, 245, 298
American Ophthalmological Society, 146, 158
American Philosophical Society, 397
American Public Health Association, 174
American Review of Tuberculosis, 241
American Social Hygiene Association (ASHA), 225
American Society of Human Genetics, 379
Amish, 53
Andrews, George Reid, 142
Anglo-Saxon Clubs, 166
Anthropometric Laboratory, 16–17
Anti-Defamation League, 382

anti-Semitism
  in Germany, 297, 302
  of Hitler, 269–70, 311
  of Ploetz, 262
Appalachia, forced sterilizations in, 3–4
Appleget, Thomas, 306, 307
Archiv für Rassen- und Gesellschaftsbiologie
  (*Archives of Race Science and Social
  Biology*), 262, 263, 265, 272, 273, 281,
  295, 308, 341
Argentina, eugenicists in, 238
Armed Forces Repository of Specimen Samples,
  430
Army, U.S.
  African-American soldiers in World War I,
  186
  intelligence tests, 80–82, 83, 84, 132, 201,
  279
  Nazi war crimes trials, 404
  number of blue-eyed recruits, 273
Arps, George, 148
Aryan superiority
  Germanization of Polish children, 331–32,
  347, 405
  Hitler's belief in, 270, 307
  master race as Nazi goal, 299, 370
  Nazi breeding program, 405, 406
  as point of agreement between American
  and German eugenicists, 266, 297
  See also Nazi eugenics; super race
ASHA. *See* American Social Hygiene
  Association
Asian Americans, bans on interracial marriages,
  146, 175–76
Association of Agricultural Colleges and
  Experimental Stations, 38
Association of Medical Officers of American
  Institutions for Idiotic and
  Feebleminded Persons, 250
Athena Diagnostics, 438
Atwood, Edith, 96
Auschwitz
  blood samples taken from prisoners,
  363–66, 376
  deaths in, 338
  gassing, 337
  liberation of, 371
  medical experiments, 331, 338, 358–59,
  361, 362–63, 376, 379
  selection of arriving Jews, 337–38
  survivors, 355–56, 359

  twins camp, 354–61, 362–63, 365–66
  See also Mengele, Josef
Austin, O. P., 57
Australia
  DNA databanks, 429, 440
  eugenicists, 238
Austria
  DNA databanks, 429
  genetic discrimination legislation, 440
  Nazi annexation, 313
Autogen, 439–40
aviation medicine, 367, 381–82
Azores, 292–93

Bach, James, 150
Baltimore Sun, 200, 202
Banker, Howard, 148
Barker, Lewellys F., 281
Barker, Olin, 150
Bates College, 75
Bateson, William, 26, 27, 28, 411–12
Baur, Erwin, 270–73
  articles in American journals, 281
  eugenic views, 281–82
  *Foundation of Human Heredity and Race
  Hygiene*, 270, 272–73, 296–97, 351
  at international meetings, 287
Belgian Eugenics Society (Société Belge
  d'Eugénique), 240–41
Belgium
  eugenicists, 195, 235, 238, 240–41, 255
  eugenic screening of potential U.S. immi-
  grants, 195, 198, 205
  Nazi occupation, 407
  occupation of German lands after World
  War I, 267, 269, 271
Bell, Alexander Graham
  discomfort with eugenics goals, 89–90
  distancing from eugenics movement, 104–5
  Eugenics Record Office and, 89–90, 94,
  101, 159
  at international meetings, 71, 213, 237
  rejection of charity, 210
  work with Davenport, 44
Bell, J. H., 116
Bellamy, Raymond, 291
Beringer, Kurt, 369
Bethnal Green (London), 221, 222
Betts, Jane, 396
Beverly, Pal S., 170–71
Bible, 9–10, 13

Bibliography of Hereditary Eye Defects, 147
Bigelow, Maurice, 422
Biggs, Hermann M., 153
Billings, John, 31, 40–41
Binet, Alfred, 76, 82
Binet-Simon test, 76, 78
bioethics, 443
    See also genetic discrimination
Biometric Laboratory, 220
biometrics, 72, 345, 430
Birkenau concentration camp, 367
Birkett, H. S., 151
birth control
    for all women, 135
    compulsory, 60
    opponents, 127
    See also birth control movement
Birth Control Clinical Research Bureau, 134
Birth Control Federation of America, 134
birth control movement, 125
    American Birth Control League, 132, 133,
        138–40
    conferences, 134, 136–37
    current activities, 426
    distinction from eugenics, 136
    eugenic beliefs, 127, 128–33, 135–37, 138
    eugenicists involved in, 133–35, 136–39
    name, 128
    organizations, 140
    origins, 126–27, 128
    Planned Parenthood, 127, 144, 426
    proposed alliances with eugenics groups,
        137–42
    support of lethal chambers, 251
    See also Sanger, Margaret
Birth Control Review, 132, 138–40, 301
Bismarck, Otto von, 266
Blacker, C. P., 300
Blacks. *See* African-Americans
The Black Stork, 257–58
Blake, John R., 175
Blakeslee, Albert F., 389
Blakeslee, Alfred, 271
blind individuals
    costs of maintaining in institutions, 147,
        153
    data collected from, 148, 152–54
    defective family members, 152–53
    eugenic legislation, 149–52, 153, 154, 155
    family pedigrees, 148
    institutions, 148, 153

Jews, 301
    marriage restrictions, 145, 146, 149–52,
        154, 155, 157
    number in United States, 147, 149, 160
    preventing reproduction, 58
    segregation of, 145, 146, 149, 153
    sterilizations, 145, 146, 148, 149, 299
blindness
    hereditary, 145, 147, 149, 301, 341
    prevention of hereditary, 145, 147–54, 237
    as test case for eugenic legislation, 152, 157
Blitch, J. S., 291
Bliven, Bruce, 306
blood groups of twins, 364, 365
Bluhm, Agnes, 268
Bobbitt, John Franklin, 29
Boeters, Gustav, 261
Bolley, Henry, 292
Bollinger, Allan, 252–54
Bollinger, Anna, 252–54
Bolton, Kate, 5
Bradby, William, 179
brains
    Hallervorden-Spatz Syndrome, 369,
        382–83
    Nazi research on, 382–83
    remains of Holocaust victims, 383–84
    See also Kaiser Wilhelm Institute for Brain
        Research
Brandeis, Louis, 120
Brandenburg State Hospital, 369
Brazil
    eugenicists, 238
    marriage restrictions, 245
Breeders Associations (Britain), 224
breeding
    of animals, 36, 44
    feminist view of, 127–28
    history of, 13
    of humans, 21–22, 28, 367
    of plants, 13, 26
    See also super race
Brigham, Carl, 82–83, 84, 85, 90
Britain. *See* England
British Columbia, Canada, sterilization law, 242
British Medical Association, 226, 434
British Medical Journal, 434
Brock, Sir Lawrence, 231, 233
Brock Commission, 233–34, 249
Brooklyn Daily Eagle, 132
Brooklyn Institute of Arts and Sciences, 34, 36

Brooks Air Force Base, 382
Brower, Daniel R., 63
Brown, Louise, 427
Brunn, James, 398
Brush Mountain, Virginia, 3–4
Bryan, W. E., 291
Bryn, Halfdan, 292
Buchenwald concentration camp
    ancestral research barrack, 331
    cruelty in, 326–27, 329
    deaths in, 324, 329
    doctors, 325–26
    functions, 324
    Katzen-Ellenbogen at, 319–21, 325–31,
        332–35
    Little Camp, 319–21, 326, 327–28, 333
    medical experiments, 324, 329–30, 331,
        367
    work details, 329
Buck, Carrie, 108, 109–10, 112, 113–17,
    120–22, 167, 400, 401–2
Buck, Emma, 108–10, 115, 116, 122
Buck, Vivian, 113, 114–15, 122
Buck v. Bell, 113–17, 119, 120–22, 123, 315,
    401–2, 409
Buffalo Eye and Ear Infirmary, 145
Bumke, Oswald, 369
Burbank, Luther, 133
Bureau of Vital Statistics, Virginia, 161, 163,
    164–65, 166, 168–69, 173–74, 175, 177,
    181
Burlington Santa Fe, 437–38
Bush, Vannevar, 393–94
Butler, Pierce, 120

California
    ban on interracial marriages, 146
    eugenicists, 277
    governor's apology for sterilizations, 400
    sterilization law, 68, 122
    sterilizations performed, 7, 69, 122, 123,
        228, 233, 277, 315, 342, 398
California School for the Deaf and Blind, 147
Cambridge University Eugenics Society, 215
Campbell, Clarence, 314–15
Campbell, E. Taylor, 398
Canada
    DNA databank, 429
    eugenicists, 238, 241–42
    genetic testing issues, 435

immigrants, 241
    insurance industry, 435
    sterilization laws, 242
cancer, genetic markers, 430, 431
Cannon, W. B., 57
capital punishment, 250, 258
Carnegie, Andrew, 31, 37, 56, 72, 140
Carnegie Endowment, 296
Carnegie Institution
    concerns about Laughlin, 191–92, 193–94,
        198
    criticism of, 101–2, 201, 202
    Department of Genetics, 388, 412
    endowments, 31, 56
    establishment, 31
    eugenics research funding, 7, 32, 296, 387
        in Europe, 245, 298
        in Germany, 258
        global goals, 235
        international conferences, 238
        lessening interest in, 385, 391, 401
    genetics research funding, 421
    medical education survey, 283
    purpose, 31
    relations with Pearson, 220
Carnegie Institution, Station for Experimental
    Evolution (Cold Spring Harbor)
    buildings, 40
    corresponding scientists, 43–44, 263, 272
    criticism of, 392
    Davenport at, 40, 105
    Davenport's retirement, 386, 388
    early years, 43
    establishment, 40, 43
    foreign scholars at, 240, 241, 244, 314, 419,
        420
    merger with Eugenics Record Office, 105
    postwar changes, 426
    proposal, 36–38
    relations with German eugenics move-
        ment, 262–63, 268–69, 270–72, 273,
        392–93
    reviews of activities, 389–90
    staff, 43
    summer biology courses, 46, 51
Carnegie Steel Company, 72
carpal tunnel syndrome, 437–38
Carr, Wilbur, 200
Carrel, Alexis, 57
Castellino, Vittorio, 96
Castle, W. E., 412

Catholic Church
  charity, 10
  opposition to eugenics, 232–33, 293
  opposition to sterilization, 70, 232, 233
Cattell, James, 76
cemeteries, white, 172–73
census, German, 309, 311
Census, U.S.
  of 1880, 41
  of 1890, 202–3, 204, 205
  of 1910, 190
  of 1920, 159, 186, 205
  information requested by eugenics
    researchers, 159, 160–61
  use of Hollerith machines, 289
Census Bureau, 159, 203
  census of state institution inhabitants,
    159–60
  data on Indians, 176–77
  defective, dependent, and delinquent popu-
    lation, 159, 160
  number of blind people, 147
  relations with eugenics researchers, 159–61
Chamberlain, George, 67
Chamberlain, J. P., 154, 155
Chamberlain, Neville, 227, 228, 233
Chambers, Robert, 247
charity
  dysgenic effects, 251
  in England, 10–11, 214, 218
  historical background, 9–10
  Hitler's rejection of, 275
  negative views of, 12, 127, 129–30, 138,
    210, 222–23
  seen as perpetuating poverty, 12, 29
Cheatham, Mrs. Robert H., 169
Cheever, Ezekiel, 200–202
chemical weapons, 258, 276
Chicago
  defective baby killed in, 252–54
  Municipal Court, 113, 256
Chicago American, 254
Chicago House of Corrections, 79
Chicago Tribune, 253
Childs, Prescott, 293
China, DNA databanks, 429
Chinese Exclusion Act, 22
Chloupek, Joseph, 106
Chloupek, Mary Sullivan, 106
Church, A. G., 233
Churchill, Winston, 71, 215, 403

Civil War, 117–18, 161, 273
Clark, Frank C., 172
Clark, Mrs. Frank C., 171–72
Clark, W. H., 169–70
Clement VII, Pope, 10
Cleveland
  multiple births, 350
  public schools, 147
clones, 427, 443
CODIS. See Combined National DNA Index
  System
Cold Spring Harbor
  Brooklyn Institute of Arts and Sciences
    biological station, 34
  See also Carnegie Institution, Station for
    Experimental Evolution; Eugenics
    Record Office
Cole, Leon J., 251
College Board, 83
Collier's, 133
Colombia, eugenicists in, 238
colonies, 156, 215, 218
Colony for Epileptics and the Feebleminded
  (Lynchburg, Virginia), 4, 5–6, 109–10,
  113–14, 122, 400
Columbia, Missouri, twins in, 350
Columbia University, 152, 154, 155
Combined National DNA Index System
  (CODIS), 429
Commission for Relief in Belgium Educational
  Foundation, 241
Commission on Feeblemindedness (Virginia),
  108–9, 111
Committee for Legalising Sterilization
  (Britain), 228
Committee on Selective Immigration, 192,
  198–99
Committee to Prevent Hereditary Blindness,
  151, 153, 157
Communism, 186
Comstock, Ada, 99
concentration camps, 313
  international awareness of, 343
  medical experiments, 331, 365, 380–81
  Ravensbrück, 406
  twin studies in, 354–61
  See also Auschwitz; Buchenwald; Dachau
conferences
  First International Congress on Eugenics,
    70–73, 207, 213, 217, 235, 263, 264
  Second International Congress on

Eugenics, 236–38, 241, 244, 268, 351,
364, 422
Third International Congress of Eugenics
(New York), 245, 298, 419
Congress
House Committee on Immigration and
Naturalization, 187–88, 192–93, 199,
311–12
Laughlin as "Expert Eugenics Agent,"
188–92, 194
racial origins of senators, 204
Senate Immigration Committee, 393–94
congresses, international. See conferences
Connecticut
Biometric ID project, 430
Department of Social Services, 430
institutions, 67–68, 69
sterilization law, 67–68
sterilizations performed, 69
Constitution, U.S., 107–8
Constitutional Convention, 204
constructive eugenics, 135
See also positive eugenics
consuls. See State Department, U.S.
Conti, Leonardo, 339
contraception. See birth control
Coolidge, Calvin, 202, 203–4
Cox, Earnest S., 165–67
Crick, Francis, 426
criminality
environmental conditions and, 24
hereditary, 23–25, 96
criminals
DNA databanks, 429, 439
executions, 250, 258
family histories, 96
names collected in Virginia, 165
in Nazi concentration camps, 367
preventing reproduction, 58
sterilizations, 59, 63–64, 65–67, 68, 69,
122, 211
See also prisons
Crismond, A. H., 175
Cuba, eugenicists in, 238
Czechoslovakia
Aryan children kidnapped by Germans,
406–7
eugenicists, 238
eugenic screening of potential U.S. immigrants, 205

Nazi invasion, 313, 323
persecution of Jews, 323

Dachau concentration camp
medical experiments, 367, 381–82
opening of, 299
war crimes trials held at, 320, 328, 329–30,
332, 333–35
Dallas, John T., 99
Danforth, C. H., 342, 344
Danielson, Florence, 54
Danvers State Hospital (Massachusetts), 321,
327
Dartmouth Medical School, 425
Darwin, Charles, 12, 13, 15, 26, 45, 71, 74, 251
Darwin, Leonard, 71, 213, 215, 216–17
data processing. See Hollerith data processing
machines
Davenport, Amzi Benedict, 32, 33
Davenport, Charles
audience with Mussolini, 293
books, 33, 35, 73–75, 255, 272, 349, 386
career, 33
Cold Spring Harbor laboratory, 36–38, 40,
43, 105
criticism of, 99–101, 102–4
data collected, 44–45, 54, 88
death, 387
English eugenicists and, 212, 230
Eugenical News and, 305
Eugenics Record Office and, 89, 105, 106,
412
eugenic views, 36, 37, 75, 416–17
on carriers of defects, 58
on euthanasia, 254
on germ plasm, 364, 386
on heredity, 105–6
identification of unfit, 78
on immigration restrictions, 187
on marriage restrictions, 147, 216
on segregation of unfit, 39, 60, 75
on skin color, 164
on twins, 349, 350
family, 32–33, 385–86
files, 397
Fisher and, 134
Galton and, 34–35, 37, 43–44
in Germany, 287–88
influence on Hitler, 259
Laughlin and, 51, 193–94, 196, 197, 202,
388, 395

Davenport, Charles (*continued*)
  leadership of American eugenics, 32,
    34–38, 48, 385
  organizational activities
    American Breeders Association, 97
    American Eugenics Society, 137
    Eugenics Research Association, 90, 91
    international cooperation, 240, 243, 244
    international meetings, 71, 73, 213, 217,
      230, 236, 237–38, 279–80, 287
    Permanent International Commission
      on Eugenics, 238, 239–40
  personality, 32, 34, 385
  physical appearance, 34
  relations with German eugenics move-
    ment, 258, 262–63, 267, 268–69,
    270–72, 273, 286, 291, 298, 313, 344,
    385
  research, 55, 56, 88, 255, 263, 386–87
  retirement, 386–87, 388
  Rockefeller Foundation and, 93
  Sanger and, 136, 139, 140
  survey of mixed race individuals, 288–89,
    290, 350
  tribute to Ploetz, 295
  whaling museum, 387
Davenport, Charlie, 385–86
Davenport, Gertrude Crotty, 34
Davenport, Jane, 33
Davis, James J., 193, 194, 195, 199, 200, 203,
  239
Davis, J. S., 108, 175
deaf individuals
  institutions, 147
  marriages, 210
  number in United States, 160
  preventing reproduction, 58
  sterilizations in Germany, 299
Deavitt, Sadie, 54–55
deCODE Genetics, 439
DeJarnette, Joseph, 7, 112, 114–15, 277
Delaware, marriage restrictions, 146
Delia (Plecker family servant), 162
Denmark
  eugenicists, 235, 238, 243–44, 418–20
  eugenic screening of potential U.S. immi-
    grants, 195, 205
  genetic discrimination legislation, 440
  Institute for Human Genetics, 418, 420–21
  marriage restrictions, 243

Nazi occupation, 407
  sterilization law, 243
  sterilization proposals, 420
Dennis, David, 150–51
Denson, William, 334
designer babies, 442
detention, forcible. *See* segregation of unfit
Deutsche Forschungsgemeinschaft. See German
  Research Society
Dickens, Charles, 11
Dight Institute, University of Minnesota, 397
Dina (Auschwitz inmate), 357
Ding-Schuler, Erwin, 329–30
diplomats. *See* State Department, U.S.
disabilities, people with
  preventing reproduction, 58
  See also blind individuals; deaf individuals;
    unfit
DNA
  analysis in criminal investigations, 429–30,
    439
  databanks, 429–30, 439–40
  double helix, 426
  fingerprints of soldiers, 430
  uses of information, 429–32, 439
  See also genetics
Dobbs, J. T., 109, 113, 122
Dodge, Cleveland H., 31, 238
Dodge Foundation, 422
Dolly (cloned sheep), 427
Donald, Mary, 6
Dora works, 329
Dow Chemical, 437
Drange-Graebe, Mary, 53
Draper, John, 351–52
Draper, Wickliffe, 288–89
Draper Fund, 422
Dugdale, Richard, 24, 25, 64, 272
Dumböck (SS lieutenant), 326–27
Dunn, L. C., 391–92, 412, 412–13
Duvall, John C., 138

East, Edward, 138, 228
Eaton, Amey, 53
Eberhard-Karls University, Tübingen, 383–84
Eliot, Thomas D., 102–3
Elizabeth City County, Virginia, 162–63
Ellinger, Tage U. H., 354, 414–16
Ellis Island, 23, 78

Emergency Fund for German Science
(*Notgemeinschaft der Deutschen
Wissenschaften*), 295, 307, 364, 365
See also German Research Society
employers, genetic testing by, 437–38, 440
Energy, Department of, 437
England
charitable institutions for poor, 10–11, 218
costs of maintaining unfit, 226
DNA analysis, 430
DNA databanks, 429, 440
eugenic screening of potential U.S. immigrants, 196, 198, 205, 222
genetic testing and discrimination, 433–35
geographic distribution of pretty women,
442
illegal sterilizations, 211, 230–32
immigrants, 22, 207
insurance industry, 433–35, 436
intelligence testing, 76, 226
Jews in, 196
laws
Defective and Epileptic Children Act,
214
Idiots Act, 214
inheritance, 212
Lunacy Act, 214
Mental Deficiency Act, 214, 215,
216–18, 227
poor laws, 10–11, 214, 220–21, 222
lethal chambers in, 247–49
Ministry of Health, 211, 227, 228, 230,
231, 232, 233
National DNA Database, 429
number of feebleminded, 132
number of unfit, 226–27
primogeniture, 212
social problems, 207, 209
England, eugenics in
American influence, 207, 208–10, 212,
224–26
comparison to American movement, 208,
219
debate on murder of unfit, 248–49
family investigations, 211–12, 216, 220,
221
international cooperation, 238
lack of public support, 211–12, 226, 234
lack of scientific foundation, 28–29
marriage restriction proposals, 211, 216,
217, 218

postwar changes, 425
preventing reproduction of unfit, 210–11
research, 212–13, 219–22
segregation proposals, 210, 211, 214,
215–18, 226
sterilization proposals, 210, 214–15,
233–34
debates on, 226
illegality, 211, 230
lack of public support, 232, 300
lobbying for law, 227–29
opposition to, 211, 226, 227, 232, 233,
234
successors of Galton, 207, 209
view of American eugenics, 99–100
See also Eugenics Education Society;
Eugenics Society; Galton, Francis J.
epilepsy
causes, 55
definitions, 56
hereditary, 55, 322
Katzen-Ellenbogen's research, 91, 321–22
link to feeblemindedness, 55
Nazi research, 367–68
epileptics
data collected on, 54–56
deaths in institutions, 256
efforts to identify and prevent reproduction, 56
institutions, 4, 54–55, 91, 98, 108, 256,
322, 327
Laughlin as, 395
marriage restrictions, 146
names collected in Virginia, 165
preventing reproduction, 58
seen as feebleminded, 55, 108
sterilizations, 5, 231, 299, 322
equality, 237, 238
ERA. See Eugenics Research Association
Der Erbarzt (*The Genetic Doctor*), 339, 346, 347,
362
Erbkrank (*The Hereditarily Diseased*), 315–16,
393
ERO. See Eugenics Record Office
Estabrook, A. H., 53, 238
Estonia Genome Project, 439
ethnic groups
conflicts, 22, 186–87
eugenic ratings, 282
immigration quotas, 202–5
melting pot concept, 22, 134

ethnic groups (*continued*)
 seen as inferior, 29–30, 35, 44–45, 134
 stereotypes of, 35
 See also race
Eugenical News
 advisory committee, 344
 articles, 105, 180, 256–57, 267
  book reviews, 166
  on eyes, 361
  by German eugenicists, 282
  German sterilization law, 300
  on hereditary blindness, 147
  on Hitler, 298
  on interstate deportation, 157
  on marriage restriction laws, 155
  on Nazi eugenics, 304–6, 314–15,
   341–42, 352–53, 388, 389, 392–93
  news from Germany, 281, 282–83, 286,
   295, 297, 304–6, 364
  news from other countries, 240–41, 243,
   244, 266
  on racial integrity laws, 174, 175, 178
  reviews of German books, 273, 282, 297,
   302, 343
  on sterilization, 211
  on twin research, 349, 350–51, 365
 criticism of Nazis, 418
 distancing of Carnegie Institution, 391,
  394
 editorial committee, 305
 editorial on Sanger, 135
 focus on race science, 280–81
 foreign readership, 305–6
 on future of eugenics, 416–17
 Howe obituary, 158
 Laughlin as editor, 98, 280–81, 305, 389
 Laughlin's obituary, 395
 launching of, 98
 name change, 425
 postwar changes, 417–18
 publication by American Eugenics Society,
  394, 424
 racial references to Jews, 282–83
 subtitles, 280–81, 300, 412
 topics covered, 98–99
 during World War II, 422
eugenics
 definitions, 18
 distinction from genetics, 391–92
 Galton's contributions, 15–17
 goals, 7

historical background, 9–13
international cooperation, 70, 235–40
lack of scientific foundation, 27, 99–100,
 296, 391, 414, 423
term, 16
transition to genetics, 411–14, 417–18,
 422–26
See also American eugenics; England,
 eugenics in; Nazi eugenics; negative
 eugenics
Eugenics (journal), 138–40, 174
eugenics curriculum, 75–76
Eugenics Education Society (EES)
 activities, 212, 214–18, 219, 220–22, 226
 debate on murder of unfit, 248–49
 focus on negative eugenics, 70
 founding, 210
 invitations to international congresses, 71
 lack of funds, 212, 223–24
 Laughlin's visit, 196
 leaders, 213
 lobbying by, 214–15, 226, 227–29
 mission, 212
 publications, 223
 relations with American organizations,
  224–25
 research, 223–24
 warning on illegality of sterilizations, 211
 See also England, eugenics in; Eugenics
  Society
Eugenics Laboratory, 27, 60
Eugenics Quarterly, 425
Eugenics Record Office (ERO)
 Board of Scientific Directors, 89–90,
  93–94, 104–5
 buildings, 51, 396
 Carnegie Institution and, 47–48
 closing of, 395–96, 413
 criticism of, 99–101, 388, 390–91
 data collected, 76, 388, 390–91, 396–97,
  425
 data collection process, 52–55, 56, 106–7,
  165, 289, 291
 Davenport's office, 386
 draft laws on people with vision problems,
  149–50
 establishment, 45
 estimated number of unfit, 216
 family pedigrees
  of blind people, 148
  collection of, 105, 106, 290, 396–97, 398

family folders, 106
family trait booklets, 106
forms, 44, 239–40, 398
field investigators, 52–55, 80, 96–97, 179–80
foreign scholars at, 419, 420
funding, 46–47
global goals, 61
Harriman funding, 47, 48, 51, 56–57, 94–95, 96, 105, 238
hereditary tuberculosis data, 255
identification of unfit Americans, 52–55, 56
influence abroad, 213
Laughlin as superintendent, 51, 52, 390
letters received after closing, 396, 397–98
marriage restriction laws, 147, 156
political advocacy activities, 46
purpose, 45, 48
reforms, 388
relationship to Carnegie Institution, 105, 388–97, 412, 413
relations with German eugenics movement, 265, 267, 272, 313, 315–16, 342
review of activities, 390–91
songs, 123
study of sterilization in foreign countries, 235
twin studies, 350, 361–62
unscientific methods, 106–7
See also Laughlin, Harry Hamilton
Eugenics Research Association (ERA)
birth control movement and, 137–38
census data requested, 160
charter members, 90, 91, 322
establishment, 90
Eugenics Record Office and, 388
meetings, 266–67
members, 412
political advocacy activities, 90
presidents, 146, 158, 193, 257, 314–15, 343, 413–14
publications, 389
relations with German eugenics movement, 313, 342
survey of mixed race individuals, 289
Eugenics Review, 216–18, 225, 267
Eugenics Society
activities, 226, 233–34
endowments, 229
lobbying by, 226, 227–29
mission, 212
name change, 225
postwar changes, 425
relations with American organizations, 225, 229–30
sterilization proposals, 233, 300
See also England, eugenics in; Eugenics Education Society
eugenics textbooks, 53, 255
by Davenport, 73–75
by Dunn, 412
German, 270, 272–73, 296–97, 351
in high schools, 75–76
by Johnson and Popenoe, 137, 255, 258, 351
read by Hitler, 259, 272–73
twin information, 349, 351
in universities, 75
euthanasia
debates on, 248–52
in Nazi Germany, 317
by neglect and abuse, 255–56
support for, 60, 247, 248
See also murders of unfit
Evian Conference, 343, 393
evolution, 12, 13
executions, 250, 258
See also murders of unfit
Extermination by Labor, 329
eyes
colors, 273, 359
hereditary defects, 361
medical experiments in concentration camps, 331, 359, 376
of twins, 361–63
vision problems, 149–52
See also blindness

Fairchild, Henry Pratt, 134, 137, 141
families
carriers of defects, 58, 296, 308, 363–66
segregation of, 215–18
superior, 46, 135, 137, 139, 301
of unfit, 24–25, 58, 59, 152–53, 215–18
See also heredity
family investigations
by Plecker, 169–72, 177–79, 180–81
of poor, 220, 221
of potential U.S. immigrants, 187, 188, 189, 192, 193, 194–99, 205
Family Life, 379

family pedigrees
  of blind people, 148
  collected by Eugenics Record Office, 105, 106, 290, 396–97, 398
  collected by Nazis, 331, 342–43, 405
  of criminals, 96
  eugenic registries, 88, 165
  Eugenics Record Office forms, 44, 239–40, 398
  genetic profiles, 431, 432
  of German Jews, 311, 316, 407–9
  of immigrants, 189
  medical histories, 431–32, 435–36, 438
  of superior families, 46
  in Tasmania, 440
  use of genetic information, 432–37
Federal Bureau of Investigation (FBI), 429
Federal Council of the Churches of Christ in America, 307
feebleminded individuals
  British policies, 214
  in concentration camps, 367
  conferences on, 80
  costs of maintaining, 39, 153, 226, 228–29
  executing, 250, 251
  identification, 78, 196, 201
  institutions, 4, 11, 254–55, 256, 257
  life expectancies, 257
  marriage restrictions, 146
  morons, 78, 79, 85, 189
  names collected in Virginia, 165
  Nazi gassing of (T-4 project), 312–13, 317, 339, 369, 382–83
  number in England, 226–27
  number in United States, 132
  preventing reproduction, 58
  segregation of, 214, 215
  See also sterilizations, of feebleminded
feeblemindedness
  definitions, 77, 78–79, 94, 96, 257
  hereditary, 77
  link to epilepsy, 55, 108
feminists
  Sanger as, 142–43
  support of eugenics, 22, 127–28
Field, James, 57
Fifth International Congress on Genetics (Berlin), 287
Filipinos, 175–76
films
  The Black Stork, 257–58

Erbkrank (The Hereditarily Diseased), 315–16, 393
fingerprints, 15, 431
Finland
  eugenicists, 245
  genetic discrimination legislation, 440
  sterilizations, 245
First American Birth Control Conference, 134, 136
First International Congress on Eugenics (London), 70–73, 207, 213, 217, 235, 263, 264
First National Conference on Race Betterment, 88–89, 249, 251
Fischer, Eugen, 263, 265–66, 272
  articles in American journals, 281
  audience with Mussolini, 293
  Davenport and, 287, 291, 294, 298
  Eugenical News advisory committee, 344
  Foundation of Human Heredity and Race Hygiene, 270, 272–73, 296–97, 351
  at international meetings, 279, 280
  involvement in Nazi eugenics, 300, 316–17, 339, 347, 354
  at Kaiser Wilhelm Institute for Anthropology, Human Heredity, and Eugenics, 286, 288
  national anthropological survey, 294–95
  postwar attacks on, 378
  retirement, 347
  at Society for Racial Hygiene, 302
  support for sterilizations, 297
Fisher, Agnes, 399
Fisher, Irving, 57, 89, 103–4, 134, 136–37, 224
Fisher, Ronald A., 221–22, 223, 351
Fletcher, Robert, 25
Flexner, Abraham, 283–84, 285
Florida State Reformatory, 291
Ford, Henry, 50–51
Forel, Auguste, 242, 243, 257
Fosdick, Raymond, 313, 314, 365
France
  African soldiers in army, 267, 271, 275, 305
  eugenicists, 235, 238, 239
  eugenic screening of potential U.S. immigrants, 196–97
  genetic discrimination legislation, 440
  Nazi occupation, 323, 407
  occupation of German lands after World War I, 267, 269, 271, 305
Frank, Hans, 404

Frankenfish, 443
Frankfurt University, 340
Frazer, Robert, 197
Frazier, Charles, 398
"Free Issue" Negroes, 170, 171
Free World, 403–4
French Canadians, 241, 242
Frick, Wilhelm, 304, 305, 314, 353, 354, 394,
    404
Fritzche, Hans, 404

Gailor, Thomas F., 99
Galton, Francis J., 73, 207
    Bateson and, 27, 28
    books, 15, 16, 18, 28
    counting by, 14–15
    data collected, 16–17
    Davenport and, 34–35, 37, 43–44
    death, 19, 70
    distribution of pretty women in England,
        442
    education, 14
    Eugenics Education Society and, 212
    eugenics work, 15–17, 18, 28, 70, 72, 209,
        219–20, 411, 418
    fingerprints, 15, 431
    German eugenicists and, 264
    Hereditary Genius, 15, 18
    on House of Lords, 212
    on human breeding, 28
    influence, 35
    intelligence testing, 76
    marginalization, 208, 209, 213
    on marriage restrictions, 18–19, 28, 211
    physical appearance, 14
    research, 15
    twin studies, 348–49, 351
Galton Institute, 425
Galton Laboratory, 99, 212–13, 220, 425
Gaupp, Robert, 369
genealogies. *See* family pedigrees
genelining, 436, 442
General Accounting Office, 400
genes
    defective, 296
    early study of, 412
    See also DNA; genetics; heredity
Genetic Anti-Discrimination Bill, 440
genetic counseling, 411, 421, 423, 431
genetic discrimination, 433–35, 436–38,
    440–41, 443–44

genetic identity, 429–32, 439–40
genetics
    designer babies, 442
    distinction from eugenics, 391–92
    eugenics transformed into, 411–14,
        417–18, 422–26
    future of, 428–29, 431, 441–44
    history of field, 26, 411–12
    medical, 421–22, 423, 424, 425
    See also DNA
Genetics Record Office, 396
    See also Eugenics Record Office
genetics research
    cloning, 427, 443
    corporate involvement, 428
    current, 426, 427–28
    funding, 424–25
    Human Genome Project, 437, 441
    Rockefeller funding, 370
    transgenic creatures, 443
genetic testing
    access to results, 434
    discrimination based on, 433–35, 436–37,
        440–41
    by employers, 437–38, 440
    privacy of information, 437
    of twins in Nazi Germany, 353–54
genetic therapies, 441
genocide
    definition, 404–5
    international treaty, 404–5
    Nazi eugenics as, 405–9, 411
    techniques, 403–4
    term coined by Lemkin, 402
German-American Hospital, Chicago, 252–54
German eugenics
    American financial support, 7
    definitions of Jews, 310–12
    international cooperation, 235
    isolation after World War I, 267–68
    leadership of international movement, 230
    national anthropological survey, 294–95
    negative eugenics, 261–62
    negative view of Jews, 282–83, 294, 295,
        296–97, 338–39
    Nordic superiority, 70, 239
    organizations, 262, 263–64
    publisher, 273
    race biological index, 282
    race biology, 259
    racial hygiene, 262

German eugenics (*continued*)
  refusal to cooperate with French and
    Belgians after World War I, 238, 239,
    258, 267–69
  relations with Nazis, 297–98
  research, 224, 229, 283, 419
  Rockefeller funding, 7, 297, 298, 302, 313,
    364–65
  sterilization proposals, 261, 265, 339
  textbooks, 270, 272–73, 296–97, 351
  twin studies, 297, 350, 352–54
  See also American eugenics, relations with
    German eugenics movement; Nazi
    eugenics
German Medical Association, 339
German Psychiatry Institute, 285, 286
German Research Society (*Deutsche
  Forschungsgemeinschaft*), 346, 353, 354,
  355, 360, 362, 364, 365, 367, 369
  See also Emergency Fund for German
    Science
Germans, stereotypes of, 35
Germany
  allied occupation, 376, 377–78
  anti-Semitism in, 297, 302
  DNA databanks, 429
  eugenic screening of potential U.S. immi-
    grants, 205
  medical education in, 283, 340
  Ministry of the Interior, 285, 287, 304, 311
  mixed race individuals, 267
  nationalists, 266, 267, 338
  Treaty of Versailles, 238, 267, 269, 271
  war reparations, 267, 268, 271
  Weimar Republic, 266–67, 268, 269
  See also German eugenics; Jews, in
    Germany; Nazis
germ plasm, 17, 18, 25, 73
  carriers of defects, 58, 363–66
  defective, 58, 386
Gesellschaft für Rassenhygiene. See Society for
  Racial Hygiene
Gestapo, 316, 323
Gildon, Mary, 169
Giuliani, Rudolph, 439
Goddard, Henry, 76–79
  classification of feeblemindedness, 77,
    78–79, 94, 257
  estimated number of feebleminded, 132
  Eugenics Record Office and, 94, 150
  intelligence testing, 78–79, 80, 90, 226, 243

The Kallikak Family, 76–77, 107, 243, 250,
    272
  on morons, 78, 85
  relations with German eugenics move-
    ment, 342
  at Vineland, 76, 91, 243
Goebbels, Joseph, 299, 318
Goethe, C. M., 277, 315, 343, 344, 379
Goldsborough, Phillip L., 71
Goodman, Aileen, 178
Goodrich, C. L., 97
Gordon, Alfred, 351
Gosney, E. S., 277, 342
Govaerts, Albert, 195, 241, 255
Grant, Madison, 30, 31, 83, 167
  Committee on Selective Immigration, 192
  Eugenics Research Association and, 90–91
  influence in Germany, 259, 273, 274–75,
    298
  international eugenics congresses, 236
  Laughlin and, 305
  on Nordic race, 29
  The Passing of the Great Race, 82, 90,
    251–52, 259, 266, 273, 274–75
  reaction to alliance with American Birth
    Control League, 139–40
  support of immigration restrictions, 188
Great Britain. *See* England
greenlining, 436
Gregg, Alan, 308
Greunuss, Werner, 331
Gypsies
  in concentration camps, 358, 359, 363, 366,
    367
  eye colors, 363
  medical experiments, 366
  Nazi persecution of, 316
  twins, 358, 359

Haase, Irmgard, 366
Haber, Fritz, 285
Hahn, Otto, 376
Haiselden, Harry, 252–54, 255, 256, 257
Hale, Howard, 4
Hall, Gertrude, 96
Hallervorden, Julius, 369, 382–83
Hallervorden-Spatz Syndrome, 369, 382–83
Hallervorden-Spatz Syndrome Association,
    382–83
Hamilton, Mascott, 171
Hanley, J. Frank, 66–67

Hansen, C. C., 293
Hansen, Sören, 243
Harding, Warren G., 193
Harrell, D. L., Jr., 5
Harriman, E. H., 46
Harriman, Mary, 46, 47
Harriman, Mary (Mrs. E. H.)
   criticism of, 101–2
   Davenport and, 46–47, 48, 49, 140
   funding of Eugenics Record Office, 47, 48,
      51, 56–57, 94–95, 96, 105, 238
   international eugenics congresses, 236
Harris, J. Arthur, 191
Harvard University, 33, 117, 119
   eugenics courses, 75
   Howe Laboratory for Ophthalmology, 146
Havemann, Robert, 376, 379
Hays, Willet M., 39–40, 47
Healy, William, 53
Hearst newspapers, 101–4, 254
Hebberd, Robert, 95
Heidelberg, University of, 312, 342, 383
Heidinger, Willi, 309–10
Henry VIII, King of England, 10
hereditary disorders
   carpal tunnel syndrome, 437–38
   classification, 58
   epilepsy, 55, 322
   feeblemindedness, 77
   genetic counseling, 431
   insanity, 295–96
   insurance industry records of, 431–32
   tuberculosis, 241, 255
   vision problems, 361
   See also blindness; genetic discrimination
heredity
   carriers of defects, 58, 296, 308, 363–66
   of character traits, 17, 74, 105–6
   of criminality, 23–25, 96
   dominant and recessive traits, 26
   environmental factors and, 27, 339, 348–49
   Galton's views, 15
   Mendelian theories, 13, 25–26, 36, 39, 101,
      208, 411
   of physical traits, 17, 27
   role of germ plasm, 17, 18, 25
   study of, 13
   theories, 17–18
   See also genetics
Herndon, C. Nash, 421
Heron, David, 99–100

Hess, Rudolf, 270, 318
HGC. See Human Genetics Commission
High Teams institution (London), 230–32
Hill, Joseph, 159
Himmler, Heinrich, 354, 355, 366
Hindenburg, Paul von, 299
Hindus, 176
Hirschhorn, Kurt, 379–80
Hitler, Adolf
   American admirers, 297, 298
   anti-Semitism, 269–70, 311
   Beer Hall Putsch, 269, 273–74
   eugenics texts read by, 259
   eugenic views, 259–60, 269–70, 274–77,
      280, 297, 307, 318, 367, 403
   imprisonment, 259–60, 269
   influence of American eugenicists, 259,
      274–76, 277, 298
   meeting with Stoddard, 318
   Mein Kampf, 270, 274, 276, 318, 354, 394
   as national Physician, 309
   rise to power, 277, 297, 298–99
   suicide, 375
   in World War I, 276
   See also Nazis
Hitlerschnitte (Hitler's cut), 304
Hodson, Cora, 223, 224–25, 226, 229–30
Hofmann, Otto, 405, 407–9
Holland
   DNA databanks, 429
   eugenicists, 238, 240, 245, 279
   eugenic screening of potential U.S. immi-
      grants, 205
   Nazi occupation, 407
   Hollerith data processing machines,
      289–90
   proposed eugenics use, 293–94, 310
   use in Jamaica survey, 290, 291
   use in Nazi Germany, 309, 311
   See also IBM
Holmes, Oliver Wendell, 209
Holmes, Oliver Wendell, Jr., 117–22, 315,
   401–2, 409
Holmes, Samuel J., 224
Holocaust. See concentration camps; Jews, in
   Germany; Nazi eugenics
Holt, Henry, 181
homeland security, 430–31
homosexuals, in concentration camps, 324, 367
Hong Kong, genetic discrimination, 438
Hoover, Herbert, 203, 205

Hoover, J. Edgar, 187
Hottentots, 263, 265–66
House Committee on Immigration and
    Naturalization, 187–88, 192–93, 199,
    311–12
Howe, Lucien, 145–46
    Committee on Selective Immigration and,
        192
    death, 158
    eugenics work, 146, 158
    interstate deportation of unfit, 156–57
    laboratory, 344
    prevention of hereditary blindness, 147–49,
        151, 152–55, 237
    support of immigration restrictions, 187
    widow, 396
Hughes, Charles Evans, 200
Human Betterment Foundation, 277, 281, 342
Human Betterment League of North Carolina,
    425
Human Genetics Commission (HGC), 434
Human Genetics League of North Carolina,
    425
Human Genome Project, 437, 441
Hungary, eugenicists in, 245
Hunter, George William, 75–76
Huntington's chorea, 54, 299, 431
Hurty, J. N., 65–66
hygiene and sanitary movement, 262

IBM
    German subsidiary, 309–10
    Iceland DNA databank project, 439
    systems designed for Nazi Germany,
        290–91, 308–10, 311, 339
    See also Hollerith data processing machines
Icarian Colony, 261
Iceland, national DNA databank, 439
IFEO. See International Federation of Eugenic
    Organizations
Illinois Institution for the Feebleminded
    (Lincoln, IL), 254–55, 256, 257
Illinois sterilization law, 67
immigrants
    to England, 22, 207
    eugenic selection of, 239
    to United States
        ethnic groups seen as inferior, 29–30
        eugenic screening in home countries,
            187, 188, 189, 192, 193, 194–99, 205
        family pedigree investigations, 189,
            194–98, 199
        feebleminded children, 132
        intelligence testing, 78, 83, 196
        Irish, 30, 35, 78, 95
        Italian, 83, 96, 187, 202
        Jews, 74, 78, 83, 95, 191, 273
        melting pot concept, 22, 35
        Nordic race preferred, 188, 192, 203
        number of blue-eyed recruits, 186
        opposition to immigration restrictions,
            191–92, 203
        southern and eastern European, 29–30,
            74, 78, 190, 192, 202
        undesirable, 185
Immigration Act of 1924, 202–5, 275
Immigration Restriction League, 188
immigration restrictions
    in foreign countries, 185
    in United States
        Chinese Exclusion Act, 22
        debates on, 185–86
        easing in 1952, 205
        eugenic principles in, 185
        on Jews, 311–12, 393, 394
        Laughlin's support of, 88, 187, 188–89,
            311–12, 387–88, 393, 394
        opposition to, 191–92, 203, 204
        quotas, 189, 190, 192–93, 202–5, 275,
            311–12, 394
        reactive, 188
        support for, 22–23, 45, 134, 187
The Independent, 254
Indiana
    Committee on Mental Defectives, 96–97
    legislature, 96–97
    poor and homeless, 25, 53, 64–65
    State Board of Health, 65
    sterilizations, 63–64, 209, 211
    Tribe of Ishmael, 25, 53, 65, 237
Indiana Reformatory, 63–64, 65–67, 148
Indian Citizenship Act, 177
Indian Health Service, 400
Indians. See American Indians
Industrial Alliance, 435
iNeurology, 382
inherited diseases. See hereditary disorders
inmates. See prisons
Innes, Alfred Mitchell, 71
insane
    census statistics, 159

costs of maintaining, 39
executing, 250
institutions, 52
marriage restrictions, 146
names collected in Virginia, 165
preventing reproduction, 58
insanity
conferences on, 80
epilepsy and, 55
hereditary, 295–96
preventing, 80
Institute for Hereditary Biology and Racial
Hygiene (Frankfurt), 341–43, 344, 345,
346, 352–53
Institute for Heredity Research (Potsdam), 267
Institute for Human Genetics (Copenhagen),
418, 420–21
Institute for Racial Hygiene (Munich), 316
Institution for Race Biology (Switzerland),
242–43
Institution Quarterly, 255
institutions
for blind, 148, 153
census of inhabitants, 159–60
conditions in, 254–55, 256
costs, 39, 130, 147, 153
data gathered in, 52, 161
deinstitutionalization, 227
in England, 227
for epileptics, 4, 54–55, 91, 98, 108, 256,
322, 327
mortality rates, 255, 256, 257
national origins of residents, 190–91
surveys of, 201
See also segregation of unfit
insurance industry
access to genetic information, 434, 435,
436
asymmetrical information, 432–33
Medical Information Bureau, 431–32,
435–36
use of genetic information, 432–37, 440
intelligence tests
Army tests, 80–82, 83, 84, 132, 201, 279
Binet-Simon test, 76, 78
criticism of, 80, 84–85
development of, 76
in England, 226
of Goddard, 78–79, 80, 90, 226, 243
identifying feebleminded, 196, 201
immigrant screening, 78, 83, 196

improving scores, 76
IQ, 82, 83, 84
racial differences, 79, 81, 82–84, 85
social use of, 83–84
spread of, 79
Stanford-Binet test, 81, 82, 84
Yerkes-Bridge Point Scale, 80
International Actuarial Association, 436
International Commission on Eugenics, 134
International Congress Against Alcoholism, 285
International Congress of Genetics
(Edinburgh), 346
International Congress of Hygiene, 267
international cooperation, 70, 235–40
See also conferences; International
Eugenics Commission; International
Federation of Eugenic Organizations
International DNA Users Conference, 430
International Eugenic Committee, 235–36, 238
International Eugenic Congress, Committee on
Immigration, 146
International Eugenics Commission, 238–40,
243, 264, 268–69, 272
International Federation of Eugenic
Organizations (IFEO), 240, 244, 258,
264
Committee on Race Crossing, 279, 291,
292, 293, 294
Committee on Racial Psychiatry, 295
Germans included in, 279, 286, 288, 302,
341
Munich meeting, 286
Rome meeting, 279–80, 293
Zurich meeting, 304
International Health Exhibition (London),
16–17
International Hygiene Exhibition, 264
International Planned Parenthood Federation,
144
International Society for Race Hygiene, 264
International Space Hall of Fame, 382
Interpol, 430
interracial marriages
of American Indians, 146, 168, 176, 177–79
preventing, 163
restrictions on, 146, 165, 174–76, 400–401
See also mixed race individuals
interstate deportation, 156–57
in vitro fertilization, 427
Iowa
Icarian Colony, 261

Iowa (continued)
    legislature, 250
    sterilization law, 68
IQ tests, 82, 83, 84
    See also intelligence tests
Ireland
    eugenic screening of potential U.S. immigrants, 205
    immigrants from, 30, 35, 78, 95
Ishmael, Tribe of, 25, 53, 65, 237
Italy
    eugenicists, 235, 238, 245
    eugenic screening of potential U.S. immigrants, 195–96, 198, 205
    immigrants from, 83, 96, 187, 202
    Mussolini's rule, 293

Jacob, 13
Jacobson, Bertha, 290
JAMA. See Journal of the American Medical Association
Jamaica, survey of mixed race individuals, 288–89, 290, 291, 350
Java, mixed race individuals, 279
Jeffreys, Sir Alec, 439
Jehovah's Witnesses, 367
Jesus Christ, 10, 13
Jeter, Mildred, 400–401
Jewish question, 297, 340
Jews
    in Africa, 295
    blindness, 301
    in England, 196
    German eugenicists' views of, 282–83, 294, 295, 296–97, 338–39
    in Germany
        Christian converts, 323
        definition of, 310–12, 315
        from eastern Europe, 275
        elimination as Nazi goal, 318, 347, 415
        family pedigrees, 311, 316, 407–9
        Final Solution, 407–8
        Hitler's view of, 270
        IBM system to collect data, 290–91, 308–10, 311, 340
        international criticism of Nazi persecution, 304
        international reaction to persecution, 311
        Kristallnacht, 316, 344, 393

Mischling (mixed breeds), 311
    Nazi persecution of, 277, 299, 304, 306, 308–9, 311, 316–17, 318, 343, 407–9
    Nuremberg Laws, 311, 312, 323, 339, 345, 392–93
    physicians, 306, 307, 308
    Ploetz's view of, 262
    refugees, 299, 304, 305, 306, 393
    registration of, 340
    sterilizations, 407–8
    twins at Auschwitz, 355–61
hereditary disorders, 343
marriages
    consanguineous, 282
    to non-Jews, 280
Mengele's identification of, 345
physical characteristics, 415–16
    in Poland, 395
    positive views of, 262
    racial stereotypes of, 316
    in United States
        immigrants, 74, 78, 83, 95, 191, 273
        immigration restrictions, 311–12, 393, 394
        intelligence testing, 83
    See also anti-Semitism; concentration camps
Johnson, Albert, 187–90, 191, 192–93, 194, 199, 201, 202, 203, 311–12
Johnson, Roswell, 137, 139
    Applied Eugenics, 137, 255, 258, 351
Johnstone, E. R., 250
Jordan, David Starr, 65, 70, 71, 210, 213, 222, 223
Jost, Alfred, 262
Journal of Delinquency, 78
Journal of Heredity
    articles, 161, 265–66
    on blindness, 147, 149
    on German eugenics, 281, 414–16
    on immigration restrictions, 188
    news from Germany, 286
    reviews of German books, 273, 341, 352
    on twin research, 350, 351, 352, 353, 354
    current mission, 425–26
    topics covered, 412
Journal of Sociology, 160
Journal of the American Medical Association (JAMA), 63, 73, 211, 281–82
    criticism of Nazis, 313
    Nazi propaganda published in, 301–2

news from Germany, 286, 341, 342–43, 352
Juhan, Joseph, 398–99
Jukes family, 24, 25, 64, 281
Julius Klaus Foundation for Heredity Research, Social Anthropology and Racial Hygiene, 242–43
Juvenile Psychopathic Institute (Chicago), 53

Kahler, Hans, 375
Kaiser Wilhelm Institute for Anthropology, Human Heredity, and Eugenics
  cover-up of wartime activities, 375
  Ellinger's visit, 414–16
  establishment, 286
  exhibits at international conferences, 298
  Fischer as director, 286, 288
  funding, 283, 286–87
  involvement in mass sterilization, 300
  Jewish doctor at, 308
  medical research, 367–68
  Nazi control, 316
  Rockefeller funding, 297, 302
  twin studies, 297, 302, 354, 359, 360–61, 364, 414, 415, 420
  Verschuer at, 339–40, 341, 347–48, 420
Kaiser Wilhelm Institute for Biology, 297–98, 369
Kaiser Wilhelm Institute for Brain Research, 283, 288, 302–3, 368–69, 379, 382, 419–20
Kaiser Wilhelm Institute for Physical Chemistry and Electrochemistry, 369–70
Kaiser Wilhelm Institute for Physics, 283
Kaiser Wilhelm Institute for Psychiatry, 283, 285, 286, 296, 306–7, 308, 369
Kaiser Wilhelm Institutes
  American funding, 283, 288
  development of, 283, 284
  Nazi control, 302–3, 306–7, 316
  Rockefeller funding, 285, 296, 306–7, 313, 369–70, 419–20
Kaiser Wilhelm Society, 283, 284
Kallikak family, 76–77, 107
Kansas, sterilizations in, 63, 122, 209
Kansas Home for the Feebleminded, 63
Katzen-Ellenbogen, Edwin, 91–93, 319–24
  at Buchenwald, 319–21, 325–31, 332–35
  education, 321
  expertise in fake symptoms, 321, 332
  as hypnotist, 331

Jewish ancestry, 92, 321
  murders by injection, 330
  as prisoner, 323–24
  as psychologist, 92, 326–27
  son, 92
  study of epilepsy, 91, 321–22
  testimony against other doctors, 332–33
  trial, 328, 329–30, 332, 333–35
Katzen-Ellenbogen, Marie Pierce, 91–92, 321
Keeler, Clyde, 344
Kellogg, Frank, 203
Kellogg, John Harvey, 88, 238
Kellogg, V. L., 75
Kellogg, Will, 88
Kellogg Company, 88
Kemp, Tage, 243–44, 418–20
Kennedy, Helena, 434
Kenya, 293
Kidder, A. V., 389–90
Kiep, Otto, 300
Kirksville, Missouri, 49, 50, 395
Klaus, Julius, 242
Klein, Zvi, 355–56
Knorr, George W., 97
Knox, P. C., 71, 72
Kogon, Eugen, 329–30, 333
Kommando 22a, 331
Kopp, Marie, 315
Kranz, Heinrich, 353
Kriegel, Vera, 362
Kristallnacht, 316, 344, 393
Kühn, Alfred, 308
Ku Klux Klan, 186
Kupas, Eva, 357

Labor, Department of, 193, 194, 197, 198, 199
labor camps, Nazi, 324, 329, 347
  See also concentration camps
Lait, Jack, 257
Lambert, Robert, 302–3
Lancashire Asylums Board, 210
Lancaster County, Pennsylvania, 53
Lancet, 382
Lang, Theodore, 307
Laski, Harold J., 120
Laughlin, Deborah, 49
Laughlin, George, 49
Laughlin, Harry Hamilton
  campaign to prevent blindness, 152–53, 157
  career, 49–50

Laughlin, Harry Hamilton (*continued*)
  census and, 160, 161
  concerns about speeches and beliefs, 185,
    191–92, 193–94, 198, 200–202,
    387–88
  congressional testimony, 189, 191, 199,
    202, 393–94
  as consultant to Municipal Court of
    Chicago, 256
  data collected, 53, 106, 361–62
  Davenport and, 51, 193–94, 196, 197, 202,
    388, 395
  death, 395
  deposition in Buck case, 115–16
  epilepsy bulletin, 55
  epilepsy of, 395
  as *Eugenical News* editor, 98, 280–81, 305,
    389
  at Eugenics Record Office, 51, 52
  eugenic views, 88–89, 193
    on carriers of defects, 58
    on immigration restrictions, 88, 187,
      188–89, 311–12, 387–88, 393, 394
    on marriage restrictions, 155, 156
    on Nazi eugenics, 304–5
    racial classifications, 161–62, 190
    social inadequacy concept, 159–60, 189,
      225
    on sterilization, 88
  European tour, 194–98, 199, 222, 238–39,
    241
  as "Expert Eugenics Agent" for Congress,
    188–92, 194
  family, 49
  Fisher and, 134
  Harriman and, 94–95
  influence in Germany, 312, 342
  model sterilization law, 113, 121
  organizational activities
    ABA Eugenic Section, 57, 250–51
    American Eugenics Society, 137
    Committee on Selective Immigration,
      192, 198–99
    Eugenics Research Association, 90, 160,
      266–67, 389
    international cooperation, 238–39, 240,
      241
    International Federation of Eugenic
      Organizations, 286
    at international meetings, 236
  Plecker and, 174, 175

  racial origins of senators investigated, 204
  racism, 191–92, 201, 387–88, 394
  rejection of charity, 222–23
  relations with German eugenics move-
    ment, 265, 266–67, 315–16, 342, 388,
    392, 394
  removal of, 388–92
  residence, 51, 52, 390
  retirement, 395
  ridicule of, 200–202, 387–88
  Sanger and, 136–37
  self-promotion, 50–51
  speeches in England, 225
  state sterilization laws compilation, 113,
    243
  study of hereditary blindness, 147, 148
  survey of state institutions, 159–60, 201
  writings, 50–51
  See also Eugenics Record Office
Laughlin, Pansy, 50, 52, 195, 390
law enforcement, DNA analysis and databanks,
    429–30, 439
Lawton, Henry, 231
League of Nations, 68–69
Lebensborn, 405
Lehmann, Julius, 273–74, 295
Lehmanns Verlag, 273, 282, 292, 302
Leipzig University, 345
Lemkin, Raphaël, 402–4
Lenz, Fritz, 272
  articles in American journals, 281, 282
  *Foundation of Human Heredity and Race
    Hygiene*, 270, 272–73, 296–97, 351
  involvement in Nazi eugenics, 316, 317,
    347
  relations with American eugenicists, 394
  research, 308
lethal chambers
  animals euthanized in, 247
  criminal executions, 258
  in England, 247–49
  support in birth control movement, 251
  use on unfit, 276
Libby, G. F., 151
Lidbetter, Ernest J., 221, 222, 223, 224, 229
Lidice, Czechoslovakia, 406–7
Lincoln. *See* Illinois Institution for the
    Feebleminded
Lincoln, Abraham, 118
Lippmann, Walter, 84
Lithuania, marriage restrictions, 245

Little, Clarence C., 137
Loeb, Clarence, 151, 285
London
    Battersea Dogs Home, 247
    First International Congress on Eugenics,
        70–73, 207, 213, 217, 235, 263, 264
    High Teams institution, 230–32
    poverty in, 221, 222
London Sociological Society, 209
Lorimer, Frank, 423
Lorinczi, Lea, 355
Los Angeles, ethnic groups, 292
Loving, Mildred Jeter, 400–401
Loving, Richard, 400–401
Loving v. Virginia, 401
Luftwaffe Institute for Aviation Medicine, 381
Lundborg, Hermann, 237, 244
lynchings, 23, 84

Magdalen Home for the Feebleminded, 80
Magnussen, Karin, 362–63
Maine
    sterilizations, 122
    Workshop for the Blind, 147
Mainz Academy of Sciences and Literature, 379
Mallory, George, 110–12
Mallory, Jessie, 111–12, 114
Mallory, Nannie, 111–12
Mallory, Willie, 110–12, 114
Malthus, Thomas, 11–12, 15, 120, 128
Manchuria, 293
marriage
    among Amish, 53
    Catholic Church views, 232–33
    consanguineous
        among elite, 74
        effects, 53, 54, 282
        of Jews, 282
    government screening, 175–76, 421
    of Jews and non-Jews, 280
    polygamy, 60
    registration in Virginia, 164
    regulations proposed by Galton, 18–19, 28
    selection of spouses, 15, 60, 74
    See also interracial marriages
marriage restrictions
    for blind, 145, 146, 149–52, 154, 155, 157
    bonding proposal, 154–56, 157
    Catholic opposition, 232
    current, 409
    Davenport's support of, 75

English proposals, 211, 216, 217, 218
found unconstitutional, 401
history of, 146
interracial marriages prohibited, 146, 165,
    174–76, 400–401
model legislation, 156
in Nazi Germany, 341, 407
in other countries, 243, 245
state laws, 146–47, 165–72, 174–76, 208,
    400–401
for unfit, 60, 146
Virginia Racial Integrity law, 165–72,
    174–76, 177, 181, 400–401
Marshall, Louis, 57, 59–60
Maryland, ban on interracial marriages, 146
Massachusetts
    institutions, 54, 321, 327
    sterilizations performed, 209
master race, 270
Max Planck Institute for Brain Research, 383
McCarran-Walter Immigration and
    Naturalization Act, 205
McCreary, James, 71
McCulloch, Oscar, 25, 64–65
McKim, W. Duncan, 248
McReynolds, James Clark, 120
medical education
    genetics, 421–22, 423
    in Germany, 283, 340
medical experiments, Nazi, 367
    at Auschwitz, 338, 358–59, 361, 362–63,
        376, 379
    aviation medicine, 367, 381
    at Buchenwald, 324, 329–30, 367
    in concentration camps, 331, 365, 380–81
    at Dachau, 367, 381–82
    funding, 365
    on Gypsies, 366
    by Mengele, 338, 359, 361, 376, 379
    postwar coverup, 375, 376–80
    real medical advances, 380–83
    retention of victims' remains, 383–84
    See also twin studies
Medical Information Bureau (MIB), 431–32,
    435–36
medical profession. See physicians
Medico-Legal Society, 227–28
Mein Kampf (Hitler), 270, 274, 276, 318, 354,
    394
Melchers, George, 363
Mencken, H. L., 200–202

Mendel, Gregor
    at Auschwitz, 355–61
    Howe and, 152
    influence on Davenport, 36, 101
    memorial to, 263
    pea breeding, 26
    publication of research, 13, 15
    rediscovery of theories, 25–26, 27, 31, 36,
        39, 208, 411
Mengele, Josef
    at Auschwitz, 337, 348, 363, 370, 376
        assistants, 356–57
        data collected, 359–61, 375
        escape, 371, 375
        medical experiments, 338, 359, 361, 376,
            379
        prisoners murdered by, 358–59
        twin studies, 353, 357, 358, 362, 365–66
    career, 344–47
    character, 357–58
    eugenic views, 344, 356
    Jewish ancestry, 345
Mental Deficiency Act, 214, 215, 216–18, 227
Mental Deficiency Committee (Britain), 226–27
mental illness. See insanity
mental retardation. See feebleminded individu-
    als
mental tests. See intelligence tests
Merriam, C. Hart, 46
Merriam, John C., 31, 185, 191, 193–94, 198,
    200, 387–88, 392
meteorology, 15
Metropolitan Relieving Officer's Association,
    221
Mexicans, 176, 387
Mexico, eugenicists in, 238
Meyer, Adolf, 90, 138, 150
MIB. See Medical Information Bureau
Michigan
    legislature, 66
    sterilizations, 123
Miles, Walter, 220
military
    DNA fingerprints of soldiers, 430
    See also Army, U.S.
Milledgeville State Hospital (Georgia), 399
Minnesota
    marriage restriction law, 208
    sterilizations, 122–23
Minnesota Historical Society, 397
miscegenation. See mixed race individuals

mixed race individuals
    in Africa, 263, 265–66
    attempts to prevent passing as white, 164
    in Canada, 242
    children of African soldiers in Rhineland,
        267, 271, 275, 305, 316
    consuls' reports on, 292–93
    efforts to eliminate, 163–82, 279, 288
    in Germany, 305, 340
    global survey of, 288–89, 291–94
    Indians, 177–80
    Jamaica survey, 288–89, 290, 291, 350
    negative views of, 30, 31
    research on, 179–80
    sterilizations, 305, 316
    study of, 89
    survey in United States, 291
    twins, 350
    in Virginia, 163, 164, 165, 169–70, 177–80
    Win Tribe, 179–80
Mjøen, Jon Alfred, 71, 244, 279
Mollison, Theodor, 344–45, 420
Monacan Indians, 177, 178
mongrelization. See mixed race individuals
Mongrel Virginians, 179–80
Monson State Hospital for Epileptics
    (Massachusetts), 54
Montana, ban on interracial marriages, 146
Montgomery County, Virginia, 3–4, 5
Moore, Elizabeth, 99
Moral Education League, 210
Morgan, J. P., 31
Morgan, William, 151
morons, 78, 79, 85, 189
    See also feebleminded individuals
moths, 308
Mozes, Eva, 355
Muller, Hermann, 351
Muller, Hermann Joseph, 302–3, 351, 379
Müller-Hill, Benno, 380
Municipal Court of Chicago, 113, 256
murders of unfit
    capital punishment, 258
    defective babies, 252–54, 257–58
    denial of treatment, 252–54, 327–29
    in Nazi Germany, 317
    Nazi T-4 project, 312–13, 339, 369,
        382–83
    opposition to, 250–52
    right of state, 262
    Sanger's views, 133, 251

support for, 247, 248–49, 250–52
See also euthanasia; lethal chambers
Mussolini, Benito, 279, 293

Nachtsheim, Hans, 367–68, 375
Nassau County, New York, 95
National Academy of Sciences, 236
National Association for the Study of Epilepsy,
    91, 322
National Committee for Federal Legislation on
    Birth Control, 134
National Committee for the Prevention of
    Blindness, 146
National Committee on Prison and Prison
    Labor, 96
National Conference on Charities and
    Correction, 248–49, 255
National Institutes of Health, 383, 425
National Origins Act, 202–5, 275
National Prison Association, 250
National Research Council, 236
National Socialism. *See* Nazis
National Socialist Doctors' Association, 307
Native Americans. *See* American Indians
natural selection, 12, 251
Nature, 198
Nazi doctors, 277
    at Auschwitz, 337
    at Buchenwald, 325–26, 329–31, 332
    political takeover of medicine, 313, 316
    war crimes trials, 320, 332–33, 381
    See also Katzen-Ellenbogen, Edwin; med-
        ical experiments; Mengele, Josef
Nazi eugenics, 316–17
    American admiration of, 7, 280, 313
    cover-up of wartime activities, 375, 379,
        380
    criticism of, 312
    depopulation strategy, 403
    elimination of Jews, 318, 347, 415
    Eugenic Courts, 318, 345
    eugenic laws, 277
    family pedigrees, 331, 342–43, 405
    films, 315–16, 393
    as genocide, 405–9, 411
    Germanization of Polish children, 331–32,
        347, 405
    increased birth rate among Germans, 403
    influence on medicine, 380–83
    international conferences, 346

racial examinations, 406, 415–16
research bodies, 339–40
revulsion of other eugenicists, 313
sterilization law, 299–301, 339
sterilizations performed, 277, 304, 315,
    316, 317, 324, 342, 406, 407–8
T-4 project (gassing of feebleminded),
    312–13, 317, 339, 369, 382–83
techniques, 403–5
twin camps, 352, 354–61, 362–63
See also American eugenics, relations with
    German eugenics movement; Aryan
    superiority; Jews, in Germany; med-
    ical experiments
Nazi Heredity Courts, 315
Nazis
    American sympathizers, 313
    Beer Hall Putsch, 269, 273–74
    census, 309, 311
    defeat, 375
    international knowledge of atrocities,
        299, 303, 311, 343, 402–4
    labor camps, 324, 329, 347
    Luftwaffe, 367, 381
    marriage restrictions, 341
    persecution of Gypsies, 316
    propaganda, 299, 315–16, 393
    registration of citizens, 310, 339
    scientists brought to United States,
        381–83
    seizure of power, 298–99
    use of IBM technology, 290–91, 308–10,
        311, 339
    See also concentration camps; Hitler,
        Adolf; Nazi eugenics
NBIA Disorders Association, 382–83
Nebraska, sterilizations in, 122
negative eugenics, 19
    Carnegie Institution funding, 32
    criticism of, 417
    in England, 209
    in Germany, 261–62
    Sanger's views, 131, 135
    term replaced by medical genetics, 424
Netherlands. *See* Holland
Die Neue Zeitung, 405
Nevada
    ban on interracial marriages, 146
    sterilization law, 68
newgenics, 428–29

New Jersey
    State Village for Epileptics (Skillman),
        54–55, 91, 98, 322, 327
    sterilization law, 68, 322
    Vineland Training School for
        Feebleminded Girls and Boys, 76, 80,
        91, 213, 243, 322, 327
New York American, 103–4
New York City
    health conditions, 126
    Second International Congress on
        Eugenics, 236–38, 241, 244, 268, 351,
        364, 422
    Third International Congress of Eugenics
        (New York), 245, 298, 419
New York City Police Department, 96, 98
New York Public Library, 54, 397
New York (state)
    county surveys of unfit, 95–96
    criminal families, 23–24
    eugenics agency, 95–96
    institutions, 52, 53, 130
    legislature, 93, 151–52, 154, 155
    marriage restrictions for blind, 151–52,
        154, 155
    Senate Commission to Investigate
        Provision for the Mentally Deficient,
        96
    sterilization law, 69
    sterilizations performed, 123
New York State Asylum (Matteawan), 52, 53
New York State Board of Charities, Bureau of
    Analysis and Investigation, 95, 96
New York State Board of Health, 152, 153
New York State Bureau of Industries and
    Immigration, 95–96
New York State Chamber of Commerce,
    Special Committee on Immigration and
    Naturalization, 394
New York State Department of Labor, 96
New York Times, 199, 297, 304, 311, 314
New York University
    College of Dentistry, 351
    eugenics courses, 75
New Zealand, eugenicists in, 238
Nordic race
    histories of, 30
    intelligence test scores, 82
    Nazi breeding program, 405
    preferences for immigrants of, 188, 192,
        203

    superiority seen, 29–31, 195, 213, 239, 244,
        266, 315
    See also Aryan superiority; super race
North Carolina
    eugenics program, 421–22
    marriage screening proposal, 421
    sterilization law, 400
    sterilizations performed, 123, 400, 421, 422
North Dakota
    ban on interracial marriages, 146
    ethnic groups, 292
Northwestern University, 75
Norway
    eugenicists, 71, 235, 238, 240, 244
    eugenic screening of potential U.S. immi-
        grants, 205
    International Society for Race Hygiene
        branch, 264
    mixed race individuals, 279, 292
    Nazi occupation, 407
    sterilization law, 244
    sterilizations performed, 244
Norwich Hospital (Connecticut), 67–68, 69
Norwich Union Insurance, 434
Notgemeinschaft der Deutschen Wissenschaften. See
    Emergency Fund for German Science
number names, 40
Nuremberg Laws, 311, 312, 323, 339, 345,
    392–93
Nuremberg war crimes trials, 325, 332–33, 381,
    404, 405–9
Nyiszli, Miklos, 356–57, 358, 359, 360–61

O'Brien, Daniel, 419
O'Brien, Stephen, 425–26
Ochsner, Albert John, 63
Offer, Moshe, 356
Ohio legislature, 250
Ohio State University, 382
Oliv (prisoner), 319–20
Olshansky, S. Jay, 425
Olson, Harry, 79, 96, 113, 195, 222, 243,
    256–57
Oneida Community, 21
Operation Paperclip, 381–83
ophthalmology, 145, 150–51
    See also eyes
Oregon
    sterilization law, 67
    sterilizations performed, 122, 400
The Origin of the Species (Darwin), 12

Osborn, Frederick, 418, 422–24
Osborn, Henry Fairfield, 37, 136, 236, 237, 238, 244, 422
Ossining, New York, Sing Sing prison, 52, 96
Owen, Teresa, 95
Owens-Adair, Bethenia, 67

Parliament, British, 211, 214–15, 217
    delegation to Nazi concentration camps, 325
    genetic testing issue, 434
    House of Lords, 212
    opposition to sterilization, 233
pathologists, at Auschwitz, 356–57, 360
Paton, Stewart, 57
Paul, Eden, 248
pauperism. See poverty
Pearl, Raymond, 150, 412
Pearson, Karl
    criticism of, 60–61, 72
    on House of Lords, 212
    relations with Carnegie Institution, 220
    research, 27, 35, 93, 212–13, 219–20
Pegram, Richard, 231
Peiffer, Jürgen, 383–84
Pennsylvania sterilization law, 66
Pennsylvania Training School for the Feebleminded, 66
Pennypacker, Samuel, 66
Perkins, Harry, 141, 142
Permanent International Commission on Eugenics. See International Eugenics Commission
Peter, W. W., 304
Peyton, David C., 148
physicians
    concentration camp inmates, 356–57, 362
    interest in eugenics, 73
    Jewish, in Germany, 306, 307, 308
    See also medical education; Nazi doctors
Pilcher, F. Hoyt, 63, 66
Pitt-Rivers, George, 279
Pius XI, Pope, 232
Plague, 10
Planck, Max, 303
Planned Parenthood, 127, 144, 426
    See also birth control movement
Plecker, Walter Ashby, 161–65
    data collection, 164–65
    goals, 165
    investigations of racial backgrounds,
        169–72, 177–79, 180–81
    Laughlin and, 163–64
    marriages prohibited by, 175–76, 400
    racism, 162, 166, 173–74, 175, 181
    registration of Virginians by race, 167, 168–72, 177–79, 180–82
Ploetz, Alfred
    anti-Semitism of, 262
    Archiv für Rassen- und Gesellschaftsbiologie, 262, 273
    Davenport and, 294, 295
    definition of racial hygiene, 341
    Eugenical News and, 306
    in international eugenics organizations, 71, 164
    relations with American eugenicists, 264, 394
    Rüdin and, 285
    Society for Racial Hygiene, 262, 263–64, 302, 339
    tribute to, 295
    in United States, 261–62
    during World War I, 266
Pocahontas, 168, 176, 177, 181
Poland
    eugenic screening of potential U.S. immigrants, 205
    German invasion, 317
    Germanization of children, 331–32, 347, 405
    marriage restrictions under German occupation, 403
    Nazi labor camps, 347
    persecution of Jews, 395
    See also Auschwitz
Poll, Heinrich, 284–85, 349
Pond, Clara, 96, 98
poor
    pauper class, 11, 12
    preventing reproduction, 58
    See also poverty
Popenoe, Paul, 424
    Applied Eugenics, 137, 255, 258, 351
    influence on Hitler, 259
    reaction to alliance with American Birth Control League, 139, 142
    relations with German eugenics movement, 272, 281, 342, 377–79
    reviews of German books, 273
    support for euthanasia, 251
    tribute to Ploetz, 295

Popenoe, Paul (*continued*)
twin studies, 351, 353
population control, 11–12, 142–43
Population Council, 422, 426
Population Reference Bureau, 426
positive eugenics, 18–19, 131, 135, 209
See also super race, building of
poverty
in England, 11, 207, 211, 218, 221, 222
history, 10
pauper class, 11, 12
seen as genetic defect, 5, 25, 65, 74, 130, 211, 220
Powell, John, 165, 166, 167, 169, 175, 181
Powhatan, 176
Priddy, Albert, 110, 111–12, 113, 114–15, 116
prisons
costs per inmate, 39, 153
data gathered in, 52
family histories of criminals collected, 96
inmates counted in censuses, 159
sterilizations in, 63–64, 65–67, 211
See also criminals
prostitution, 284, 419
psychiatry. *See* Kaiser Wilhelm Institute for Psychiatry
psychologists. *See* Katzen-Ellenbogen, Edwin
Public Health Service, U.S., 196
punch cards. *See* Hollerith data processing machines
Puzyna, Martina, 356, 375

Quebec, genetic testing issues, 435
Quota Board, 203–4
quotas. *See* immigration restrictions

race
classifications, 161–62, 168, 190
concepts, 65
conflicts related to, 186–87
eugenic ratings, 282
intelligence test scores, 79, 81, 82–84, 85
racism, 22, 269–70
registration of citizens by, 163, 165, 166, 167, 168–73, 177–79
segregation, 21, 171, 172–73
See also African-Americans; interracial marriages; mixed race individuals; whites
Race Betterment Foundation, 87–89, 238
race cards, 282
Race Policy Office of the Nazi Party, 315

race science, 31
race suicide, 22–23, 31, 209
racial hygiene (*Rassenhygiene*), 262, 280–81, 341
Racial Integrity Act (Virginia), 165–72, 174–76, 177, 181, 400–401
Rascher, Sigmund, 381
Rassenhygiene. *See* racial hygiene
Ratner, Brett, 106–7
Ravensbrück concentration camp, 406
Reconstruction, 162
redlining, 436
Red Scare, 186
Reed, Horace, 67
Reed, Sheldon, 397, 423–24
Reeves, Helen, 54
"Rehoboth bastards," 263, 265–66
Reichenberg boys, 358
Reich Research Fund, 365
Reich Statistics Bureau, 353–54
Rentoul, Robert, 208–10, 241–42, 250
Race Culture; Or, Race Suicide?, 146–47, 209, 248
reproduction. *See* birth control; breeding; sterilizations
Richardson, Benjamin Ward, 247, 262
Richmond Times-Dispatch, 166, 167, 277
Riddell, Lord George Allardice, 215–16, 227–29
Ripley, William, 82
Risk Management, 434
Riverview Cemetery, 172–73
Robertson, John Dill, 252–53
Robinson, William, 251
Rockbridge County, Virginia, 177, 178
Rockefeller, John D., 57, 93
Rockefeller, John D., Jr., 93, 284
Rockefeller Foundation
establishment, 93
eugenics research funding
avoidance of, 296, 313–14, 365, 370
criticism of, 101–2
Eugenics Record Office, 93, 94, 95
in Europe, 243–44, 245, 418–21
in Germany, 7, 258, 294–95, 297, 298, 302, 313, 364–65
move to genetics from, 244
twin studies, 349, 364
German research funding
criticism of, 306–7
eugenics research, 7, 258, 294–95, 297, 298, 302, 313, 364–65

Kaiser Wilhelm Institutes, 283, 285, 296,
306–7, 419–20
protection of German scientists, 302–3
scientific research, 283–85, 288, 295,
296, 306–8, 313, 365, 369–70
goals, 364–65
Paris office, 302–3, 419
political neutrality, 302, 364
Population Council funding, 422
Rodenwaldt, Ernst, 312
Rogers, Samuel, 160–61
Rolfe, John, 176
Roman Empire, 10
Romania, eugenicists in, 245
Roosevelt, Franklin D., 403
Roosevelt, Theodore, 46, 99, 118, 209
Root, Elihu, 31
Rosanoff, A. J., 100
Ross, E. A., 23, 209
Royal Commission on the Care and Control of
the Feebleminded, 214, 217, 226
Royal Commission on the Poor Law, 220–21
Royal Horticultural Society, 39, 411
Royal Statistical Society, 60
Rucker, W. C., 147
Rüdin, Ernst
Eugenical News advisory committee, 344
eugenic views, 285
family profiles collected by, 285, 286
at German Psychiatry Institute, 285
International Federation of Eugenic
Organizations and, 286, 295, 302
involvement in Nazi eugenics, 316, 364
at Kaiser Wilhelm Institute for Psychiatry,
307, 369
Nazis and, 306–7
relations with American eugenicists, 286,
394
research, 286, 295–96, 307, 308, 419
Rockefeller funding, 307
sterilization law, 299
sterilization proposals, 295–96, 301
RuSHA. See SS Race and Settlement Office
Ruttke, Falk, 344

Sachs, Jake and Saide, 126
Saleeby, Caleb, 213, 216, 217, 218, 224, 248–49
San Francisco Daily News, 101–3
Sanger, Margaret, 125–27
ancestors, 128
conferences, 134

eugenicists and, 315
eugenic views, 127, 128–33, 135–37, 139,
143–44, 156
feminism of, 142–43
on large families, 132–33, 139, 301
legacy of, 426
opposition to charity, 127, 129–30
organizations, 140, 142
Pivot of Civilization, 129–30, 131, 251
rejection of euthanasia, 251
support for sterilizations and segregation,
131–32
Woman and the New Race, 132–33
See also birth control movement
SAT. See Scholastic Aptitude Test
Saure, Philip N., 175–76
Scandinavia, eugenics in, 70, 224, 240
See also Denmark; Finland; Iceland;
Norway; Sweden
Schiedlausky, Gerhard, 326, 330–31
Schlaginhaufen, Otto, 242, 243
Schneider, Carl, 312
Scholastic Aptitude Test (SAT), 83
schools, racially segregated, 171, 173
Schultz, A. P., 203
Schultz, Bruno, 419
Science Magazine, 202
Second International Congress on Eugenics
(New York), 236–38, 241, 244, 268, 351,
364, 422
Second Race Betterment Conference, 89, 134
segregation, racial, 22, 171, 172–73
segregation of unfit
of alcoholics, 285
of blind, 145, 146, 149, 153
colonies, 156, 215, 218
English proposals, 210, 211, 214, 215–18,
226
of feebleminded, 214, 215
including extended families, 215–18
interstate deportation, 156–57
support for, 52, 60, 78, 280, 285
by Davenport, 39, 60, 75
by Laughlin, 88–89
by Sanger, 131–32
See also institutions
Senate, U.S.
Immigration Committee, 393–94
racial origins of senators, 204
sexology, 243
Sexual Sterilization Act (Alberta), 242

Shackleford, C. D., 108–10
Sharp, Harry Clay, 63–64, 65–67, 69, 148, 211, 273
Shaw, George Bernard, 28, 248
sheep, clones of, 427
Shelton, Robert, 113, 114
Sherlock, E. B., 250
shiftlessness, 74, 108, 189
ship captains, 105
Siemens, H. W., 281
Simon, Theodor, 76
Sing Sing prison, 52, 96
Sixth International Neo-Malthusian and Birth Control Conference, 134, 136–37
Skinner, Robert, 197
Slaughter, Louise, 440, 441
slaves, breeding of, 21
Smith, Buck, 5–6
Smith, D. F., 248
Smith, John, 176
Snyder, Laurence, 413–14
Social Biology, 425
social Darwinism, 12–13, 119, 129, 262, 275
social inadequacy, 159–60, 189, 225
socialists, 367
social reform movements, 125, 262
    See also birth control movement
Société Belge d'Eugénique (Belgian Eugenics Society), 240–41
Society for Racial Hygiene (Gesellschaft für Rassenhygiene), 262, 263–64, 274, 285, 302, 316, 339, 341
Society for the Prevention of Blindness, 315–16
Society for the Study of Social Biology, 425
Solvay Institute, 195, 241
Southard, E. E., 89
South Carolina, sterilizations in, 400
South Dakota, sterilizations in, 122
Soviet Union
    liberation of Auschwitz, 371
    mixed race individuals, 292
Sparta, 251
Spatz, Hugo, 369, 382–83
Spencer, Herbert, 12, 13, 15, 26, 119, 129, 262
Spiegel, Magda, 355
Spielmeyer, Walther, 306, 308
SS, 316, 326–27, 329, 337
SS Race and Settlement Office (RuSHA)
    Genealogical Section, 347
    Hollerith data processing machines, 290–91

Mengele at, 347, 348
persecution of Jews, 407–9
Polish children kidnapped by, 331–32, 405
racial examinations, 406
Stalin, Josef, 403
Stanford, Leland, 65
Stanford-Binet test, 81, 82, 84
Stanford University, 65, 75
State Department, U.S.
    consul in Jamaica, 289
    consuls involved in immigrant screening, 195, 196, 197, 199, 205
    consuls involved in survey of mixed race individuals, 292–93
    invitations to eugenics meetings sent by, 71–72, 236, 268
    Jewish refugees refused visas by, 393
    reluctance to cooperate with other departments, 196–97, 199–200
Steggerda, Morris, 305
sterilization laws
    amending to include blind, 149
    British, 227–29
    Canadian, 242
    constitutionality, 59–60, 64, 113
    Danish, 243
    due process safeguards, 113
    model, 113, 121
    Nazi, 299–301, 339
    remaining on books, 400
    of states, 66–69, 89, 112–13, 122, 208, 209, 300, 408
    Swiss, 243
sterilizations
    of alcoholics, 285
    of American Indians, 5, 400
    of blind, 145, 146, 148, 149, 299
    in Canada, 242
    of criminals, 59, 63–64, 65–67, 68, 69, 122
    criticism of, 63
    Davenport's support of, 75
    of deaf, 299
    in Denmark, 420
    of epileptics, 5, 231, 299, 322
    of family members of unfit, 225, 295–96
    of feebleminded
        Buck case, 108–10, 113–17, 119, 120–22, 167, 315, 401–2
        in California, 68, 122
        in Canada, 242
        debates, 63, 66

in England, 227–29
in New York, 96
supporters, 77–78
in Virginia, 3–7
in foreign countries, 235
in Germany
by Nazis, 277, 299–301, 304, 315, 316,
317, 324, 342, 406, 407–8
proposals, 261, 265, 297, 339
governors' apologies for, 400
grounds for, 4, 123
Hitler on, 276
illegal, 63–64, 65–67, 110–12, 211, 230–32
increasing number, 122–23
lack of public support, 66, 69–70
Laughlin's support of, 88
lawsuits by victims, 398–400
lobbying for legislation, 64, 66–67
Mallory case, 110–12
of mixed race individuals, 305, 316
model law, 113, 300
number in United States, 7, 234, 301, 398
opposition to, 70, 87, 101–4
press coverage of plans, 101–4
recent, 400, 409
Sanger's support of, 131–32, 143
therapeutic, 110
of unfit, 3–7, 39, 52
voluntary, 60, 228
See also England, eugenics in
Stern, Hedvah, 356, 366
Stern, Leah, 356, 366
Stern, William, 82
Stimson, Henry, 72
Stoddard, Lothrop
Into the Darkness, 317–18
Eugenics Research Association and, 90, 91
Hitler and, 298, 318
on immigration, 29–30, 35
involvement in birth control movement,
133–34, 167
support of Virginia Racial Integrity Act,
167
Streicher, Julius, 404
Stribling, Francis T., 4
Strode, Aubrey, 113, 114, 115, 116, 117
Strughold, Hubertus, 381–82
Stubbs, Walter, 71
Stypanovitz, Eva, 95
Sullivan, Mary, 106
super race, building of

breeding Aryan babies, 367
Davenport's goals, 36, 37, 416–17
eliminating lower levels of population, 59
encouraging reproduction in superior fam-
ilies, 46, 135, 137, 139, 301
Galton on, 15
as goal of American eugenics, 7, 88
Grant on, 90
Hitler on, 270, 276, 367
Mengele's goals, 356
Nazi goals, 370
potential in future, 441–44
role of twin studies, 351, 354
See also Aryan superiority; Nordic race
Supreme Court, U.S.
Buck v. Bell, 117, 119, 120–22, 123, 315,
401–2, 409
Holmes as justice, 118–19
Loving v. Virginia, 401
survival of fittest, 12, 29
Sweden
emigration discouraged, 195
eugenicists, 237, 238, 244
eugenic screening of potential U.S. immigrants, 195, 198, 205
genetic discrimination legislation, 440
International Eugenics Commission meeting, 238–39, 268–69
International Society for Race Hygiene
branch, 264
State Institute of Race-Biology, 195, 244
sterilization law, 244–45
Swift, Walter, 98
Switzerland
eugenicists, 242–43
sterilization law, 243
syphilis, 146

T-4 project (gassing of feebleminded by Nazis),
312–13, 317, 339, 369, 382–83
Taft, William Howard, 120
Tasmania, DNA databank, 440
Terman, Lewis, 80, 81, 82, 84, 90, 150, 226
terrorism, 430–31
test-tube babies, 427
Thackera, A. M., 196–97
thalassophilia, 105
Thayer, Ethel, 97
Third International Conference on Genetics,
412

Third International Congress of Eugenics (New York), 245, 298, 419
Third Reich. *See* Nazis
Thomas, Elsie M., 175–76
Timoféeff-Ressovsky, Nikolai, 288
Tisdale, Walter, 308
Tonga, national DNA databank, 439–40
Town, Clara, 257
Trades Union Congress, 234
transgenic creatures, 443
Treaty Against Genocide, 404–5
Treaty of Versailles, 238, 267, 269, 271
Tredgold, Arthur F., 249, 256
Tribe of Ishmael, 25, 53, 65, 237
Trinkle, E. Lee, 175
Trowbridge, Augustus, 297
tuberculosis
    in cattle, 254–55
    hereditary, 241, 255
    mortality from, 255, 256
    Nazi research, 368
    relationship to racial mixing, 279
    research on, 241
    twins and, 352, 358
    in Virginia, 280
Tucker, J. R., 181
twins
    blood groups, 364, 365
    eyes, 361–63
    mixed race, 350
    in Nazi Germany, 352, 353–54
twin studies
    anatomical, 351–52
    in concentration camps, 354–61, 362–63
    criminal behavior, 353
    deaths in, 354, 359
    difficulty of, 349
    by Galton, 348–49
    in Germany, 350, 352–54, 362–63, 364, 365–66, 414, 415
    journals, 353
    at Kaiser Wilhelm Institute for Anthropology, Human Heredity, and Eugenics, 297, 302, 354, 359, 360–61, 364, 414, 415, 420
    by Mengele, 353, 357, 358, 362, 365–66
    racial differences in twin frequency, 350
    Rockefeller funding, 349, 364
    in United States, 361–62
    value to eugenicists, 348
Twitchen, Henry, 229

UK Biobank, 440
UK Forum for Genetics and Insurance, 436
Ulster County, New York, criminals, 23–24
unfit
    borderline, 58–59
    categories, 58
    costs of maintaining, 39, 130, 147, 153
    county surveys in New York, 95–96
    elimination of, 12, 29, 39–40, 56, 61
    in England, 214, 227
    extended families of, 24–25, 58, 59, 152–53, 215–18
    financial responsibility for, 156
    global cataloging of, 279, 280
    identification, 52–55, 56, 95, 96–97, 107
    interstate deportation, 156–57
    lack of standards to identify, 76
    number in United States, 58–59, 216
    preventing reproduction, 18, 21, 46, 57–61, 210–11, 403
    seen as subhuman, 21
    See also epileptics; feebleminded individuals; insane; marriage restrictions; murders of unfit; segregation of unfit; sterilizations
United Kingdom. *See* England
United Nations, 403–4
United States
    knowledge of Nazi atrocities, 299
    See also American eugenics
universities
    entrance exams, 83
    eugenics classes, 75, 103–4, 412
University of California, Berkeley, 75
    Lawrence Berkeley Lab, 437
University of Chicago, 33, 75
University of Copenhagen, 243, 420–21
University of Heidelberg, 312, 342, 383
University of Indiana, 65
University of London, 425
University of Minnesota, Dight Institute, 397
University of Münster, Institute of Human Genetics, 379–80
urbanization, 186
U.S. News & World Report, 437
Utah
    investigation of feebleminded, 97
    sterilizations, 122
utopian societies, 261

vagrancy, 10–11
Van Guilder, Elsie, 398
van Wagenen, Bleeker, 213
vasectomies. *See* sterilizations
Vatican, 232, 293
    See also Catholic Church
Vaud, Switzerland, sterilization law, 243
Vaughan, Victor, 151
Vekler, Jancu, 362
Venezuela, eugenicists in, 238
Venter, J. Craig, 443
Vermont, survey of unfit, 107
Verschuer, Otmar Freiherr von, 338–44
    bibliographies, 380–81
    blood research, 363–66, 376
    books, 343, 352
    death, 380
    Institute for Hereditary Biology and Racial
        Hygiene, 341–43, 344, 345, 346,
        352–53
    investigation of, 380
    involvement in Nazi eugenics, 375, 380
    at Kaiser Wilhelm Institute for
        Anthropology, Human Heredity, and
        Eugenics, 339–40, 341, 347–48, 420
    Mengele and, 345, 346, 347, 348, 366, 376,
        380
    postwar rehabilitation efforts, 376–80
    relations with American eugenicists, 394
    research, 339–40, 368
    twin studies, 352–53, 354, 355, 359,
        360–61, 362–63, 364, 365–66, 420
Vineland Training School for Feebleminded
    Girls and Boys (New Jersey), 76, 80, 91,
    213, 243, 322, 327
Virginia
    Buck sterilization case, 108–10, 113–17,
        119, 120–22, 167, 315, 401–2
    Bureau of Vital Statistics, 161, 163,
        164–65, 166, 168–69, 173–74, 175,
        177, 181
    colonial settlers, 162, 168
    descendants of Pocahontas, 168, 176, 181
    eugenic registry, 165
    governor's apology for sterilizations, 400
    institutions, 108
    legislature, 112, 167
    marriage restriction laws, 165–72, 400–401
    mixed race individuals in, 163, 165,
        169–70, 177–80

Racial Integrity Act, 167–68, 174–76, 177,
    181, 400–401
registration of citizens by race, 163, 165,
    166, 167, 168–73, 177–79
sterilization law, 68, 113, 114–17, 121, 167
sterilizations performed, 3–7, 108–12,
    113–17, 122, 123
tuberculosis in, 280
Virginia Department of Health, 168
Virginia State Health Bureau, 163
vision problems
    eugenic legislation, 149–50
    marriage restrictions, 149–52
    See also blindness
Vogt, Oskar, 288, 302–3
von Hofmann, Géza, 264–65, 266, 273

Wagner, Werner, 369
Wake Forest Medical School, Department of
    Medical Genetics, 421–22
Walker, Francis, 22–23
Walsh, Catherine, 252
Wannsee Protocol, 407–8
war crimes trials
    at Dachau, 320, 328, 329–30, 332, 333–35
    of Katzen-Ellenbogen, 320, 328, 329–30,
        332, 333–35
    of Nazi doctors, 320, 332–33, 381
    Nuremberg, 325, 332–33, 381, 404, 405–9
    preparations, 404
    threats of, 402, 403
Ward, Robert DeCourcy, 57, 187, 192, 202
Warren, Earl, 401
wars, eugenic effects, 222–23
Washington (state)
    sterilization law, 67
    sterilizations performed, 69
Watson, James, 426, 442–43
Watson, Thomas J., 290, 309–10
wealthy benefactors of eugenics movement,
    56–57, 87, 93–95, 140, 237–38
    See also Harriman, Mary (Mrs. E. H.);
        Rockefeller Foundation
Weaver, Warren, 370
Wehrmacht, 323–24
Weimar Republic, 266–67, 268, 269
Weismann, August, 17
Welch, William, 89, 93–94, 138
Wells, H. G., 130, 209, 248
Western State Hospital (Virginia), 4, 7

Westrope, Lionel L., 230–32
White, Arnold, 248
White, Wendell, 175
White America Society, 165
Whitehead, Irving, 114, 116, 117
whites
    cemeteries, 172–73
    death rates, 163
    definitions of, 166, 168
    intelligence test scores, 79, 81
    multiple birth frequency, 350
    racial purity, 162, 165–66
    See also Aryan superiority; Nordic race;
        race
white supremacists, 166
Whitney, Leon, 138–40, 142, 244, 259, 317
Wiggam, Albert, 99
Wigmore, John, 120
Wilhelm II, Kaiser, 284
Wilks, Michael, 434
Williams, Minnie, 398
Willis, Sir Frederick, 227
Wilson, James, 47, 98
Wilson, William George, 231
Wilson, Woodrow, 68–69, 99, 258, 320, 322
Wimmer, August, 243
Winston-Salem Journal, 422
Win Tribe, 179–80
Wisconsin
    sterilization law, 67
    sterilizations performed, 122

Wise, Stephen, 129
women
    promiscuous, 110
    See also birth control; feminists; steriliza-
        tions
Wood, Sir Arthur, 226–27
Wood, Patricia, 383
Woodhull, Victoria, 22, 128
workers, genetic testing of, 437–38, 440
World Population Congress, 314
World War I
    African-American soldiers, 186
    effects on eugenics movements, 221, 222-
        23, 236, 238, 266
    intelligence testing of American soldiers,
        80-82
    poison gas, 276
World War II
    beginning of, 317, 346
    See also concentration camps; Nazis

Yagudah, Judith, 355
Yerkes, Robert, 80, 81, 82, 90, 99, 150, 226
Yerkes-Bridge Point Scale for Intelligence, 80

Zahn, Friedrich, 310
Zangwill, Israel, 22